T0183655

Lecture Notes in Artificial Intelligence 12595

Subseries of Lecture Notes in Computer Science

Series Editors

Randy Goebel
University of Alberta, Edmonton, Canada

Yuzuru Tanaka
Hokkaido University, Sapporo, Japan

Wolfgang Wahlster
DFKI and Saarland University, Saarbrücken, Germany

Founding Editor

Jörg Siekmann
DFKI and Saarland University, Saarbrücken, Germany

Chee Seng Chan · Hong Liu ·
Xiangyang Zhu · Chern Hong Lim ·
Xinjun Liu · Lianqing Liu ·
Kam Meng Goh (Eds.)

Intelligent Robotics and Applications

13th International Conference, ICIRA 2020
Kuala Lumpur, Malaysia, November 5–7, 2020
Proceedings

 Springer

Editors
Chee Seng Chan
University of Malaya
Kuala Lumpur, Malaysia

Hong Liu
Harbin Institute of Technology
Harbin, China

Xiangyang Zhu
Shanghai Jiao Tong University
Shanghai, China

Chern Hong Lim
Monash University
Selangor, Malaysia

Xinjun Liu
Tsinghua University
Beijing, China

Lianqing Liu
Shenyang Institute of Automation
Shenyang, China

Kam Meng Goh
Tunku Abdul Rahman University College
Kuala Lumpur, Malaysia

ISSN 0302-9743   ISSN 1611-3349 (electronic)
Lecture Notes in Artificial Intelligence
ISBN 978-3-030-66644-6   ISBN 978-3-030-66645-3   (eBook)
https://doi.org/10.1007/978-3-030-66645-3

LNCS Sublibrary: SL7 – Artificial Intelligence

This Springer imprint is published by the registered company Springer Nature Switzerland AG
The registered company address is: Gewerbestrasse 11, 6330 Cham, Switzerland

Preface

On behalf of the Organizing Committee, we welcome you to the proceedings of the 13th International Conference on Intelligent Robotics and Applications (ICIRA 2020), organized by the IEEE Computational Intelligence Society (Malaysia Chapter), technically sponsored by Springer, and financially sponsored by Malaysia Convention & Exhibition Bureau. ICIRA 2020 with the theme of "Cloud Robotics" offered a unique and constructive platform for scientists and engineers throughout the world to present and share their recent research and innovative ideas in the areas of robotics, automation, mechatronics, and applications.

ICIRA 2020 was a successful event this year in spite of the COVID-19 pandemic. It was held virtually and attracted 66 submissions regarding state-of-the-art developments in robotics, automation, and mechatronics. The Program Committee undertook a rigorous review process to select the most deserving research for publication. Most of the submissions were of high quality; 45 submissions were selected for publication in Springer's Lecture Notes in Artificial Intelligence, a subseries of Lecture Notes in Computer Science. We sincerely hope that the published papers of ICIRA 2020 will prove to be technically beneficial and constructive to both the academic and industrial community in robotics, automation, and mechatronics. We would like to express our sincere appreciation to all the authors, the participants, and the distinguished keynote and invited speakers.

The success of the conference is also attributed to the Program Committee members and invited peer reviewers for their thorough review of all the submissions, as well as to the Organizing Committee and volunteers for their diligent work. Special thanks are extended to Alfred Hofmann, Anna Kramer, Ronan Nugent, and Celine Chang from Springer for their consistent support.

October 2020

Chee Seng Chan
Hong Liu
Xiangyang Zhu
Chern Hong Lim
Xinjun Liu
Lianqing Liu
Kam Meng Goh

Organization

General Chairs

Chee Seng Chan	University of Malaya, Malaysia
Hong Liu	Harbin Institute of Technology, China
Xiangyang Zhu	Shanghai Jiao Tong University, China

Program Chairs

Chern Hong Lim	Monash University, Malaysia
Xinjun Liu	Tsinghua University, China
Lianqing Liu	Chinese Academy of Sciences, China

Keynote and Panel Chairs

Honghai Liu	University of Portsmouth, UK
Jangmyung Lee	Pusan National University, Republic of Korea

Finance Chairs

Ven Jyn Kok	National University of Malaysia, Malaysia
Mei Kuan Lim	Monash University, Malaysia

Awards Chairs

Naoyuki Kubota	Tokyo Metropolitan University, Japan
Kok-Meng Lee	Georgia Institute of Technology, USA

Special Session Chairs

Dalin Zhou	University of Portsmouth, UK
Xuguang Lan	Xi'an Jiaotong University, China

Demo/Industry Chairs

Zati Hakim Azizul Hasan	University of Malaya, Malaysia
Zhaojie Ju	University of Portsmouth, UK

Publication Chairs

Kam Meng Goh	Tunku Abdul Rahman University College, Malaysia
Jiangtao Cao	Liaoning Shihua University, China

Local Arrangements Chairs

Wai Lam Hoo   University of Malaya, Malaysia
Sim Ying Ong   University of Malaya, Malaysia

European Liaison

Qinggang Meng   Loughborough University, UK
Serge Thill   Radboud University, The Netherlands

North and South America Liaison

Ning Jiang   University of Waterloo, Canada
Rodney Roberts   Florida State University, USA

Contents

Human-Robot Interaction

Mobile Robots and Intelligent Autonomous System

Recent Trends in Computational Intelligence

Robot Design, Development and Control

Robotic Vision, Recognition, and Reconstruction

Soft Actuators

Advanced Measurement and Machine Vision System

An Efficient Calibration Method for 3D Nonlinear and Nonorthogonal Scanning Probe in Cylindrical CMM

Kangyu Yang[1], Xu Zhang[2], Jinbo Li[2], and Limin Zhu[1]($\boxtimes$)

[1] School of Mechanical Engineering, Shanghai Jiao Tong University,
Shanghai 200240, People's Republic of China
`zhangxu@hust-wuxi.com`
[2] Huazhong University of Science and Technology, Wuxi Research Institute,
Wuxi 214174, People's Republic of China

Abstract. In terms of the measurement accuracy of traditional CMM, a three-dimensional (3D) probe plays a great part in machine performance. However, there is few models and calibrations for 3D nonlinear and nonorthogonal scanning probe to have been published in cylindrical coordinate measuring machine (Cylindrical CMM). In this paper, a proper nonlinear model on the 3D probe is established which is based on Taylor series expansion. Moreover, an efficient calibration method is proposed to compensate not only 3D deformation of the probe but also radius of the probe tip. A high-accuracy sphere is selected as an artefact and several practical and feasible scanning paths are designed. During the calibration, optimization of the third-order model benefits from the second-order model for providing the approximate coefficient magnitudes. The proposed calibration method was experimentally carried out on an assembled four-axis CMM. Although both of the two models can reach a micron level, a compensation of the third-order model is more accurate. The correctness of the model and the efficiency of the calibration method were successfully demonstrated by probing another high-accuracy ball and the uncertainty was also analyzed.

Keywords: 3D scanning probe · Nonlinear calibration · Cylindrical CMM

1 Introduction

With the rapid development of industrialization, quality control and accuracy demands are significant for a large number of machining components such as blades [1]. The CMM plays a striking role in manufacturing processes. At present, an extra rotary table is added to improve the measurement efficiency of rotary parts. When a linear axis is fixed, it becomes a Cylindrical CMM.

In order to improve the accuracy of linear axes, there are two ways: precision design and error compensation [2]. The former is adopted during the design phase and the latter is an effective solution to reduce measurement errors after a CMM

© Springer Nature Switzerland AG 2020
C. S. Chan et al. (Eds.): ICIRA 2020, LNAI 12595, pp. 3–15, 2020.
https://doi.org/10.1007/978-3-030-66645-3_1

is assembled. During the measurement, some factors affect the accuracy such as geometric error, thermal error and kinematics error. Among them, geometric error is one of the most source of inaccuracy [3] and geometric error techniques can be divided into three steps: establishing an error model, measuring errors and compensating the machine [4]. Nowadays, they have been able to standardize measurement by a laser interferometer and a set of optics [5].

The identification and compensation of geometric errors of rotary axis, which can be divided into two groups [6,7], i.e. position-independent geometric errors (PIGEs) and position-dependent geometric errors (PDGEs), are also crucial. Focusing on PDGEs, Huang et al. brought up and analyzed two different models, which were named as "Rotary axis component shift" and "Rotary axis line shift" [8]. Later, they successfully obtained PDGEs by using a touch-trigger probe [9]. However, a test piece was installed and pre-machined before on-machine measurement so that it is not suitable in CMM. Chen et al. [10,11] proposed a series of methods to identify PIGEs and PDGEs by a touch-trigger probe and high-accuracy spheres. Consequently, Chen's methods can be also performed on the CMM with a rotary table.

The contact probe is generally classified into touch-trigger probe and scanning probe [12]. As the higher efficiency is increasingly demanded, 3D compensation concerning a probe is gradually utilized and the 3D scanning probe such as SP25M produced by Renishaw plays an important role. In traditional CMM, it has been solved perfectly by Renishaw. Nevertheless, the calibration provided by Renishaw cannot be used in Cylindrial CMM. Unfortunately, few researchers have addressed the problem of 3D compensations although they are vital and worthwhile.

In this paper, a kinematic error model of the machine is built by means of the homogeneous transformation matrix (HTM). The model is established to depict the relationship between 3D deformation of the probe and voltage output of the sensor, which is based on Taylor series expansion mathematically. A sphere with high accuracy is selected as an artefact and practical paths are designed to calibrate the probe. It achieved micron accuracy and the uncertainty was estimated. The main novel contributions of this work are the establishment of nonlinear model, the estimation of coefficient magnitudes and the compensation of 3D deformation and radius of the probe tip.

2   Kinematic Error Model Establishment

2.1   Coordinate System Establishment

In this study, a four-axis CMM configuration with a rotary table (as shown in Fig. 1) is investigated, which can be considered as a traditional CMM when the rotary table is fixed and a Cylindrical CMM when a linear axis is fixed. The machine coordinate system (MCS) is set at the initial position of the machine. The Y-axis (YCS), X-axis (XCS) and Z-axis coordinate system (ZCS) overlap each other when the machine is at the initial position and they are aligned with the Y-, X- and Z-axis of the MCS. The C-axis coordinate system (CCS) is defined at the center of the rotary table. The 3D probe is mounted on the Z-axis of the CMM. Accordingly, the probe coordinate system (PCS) is built at the ruby center of the probe which locates at the position (t_x, t_y, t_z) with respect to the ZCS.

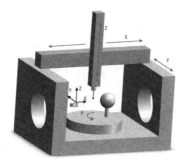

Fig. 1. Machine structure.

2.2 Kinematic Error Model of the CMM

The kinematic error model can be established by applying HTM under the assumptions of small angular motions. the ruby center positions regarding the MCS in ideal and real case can be acquired by using a kinematic chain:

$$
\begin{bmatrix} x^i \\ y^i \\ z^i \\ 1 \end{bmatrix} = {}^M_Y T \, {}^Y_X T^i \, {}^X_Z T^i \, {}^Z_P T \begin{bmatrix} 0 \\ 0 \\ 0 \\ 1 \end{bmatrix} = \begin{bmatrix} x_c + t_x \\ y_c + t_y \\ z_c + t_z \\ 1 \end{bmatrix} \tag{1}
$$

$$
\begin{bmatrix} x^r \\ y^r \\ z^r \\ 1 \end{bmatrix} = {}^M_Y T^r \, {}^Y_X T^r \, {}^X_Z T^r \, {}^Z_P T \begin{bmatrix} 0 \\ 0 \\ 0 \\ 1 \end{bmatrix} \tag{2}
$$

where x_c, y_c and z_c represent the commands of X-, Y- and Z-axis, respectively. ${}^M_Y T^i$ denotes the ideal transfer matrix form YCS to MCS and others are similar.

The model of the rotary table is established according to the theory of "Rotary axis component shift" [8]. The ideal and real transformation are shown as follows.

$$
{}^{\theta_0}_C T^i = \begin{bmatrix} \cos\theta & -\sin\theta & 0 & 0 \\ \sin\theta & \cos\theta & 0 & 0 \\ 0 & 0 & 1 & 0 \\ 0 & 0 & 0 & 1 \end{bmatrix} \tag{3}
$$

$$
{}^{\theta_0}_C T^r = \begin{bmatrix} \cos\theta - \varepsilon_{zc}\sin\theta & -\sin\theta - \varepsilon_{zc}\cos\theta & \varepsilon_{yc} & \delta_{xc} \\ \varepsilon_{zc}\cos\theta + \sin\theta & -\varepsilon_{zc}\sin\theta + \cos\theta & -\varepsilon_{xc} & \delta_{yc} \\ -\varepsilon_{yc}\cos\theta + \varepsilon_{xc}\sin\theta & \varepsilon_{yc}\sin\theta + \varepsilon_{xc}\cos\theta & 1 & \delta_{zc} \\ 0 & 0 & 0 & 1 \end{bmatrix} \tag{4}
$$

where θ is the command of the rotary table. Accordingly, the relations from CCS to MCS are calculated in Eqs. (5) and (6).

$$
\begin{bmatrix} x^i \\ y^i \\ z^i \\ 1 \end{bmatrix} = {}^{M}_{\theta_0}T\,{}^{\theta_0}_{C}T\,T^i \begin{bmatrix} x_{rot} \\ y_{rot} \\ z_{rot} \\ 1 \end{bmatrix} \tag{5}
$$

$$
\begin{bmatrix} x^r \\ y^r \\ z^r \\ 1 \end{bmatrix} = {}^{M}_{\theta_0}T\,{}^{\theta_0}_{C}T\,T^r \begin{bmatrix} x_{rot} \\ y_{rot} \\ z_{rot} \\ 1 \end{bmatrix} \tag{6}
$$

where ${}^{M}_{\theta_0}T$ is a constant matrix that defines the position and orientation of CCS in terms of MCS at $\theta = 0$ and $(x_{rot},\ y_{rot},\ z_{rot})$ is the position in CCS.

2.3 Forward and Inverse Kinematics

When X-axis is fixed at the proper position, which means the x coordinate value of the CCS in terms of MCS at $\theta = 0$, it becomes a Cylindrical CMM. Therefore, the first row and the fourth column of ${}^{M}_{\theta_0}T$ is 0 in the Cylindrical CMM.

The forward kinematics is that $(x_{rot},\ y_{rot},\ z_{rot})$ is obtained at the given commands of Y-, Z- and C-axis. It is easy to get the result according to Eqs. (1–6):

$$
\begin{bmatrix} x_{rot} \\ y_{rot} \\ z_{rot} \\ 1 \end{bmatrix}^i = \left({}^{M}_{\theta_0}T\,{}^{\theta_0}_{C}T\,T^i \right)^{-1} \begin{bmatrix} 0 + t_x \\ y_c + t_y \\ z_c + t_z \\ 1 \end{bmatrix} \tag{7}
$$

$$
\begin{bmatrix} x_{rot} \\ y_{rot} \\ z_{rot} \\ 1 \end{bmatrix}^r = \left({}^{M}_{\theta_0}T\,{}^{\theta_0}_{C}T\,T^r \right)^{-1} {}^{M}_{Y}T\,T^{rY}_{X}T\,T^{rX}_{Z}T\,T^{rZ}_{P}T \begin{bmatrix} 0 \\ 0 \\ 0 \\ 1 \end{bmatrix} \tag{8}
$$

The inverse kinematics is just the opposite. Apparently, it is hard to get the real commands of Y-, Z- and C-axis directly on account of geometric errors. An iteration based on the ideal inverse kinematics is implemented. Because x_c is 0, an equation containing only the unknow θ can be acquired from Eq. (7).

$$
\begin{aligned}
&t_x - \left({}^{M}_{\theta_0}T \right)_{13} z^i_{rot} - \left({}^{M}_{\theta_0}T \right)_{14} \\
&= \left(\left({}^{M}_{\theta_0}T \right)_{12} x^i_{rot} - \left({}^{M}_{\theta_0}T \right)_{11} y^i_{rot} \right) \sin\theta + \left(\left({}^{M}_{\theta_0}T \right)_{11} x^i_{rot} + \left({}^{M}_{\theta_0}T \right)_{12} y^i_{rot} \right) \cos\theta
\end{aligned} \tag{9}
$$

where $\left({}^{M}_{\theta_0}T \right)_{12}$ is the value of the first row and the second column of ${}^{M}_{\theta_0}T$, and others are similar.

There is no doubt that two results are calculated that locate at the both side of C-axis by means of the auxiliary angle formula and the normal vector of a point will determine the right result θ_c. Then, it is extremely easy to get the ideal y_c and z_c based on Eq. (7). Use the ideal θ_c, y_c and z_c as the initial iteration values and inquire all the errors, similar to the ideal case, derive an equation containing only the unknow θ from Eq. (8) and get the corresponding θ_c^k, y_c^k and z_c^k (k means number of iterations), and then repeat the process until the difference between two results is less than the threshold as presented in Eq. (10). The geometric errors are so small that it usually iterates two or three times to get the result when threshold is 10^{-4}.

$$\| \left(\theta_c^{k+1} \; y_c^{k+1} \; z_c^{k+1} \right) - \left(\theta_c^k \; y_c^k \; z_c^k \right) \| < threshold \tag{10}$$

3 Probe and Calibration

3.1 Probe Model

Up to now, these calculations are based on the assumption that the probe is not deformed. When the deformation of the probe is (Δx, Δy, Δz), the real position in MCS is given by Eq. (11).

$$\mathbf{P}^d = \begin{pmatrix} x^d \\ y^d \\ z^d \end{pmatrix} = \begin{pmatrix} x^r + \Delta x \\ y^r + \Delta y \\ z^r + \Delta z \end{pmatrix} \tag{11}$$

The deformation is related to 3D voltage output of the probe sensor noted as (p, q, r). If the voltage outputs are (p_1, q_1, r_1) and (p_0, q_0, r_0) with and without the probe deformation, it can be obtained by Eq. (12).

$$p = p_1 - p_0 \; q = q_1 - q_0 \; r = r_1 - r_0 \tag{12}$$

$$\mathbf{\Delta} = \mathbf{f}(p,q,r) = \begin{bmatrix} f_x(p,q,r) \\ f_y(p,q,r) \\ f_z(p,q,r) \end{bmatrix} \tag{13}$$

The Taylor series expansion is used to depict the nonlinear and nonorthogonal relationship and $f_x(p,q,r)$ is taken as an example in Eq. (14). As a result, they can be formulated by a third-order expansion as shown in Eq. (15). For the sake of depiction, it is called a second-order model if $\mathbf{C}$ is zero, otherwise it is a third-order model. When $\mathbf{B}$ and $\mathbf{C}$ are both zero, Eq. (15) becomes the linear case.

$$f_x(p,q,r) = \sum_{k=1}^{n} \left(p\frac{\partial}{\partial p} + q\frac{\partial}{\partial q} + r\frac{\partial}{\partial r} \right)^k f_x(0,0,0) + R_{2n} \tag{14}$$

$$
\begin{bmatrix} \Delta x \\ \Delta y \\ \Delta z \end{bmatrix} = \mathbf{A}_{3\times3} \begin{bmatrix} p \\ q \\ r \end{bmatrix} + \mathbf{B}_{3\times6} \begin{bmatrix} p^2 \\ pq \\ pr \\ q^2 \\ qr \\ r^2 \end{bmatrix} + \mathbf{C}_{3\times10} \begin{bmatrix} p^3 \\ p^2q \\ p^2r \\ pq^2 \\ pr^2 \\ pqr \\ q^3 \\ q^2r \\ qr^2 \\ r^3 \end{bmatrix} \tag{15}
$$

where $\mathbf{A}_{3\times3}$ is a matrix with 3 rows and 3 columns, and $\mathbf{B}$ and $\mathbf{C}$ are similar.

3.2 Probing Procedure

A standard sphere with high accuracy is used as an artefact. Several practical paths have to be designed and it should be feasible to carry out and similar to the actual measurement that the probe is adopted to measure a machine part after having been calibrated. Taking these two essential elements into consideration, six special scanning paths are conceived and they are blue as shown in Fig. 2 (Renishaw's paths are referenced). The sphere radius is R_1 and the ruby radius is R_2. It is specially noted that part paths need be removed to avoid interference when X-axis is fixed and the probe is horizontal.

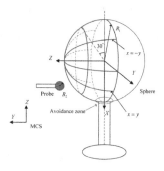

Fig. 2. Six scanning paths.

The effective deformation of the probe is usually in a small interval. Then it must be emphasized that it is vital to make sure the deformation is maintained at the reasonable range. If the interval is $[\delta_1, \delta_2]$, the radius of the sphere where paths are planned is calculated as follows:

$$
R_a = R_1 + R_2 - \delta_1 \quad R_b = R_1 + R_2 - \delta_2 \tag{16}
$$

Because of the maximum and minimum deformation, there are twelve paths in total. The position of the sphere with respect to the CCS should be determined roughly before scanning. Rotate the C-axis to three or four angles and probe several points at each angle, then calculate the corresponding coordinate $(x_{rot}, y_{rot}, z_{rot})$ according to Eq. (8). Finally the least squares method is applied to get the sphere center noted as $\mathbf{S}_0$. During probing points, it is better to ensure the deformation of the probe small to get a more accurate $\mathbf{S}_0$ by lightly contacting the sphere with a low feed speed. After getting $\mathbf{S}_0$, the coordinates of twelve scanning paths in the CCS can be obtained and the commands (θ_c, y_c, z_c) can be acquired according to the inverse kinematics of Sect. 2.3.

3.3 Calibration Algorithm

When the 3D probe has been installed and the CMM has been started, the initial value (p_0, q_0, r_0) of the voltage output can be read. During scanning, the command (θ_c, y_c, z_c) and the voltage output (p_1, q_1, r_1) corresponding to every contacting point are available in time. Based on Eqs. (6) and (11–15), the position $\mathbf{P}_i^{rot}$ of every point in the CCS, which is the function of p, q and r, can be acquired. All the points are on the sphere and its approximate radius is the sum of R_1 and R_2 because R_2 also needs to be calibrated. Consequently, the objective function is expressed as follows:

$$\min \left\{ \sum_{i=1}^{n} \left(\| \mathbf{P}_i^{rot} - \mathbf{S} \| - R \right)^2 \right\} \tag{17}$$

where $\mathbf{S}$ and R is the actual center and radius of the sphere, respectively.

Obviously, it is a typical nonlinear problem. Levenberg-Marquardt method will be used to calibrate coefficients of the 3D probe. Although Eq. (15) shows the mathematical model of the probe and the method seems to figure out it perfectly, it is arduous to get al.l the coefficients because as the order increases, the products of voltage outputs increase exponentially, and accordingly the coefficients decrease exponentially. It is not difficult to infer that the matrix $\mathbf{C}$ is much smaller than $\mathbf{A}$ in magnitude, indicating that the matrix $\mathbf{C}$ is too trivial to optimize. Hence, two adjustment coefficients K_b and K_c have to be introduced to make sure matrix $\mathbf{B}$ and $\mathbf{C}$ are approximately the same as the matrix $\mathbf{A}$ in magnitude as shown in Eq. (18).

$$\mathbf{B} = K_b \cdot \mathbf{B}^* \quad \mathbf{C} = K_c \cdot \mathbf{C}^* \tag{18}$$

It is of great importance to estimate K_b and K_c. In the second-order model, the magnitude of $\mathbf{A}$ and $\mathbf{B}$ are optimized as $\mathbf{A}_{mag}$ and $\mathbf{B}_{mag}$. Because of their approximate exponential change, K_b and K_c are roughly estimated in Eq. (19). As a result, the third-order model can be optimized. The maximum and minimum error, and the root mean square (RMS) can be obtained, which are essential to evaluate the optimization.

$$K_b = \frac{\mathbf{B}_{mag}}{\mathbf{A}_{mag}} \quad K_c = K_b \cdot K_b \tag{19}$$

4   Experimental Results

The proposed optimization method was performed on a four-axis CMM with a rotary table as shown in Fig. 1. The 3D nonlinear and nonorthogonal scanning probe is SP25M (Renishaw) whose major specifications are listed in Table 1. A 25-mm-diameter sphere whose roundness uncertainty is ±0.025 μm is the artefact. The experiment was carried out at a constant temperature room where it was maintained at 20 ± 0.5, accounting for the fact that the thermal error is negligible. Additionally, the probe was installed on the vertical axis (Z-axis) at first and the artefact was fixed on the rotary table, respectively. Moreover, the nominal radius of the ruby is 1.0000 mm. Before the experiment, the geometric error compensation of the linear axes had been completed and the PDGEs of the rotary table had been identified based on Chen's method [19].

Table 1. Specification of the 3D scanning probe.

Type	SP25M
Measurement range	±0.5 mm deflection in all directions
Overtraval range of X and Y	±2 mm
Overtraval range of +Z and −Z	1.7 mm 1.2 mm
Resolution	Capable of <0.1 μm

According to the installation of the rotary table, the X-axis was fixed at 122.0990 mm and the C-axis was approximately in YZ plane. Then the probe was rotated to the −Y direction and it became a Cylindrical CMM. Later, as stated in Sect. 3.2, the C-axis was rotated to four angles and several points were probed at each angle. The fitted sphere center S_0 was (5.65635, 15.88550, 0.00166) mm. It is important to note that there is no need to scan every path solely and some connecting points between two different paths can be added to save scanning time. Figure 3 shows one of the pictures during scanning.

The initial value of the voltage output in counts were (1221, −2831, 1953). After finishing the operation, a total of 4012 sets of data were recorded when the scanning speed was 4 mm/s and the sampling frequency 10 Hz. The initial values of $\mathbf{A}$, $\mathbf{B}$ and $\mathbf{C}$ were zero matrices. Firstly, the second-order model were optimized and $\mathbf{A}_{mag}$ and $\mathbf{B}_{mag}$ were about 10^{-5} and 10^{-10}. Admittedly, K_b and K_c could be 10^{-5} and 10^{-10}, respectively. Eventually, the results of the third-order model were able to be acquired. Part of them are displayed in Table 2 since the whole coefficients are considerable. It is compelling that the magnitudes of coefficient matrices are about 10^{-5}, 10^{-10} and 10^{-14} regardless of their not presentation in Table 2, which indicates the estimation of magnitudes from the second-order model is feasible.

As presented in Table 2, comparing with the second-order model, the third-order model has the superiority of higher accuracy in that the RMS reduces

Fig. 3. Experimental scene.

Table 2. Results of optimization.

Item	Second-order model	Third-order model
Sphere center S (mm)	$(5.65637, 15.88773, 0.00329)$	$(5.65553, 15.88427, 0.00339)$
Ruby radius R_2 (mm)	1.00018	0.99991
Maximum error (μm)	6.61	4.59
Minimum error (μm)	−7.84	−4.32
RMS (μm)	1.14	0.77

from 1.14 μm to 0.77 μm. Simultaneously, the maximum and minimum error significantly decrease to 4.59 μm and −4.32 μm so that its accuracy can reach a micron level. Although the RMS of the second-order model is just 1.14 μm that is also acceptable, the maximum and minimum error are 6.61 μm and −7.84 μm. Figure 4 denotes the error distributions and it is clear that majorities of errors are within ±2 μm . Specifically, the rates of the second-order and third-order model are 97.53% and 98.70%.

5   Verification of Accuracy

5.1   Verification of the Compensation

The third-order model should be applied to the actual measurement for a better micron accuracy although the second-order model can also reach the micron level. Another high-accuracy ball with a diameter of 44.9613 mm and a roundness of ±0.05 μm was probed to verify its accuracy and it was implemented at the same constant temperature room as shown in Fig. 5. The results are listed in Table 3.

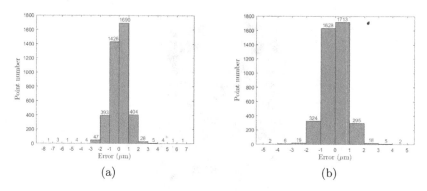

Fig. 4. Distribution of errors. (a) Errors of second-order model. (b) Errors of third-order model.

Fig. 5. Probing another ball.

As shown in Table 3, the accuracy is virtually the same when it comes to RMS. However, the radius of using compensation of the third-order model is closer to the ground truth value 22.48065 mm and its error is just 2.90 μm, while the radius error of the second-order model is 4.41 μm. In conclusion, two compensations can both reach a micron level and the compensation of a third-order model is better. Theoretically, the higher order Taylor expansion we take, the higher accuracy we acquire. However, it will undoubtedly increase the complexity of the probe model and it is definitely harder to optimize. On the contrary, it is almost impossible to upgrade measurement accuracy rapidly. Taking the model complexity and measurement accuracy into account, the third-order model is satisfied with a great deal of high-accuracy demand in industry.

Table 3. Results of the ball measurement.

Number	The second-order model		The third-order model	
	Radius (mm)	RMS (μm)	Radius (mm)	RMS (μm)
1	22.48548	1.77	22.48304	1.53
2	22.48547	1.65	22.48388	1.70
3	22.48482	1.79	22.48372	1.68
4	22.48526	1.79	22.48296	1.82
5	22.48503	1.73	22.48314	1.65
6	22.48432	1.66	22.48356	1.71
Average	22.48506		22.48355	
ΔR (mm)	0.00441		0.00290	
Std deviation	0.00045		0.00038	

5.2   Uncertainty Analysis

An uncertainty analysis to investigate compensations of the two models is executed based on the expression described in [13]. The uncertainty is given by Eq. (20).

$$u_c = \sqrt{u_a^2 + u_b^2} \tag{20}$$

where u_a and u_b are the standard deviation of Type A and Type B.

The standard deviation of Type A is acquired by means of the measurement data presented in Table 3. The results of the second-order model and the third-order model are:

$$u_{a2} = \frac{0.00045}{\sqrt{6}} = 0.00018 \quad u_{a3} = \frac{0.00038}{\sqrt{6}} = 0.00015 \tag{21}$$

The standard deviation of Type B is estimated based on the resolution of the probe (0.1 μm in Table 1) and the roundness uncertainty of the sphere artifact (± 0.05 μm). According to a uniform distribution, it is estimated as 0.06 μm by Eq. (22).

$$u_b = \sqrt{u_{probe}^2 + u_{sphere}^2}$$
$$u_{probe} = \frac{0.1}{\sqrt{3}} \quad u_{sphere} = \frac{0.05}{\sqrt{3}} \tag{22}$$

As a result, the uncertainties of the second-order and third-order model are 0.19 μm and 0.16 μm. Accordingly, the expanded uncertainties (coverage factor $k = 2$) are 0.38 μm and 0.32 μm , which have a level of confidence of 95%.

6   Conclusion

Contact measurement has been widely used in a mass of applications of high-accuracy manufacture in which probes play a great role. In this paper, a nonlinear and nonorthogonal model is established and a systematic optimization method is proposed to figure out the problem. Taylor series expansion makes sure of its correctness so that the probe model is successfully built. A high-accuracy sphere with ± 0.025 μm roundness uncertainty is the artefact and several practical and feasible scanning paths are designed to calibrate the probe. During the calibration, the optimization of the third-order model is carried out with the help of two estimated adjustment coefficients derived from the former second-order optimization. Compared with the second-order model, the third-order model has the merit of higher accuracy because RMS is just 0.77 μm and errors in ± 2 μm account for 98.70%. In the case of another ball measurement, the measurement accuracy is verified with a more accurate radius whose error is just 2.90 μm and the expanded uncertainty is 0.32 μm. As a result, it is obviously demonstrated that the calibration method is effective and precise.

Acknowledgement. This research was supported by the key research project of Ministry of Science and Technology (Grant No. 2017YFB1301503).

References

1. Li, Y.D., Gu, P.H.: Free-form surface inspection techniques state of the art review. Comput. Aid. Des. **36**(13), 1395–1417 (2004)
2. Li, J., Xie, F., Liu, X.-J., Li, W., Zhu, S.: Geometric error identification and compensation of linear axes based on a novel 13-line method. Int. J. Adv. Manuf. Technol. **87**(5), 2269–2283 (2016). https://doi.org/10.1007/s00170-016-8580-x
3. Zhu, S., Ding, G., Qin, S., Lei, J., Zhuang, L., Yan, K.: Integrated geometric error modeling, identification and compensation of CNC machine tools. Int. J. of Mach. Tools Manufact. **52**(1), 24–29 (2012)
4. Okafor, A.C., Ertekin, Y.M.: Derivation of machine tool error models and error compensation procedure for three axes vertical machining center using rigid body kinematics. Int. J. Mach. Tool Manuf. **40**(8), 1199–1213 (2000)
5. Sun, G., He, G., Zhang, D., Sang, Y., Zhang, X., Ding, B.: Effects of geometrical errors of guideways on the repeatability of positioning of linear axes of machine tools. Int. J. Adv. Manuf. Technol. **98**(9), 2319–2333 (2018). https://doi.org/10.1007/s00170-018-2291-4
6. Ibaraki, S., Oyama, C., Otsubo, H.: Construction of an error map of rotary axes on a five-axis machining center by static R-test. Int. J. Mach. Tool Manuf. **51**(3), 190–200 (2011)
7. Xiang, S., Yang, J.: Using a double ball bar to measure 10 position-dependent geometric errors for rotary axes on five-axis machine tools. Int. J. Adv. Manuf. Technol. **75**(1), 559–572 (2014). https://doi.org/10.1007/s00170-014-6155-2
8. Huang, N.D., Bi, Q.Z., Wang, Y.H.: Identification of two different geometric error definitions for the rotary axis of the 5-axis machine tools. Int. J. Mach. Tool Manuf. **91**, 109–114 (2015)

9. Bi, Q.Z., Huang, N.D., Sun, C., Wang, Y.H., Zhu, L.M., Ding, H.: Identification and compensation of geometric errors of rotary axes on five-axis machine by on-machine measurement. Int. J. Mach. Tool Manuf. **89**, 182–191 (2015)

10. Chen, Y.-T., More, P., Liu, C.-S.: Identification and verification of location errors of rotary axes on five-axis machine tools by using a touch-trigger probe and a sphere. Int. J. Adv. Manuf. Technol. **100**(9), 2653–2667 (2018). https://doi.org/10.1007/s00170-018-2863-3

11. Chen, Y.-T., More, P., Liu, C.-S., Cheng, C.-C.: Identification and compensation of position-dependent geometric errors of rotary axes on five-axis machine tools by using a touch-trigger probe and three spheres. Int. J. Adv. Manuf. Technol. **102**(9), 3077–3089 (2019). https://doi.org/10.1007/s00170-019-03413-x

12. Bastas, A.: Comparing the probing systems of coordinate measurement machine: scanning probe versus touch-trigger probe. Measurement **156**, 107604 (2020)

13. Evaluation of measurement data–guide to the expression of uncertainty in measurement. https://www.bipm.org/en/publications/guides/

A Matching Algorithm for Featureless Sparse Point Cloud Registration

Zeping Wu[1], Yilin Yang[1], Xu Zhang[1(✉)], and Lin Zhang[2]

[1] School of Mechatronic Engineering and Automation, Shanghai University,
Shanghai, China
xuzhang@shu.edu.cn
[2] HUST-Wuxi Reasearch Institute, No 329 Yanxin Road,
Huishan District, WuXi 214100, China

Abstract. Be confronted with the challenges of efficiency and accuracy, point cloud registration, as a universal technique adopted in vision system, has always been used for large-dimension workpieces measurement. In this paper, we present a matching algorithm for determining the transformation relation between local point cloud and global point cloud with corresponding points unknown. First, multilinked lists of distance for point clouds are constructed with k-D tree. Then, a closed-traversal matching algorithm is proposed, which uses subgraph isomorphism to find possible matching results. The possible results still need further verification by recursive to get a credible matching result. In the end, a method is designed to solve and verify the transformation matrix by singular value decomposition. The performance of the algorithm is evaluated with the actual data obtained from vision measuring system. The experiments show that the algorithm is of high performance and efficiency and can be applied to practical problems of point cloud registration.

Keywords: Point cloud registration · Vision measuring system · k-d tree · Subgraph isomorphism

1 Introduction

In the field of industrial manufacturing, it's difficult to comprehensively measure large workpieces, especially those with complex curved surfaces, at only one viewpoint by a binocular structure-light scanner. On the other hand, when scanning from multiple viewpoints, due to the fact that the camera poses are unknown, there is an essential procedure to unify the results into a general coordinate system, which is also known as the automatic registration technique. Nowadays, though lots of solutions of point cloud registration have been proposed, they are still not accurate and efficient enough to obtain the precise 3D point cloud simultaneously during measurement.

The registration problem is comprised of two related sub-problems: corresponding points matching and motion estimation. The former means finding the corresponding points from different point cloud pieces. The latter can also

© Springer Nature Switzerland AG 2020
C. S. Chan et al. (Eds.): ICIRA 2020, LNAI 12595, pp. 16–27, 2020.
https://doi.org/10.1007/978-3-030-66645-3_2

be regarded as minimization of the distance between the corresponding points. After achieving corresponding points matching of high accuracy, iterative closest point [1] and singular value decomposition (SVD) [2] can be used to obtain accurate motion estimation. An efficient registration method is fundamental for 3D reconstruction.

Some researchers have studied and improved the problem in many aspects. Li. et al. [3] proposed a method to unify coordinate systems, and a multiaxial 3D laser scanning system consisting of a portable 3D laser scanner, an industrial robot and a turntable is constructed. Y. Ye et al. [4] developed a turntable-base structured light system for the automatically 3ED scanning, which could get the center and the direction of the axis of rotation quickly and accurately. Both of the above methods rely on specialized equipment to realize automatic registration, which will also bring great limitations in practical operation for immobile or large size objects.

Some researchers adopted feature information of objects to splice point clouds, such as 3D feature descriptors or 2D image features. Kim et al. [5] proposed a registration method based on principal component analysis (PCA). The eigenvector of point cloud is calculated with the covariance matrix and then three direction of the feature point cloud is obtained. Rusu R B et al. [6] proposed fast point feature histograms for 3D registration. The method results in a new type of local features, which retains most of the discriminative power of point feature histograms. 2D image feature matching method aims to solve the model-to-image registration problem. Joon Kyu Seo et al. [7] presented a new way for 3D distance data registration based on 2D local photometric characteristics. Zhang et al. [8] proposed a method of combining coded and non-coded marked points to match corresponding points. As a classical algorithm, SoftPOSIT [9] is also widely used for solving the registration problem, which is to treat all possible matches identically throughout the search for an optimal pose. The above-mentioned registration methods depend on the shape information of the objects or the strict experimental conditions, which is difficult to be popularized in practice.

To solve the model-to-model registration problem, an algorithm that realized by searching the corresponding points between local point cloud and global point cloud with non-coded marked points is proposed in this paper. We assume that there is no additional information constraining the pose of the object or the correspondences except for the 3D positions of non-coded marked points in the global and local coordinates. This paper uses the k-D tree structure to process point cloud data as well as the principle of subgraph isomorphism to match corresponding points. With the aid of high-precision vision measuring system, the algorithm shows satisfactory performance.

The paper is organized as follows. Section 2 introduces multilinked lists of distance. Section 3 expounds the detailed matching algorithm based on closed-traversal matching. Section 4 introduces a method for solving and verifying the transformation matrix. Section 5 presents the experiment results of the proposed algorithm and Sect. 6 concludes.

2   Multilinked List of Distance

Starting from the head of each linked list, the distances from other points to the current point along with their serial numbers are stored according to the order from near to far. The information can be recorded by adjacency matrix. In order to save space and search time, it is not necessary to calculate all the distances between each other but only an appropriate amount in each linked list, which is sufficient to ensure corresponding points matching. Multilinked list of distance is adopted to record the distance information, in which each single point constitutes the head of a separate linked list. Those suitable points with the shortest distance will be selected by a k-D binary tree. Compared with the time complexity of adjacency matrix $O(n^2)$, the method shows better real-time performance with its time complexity reduced to $O(nlogn)$.

The k-D tree is a generalization of a transformation from k dimensions to the same one-dimensional structure which is fundamentally a binary tree [10]. In our case, k is set 3. A 3-D tree is constructed by the following steps. Firstly, a plane perpendicular to one dimension (one of the x, y, z axes) is chosen as the segmentation interface to split the whole space into two subspaces such that there are approximately equal numbers of points in either subspace. We store the median point as the root node and then obtain a left subtree and a right subtree. Each subtree is further split into two smaller subspaces in another dimension. We generally loop x, y, z axis by default. Above operations will be repeated until the data can no longer be divided. The searching for the closest point cloud pieces is conducted by binary search and verified backtracking. Each node that is relatively close is traversed by comparison in each dimension, and the most probable points with the closest distance will be found by the backtracking method. Once the nearest point is found, it will be recorded and removed, then search again to find the next nearest point. The process repeats continually until all the required points are found. By constructing and querying k-D tree, the closest neighbor point cloud pieces are searched in $O(nlogn)$ time complexity with $O(n)$ storage complexity, which are both optimal [11].

Algorithm 1. Algorithm for multilinked list of distance

Input: The set of 3D points $P_i = (X_i, Y_i, Z_i), i = 1, ..., n$;
Output: Distance multiple linked list L;
1: **function** CreateDistanceMultipleLinkedList (P)
2:   create the k-d tree T of P
3:   **for** i = 1 **to** n **do**
4:      search for 10 closest points to P_i in T, and save ids in list v and distances in list d
5:      create a new list l, and update $L(i) = l$
6:      **for** j = 1 **to** 10 **do**
7:         $l \rightarrow val = \{v(j), d(j)\}, l = l \rightarrow next$
8:   **return** L;
9: **end function**

The multilinked list of distance is created based on the k-d tree. The pseudocode is shown in Algorithm 1. P is the set of 3D coordinate points of a point cloud. The number of the point is n. L represents Multilinked list of distance. The k-d tree data structure is provided in OpenCV.

3   Closed-Traversal Matching Algorithm

Referring to the distribution of point clouds in space, a matching method using subgraph isomorphism is proposed. In order to simplify the problem, we assume that the distance between each two points in the local point cloud is the same as that of the corresponding two points in the global point cloud, which means that there is no accuracy loss in the process of 3D reconstruction.

In this framework, each possible matching result is successively generated by traversal. We don't worry about getting bad results because the backtracking method has a good chance of correctly telling apart right matching results from false ones and, by inference, there is smaller risk of inadvertently selecting the erroneous structure for a subgraph. Moreover, as shown in Algorithm 1 multilinked list only stores the closest point cloud pieces so that a number of steps that can be skipped, and that helps to improve the efficiency of matching.

3.1   Subgraph Isomorphism

Graph is one of the most common data structures which is an effective method to find the contact between data. Subgraph isomorphism is able to find a copy of some previously described smaller graph. In this paper, our intention is to determine the corresponding points with equal distances in two graphs distances by subgraph isomorphism.

The subgraph isomorphism problem is, given two multilinked lists of distance M and N, we will find whether there is a subgraph isomorphism from N to M. If such corresponding relationship exists, we say that N is a copy of a part of M. Figure 1. shows an example of subgraph isomorphism for global and local point cloud.

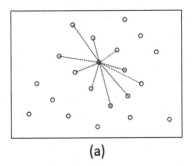 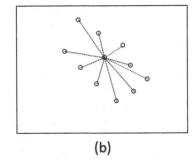

(a)   (b)

Fig. 1. Subgraph isomorphism for local point cloud and global point cloud. (a) Global point cloud. (b) Local point cloud. The dots connected by the dotted lines represent the corresponding part of the point clouds, which is the subgraph isomorphism.

The data would have been traversed to match corresponding points, because of the mutual connection with each point, however, we can determine the next corresponding point with the aid of the distance information of the last matched point instead of repeatedly calculating all points' distance information. In other words, the adjacent points with the same distance away from one of a pair of corresponding points that have been identified in the two graphs can also match. Therefore, all point correspondences can be deduced as soon as only one pair of corresponding points is obtained.

As the pseudocode in the first function of Algorithm 2 shows, the process takes place mainly in two loops, along with the traversal that selecting the corresponding points and determining the relationship between adjacent points by Euclidean distance. After creating the multilinked lists of distance from two pieces of point clouds, we start to compare each point. Considering the error of point cloud reconstruction in practice, it is unlikely to be a perfect matching so that we adopt two threshold values to match the corresponding point: the threshold of distance error T1 and the threshold of matching numbers T2. If T1 is too small or T2 is too large, little points will be matched, which leads to the failure to find a result. If T1 is too large or T2 is too small, points of false positives will contaminate the result, which easily leads to mismatch.

At the end of the traversal, a void return value of the function signifies there are no corresponding points. If a pair of possible corresponding points is found, the procedure will immediately suspend and then moves on to the next step.

3.2 Verification of Matching Correctness

A recursive process is used to verify the correctness of the previous possible solution, where the backtracking algorithm proposed by Bron [12] is applied. The algorithm relies on the depth-first search which explores the solution set in depth from a possible solution. The program will backtrack to the original traversed location as soon as it searches downward for a mismatch.

A matching pair is selected from the possible solution at random at the beginning, aiming at verifying whether the subsequent iterations meet the expectations: we'll come to a potential solution if the times of iteration is sufficient. Still based on the distance error threshold and the matching numbers threshold, we use recursion to solve the problem. The procedure of comparison is the same as that in the previous subsection apart from marking the times of iteration. What's more, appropriate thresholds should be adopted according to the actual situation in order to ensure high probability of perfect matching. This part is summarized in the second function of Algorithm 2.

Those typical methods mentioned in the introduction that focus on object characteristics and turntable control exist computational difficulties and scene constrains. In addition, point cloud registration without pairs of definite corresponding points is inaccurate. A slew of coded values captured in each perspective, however, will bring some trouble for coding and decoding. The matching algorithm proposed in this paper shows two significant advantages over conventional methods. Firstly, the problem of complex surface registration is transformed into the problem of matching non-coded marked points. Secondly, it does not involve the calculation of complex formulas and will not fall into an infinite loop, which means that the obtained result actually depends on the accuracy of the device.

Algorithm 2. Algorithm for closed-traversal matching

Input: the set of global point cloud $M_i = (X_i, Y_i, Z_i), i = 1,, m$; the set of local point cloud $N_i = (x_i, y_i, z_i), i = 1,, n$; the threshold of distance error T1; the threshold of matching number $T2$;

Output: the list of matching pair R;

1: **function** SubgraphMatching(M, N)

2: L_M = CreateDistanceMultipleLinkedList(M)

3: L_N = CreateDistanceMultipleLinkedList(N)

4: **for** i←1 **to** m **do**

5:     **for** j←1 **to** n **do**

6:         $n_{condition} = \sum\limits_{k=0}^{9} \sum\limits_{h=0}^{9} (|L_M(i) \to next(k) \to val.d - L_N(j) \to next(h)$

7:         $\to val.d| < T1)$

8:         $\mathbf{R} = \{(L_M(i) \to next(k) \to val.v, L_N(i) \to next(k) \to val.v)$

9:         $: (|L_M(i) \to next(k) \to val.d - L_N(j) \to next(h) \to val.d| < T1,$

10:         $k = 0, \ldots, 9, \ h = 0, \ldots, 9$

11:         **if** $n_condition > T2$

12:             randomly select a matching pair (u,v) from $\mathbf{R}$

13:             $t \leftarrow 0, \ flag \leftarrow VerifyMatchCorrection(L_M, L_N, u, v, t)$

14:             **if** flag is true

15:                 **Return R**

16:     **return** void

17: **end function**

18:

19: **function** VerifyMatchCorrection(D_M, D_N, u, v, t)

20:     **if** t>4

21:         **return** true

22:     $n_{condition} = \sum\limits_{k=0}^{9} \sum\limits_{h=0}^{9} (|L_M(i) \to next(k) \to val.d - L_N(j) \to next(h) \to val.d| <$ $T1)$

23:     $r = \{| (L_M(i) \to next(k) \to val.v, L_N(i) \to next(k) \to val.v) : L_M(i) \to$

24:     $next(k) \to val.d - L_N(j) \to next(h) \to val.d| < T1, k = 0, \ldots, 9, \ h = 0, \ldots, 9\}$

25:     **if** $n_{condition} > T2$

26:         t=t+1

27:         randomly select a matching pair (u,v) from r

28:         flag←VerifyMatchCorrection(L_M, L_N, u, v, t)

29:     **else**

30:         flag←false

31:     **return** $flag$;

32: **end function**

---

4   Point Cloud Registration

After matching corresponding points of the global and local point clouds, the last process is to compute the transformation matrix which is the best fitting of the two 3D point sets. Our process consists of the following two steps: solve and verify the transformation matrix. The transformation matrix can be quickly solved by SVD. In order to further confirm the correctness of the matching, it is necessary to calculate the cost function to determine whether the transformation matrix is right.

According to the previous matching relationship, two corresponding 3D point sets are obtained, the global point set G_i and the local point set $L_i = 1, ..., K$ (G_i and L_i are 3 × 1 column matrices). The transformation matrix can be written as

$$G_i = RL_i + T \tag{1}$$

where R is a 3 × 3 rotation matrix and T is a 3 × 1 translation vector. R and T constitute the transformation matrix. Then the problem can be formulated as minimization of the cost function

$$J^2 = \sum_{i=1}^{K} G_i - (RL_i + T)^2 \tag{2}$$

If the centroids of the two datasets are moved to the origin, we only need to calculate the rotation matrix. The set of points with the centroid at the origin are

$$g_i = G_i - c \tag{3}$$

$$l_i = L_i - c' \tag{4}$$

where

$$c = \frac{1}{K} \sum_{i=1}^{K} G_i \tag{5}$$

$$c' = \frac{1}{K} \sum_{i=1}^{K} L_i \tag{6}$$

The cost function can be rewritten as

$$J^2 = \sum_{i=1}^{K} g_i - Rl_i^2 \tag{7}$$

To minimize this objective function, we use SVD method to solve the problem

$$[U, \ S, V] = SVD(H) \tag{8}$$

where

$$H = \sum_{i=1}^{K} g_i l_i \tag{9}$$

Then the optimal rotation matrix R' can be computed

$$R' = VU^T \tag{10}$$

The optimal translation matrix T' can be returned as

$$T' = -R' * c + c' \tag{11}$$

Finally, the matrices R' and T' can be substituted back into the Eq. (2). We deem that the transformation matrix is desired if the computational cost J is lower than the expected value. Supposing that the outliers are mixed in the sets of points, the RANSAC technique would be used to combat against outliers which uses 3 points at a time [13].

5   Experiments

One of the most important factors affecting the algorithm performance is the success rate of matching, and that is the key issue this section will focus on. In the meanwhile, we also care about the actual running time of the algorithm.

5.1   Results and Analysis

The algorithm has been evaluated in terms of point clouds registration. The 3D data point clouds come from photogrammetric system [14] and line structured-light scanner [15], and the accuracy of the two devices are within $\pm 0.02\,\text{mm/m}$ and $\pm 0.04\,\text{mm/m}$ respectively. The point cloud obtained from photogrammetric system is called global point cloud and the point cloud obtained from stereo vision system is called local point cloud. To be more specific, the coordinates of these point clouds represent the positions of the reflective non-coded marked points that attached to the measured object. In our experiment, the number of the global point cloud is 57. Our matching pipeline is entirely written in C++, and all the experiments are conducted on a PC with 2.5 GHz CPU.

We consider a matching result to be good when the global point cloud can be accurately projected onto the image obtained from the stereo vision system by a computed transformation matrix. If the projected points cannot match the reflective marked points on the image, the algorithm declares failure. Figure 2. shows an example of the successful matching result.

Figure 3 shows two experimental results before and after point cloud registration. The point cloud in white represents scanning point cloud, while the point cloud in red represents local point cloud and the point cloud in green represents global point cloud. For local parts of the global point cloud, this algorithm can always find the right corresponding points. Of special attention is that it makes no difference to the outcome even if there are a few outliers, which means some points are away from or absent from the point cloud in red. Under the consideration, appropriate thresholds should be set according to experimental conditions.

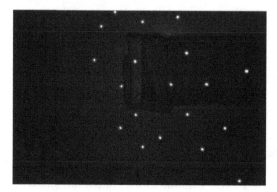

Fig. 2. A good matching result with projected points overlapped. The white dots are the reflective marked points and the red circles are projections of the global point cloud.

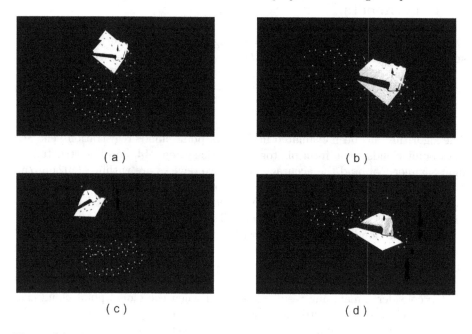

Fig. 3. (a) The initial point cloud in the first experiment. (b) The point cloud after point cloud registration in the first experiment. (c) The initial point cloud in the second experiment. (d) The point cloud after point cloud registration in the second experiment (Color figure online).

Figure 4 shows the pass rate and success rate of the algorithm function with fixed matching number threshold and that with fixed distance error threshold. The pass rate represents the proportion of valid results and the success rate represents the proportion of the results confirmed to be good. Obviously, the looser the threshold, the higher the pass rate, but the lower the success rate after reaching a certain limit. This is because points are easily mismatched in a looser threshold. In our experiment, the success rate could achieve a high level with the threshold of matching numbers fixed at 4. Considering the overall accuracy of the devices, the algorithm could perform well if the threshold of distance error set a little looser than that above.

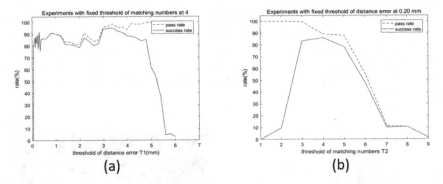

(a)   (b)

Fig. 4. Pass and success rate obtained by the algorithm operation with fixed thresholds. (a) Fix the threshold of matching numbers at 4. (b) Fix the threshold of distance error at 0.20 mm.

The time complexity of our algorithm is $O(MN)$, where M is the number of global point cloud and N is the number of the local point cloud. Comparison experiments are performed to test the efficiency of the proposed method. Our method shorten time in half compared to ICP. The comparison experiments of time consumption are shown in Fig. 5.

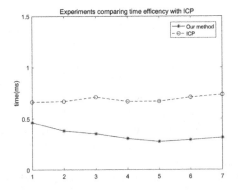

Fig. 5. The efficiency comparison with ICP algorithm.

5.2   Application

In the 3D measurement of large dimension workpieces based on point cloud registration, the workpieces need to be scanned for many times. In this subsection,
the matching algorithm is integrated into the industrial vision measuring system for large dimension workpieces. As shown in Fig. 6. (a), reflective non-coded
marked points are used to splice the point clouds obtained by vision measuring
system. The vision measuring system is controlled by a robotic arm to move to
scan different areas. The 3D coordinates of the non-coded marked points have
been measured in advance by photogrammetry. The measurement result is shown
in Fig. 6. (b). The result was good, and this algorithm was proved to be effective
for point cloud registration.

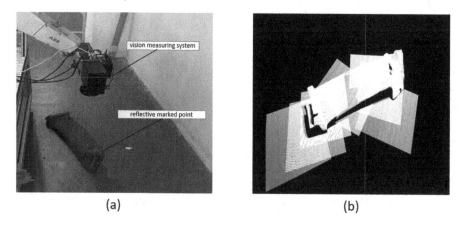

Fig. 6. (a)Experiment scene. (b)Measurement result.

6   Conclusions

In this paper, we have proposed and tested the matching algorithm to determine
the corresponding points between local point cloud and global point cloud with
a small amount of featureless points. This algorithm can be used as a component in industrial vision measuring system for large dimension workpiece. Our
experiment indicates that the algorithm performs well on real data and can meet
the real-time requirement of measurement. Another advantage of this algorithm
is that there is no demand for coded marked points, which is no limit to coded
values.

Acknowledgments. This research was partially supported by the key research
project of the Ministry of Science and Technology (Grant No. 2017YFB1301503) and
the National Natural Science Foundation of China (Grant No. 51975344).

References

1. Besl, P.J., Mckay, N.D.: A method for registration of 3-D shapes. In: Sensor Fusion IV: Control Paradigms and Data Structures. IEEE Computer Society (1992)
2. ARUN, K. S. Least-squares fitting of two 3-D point sets. IEEE Trans. Pattern Anal. Mach. Intell. (1987)
3. Li, J., Chen, M., Jin, X., et al.: Calibration of a multiple axes 3-D laser scanning system consisting of robot, portable laser scanner and turntable. Optik - Int. J. Light Electron Opt. **122**(4), 324–329 (2011)
4. Ye, Y., Song, Z.: An accurate 3D point cloud registration approach for the turntable-based 3D scanning system. In: IEEE International Conference on Information & Automation. IEEE (2015)
5. Kim, S.H., Seo, J., Jho, C.W.: Automatic registration of 3D data sets from unknown viewpoints. Proc. Workshop Front. Comput. Vers. **12**(2), 155–159 (2003)
6. Rusu, R.B., Blodow, N., Beetz, M.: Fast point feature histograms (FPFH) for 3D registration. In: IEEE International Conference on Robotics & Automation. IEEE (2009)
7. Seo, J.K., Sharp, G.C., Lee, S.W.: Range data registration using photometric features. In: IEEE Computer Society Conference on Computer Vision & Pattern Recognition. IEEE (2005)
8. Zhang, W., Zhang, L., Zhang, H., et al.: 3D reconstruction from multiple perspective views with marked points. China Mech. Eng. **17**(16), 1711–1715 (2006)
9. David, P., Dementhon, D., Duraiswami, R., et al.: SoftPOSIT: simultaneous pose and correspondence determination. In: European Conference on Computer Vision. Springer, Berlin, Heidelberg (2002)
10. Bentley, J.L.: Multidimensional binary search trees used for associative searching. Commun. ACM **18**(9), 509–517 (1975)
11. Preparata, F.P, Shamos, M.I.: Computational Geometry (1985).https://doi.org/10.1007/978-1-4612-1098-6
12. Bron, C., Kerbosch, J.: Algorithm 457: finding all cliques of an undirected graph. Commun. ACM **16**(9), 575–577 (1973)
13. Martin, A., Fischler, et al. Random sample consensus: a paradigm for model fitting with applications to image analysis and automated cartography. Commun. ACM, **24**(6), 381–395 (1981)
14. Luhmann, T., Robson, S., Kyle, S.A, et al.: Close Range Photogrammetry: Principles, Techniques and Applications. Blackwell Publishing Ltd, Hoboken (2006)
15. Ozturk, A.O., Halici, U., Ulusoy, I., et al.: 3D face reconstruction using stereo images and structured light. In: IEEE 16th Signal Processing and Communications Applications Conference, Didim, Turkey, vol. 1–2, pp. 878–881. IEEE (2008)

Iterative Phase Correction Method and Its Application

Li Chen[1(✉)], Jin Yun[2], Zhang Xu[2], and Zhao Huan[2]

[1] School of Mechanical Science and Engineering, Huazhong University of Science and Technology, Wuhan, China
lichenhaod@126.com
[2] School of Mechatronic Engineering and Automation, Shanghai University, Shanghai, China
xuzhang@shu.edu.cn

Abstract. The iterative Gaussian filter method is proposed to eliminate the phase error of the wrapped phase (which is recovered from the low-quality fringe images). The main approach is regenerating the fringe images from the wrapped phase and performed the iterative Gaussian filter. Generally, the proposed iterative Gaussian filter method can filter the noise without interference from reflectivity, improve the measurement accuracy and recover the wrapped phase information from the low-quality fringe images. The proposed method is verified by the experiment results. For the binocular system, the proposed method can improve the measurement accuracy (the root mean square (RMS) deviations of measurement results can reach 0.0094 mm).

Keywords: Iterative phase correction · Phase recovery · Binocular structured · Measurement accuracy · Phase shift image

1 Introduction

Optical three-dimensional (3D) measurement methods, are playing an increasingly important role in modern manufacturing, such as structured light method [1] and phase measurement deflectometry [2]. Where, the phase shifting technique [3] is used to determine the phase information. The accuracy of phase information directly determines the measurement accuracy. Generally, the quality of the phase shift images (contains: noise, non-linear intensity and surface reflectivity), the number of phase shift steps and the intensity modulation parameter are the main reasons that affect phase accuracy. Increasing the number of phase shift steps will greatly reduce the measurement speed. And generally, the adjustment range of intensity modulation parameter is limited. In summary, it is an effective way to improve the accuracy of phase recovery by suppressing the phase error caused by the low-quality of the fringe images.

Filtering methods are used to inhibit the noise in captured fringe images. For instance, Gaussian filtering [4] is used to inhibited the phase errors caused by Gaussian noise in the captured fringe images. The fuzzy quotient space-oriented partial differential equations filtering method is proposed in literature [5] to inhibit Gaussian noise

© Springer Nature Switzerland AG 2020
C. S. Chan et al. (Eds.): ICIRA 2020, LNAI 12595, pp. 28–37, 2020.
https://doi.org/10.1007/978-3-030-66645-3_3

contrary to literature [4]. The median filtering [6] is used to preprocess the captured images and filter out the invalid data by the masking. The wavelet denoising method and Savitzky-Golay method are proposed in literature [7, 8] and literature [4], respectively. The captured images are converted to the frequency domain for filtering [9] the captured fringe images.Thereby the influence of noise is inhibited and the accuracy of phase recovery is improved. The gamma value of the light source [10, 11] is calibrated to correct the non-linear intensity. Tn literature [12], the measurement error (caused by Gamma) is inhibited by the Gamma calibration method expressed as Fourier series and binomial series theorem. A robust gamma calibration method based on a generic distorted fringe model is proposed in literature [13]. In literature [14], a gamma model is established to inhibit phase error by deriving the relative expression.

The multi-exposure and polarization techniques are applied to solve the measurement of objects with changes in reflectivity. For instance, multi-exposure technique [15] is proposed to measure objects with high reflectivity. Among them, reference image with middle exposure is selected and used for the slight adjustment of the primary fused image. High signal-to-noise ratio fringe images [16, 17] are fused from rough fringe images with different exposures by selecting pixels with the highest modulated fringe brightness. In literature [18], the high dynamic range fringe images are acquired by recursively controlling the intensity of the projection pattern at pixel level based on the feedback from the reflected images captured by the camera. The absolute phase is recovered from the captured fringe images with high dynamic range by multi-exposure technique. Spatially distributed polarization [19] state is proposed to measure objects with high contrast reflectivity. Generally, the degree of linear polarization (DOLP) is estimated, and the target is selected by DOLP, finally the selected target is reconstructed. The polarization coded can be applied for the target enhanced depth sensing in ambient [20]. But, the polarization technique is generally not suitable for the measurement of complex objects.

In general, the factors (noise, non-linear intensity and surface reflectance changes) that affect the quality of the captured phase shift fringe images are comprehensive and not isolated. When the noise is suppressed by the filtering method, the captured images are distorted by the interference surface reflectivity. And the multi-exposure [15–18] limited in the measurement speed for its large number of required project images. in order to improve the phase accuracy from the low-quality fringe images (affected by noise, non-linear intensity and surface reflectance changes), an iterative Gaussian filter method is proposed. The main approach is regenerating the fringe images from the wrapped phase and performed the iterative Gaussian filter. Generally, the proposed iterative Gaussian filter method can filter the noise without interference from reflectivity, improve the measurement accuracy and recover the wrapped phase information from the low-quality fringe images.

2 Principle of Iterative Phase Correction Method

For the optical 3D measurement methods, the standard phase shift fringe technique is widely used because of its advantages of good information fidelity, simple calculation and high accuracy of information restoration. A standard N-steps phase shift algorithm

[21] with a phase shift of $\pi/2$ is expressed as

$$I_n(x, y) = A(x, y) + B(x, y) \cos[\phi(x, y) + \frac{(n-1)}{N}2\pi], \quad n = 1, 2, \cdots N, \quad (1)$$

where $A(x, y)$ is the average intensity, $B(x, y)$ is the intensity modulation, N is the number of phase step, and $\phi(x, y)$ is the wrapped phase to be solved for. $\phi(x, y)$ can be calculated from the Eq. (1).

$$\phi(x, y) = \arctan\left\{ \frac{\sum\limits_{n=1}^{N} I_n \sin\left(\frac{2\pi(n-1)}{N}\right)}{\sum\limits_{n=1}^{N} I_n \cos\left(\frac{2\pi(n-1)}{N}\right)} \right\}. \quad (2)$$

The phase error in wrapped phase (ϕ) is propagated from the intensity (I_n). The relationship between the intensity standard variance σ_{I_n} and the phase standard variance σ_ϕ is calculated by the principle of the error propagation and described as

$$\sigma_\phi^2 = \sum_{n=1}^{N} \left[\left(\frac{\partial \phi}{\partial I_n}\right)^2 \sigma_{I_n}^2 \right]. \quad (3)$$

$\frac{\partial \phi}{\partial I_n}$ (The partial derivatives of ϕ to I_n) is derived from Eq. (2).

$$\frac{\partial \phi}{\partial I_n} = \frac{1}{1 + \left(\frac{\sin(\phi)}{\cos(\phi)}\right)^2} \left\{ \frac{\left[\sum\limits_{n=1}^{N} I_n \cos\left(\frac{2\pi(n-1)}{N}\right)\right] \sin\left(\frac{2\pi(n-1)}{N}\right) \left[\sum\limits_{n=1}^{N} I_n \sin\left(\frac{2\pi(n-1)}{N}\right)\right] \cos\left(\frac{2\pi(n-1)}{N}\right)}{\left[\sum\limits_{n=1}^{N} I_n \cos\left(\frac{2\pi(n-1)}{N}\right)\right]^2} \right\}$$

$$- \frac{1}{1 + \left(\frac{\sin(\phi)}{\cos(\phi)}\right)^2} \left\{ \frac{2\sin(\phi)\sin\left(\frac{2\pi(n-1)}{N}\right) - 2\cos(\phi)\cos\left(\frac{2\pi(n-1)}{N}\right)}{NB\cos^2(\phi)} \right\}$$

$$= -\frac{2}{NB} \sin\left(\phi - \frac{2\pi(n-1)}{N}\right). \quad (4)$$

The relationship between the wrapped phase and the intensity is expressed as follows.

$$\begin{cases} \sum\limits_{n=1}^{N} I_n \sin\left(\frac{2\pi(n-1)}{N}\right) = \frac{NB}{2}\sin(\phi), \\ \sum\limits_{n=1}^{N} I_n \cos\left(\frac{2\pi(n-1)}{N}\right) = \frac{NB}{2}\cos(\phi). \end{cases} \quad (5)$$

In addition, the influence of error sources to N images is not different from one to another, therefore:

$$\sigma_{I_1} = \sigma_{I_2} = \cdots = \sigma_{I_N} = \sigma_I. \quad (6)$$

In summary, the relationship between σ_{I_n} and σ_ϕ is calculated and expressed as

$$\sigma_\phi = \sqrt{\frac{2}{N}} \cdot \frac{\sigma_I}{B}. \quad (7)$$

As can be seen from the above content, reducing phase errors caused by phase shift images, increasing the number of the phase shift steps and improving the intensity modulation B are the main method to reduce the phase error in wrapped phase. In which, increasing the number of phase shift steps means to reduce the measurement speed. And, the intensity modulation parameter is limited by the measurement system (such as camera and projector) and the properties of measured objects (such as reflectivity). In summary, inhibit the phase error caused by the image error is a feasible way to improve the phase accuracy. Image noise ($n_{oise\_n}(x, y)$), non-linear intensity of light source (g_{amma}) and surface reflectivity changes ($r(x, y)$) are the main factors that affect the quality in captured fringe images.

$$I_n^c(x, y) = r(x, y)(I_n(x, y)^{gamma}) + n_{oise\_n}(x, y) \qquad (8)$$

The noise of fringe images is generally expressed as Gaussian distribution. Hence, the noise in captured fringe images can be inhibited by Gaussian filtering. And gamma value is calibrated to inhibit the phase errors caused by non-linear intensity. However, the pixels in the imaging area are affected by noise and the non-linear intensity are not uniform (Fig. 1), due to the non-uniform characteristics of the reflectivity of the object surface. The convolution operation is performed on the filter area in Gaussian filtering. For the Uneven reflectivity of objects, filtering effect is affected. Therefore, the conventional Gaussian filter method is limited in improving the wrapped phase accuracy from the objects with uneven reflectivity (Fig. 1).

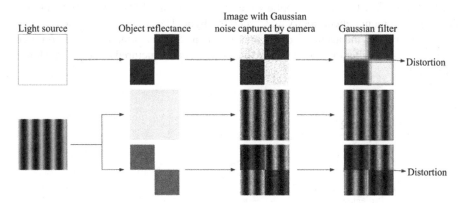

Fig. 1. Influence of uneven reflectivity on filtering effect

Hence, an iterative Gaussian filter method with iterative Gaussian filtering is proposed to improve the accuracy of wrapped phase accuracy recovered from the phase shift images (Fig. 2).

Step 1: initialing calculation ϕ^c: $\phi^c = \arctan\left\{ \dfrac{\sum\limits_{n=1}^{N} I_n^c \sin\left(\frac{2\pi(n-1)}{N}\right)}{\sum\limits_{n=1}^{N} I_n^c \cos\left(\frac{2\pi(n-1)}{N}\right)} \right\}, n = 1, 2, \cdots N,$

when the projected fringe pattern is I_n, the image captured by the camera is I_n^c;

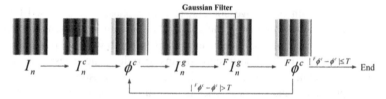

Fig. 2. Iterative Gaussian filter method

Step 2: wrapped phase ϕ^c is projected to the phase shift fringe image space, then I_n^g is generated: $I_n^g(x, y) = A(x, y) + B(x, y) \cos[\phi^c(x, y) + \frac{(n-1)}{N}2\pi]$;

Step 3: $^F I_n^g$ is determined by filtering the images I_n^g with Gaussian filter, and the wrapped phase $^F \phi^c$ is recalculated according to step 1;

Step 4: repeat step 2 and step 3. The stop condition is that the phase error between $^F \phi^c$ and ϕ^c is less than the set threshold T.

It is worth noting that the quality of fringes is improve by Gaussian filtering without changing the sine of fringes if the objects with uniform reflectivity. This is the theoretical premise that step 3 can improve the accuracy of phase recovery by projecting the phase with errors into the fringe image space and performing Gaussian filtering.

3 Application-Binocular Structured Light

Phase shift is a key technique in optical measurement methods, which is applied to the phase measurement deflectometry and structured light. The accuracy of phase recovery directly determines the accuracy of 3D measurement. The performance of the proposed iterative Gaussian filter method is verified by the binocular structured light [22], as shown in Fig. 3.

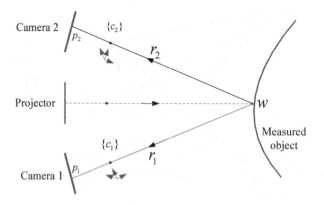

Fig. 3. The measurement principle of binocular structured light

3.1 Measurement Principle of Binocular Structured Light

The binocular structured light is established with two cameras (camera 1, camera 2) and a projector. The light projected by the projector is reflected by the object point w and imaged on the pixel p_1 of camera 1 and pixel p_2 of camera 2. $^{c1}R_{c_2}{}^{c1}T_{c_2}$ is the posed relationship between the two cameras. K_1 and K_2 are the intrinsic parameters of camera 1 and camera 2, respectively. The coordinates of the object point w are $^{c1}X_w$ and $^{c2}X_w$ in the coordinates systems of camera 1 and camera 2, respectively. The coordinates of the object point w is determined according to the reflected light r_1 determined by p_1 and reflected light r_2 determined by p_2. The correspondence between p_1 and p_2 is determined from the absolute phases calculated from the captured fringe images.

$$\begin{cases} ^{c1}X_w = k_1 \cdot r_1, \ r_1 = K_1^{-1} \cdot \begin{bmatrix} p_1 & 1 \end{bmatrix}^T, \\ ^{c2}X_w = k_2 \cdot r_2, \ r_2 = K_2^{-1} \cdot \begin{bmatrix} p_2 & 1 \end{bmatrix}^T, \\ ^{c1}X_w = {}^{c1}R_{c_2} \cdot {}^{c2}X_w + {}^{c1}T_{c2}. \end{cases} \tag{9}$$

The coefficient parameters k_1 and k_2 is determined through the Eq. (9). The camera intrinsic parameters K_1, K_2, and the posed relationship $^{c1}R_{c_2}{}^{c1}T_{c_2}$ are calibrated by the iterative calibration method [23].

3.2 Measurement Experiments

The binocular structured light is constructed by two cameras (resolution: 1280 × 1024 pixels) with 12-mm lens and a laser projector (resolution: 1280 × 1024 pixels). The multi-frequency heterodyne is used to determine the phase order of the wrapped phase. The periods are chosen as (28, 26, 24) in multi-frequency heterodyne. Befor measurement, the system parameters of binocular structured light is calibrated at first. During calibration process, the calibration board (chess board) at different positions are captured by two cameras. The iterative calibration method is applied to determine the intrinsic parameters of the two cameras and the their posed relationship, as shown in Table 1. The posed relationship (the camera 2 relative to camera 1) is $^{c1}R_{c_2}{}^{c1}T_{c_2}$. To verify the performance of the proposed iterative phase correction technique, the non-linear parameter (gamma) of the projector is not calibrated.

Table 1. The intrinsic parameters of camera 1 and camera 2

Unit/pixel	Camera 1	Camera 2
Focal length	[1761.5,1760.8]	[1748.5,1747.5]
Principle point	[637.75,499.19]	[630.88,520.77]
Distortion	[− 0.091,0.0793]	[− 0.110,0.352]
Re-projection error	[0.049,0.049]	[0.048,0.048]

$$\begin{cases} {}^{c1}\mathbf{R}_{c2} = \begin{bmatrix} 0.9044 & -0.0120 & 0.4263 \\ 0.0106 & 0.9999 & 0.0056 \\ -0.4264 & -0.0005 & 0.9045 \end{bmatrix} \\ {}^{c1}\mathbf{T}_{c2} = \begin{bmatrix} -118.6958 & -0.7826 & 26.1722 \end{bmatrix} \end{cases}$$

A Φ38.092-mm standard spherical with the surface accuracy of 0.5-μm is measured to verify the proposed iterative Gaussian filter method. As shown in Fig. 4, the absolute phase obtained by direct Gaussian filtering is less accurate than the absolute phase recovered from the proposed iterative Gaussian filter method.

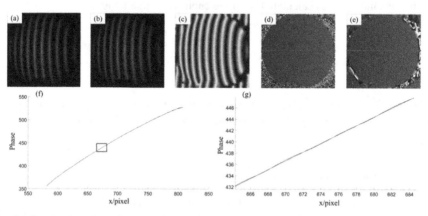

Fig. 4. The absolute phase decoding from the captured image with Gaussian filtering and iterative phase correction. (a): The phase shift fringe image captured by camera 2; (b): image obtained by applying the Gaussian filter to (a); (c):$^F I_n^g(x, y)$ determined from the absolute phase; (d): the absolute phase calculated from (a); (e): the absolute phase calculate by the iterative Gaussian filter method from (c); (f): the absolute phases of blue line in (d) and red line in (e),respectively; (g): the enlarged view of the red box in (f). (Color figure online)

Figure 5 shows the measurement results of standard spherical. The error of the reconstructed point cloud with direct Gaussian filtering is shown in Fig. 5(c). The reconstruction error are 0.13 mm and the RMS deviations is 0.0247 mm. The error of the reconstructed point cloud with proposed iterative Gaussian filter method is shown in Fig. 5(d). The less accurate parts of the reconstruction error are around 0.04 mm and the RMS deviations is 0.0094 mm.

The conventional direct filtering method will fail when reconstructing the 3D information of object with change drastically in reflectivity (for instance the reflectivity is too low or too high), as shown in Figs. 6(b), 7(b). The iterative Gaussian filter method proposed is not sensitive to surface reflectivity. Therefore, effective measurement results can still be obtained by measuring the surface of the objects whose reflectance change more drastically, as shown in Figs. 6(d), 7(c). The error of the reconstructed point cloud with direct Gaussian filtering is shown in Fig. 6(c), The reconstruction error are 0.48 mm and the root mean square (RMS) deviations is 0.042 mm. The error of the reconstructed point cloud with proposed iterative Gaussian filter method is shown in Fig. 6(d). The less

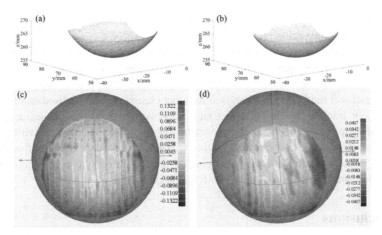

Fig. 5. The measurement results of standard spherical. (a): the point cloud determined from the absolute phase calculated from the captured images with direct Gaussian filtering; (b): the point cloud determined from the absolute phase calculated from the captured images with iterative phase correction; (c): the reconstruction error of (a); (d): the reconstruction error of (b).

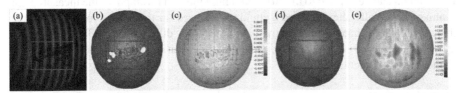

Fig. 6. The measurement results of table tennis (low reflectivity in localized areas). (a): the captured phase shift fringes; (b): the surface determined from (a) with direct Gaussian filtering; (c): the reconstruction error of (b); (d): the surface determined from (a) with iterative phase correction; (e): the reconstruction error of (d).

Fig. 7. The measurement results of box (low reflectivity in localized areas). (a): the captured phase shift fringes; (b): the surface determined from (a) with direct Gaussian filtering; (c): the surface determined from (a) with iterative phase correction

accurate parts of the reconstruction error are around 0.13 mm and the RMS deviations is 0.024 mm. When the surface reflectivity of the object decreases, the intensity of the captured phase shift fringe images also decreases. When the reflectivity of the surface of the object is very low, it is difficult to recover the effective phase from the captured fringe images by convention methods.

The experimental results show (Fig. 7) that the wrapped phase information can be recovered from the low-quality phase shift fringe images by the iterative Gaussian filter method, and then measurement accuracy can be improved. Generally, compared with conventional methods, the proposed iterative Gaussian filter method can filter the noise without interference from reflectivity, improve the measurement accuracy and recover the wrapped phase information from the low-quality fringe images (which is difficult to recover with the conventional methods). Thereby, compare with multi-exposure technique, objects (with drastically changing surface reflectivity) can be reconstructed without additional projection of phase shift fringe images.

4   Conclusions

The iterative Gaussian filter method is proposed to recover the wrapped phase information. The whole approach is regenerating the fringe images from the wrapped phase and performing the iterative Gaussian filter. The phase errors caused by the low-quality in fringe images are effectively inhibited. Therefore, the proposed iterative Gaussian filter method can be applied to the structured light method for improve the measurement accuracy. Especially, the proposed method can be used for the phase recovery from the objects with large changes in reflectivity (which is very difficult to be effectively recovery with the conventional methods). And, the effectiveness of the proposed method is verified by experiments.

Acknowledgements. This study is supported by the National Natural Science Foundations of China (NSFC) (Grant No. 51975344, No. 51535004) and China Postdoctoral Science Foundation (Grant No. 2019M662591).

References

1. Zhang, S.: High-speed 3D shape measurement with structured light methods: a review. Opt. Lasers Eng. **106**, 119–131 (2018)
2. Chang, C., Zhang, Z., Gao, N., et al.: Improved infrared phase measuring deflectometry method for the measurement of discontinuous specular objects. Opt. Lasers Eng. **134**, 106194 (2020)
3. Zuo, C., Feng, S., Huang, L., et al.: Phase shifting algorithms for fringe projection profilometry: a review. Opt. Lasers Eng. **109**, 23–59 (2018)
4. Butel, G.P.: Analysis and new developments towards reliable and portable measurements in deflectometry. Dissertations & Theses - Gradworks (2013)
5. Yu, C., Ji, F., Xue, J., et al.; Fringe phase-shifting field based fuzzy quotient space-oriented partial differential equations filtering method for gaussian noise-induced phase error. Sensors **19**(23), 5202 (2019)
6. Skydan, O.A., Lalor, M.J., Burton, D.R.: 3D shape measurement of automotive glass by using a fringe reflection technique. Meas. Sci. Technol. **18**(1), 106 (2006)
7. Wu, Y., Yue, H., Liu, Y.: High-precision measurement of low reflectivity specular object based on phase measuring deflectometry. Opto-Electron. Eng. **44**(08), 22–30 (2017)

8. Wu, Y., Yue, H., Yi, J., et al.: Phase error analysis and reduction in phase measuring deflectometry. Opt. Eng. **54**(6), 064103 (2015)
9. Yuhang, H., Yiping, C., Lijun, Z., et al.: Improvement on measuring accuracy of digital phase measuring profilometry by frequency filtering. Chin. J. Lasers **37**(1), 220–224 (2010)
10. Zhang, S.: Absolute phase retrieval methods for digital fringe projection profilometry: a review. Opt. Lasers Eng. **107**, 28–37 (2018)
11. Zhang, S.: Comparative study on passive and active projector nonlinear gamma calibration. Appl. Opt. **54**(13), 3834–3841 (2015)
12. Yang, W., Yao, F., Jianying, F., et al.: Gamma calibration and phase error compensation for phase shifting profilometry. Int. J. Multimedia Ubiquit. Eng. **9**(9), 311–318 (2014)
13. Zhang, X., Zhu, L., Li, Y., et al.: Generic nonsinusoidal fringe model and gamma calibration in phase measuring profilometry. J. Opt. Soc. Am. A Opt. Image Sci. Vis. **29**(6), 1047–1058 (2012)
14. Cui, H., Jiang, T., Cheng, X., et al.: A general gamma nonlinearity compensation method for structured light measurement with off-the-shelf projector based on unique multi-step phase-shift technology. Optica Acta Int. J. Opt. **66**, 1579—1589 (2019)
15. Song, Z., Jiang, H., Lin, H., et al.: A high dynamic range structured light means for the 3D measurement of specular surface. Opt. Lasers Eng. **95**, 8–16 (2017)
16. Jiang, H., Zhao, H., Li, X.: High dynamic range fringe acquisition: A novel 3-D scanning technique for high-reflective surfaces. Opt. Lasers Eng. **50**(10), 1484–1493 (2012)
17. Zhao, H., Liang, X., Diao, X., et al.: Rapid in-situ 3D measurement of shiny object based on fast and high dynamic range digital fringe projector. Opt. Lasers Eng. **54**, 170–174 (2014)
18. Babaie, G., Abolbashari, M., Farahi, F.: Dynamics range enhancement in digital fringe projection technique. Precis. Eng. **39**, 243–251 (2015)
19. Xiao, H., Jian, B., Kaiwei, W., et al.: Target enhanced 3D reconstruction based on polarization-coded structured light. Opt. Express **25**(2), 1173–1184 (2017)
20. Xiao, H., Yujie, L., Jian, B., et al.: Polarimetric target depth sensing in ambient illumination based on polarization-coded structured light. Appl. Opt. **56**(27), 7741 (2017)
21. Liu, Y., Zhang, Q., Su, X.: 3D shape from phase errors by using binary fringe with multi-step phase-shift technique[J]. Opt. Lasers Eng. **74**, 22–27 (2015)
22. Song, L., Li, X., Yang, Y.G,. et al.: Structured-light based 3d reconstruction system for cultural relic packaging[J]. Sensors 18(9), p. 2981 (2018)
23. Datta, A., Kim, J. S, Kanade, T.: Accurate camera calibration using iterative refinement of control points. In: 2009 IEEE 12th International Conference on Computer Vision Workshops, ICCV Workshops. IEEE (2010)

Automation

An Improved Calibration Method
of EMG-driven Musculoskeletal Model
for Estimating Wrist Joint Angles

Jiamin Zhao, Yang Yu, Xinjun Sheng$^{(\boxtimes)}$, and Xiangyang Zhu

State Key Laboratory of Mechanical System and Vibration, Shanghai Jiao Tong
University, 800 Dongchuan Road, Shanghai, China
xjsheng@sjtu.edu.cn
http://bbl.sjtu.edu.cn/

Abstract. Lumped-parameter musculoskeletal model based on surface electromyography (EMG) promises to estimate multiple degrees-of-
freedom (DoFs) wrist kinematics and might be potentially applied in
the real-time control of powered upper limb prostheses. In this study,
we proposed a new parameter calibration method based on the lumped-
parameter musculoskeletal model. Compared with the existing calibration
method in the lumped-parameter musculoskeletal model, this paradigm
used an improved method of calculating estimated joint angles in optimization and a reduced training dataset (data from only single-DoF movements) to optimize model parameters. Surface EMG signals were then
mapped into the kinematics of the wrist joint using the optimized musculoskeletal model. In the experiments, wrist joint angles and surface EMG
signals were simultaneously acquired from able-bodied subjects while performing 3 movements, including flexion/extension (Flex/Ext) only, prona-
tion/supination (Pro/Sup) only, and 2-DoF movements. The offline tracking performance of the proposed method was comparable to that of the
existing calibration method with averaged r = 0.883 and NRMSE = 0.218.
Moreover, the results demonstrated significant superiority of the proposed
method over the existing method with less amount of data for parameter tuning, providing a promising direction for predicting multi-DoF limb
motions with only single-DoF information.

Keywords: Parameter calibration · Musculoskeletal model · EMG

1 Introduction

Electromyography (EMG) based human-machine interface (HMI) promises to
enhance the quality of human-machine interaction and user experience. Neural control information extracted from EMG signals has been widely applied in
prosthetics, exoskeleton systems and robots for decades [1–4]. With regard to
myoelectric control schemes, direct control [5], pattern recognition (PR) [6,7]
and simultaneous and proportional control (SPC) [8,9] have been proposed in

© Springer Nature Switzerland AG 2020
C. S. Chan et al. (Eds.): ICIRA 2020, LNAI 12595, pp. 41–51, 2020.
https://doi.org/10.1007/978-3-030-66645-3_4

the past decades. The direct control requires the user to co-contract a pair of antagonistic muscles to switch various functions between different DoFs, which is nonintuitive. Besides, PR-based scheme and SPC with regression methods are data-driven approaches, failing to account for the complex interactions between numerous neural, muscular, and skeletal components that influence motor commands during multi-joint movements [10,11].

To this end, many researchers proposed to estimate human kinematics using the musculoskeletal model [12–15], which simulates limb movements by incorporating the coordination of bones and muscles. Crouch *et al.* [15] proposed a lumped-parameter, EMG-driven musculoskeletal model of wrist and hand, estimating multi-joint movements. Compared with complex models [16,17], lumped-parameter models reduce the computation complexity and simplify the action of several muscles in a single DoF into a pair of antagonistic muscles, which indicates that the lumped-parameter model is more applicable to the modeling of complex forearm muscles.

The size of the training dataset used in the musculoskeletal model is smaller than that of data-driven approaches. Throughout the training dataset of the previously-reported musculoskeletal model [15,18,19], it consists of data from trials of all movements (single-DoF and multi-DoF movements). Although using such a dataset for calibration proved to be feasible, further improvement of the musculoskeletal model might be facilitated by reducing the size of the training dataset further. In this study, we proposed a calibration method based on the lumped-parameter musculoskeletal model, in which the training dataset consisted of data solely from single-DoF movements (henceforth called reduced training datasets) and the method of calculating joint angles during parameter optimization was improved. The results demonstrated the feasibility of offline wrist kinematics prediction with the proposed method in multi-DoF tasks and its superiority over the existing method [15] with reduced training datasets.

2   Methodology

2.1   Subjects

Two able-bodied subjects (one male, age 27, one female, age 24, both right-handed) with no history of neuromuscular or joint diseases participated in the experiments. The experiments were in accordance with the Declaration of Helsinki.

2.2   Experimental Setup

EMG and kinematic data were recorded and synchronized using Biometrics DataLog (Biometrics Ltd., UK). With regard to kinematic data acquisition, wrist flexion/extension (Flex/Ext) and pronation/supination (Pro/Sup) joint angles were collected 100 Hz by two goniometers (SG Series and Q Series, Biometrics Ltd., UK) (Fig. 1). Based on the musculoskeletal geometry, four selected muscles generate wrist Flex/Ext and Pro/Sup joint moments in able-bodied subjects, including flexor carpi radialis (FCR), extensor carpi radialis longus (ECRL), pronator teres (PT), and biceps brachii (BB). Among them, FCR and ECRL contribute to wrist Flex/Ext, while the others contribute to wrist Pro/Sup. The EMG signals were sampled 2000 Hz. To reduce the contact impedance, the subjects' skins were cleaned with alcohol wipes. Four bipolar electrodes (SX230, Biometrics Ltd., UK) were attached to these muscles (Fig. 1).

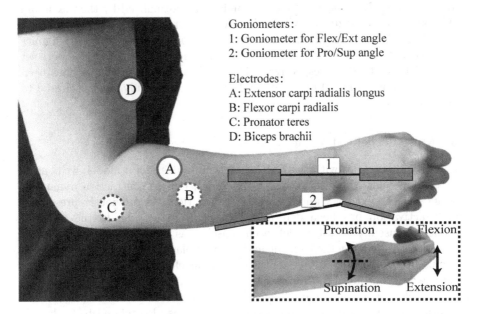

Fig. 1. Experimental setup: The positions of the surface EMG electrodes and goniometers. Circles and rectangles denote electrodes and goniometers, respectively. Dashed circles represent electrodes placed on the interior aspect of the forearm.

2.3 Experimental Protocol

The experiment consisted of 3 different types of wrist movement: 1. wrist Flex/Ext only; 2. wrist Pro/Sup only; 3. simultaneous wrist Flex/Ext and Pro/Sup. The subject performed 3 repeated trials of 20 s for each type of movement with the dominant arm.

During the experiments, the subjects were instructed to keep the hand and arm in a neutral posture with the elbow flexed to 90° (Fig. 1). First of all, the maximum EMG signal values were recorded during maximum voluntary contraction (MVC). Next, two measured joint angles were zeroed when the subjects' arms and hands were kept in the neutral posture. Finally, the subjects conducted motion tasks as aforementioned. And there was a 10 s rest between trials to avoid fatigue.

2.4 Data Processing

The kinematic data were low-pass filtered by a 4th-order zero-lag Butterworth digital filter with cut-off frequency 6 Hz. For the EMG signals, we firstly fed them into a 4th-order zero-lag Butterworth bandpass filter (20–450 Hz). Then, a notching comb filter (50 Hz and its multiplications) was used to eliminate the interferences caused by power frequency. And then, the filtered signals were rectified and low-pass filtered (4th-order zero-lag Butterworth digital filter, cut-off frequency 4 Hz). Next, the processed signals were normalized by the maximum EMG values recorded during MVC. Finally, the normalized signals were down-sampled 100 Hz and converted to muscle activations using activation dynamics [20]. In this study, the activation dynamics parameters, A, β_1 and β_2, were set to -1, -0.09 and -0.09, respectively. The electromechanical delay, d, has been reported to range from 10 ms to about 100 ms. In this study, d was selected to 40 ms [13,21].

2.5 Musculoskeletal Model

To predict joint angles with the EMG-driven musculoskeletal model, we considered 4 components: muscle activation dynamics, muscle contraction dynamics, musculoskeletal geometry and parameter calibration. As previously presented, muscle activation has been determined in *Data Processing*. In the lumped-parameter musculoskeletal model, each muscle was represented as a Hill-type actuator with a contractile element and parallel elastic element. The force (F) of a muscle was the sum of forces of the contractile element and parallel element, while the force output of contractile elements was a function of its length (l), contraction velocity (v), maximum isometric contractile element force (F_{max}) and activation (a). The parallel element force (F_{pee}) was also related to its length (l). In this study, we focused on the parameter calibration rather than muscle modeling. Therefore, we compared the estimation performance of different calibration methods based on a previously-reported lumped-parameter musculoskeletal model [15]. In our method, wrist Flex/Ext and Pro/Sup joint angles would be predicted simultaneously. More details of muscle contraction dynamics and musculoskeletal geometry in the lumped-parameter musculoskeletal model could be found in [15].

2.6 Parameter Calibration

For parameter calibration, there were two main differences between the proposed one and the existing one [15] : the method of calculating predicted joint angles used in optimization and data used for parameter tuning. Specifically, the flow chart of computing estimated joint angles in optimization is illustrated in Fig. 2. Based on the muscle contraction dynamics, joint moments generated joint rotations, caused by changes of muscle length (l), contraction velocity (v) and activation (a). Compared with the lumped-parameter model proposed in [15], measured joint angles ($\theta_{measured}$) were used to calculate not only the estimation errors ($Error$) (refer to Eq. 1) but also the fiber length (l_i) and contraction velocity (v_i) at each sampling time.

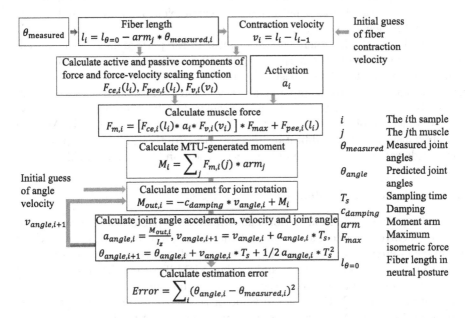

Fig. 2. The flow chart of the proposed method of calculating predicted joint angles during optimization.

The second difference is embodied in the training dataset:

(i) The training dataset I for parameter tuning in previously-reported model [15] was from 3 of the 9 trials, arbitrarily selected from each of three movement types.

(ii) The training dataset II in our method, unlike the aforementioned training dataset, was from one of the trials (2000 samples) collected in Flex/Ext and Pro/Sup, respectively for the parameter optimization of the FCR&ECRL and PT&BB related models. Therefore, the remaining data (7 trials, 14000 samples for each subject) were used for testing.

In order to keep the number of samples consistent between two training datasets, the training dataset I consisted of the first two-thirds of each selected trial. The testing dataset was composed of the data not included in the training dataset. The setup remained the same for the rest of parameter calibration except these two differences. Specifically, a global optimization using the *GlobalSearch* function in MATLAB was performed to calculate subject-specific muscle parameters, where the objective function was as follow:

$$Error_{min} = min \sum_{i=1}^{N}(angle_{i,predicted} - angle_{i,measured})^2 \qquad (1)$$

where N is the number of samples in trials of parameter tuning, $angle_{i,predicted}$ and $angle_{i,measured}$ are ith predicted and measured joint angles in specific DoF of the wrist, respectively. For each muscle, six parameters were optimized with constraints [15], including maximum isometric force, constant moment arm across joint angles, optimal fiber length, parallel elastic element stiffness and muscle length at the neutral position.

To evaluate whether there was an increase in the performance of the musculoskeletal model due to the proposed calibration method, we introduced a third calibration method with the existing method of calculating joint angles and the training dataset II, and compared its estimation performance with the other methods. In summary, there were 3 different calibration methods: previously-reported method of calculating joint angles with the training dataset II (Method 1), previously-reported calibration method [15] (Method 2) and the proposed joint angle calculation method in Fig. 2 with the training dataset II (Method 3).

2.7 Performance Evaluation

To evaluate the estimation performance of different calibration methods, the Pearson's correlation coefficient (r) and the normalized root mean square error (NRMSE) were used to measure similarities of predicted joint angles with respect to the measured angles. Larger r and smaller NRMSE reveal that the estimated angles are more approximate to the measured joint angles. For single-DoF circumstance, the r or NRMSE is the value of the actuated DoF. For 2-DoF circumstance, the r or NRMSE is the mean value over multiple DoFs.

2.8 Statistical Analysis

In the experiments, we considered 2 factors (independent variables) affecting the estimation performance, i.e., the calibration methods and the movement types. Dependent variables were r and NRMSE. A two-way analysis of variance (ANOVA) was used to analyze the influence of calibration methods and movement types on the two dependent variables.

3 Results

3.1 Estimation Performance of the Proposed Calibration Method

Figure 3 shows the measured and predicted wrist Flex/Ext and Pro/Sup joint angles using the proposed calibration method during different movements from one subject, in which movement trends and amplitudes are similar at both DoFs. In this example, r/NRMSE during Flex/Ext-only and Pro/Sup-only are 0.88/0.21 and 0.83/0.23, respectively, while r/NRMSE for Flex/Ext and Pro/Sup during simultaneous 2-DoF movements are 0.89/0.26 and 0.90/0.25, respectively.

3.2 Comparison of Estimation Performance of Different Calibration Methods

Figure 4 (a) and (b) illustrate the r and NRMSE values of different calibration methods during different movements, respectively. The r values on average across

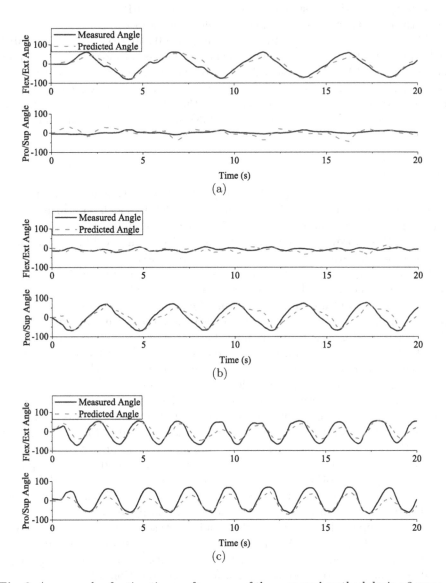

Fig. 3. An example of estimation performance of the proposed method during 3 movements from one subject (angle, unit: degree): (a) wrist Flex/Ext only, (b) wrist Pro/Sup only and (c) wrist Flex/Ext and Pro/Sup simultaneously. The blue and orange lines represent the measured and predicted joint angles, respectively. In each subfigure, the first figure refers to wrist Flex/Ext angles, while the second one is wrist Pro/Sup angles. (Color figure online)

all movements and subjects for Method 1, 2 and 3 are 0.840 ± 0.020, 0.890 ± 0.018 and 0.883 ± 0.020, respectively (Fig. 4(a)). The two-way ANOVA on r values shows a two-way interaction ($p < 0.001$) between movement types and calibration methods. According to the statistical analysis, there is a significant difference in all circumstances between Method 1 and Method 2 or 3. However, there is no statistically significant difference ($p = 0.618$, $p = 1$ and $p = 1$, respectively) between Method 2 and 3 in all circumstances.

The NRMSE values of Method 1, 2 and 3 are 0.240 ± 0.020, 0.207 ± 0.012 and 0.218 ± 0.019, respectively, when averaged across all movement types and subjects (Fig. 4(b)). The two-way ANOVA on NRMSE values shows no two-way interaction ($p = 0.143$) between movement types and calibration methods and significant difference ($p < 0.001$) between Method 1 and Method 2 or Method 3, as well as no significant difference ($p = 0.143$) between Method 2 and 3.

4   Discussion

In this study, we proposed a novel calibration method for tuning the parameters of a lumped-parameter EMG driven musculoskeletal model, in which a new optimization method and only single-DoF movement information were used. Moreover, the improved musculoskeletal model with the new calibration method was evaluated in estimating two-DoF wrist movements.

Experimental results are encouraging since the averaged r and NRMSE values are 0.883 and 0.218 (Fig. 4), comparable to those of conventional lumped-parameter model optimized by data from all circumstances. However, as shown in Fig. 3 (a) and (b), during single-DoF movements, the estimation performance of the inactivated DoF is relatively poor. Further, the cross-talk of EMG signals and the physiological relevance of selected muscles are very likely to result in this phenomenon. Thus, it is of vital importance to eliminate the mutual effect between different pairs of antagonistic muscles in future work.

To validate the effectiveness of the proposed method on reduced training datasets, we compared our method with Method 1. The training datasets of most of current calibration methods in the musculoskeletal model [15] are similar to the training dataset I, consisting of data from all the tasks or movements. Conversely, the training dataset II adopted in our proposed method is only composed of data from single-DoF movements. As shown in Fig. 4, no significant difference is recognized between the estimation performances (r and NRMSE) of Method 2 and Method 3, which indicates that the estimation performance of these two methods is comparable even though the training datasets are different. Moreover, Method 2 and 3 outperform Method 1, indicating a poor estimation performance of previously-reported optimization method with less training data. Perhaps, one possible explanation for these results is the different functions of measured joint angles during the parameter optimization (described in *Parameter Calibration*), as more accurate values of fiber lengths and contraction velocities are computed using measured joint angles. Although there are only 3 different movements in this study, the proposed method could be a potential solution for reducing the

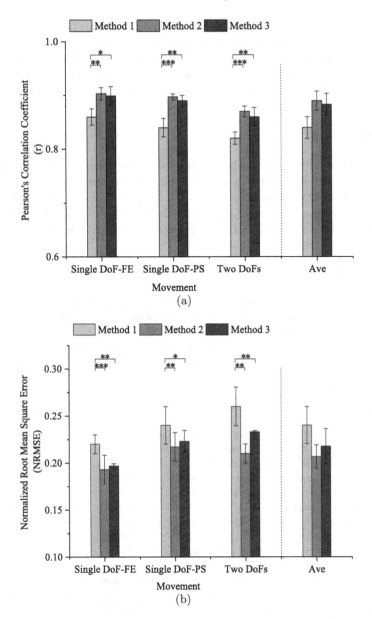

Fig. 4. Results of ANOVA. (a) Pearson's correlation coefficient (r) and (b) normalized root mean square error (NRMSE) of three calibration methods of each movement across both subjects. The blue, green and red bars illustrate the estimation results of Method 1, 2 and 3, respectively. Significant differences with a level of $(0.01 < p < 0.05)$, $(0.001 \leqslant p \leqslant 0.01)$, $(p < 0.001)$ are represented by symbols *, ** and ***, respectively. (Color figure online)

amount of data required to train a musculoskeletal model. In the future, more movement types should be conducted in order to further test the performance of the proposed method on the reduced training datasets. In addition, the feasibility of the musculoskeletal model using the proposed calibration method was only verified offline. In future work, we will also test the proposed method online.

5   Conclusion

We proposed a calibration method embodied in the lumped-parameter EMG-driven musculoskeletal model and validated its effectiveness on estimating wrist two-DoF movements offline. Compared with the previously-reported calibration method, the results indicate that the proposed method could obtain superior estimation performance with less data merely from single-DoF movements.

References

1. Farina, D., et al.: The extraction of neural information from the surface EMG for the control of upper-limb prostheses: emerging avenues and challenges. IEEE Trans. Neural Syst. Rehabil. Eng. **22**(4), 797–809 (2014)
2. Graupe, D., Cline, W.K.: Functional separation of EMG signals via ARMA identification methods for prosthesis control purposes. IEEE Trans. Syst. Man Cybern. **5**(2), 252–259 (1975)
3. Doerschuk, P.C., Gustafon, D.E., Willsky, A.S.: Upper extremity limb function discrimination using EMG signal analysis. IEEE Trans. Biomed. Eng. **30**(1), 18–29 (2007)
4. Fougner, A., Stavdahl, O., Kyberd, P.J., Losier, Y.G., Parker, P.A.: Control of upper limb prostheses: terminology and proportional myoelectric control-a review. IEEE Trans. Neural Syst. Rehabil. Eng. **20**(5), 663–677 (2012)
5. Davidson, J.: A survey of the satisfaction of upper limb amputees with their prostheses, their lifestyles, and their abilities. J. Hand Ther. **15**(1), 62–70 (2002)
6. Englehart, K., Hudgins, B.: A robust, real-time control scheme for multifunction myoelectric control. IEEE Trans. Biomed. Eng. **50**(7), 848–854 (2003)
7. Huang, H., Zhou, P., Li, G., Kuiken, T.A.: Spatial filtering improves EMG classification accuracy following targeted muscle reinnervation. Ann. Biomed. Eng. **37**(9), 1849–1857 (2009)
8. Jiang, N., Englehart, K.B., Parker, P.A.: Extracting simultaneous and proportional neural control information for multiple-DOF prostheses from the surface electromyographic signal. IEEE Trans. Biomed. Eng. **56**(4), 1070–1080 (2009)
9. Jiang, N., Dosen, S., Muller, K.R., Farina, D.: Myoelectric control of artificial limbs-is there a need to change focus? [In the spotlight]. IEEE Signal Process. Mag. **29**(5), 150–152 (2012)
10. Crouch, D.L., Huang, H.: Musculoskeletal model predicts multi-joint wrist and hand movement from limited EMG control signals. In: Engineering in Medicine and Biology Society, pp. 1132–1135 (2015)
11. Sartori, M., Durandau, G., Dosen, S., Farina, D.: Robust simultaneous myoelectric control of multiple degrees of freedom in wrist-hand prostheses by real-time neuromusculoskeletal modeling. J. Neural Eng. **15**(6), 066,026.1-066,026.15 (2018)

12. Lloyd, D.G., Besier, T.F.: An EMG-driven musculoskeletal model to estimate muscle forces and knee joint moments in vivo. J. Biomech. **36**(6), 765–776 (2003)
13. Buchanan, T.S., Lloyd, D.G., Manal, K., Besier, T.F.: Neuromusculoskeletal modeling: estimation of muscle forces and joint moments and movements from measurements of neural command. J. Appl. Biomech. **20**(4), 367–395 (2004)
14. Sartori, M., Reggiani, M., Farina, D., Lloyd, D.G.: EMG-driven forward-dynamic estimation of muscle force and joint moment about multiple degrees of freedom in the human lower extremity. PloS One **7**(12), 1–11 (2012)
15. Crouch, D.L., Huang, H.: Lumped-parameter electromyogram-driven musculoskeletal hand model: A potential platform for real-time prosthesis control. J. Biomech. **49**(16), 3901–3907 (2016)
16. Manal, K., Gonzalez, R.V., Lloyd, D.G., Buchanan, T.S.: A real-time EMG-driven virtual arm. Comput. Biol. Med. **32**(1), 25–36 (2002)
17. Chadwick, E., Blana, D., van den Bogert, A., Kirsch, R.: A real-time, 3-D musculoskeletal model for dynamic simulation of arm movements. IEEE Trans. Biomed. Eng. **56**(4), 941–948 (2009)
18. Pan, L., Crouch, D.L., Huang, H.: Musculoskeletal model for simultaneous and proportional control of 3-DOF hand and wrist movements from EMG signals. In: 8th International IEEE/EMBS Conference on Neural Engineering, NER, pp. 325–328. IEEE (2017)
19. Pan, L., Crouch, D.L., Huang, H.: Comparing EMG-based human-machine interfaces for estimating continuous, coordinated movements. IEEE Trans. Neural Syst. Rehabil. Eng. **27**(10), 2145–2154 (2019)
20. Heine, R., Manal, K., Buchanan, T.S.: Using hill-type muscle models and EMG data in a forward dynamic analysis of joint moment evaluation of critical parameters. J. Mech. Med. Bio. **3**(2), 169–186 (2003)
21. Corcos, D.M., Gottlieb, G.L., Latash, M.L., Almeida, G.L., Agarwal, G.C.: Electromechanical delay: an experimental artifact. J. Electromyogr. Kinesiol. **2**(2), 59–68 (1992)

Non-invasive Measurement of Pulse Rate Variability Signals by a PVDF Pulse Sensor

Dun Hu[1,2] , Na Zhou[3], Chenlei Xie[1,2,4] , and Lifu Gao[2(✉)]

[1] Institute of Intelligent Machines, HFIPS, Chinese Academy of Sciences, Hefei 230031, China
[2] University of Science and Technology of China, Hefei 230026, China
lifugao@iim.ac.cn
[3] The Fourth Affiliated Hospital of Auhui Medical University, Hefei 230000, China
[4] Anhui Jianzhu University, Hefei 230022, China

Abstract. Pulse rate variability (PRV) is a small change in the heart beat cycle that can be obtained from the pulse signal. PRV has important application value in clinical diagnosis, disease monitoring, and prevention. PRV can be conveniently extracted from the fingertip pulse signal obtained by a photoplethysmography (PPG) pulse sensor. However, this method requires clamping the fingertip during the measurement, which is uncomfortable for the monitored person and is not conducive to continuous PRV detection in family monitoring or in a specific environment, such as driving. Thus, in this paper, we propose a pulse sensor with a soft polyvinylidene fluoride (PVDF) piezoelectric film. The non-invasive pulse signals can be collected by lightly pressing the fingertip on the sensor. In the experiment, two PVDF pulse sensors were used to collect the pulse waves from the left wrist and left forefinger; simultaneously, an infrared PPG pulse sensor measures the pulse wave of the right forefinger. The pulse waves measured by the three methods were further filtered to extract PRV signals and compare the differences. The results show that the PRV signal obtained by the PVDF sensor pressing measurement method has good consistency with the PRV signal obtained by PPG measurement, and the PVDF pulse sensor can be conveniently applied in wearable devices and portable medical devices to obtain the PRV.

Keywords: Pulse rate variability · PVDF · Pulse sensor · PPG · Non-invasive measurement

1 Introduction

Heart rate variability (HRV) arises from small changes in the heart beat cycle. HRV is an important indicator of sympathetic and parasympathetic nerve activity and balance in the autonomic nervous system. HRV contains rich pathological information about the cardiovascular nervous system and the humoral regulation system [1], which has important application value in clinical diagnosis, disease monitoring, and prevention. HRV has been widely used as a tool to support clinical diagnosis and measurement of biological information for health purposes, such as sleep phase, stress state, and fatigue [2]. HRV is obtained from ECG signal, and ECG signal is picked up by lead mode. The

© Springer Nature Switzerland AG 2020
C. S. Chan et al. (Eds.): ICIRA 2020, LNAI 12595, pp. 52–64, 2020.
https://doi.org/10.1007/978-3-030-66645-3_5

complicated connection methods restrict the application of HRV signals in wearable devices and portable medical instruments [3, 4].

Pulse rate variability (PRV) also arises from small changes in the heart beat cycle and can be obtained from the pulse signal. Previous studies showed that PRV can replace HRV to reflect the characteristics of the heart beat [5]. Pulse signals are more easily measured compared with electrical signals. In the pulse signal, the P peak has obvious characteristics and is easy to detect, as shown in Fig. 1. The interval between adjacent P peaks is regarded as the duration of a pulse, which is called the PP interval. The slight changes in each PP interval form a time series, called the PP series and also known as the PRV signal, as shown in Fig. 1.

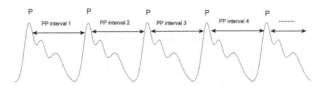

Fig. 1. Pulse rate variability (PRV) signal.

An important first step in PRV analysis is the accurate detection of the pulse wave and pulse periods [6]. In the past decade, a wide variety of sensors were studied to detect pulse signals [7, 8]. There are two main measurement methods:

1. Pressure detection method: The waveform of sensor pressure (F) with time is approximated as the waveform of the arterial pressure with time, and the ideal pulse wave waveform can be obtained by subsequent signal processing [9]. Generally, the radial artery at the wrist is more often selected as a measuring point.
2. Photoplethysmography (PPG) detection method: When light is irradiated onto human tissue, the absorption of light by muscles and bones is constant, and the rhythmic beat of blood vessels causes the expansion and contraction of blood vessels, causing the internal blood volume to change periodically. The PPG signal can be obtained by collecting the attenuated light by a photo-detector and converting it into an electrical signal [10]. Generally, the index finger or middle finger is chosen as the test point.

Most PRV measurement equipment on the market uses the PPG method, and PRV extracted from PPG finger signals can be used as an alternative to HRV [11]. The PPG method is more convenient in terms of measurement, but the performance of PPG-based algorithms can be degraded by poor blood perfusion, ambient light, and motion artifacts [12]. Most importantly, this method requires clamping the fingertip during the measurement, which is uncomfortable to the monitored person and is not conducive to continuous PRV detection in family monitoring or in a specific environment, such as driving [13].

Polyvinylidene fluoride (PVDF) is a piezoelectric material having a solid structure with approximately 50%–65% crystallinity. In this study, we propose a pulse sensor with a soft PVDF film that can measure the pulse wave. The non-invasive collection of pulse signals can be completed by lightly pressing the fingertip on the sensor. In the experiment, the PVDF pulse sensor was used to measure the pulse signal synchronously with an infrared photoelectric pulse sensor. The pulse waves measured by different measurement methods were further filtered and processed. The PRV signals were extracted and the differences in PRV between measured methods were compared to evaluate the accuracy of the PVDF sensor pressing measurement method.

2   Materials and Methods

2.1   Principle of the PVDF Film Pulse Sensor

The whole measurement system consists of three parts: (1) PVDF film sensor, (2) charge transfer and amplifier, and (3) data collection. As the human body's signal is weak, the charge output of the PVDF sensor needs to be converted into a voltage signal, which can be collected and analyzed after being amplified. The physical map of the PVDF film pulse sensor system is shown in Fig. 2.

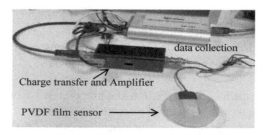

Fig. 2. The schematic diagram of the measurement system.

In this paper, MEAS's SDT1-028K film is used to collect pulse signals. SDT1-028K piezo film sensor consist of a rectangular element of PVDF film together with a molded plastic housing and $18''$ of coaxial cable. Figures 3a,b show the physical map and simplified model structure of SDT1-028K. The SDT1-028K parameters are given in Table 1.

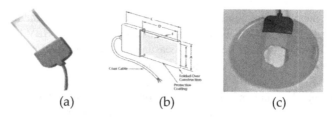

(a)   (b)   (c)

Fig. 3. (a) The physical map of SDT1-028K; (b) The simplified model structure of SDT1-028K; (c) The physical map of PVDF film sensor.

Table 1. The parameters of the SDT1-028K.

DIMENSIONS in INCHES (mm)						
Description	A Film	B Electrode	C Film	D Electrode	t (μm)	Cap (nF)
SDT1-028K	.640 (16)	.520 (13)	1.64(41)	1.18 (30)	75	2.78

SDT1-028K film is placed in the middle of two circular silicone films. The upper silicone film has a circular hole, and the fingertips contact the SDT1-028 K film through the circular holes to collect pulse signals, as shown in Fig. 3c. The role of the silicone sheets is to fix the SDT1-028K film while reducing the effects of environmental vibrations and finger motion artifacts during pulse measurement.

The amount of charge obtained from the PVDF sensor is very weak and needs to be amplified by a charge amplifier. We use the VK202 charge amplifier (Vkinging Corporation, Shenzhen, China). The amount of voltage from the charge amplifier is an analog signal and needs to be converted into a digital value, and can be stored and processed by the computer. We used a signal collection board (VK-701: Vkinging Corporation, Shenzhen, China). The digital signal obtained from the VK-701 is sent to the computer for storage and further processing.

2.2 Measurement and Analysis of Pusle Wave Signals

To verify the accuracy of the pulse measurement of the PVDF film pulse sensor, in the experiment, a PPG pulse sensor (HKG-07B, Hefei Huake Electronic Technology Research Institute, Hefei, China) was selected for synchronous measurement. HKG-07B contains amplification, filter, and other circuits. The HKG-07B is depicted in Fig. 4a.

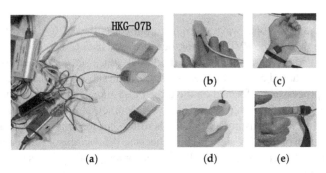

Fig. 4. Experimental scenario: (a) synchronization measurement system, (b) right forefinger is clamped by photoplethysmography (PPG) pulse sensor, (c) left radial artery is bundled by polyvinylidene fluoride (PVDF) film pulse sensor, and palm down during synchronous acquisition, (d) left forefinger is measured by pressing PVDF film pulse sensor, and (e) left forefinger is bundled by PVDF film pulse sensor.

Figure 4 shows three pulse sensors and different measurement methods used in this study. According to Yeragani et al., there is no significant difference between the right and left sides of the body for measurement of PPG signals in normal controls [14]. To facilitate the comparison of the results in the simultaneous measurement, in our experiments, the PPG pulse sensor was placed on the right forefinger, and two PVDF sensors were placed on the left wrist radial artery and the left forefinger, as shown in Figs. 4b–d. In Fig. 4e, we depict finger pulse signal collection in a bundled way to evaluate the feasibility of the PVDF pulse sensor. The pulse waves of four acquisition methods are shown in Fig. 5.

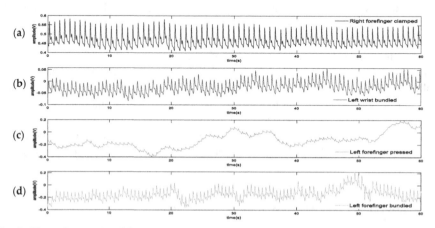

Fig. 5. The pulse waves of four measurements: (a) The blue pulse wave from right forefinger clamped by the PPG pulse sensor; (b) the red pulse wave from left wrist bundled by the PVDF film pulse sensor; (c) the green pulse wave from left forefinger pressed by the PVDF film pulse sensor; (d) the turquoise pulse wave from the left forefinger bundled by the PVDF film pulse sensor. (Color figure online)

In Fig. 5, the blue pulse wave of right forefinger collected by the PPG sensor is clearer than the data of the PVDF sensor, because the HKG-07B sensor has a filtering process inside, and the data collected by the PVDF pulse sensor are unprocessed and has baseline drift caused by low-frequency noise. In the subsequent data processing, the same filtering process is performed on the three pulse signals measured synchronously, which is convenient for PRV signal extraction and accuracy comparison. The pulse signal amplitudes of different measurement methods differ, but the amplitude value does not affect the extraction of the PRV signal. This is because the two sensor acquisition systems use different amplifier circuits, and the PVDF sensor amplitude value is related with the strength of the binding.

In Figs. 5b,d, the pulse signal of the wrist radial artery is more stable than the finger-tip pulse signal with the PVDF sensor binding measurement method, but the difference in waveform is small. This shows that the PVDF sensor is feasible for collecting finger pulses, but the fingertip pulse is measured more conveniently than the wrist pulse. By comparing Figs. 5c and 5d, the pulse wave collected by the fingertip bundling measurement method is better than the pressing measurement method, where the low-frequency noise of the pulse signal is prominent and subsequent filtering is required to extract the PRV signal. Improving the structure of the PVDF sensor and reducing the interference of finger motion artifacts require further research to ensure the measurement result of the pressing fingertip method is close to that of the bundling fingertip.

2.3   Data Collection and Processing

In the experiment, three pulse sensors were used to simultaneously collect the pulse signal for 5 min, whose 1-minute valid data were used for PRV signal extraction and analysis. The right forefinger of each participant was clamped by a PPG pulse sensor, the radial artery of the left wrist was bound by a PVDF pulse sensor, and the left forefinger was lightly pressed on the other PVDF pulse sensor.

All the data were sampled at 320 Hz. A Nyquist frequency of 320 Hz is sufficient for pulse signal frequency. Data samples were collected from four participants, two men and two women. Self-reported age ranged from 23 to 35 years, as shown in Table 2. Each participant sat on the seat for 5 min while the data of the three pulse sensors were collected. During data collection, all participants remained still.

Table 2. Participant data.

	P1	P2	P3	P4
Gender	M	M	F	F
Age(years)	35	23	26	23

From the time and frequency domain viewpoints, we analyzed the pulse waves of Participant P1 from three pulse sensors, as shown in Fig. 6. The frequency distribution of the pulse signals was in the range of 0–5 Hz, and the main frequency component of the three pulse signals was 1.533 Hz. A 0.4 Hz frequency component exists in the wrist data, which is easy to see. According to the analysis of the frequency range, the 0.4 Hz component should be caused by breathing. In the spectrum of PPG data, no amplitude at 0.4 Hz is observed, but low frequency components can be seen, which should be caused by measurement noise. Compared with PPG pulse sensors, PVDF pulse sensors may have more advantages in breathing component detection, which is worth further research in the future.

In the frequency spectrum of the left forefinger data from the PVDF sensor, a very strong low-frequency noise was observed, overwhelming the 0.4 Hz breathing component, as shown in Fig. 6b. Low frequency noise comes from the relative movement of fingertips and sensors during the measurement. Myoelectric interference, power frequency and multiplier interference, and baseline drift below 0.7 Hz are the main interference sources for body surface pulse signals [15–17]. Myoelectric interference is caused by human muscle vibration, which has a frequency range generally between 5 Hz and 2 kHz [15, 16]. In the next work, these noise signals need to be filtered to extract PRV signals.

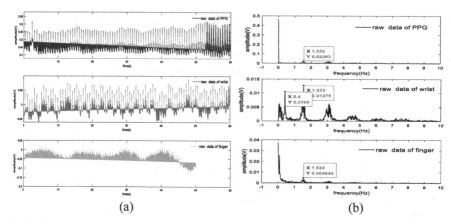

Fig. 6. The pulse waves of P1 from three pulse sensors for 1 min of data collection at 320 Hz. (a) The raw data of the right forefinger from the PPG sensor are shown in blue and the raw data of the wrist from the PVDF sensor are shown in red. The raw data of the left forefinger from the PVDF sensor are shown in green. (b) The frequency spectra of the three groups of raw data. (Color figure online)

To extract the PRV, the raw voltage output signals from three pulse sensors were first filtered using a band-pass filter with cutoff frequencies of 0.8 and 2 Hz. The cutoff frequencies would be sufficient for adults with a heart rate above 50 beats per minute (bpm) and below 120 bpm. The waves after filtering are shown in Fig. 7.

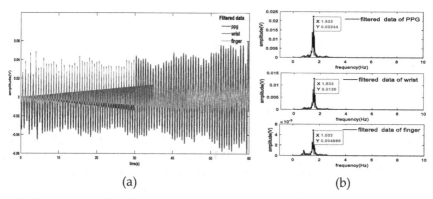

Fig. 7. The pulse data after filtering: (a) The blue wave denotes filtered PPG data. The red wave denotes filtered wrist data. The green wave indicates filtered left forefinger data. (b) The frequency spectrum of the three groups of data after filtering. (Color figure online)

In Fig. 7a, the P peak is left in the filtered data and the tidal wave is filtered out because only the P peak is needed for PRV detection. The amplitudes of the three waves are different, but their changes have a consistent trend, indicating a difference in the peak values of each pulse. This is not caused by measurement errors, but is a characteristic of the pulse signal. The wave peak at the red wrist is a little earlier than the PPG and finger data because the pulse propagates through the wrist to the end of the finger, and the results displayed by the waves are consistent with our experience.

Figure 7b shows that the frequency spectrum of the three groups of data after filtering is mainly distributed around 1.533 Hz. We estimated that P1 has a pulse rate above 90 beats per minute. The next step was to extract the PRV signal from the filtered data.

2.4 PRV Detection

The main differences between PRV and HRV are due to physiological factors and to the variability in the location of the PPG fiducial point [18]. Different fiducial points for the temporal location of each pulse wave were proposed in several studies, such as the apex, middle-amplitude, and foot points of the PPG signal; maximum of the first- and second-order derivative PPG signal; or the tangent intersection point, depending on the application. The apex point of the first derivative was considered the most promising fiducial point for use in pulse arrival time [19].

In this study, the apex point of the first derivative was used as the fiducial point of the PRV signal, and the PRV signals were extracted from the filtered data from three sensors. In Fig. 8a, the fiducial points of the PRV signal measured from left forefinger by pressing the PVDF film pulse sensor were located, and the final PRV signal was calculated through the reference points, as shown in Fig. 8b. In the figure, there were 90 PP intervals in one minute, and the duration of each interval was between 0.5 and 0.8 s. In the same process, the PP series of the right forefinger PPG pulse and the left wrist pulse were obtained.

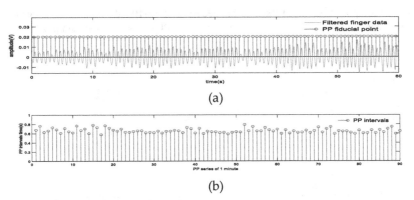

(a)

(b)

Fig. 8. PRV detection: (a) green wave from the filtered data of the left forefinger pressed on PVDF film pulse sensor. The blue points are the fiducial points of the PRV signal. (b) PP series in one minute calculated through the reference points. (Color figure online)

3   Results

Figures 9, 10, 11 and 12 show Bland–Altman plots and mountain plots of four participants, showing the discrepancies of each participant's PP series obtained from three measures and the stability across a wider range of values. Bland–Altman plots were used to indicate the discrepancies between PP series obtained from PPG signals and left forefinger data from the PVDF pulse sensor. In this graphical method the differences (or alternatively the ratios) between the two techniques are plotted against the averages of the two techniques. Horizontal lines are drawn at the mean difference, and at the limits of agreement (LoA), which are defined as the mean difference plus and minus 1.96 times the standard deviation of the differences. If the differences within mean ± 1.96 SD are not clinically important, the two methods may be used interchangeably [20]. Mountain plots allowed for comparison of the three PP series from three pulse measurement methods. The left wrist and left forefinger assay were compared with the PPG reference assay. The mountain plot provides information about the distribution of the differences between methods. If two assays are unbiased with respect to each other, the mountain is centered over zero. Long tails in the plot reflect large differences between the methods. It is a useful complementary plot to the Bland–Altman plot [21].

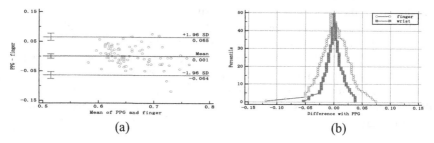

(a)   (b)

Fig. 9. The comparison of three PP series obtained from P1: (a) Bland–Altman plot of the discrepancies between the PPG PP series and the left forefinger PP series; (b) mountain plots compared with the PPG reference assay.

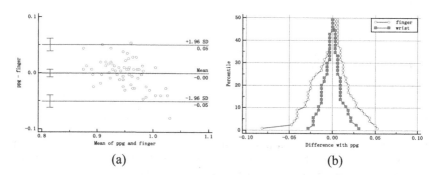

(a)   (b)

Fig. 10. The comparison of three PP series obtained from P2: (a) Bland–Altman plot of the discrepancies between the PPG PP series and the left forefinger PP series; (b) mountain plots compared with the PPG reference assay.

Figure 9a shows the Bland–Altman plot indicating the discrepancies between PPG PP series and left forefinger PP series obtained from P1, who had a total of 90 paired PP measurements were used for the analysis. The parameters of Bland–Altman plots are

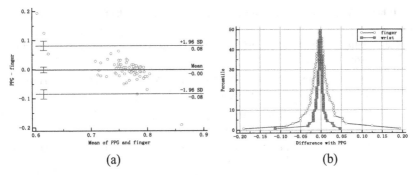

Fig. 11. The comparison of three PP series obtained from P3: (a) Bland–Altman plot of the discrepancies between the PPG PP series and the left forefinger PP series; (b) mountain plots compared with the PPG reference assay.

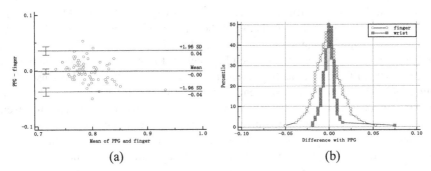

Fig. 12. The comparison of three PP series obtained from P4: (a) Bland–Altman plot of the discrepancies between the PPG PP series and the left forefinger PP series; (b) mountain plots compared with the PPG reference assay.

shown at Table 3. As shown in the Bland–Altman diagram shown in Fig. 9a, four points (4.44%) fall outside the LoA (mean difference ± 1.96 × SD values, the upper and lower dashed lines in the figure) and the other points are randomly distributed on the between horizontal line of the average difference and the LoA. The average difference of the measurement results was 0.0005556 s, and the 95% CI of the overall average difference was –0.006360 to 0.007471, including 0. There was no statistically significant differences in the measured values of the two methods. The two measurement methods have good consistency, and they may be used interchangeably.

The same analysis can be applied to the Bland–Altman plots of P2–P4, as shown in Figs. 10a, 11a, and 12a. The one minute PP sequence numbers of P2–P4 were 62, 78, and 73, respectively. Although we found differences in the bias arithmetic mean and the LoA values of the four participants, their Bland–Altman plots showed that the two methods may be used interchangeably. The measurement error of the pressing PVDF pulse sensor type is related to the finger placement method of the experimenter.

The mountain graphs in Figs. 9b, 10b, 11b, and 12b show that the mountain of solid square points is more centered over zero than the mountain of circle points. The

Table 3. The parameters of the Bland–Altman plots from the four participants.

Participants	Sample size	Arithmetic mean	95% CI	P (H0: Mean = 0)	Lower limit	Upper limit
P1	90	0.0005556	−0.006360 to 0.007471	0.8735	−0.06416	0.06527
P2	62	−0.0004536	−0.007004 to 0.006097	0.8903	−0.05101	0.05010
P3	78	−0.001042	−0.01060 to 0.008521	0.8289	−0.08417	0.08209
P4	73	−0.0006421	−0.005027 to 0.003743	0.7712	−0.03748	0.03619

solid square points are the difference between the left wrist PRV and the PPG PRV, and the circle points are the difference between the left forefinger PRV and the PPG PRV. The four participants' mountain graphs show that the PP series of the left wrist agrees more with the PP series measured by the PPG than the PP series measured from the left forefinger. The difference in the measurement of the pressing PVDF pulse sensor type is high, caused by the relative movement of the fingertip in contact with the PVDF film. When the finger is bundled to the PVDF sensor, good results are produced. However, this affects the convenience of measurement. Improving the structure of the sensor can better and more stably fit the fingertips during the measurement process, which would help improve the accuracy of pressing PVDF film measurement in the future.

4 Discussion

In this paper, we focused on the convenience and the validity of PRV signal measurement using a PVDF pulse sensor. PVDF pulse sensors were designed to non-invasively collect pulse signals measured synchronously with an infrared PPG pulse sensor for differential analysis. By analyzing the difference between the Bland–Altman plots and the mountain plots of the four participants, the results showed that the PVDF pulse sensor can obtain the pulse signal from the fingertip, providing an effective method for non-invasive PRV signal measurement. Due to the flexible nature of PVDF film, it could be widely used in wearable devices and portable medical devices, and used in home monitoring and special environments.

The signals in this study were obtained in a well-controlled environment, and motion artifacts were significantly suppressed. In reality, due to the relative movement of the fingertip in contact with the PVDF film, motion artifacts may be more disturbing, which would affect the accuracy of the PRV sequence. In future work, improving the structure of the PVDF sensor to reduce motion interference in contact with the fingertip and trying new filtering methods to extract PRV signals are important tasks to improve the accuracy of measuring PRV with the fingertip by pressing onto PVDF film.

References

1. Makivi, B., Niki, M.D., Willis, M.S.: . Heart rate variability (HRV) as a tool for diagnostic and monitoring performance in sport and physical activities. J. Exerc. Physiol. **16**(3), 103–131 (2013)
2. Baek, H.J., Shin, J.W.: Effect of missing inter-beat interval data on heart rate variability analysis using wrist-worn wearables. J. Med. Syst. **41**(10), 147 (2017)
3. Shi, B., Chen, F., Chen, J., et al.: Analysis of pulse rate variability and its application to wearable smart devices. Chin. J. Med. Instrum. **39**(2), 95–97 (2015)
4. Leonhardt, S., Leicht, L., Teichmann, D.: Unobtrusive vital sign monitoring in automotive environments—a review. Sensors 18(9), 3080 (2018)
5. Yu, E., He, D., Su, Y., et al.: Feasibility analysis for pulse rate variability to replace heart rate variability of the healthy subjects. In: 2013 IEEE International Conference on Robotics and Biomimetics (ROBIO), pp. 1065 –1070 (2013)
6. Peralta, E., Lazaro, J., Bailon, R.:. Optimal fiducial points for pulse rate variability analysis from forehead and finger PPG signals. Physiol.Meas. 40(2), p. 025007 (2019)
7. Haseda, Y., et al.: Measurement of pulse wave signals and blood pressure by a plastic optical fiber FBG sensor. Sensors-Basel 19, p. 5088 (2019)
8. Nguyen, T.-V., Ichiki, M.: MEMS-based sensor for simultaneous measurement of pulse wave and respiration rate. Sensors-Basel 19, p. 4942 (2019)
9. Wang, W., Xu, Y., Zeng, G., et al.: Extraction and dual domain analysis of pulse wave signals. Electron. Technol. Appl. (2019)
10. Ding, X.Y, Chang, Q., Chao, S.: Study on the extract method of time domain characteristic parameters of pulse wave. In: IEEE International Conference on Signal & Image Processing (2017)
11. Chuang, C.-C., Ye, J.-J., Lin, W.-C., Lee, K.-T., Tai, Y.-T.: Photoplethysmography variability as an alternative approach to obtain heart rate variability information in chronic pain patient. J. Clin. Monit. Comput. **29**(6), 801–806 (2015). https://doi.org/10.1007/s10877-015-9669-8
12. Fallow, B.A., Tarumi, T.: Influence of skin type and wavelength on light wave reflectance. J. Clin. Monit. Comput. **27**(3), 313–317 (2013). https://doi.org/10.1007/s10877-013-9436-7
13. Wusk, G., Gabler, H.: Non-invasive detection of respiration and heart rate with a vehicle seat sensor. Sensors (Basel) **18**, 1463 (2018)
14. Yeragani, V.K., Kumar, R., Bar, K.J., et al.: Exaggerated differences in pulse wave velocity between left and right sides among patients with anxiety disorders and cardiovascular disease. Psychosom. Med. **69**(8), 717–722 (2007)
15. Abulkhair, M.F., Salman, H.A., Ibrahim, L.F.: Using mobile platform to detect and alerts driver fatigue. Int. J. Comput. Appl. **123**(8), 27–35 (2015)
16. Sun, Y., Yu, X.B.: An innovative nonintrusive driver assistance system for vital signal monitoring. IEEE J. Biomed. Health Inform. **18**(6), 1932–1939 (2014)
17. Chu, Y., Zhong, J., Liu, H., et al.: Human pulse diagnosis for medical assessments using a wearable piezoelectret sensing system. Adv. Funct. Mater. **28**(40), 1803413.1–1803413.10 (2018)
18. Gil, E., et al.: Photoplethysmography Pulse Rate Variability as a Surrogate Measurement of Heart Rate Variability during Non-Stationary Conditions. Physiol. Meas. **31**(9), 1271–1290 (2010)
19. Rajala, S., et al.: Pulse arrival time (PAT) measurement based on arm ECG and finger PPG signals - comparison of PPG feature detection methods for PAT calculation. In: 2017 39th Annual International Conference of the IEEE Engineering in Medicine and Biology Society (EMBC), vol. 2017, pp. 250–253 (2017)

20. Bland, J.M., Altman, D.G.: Measuring agreement in method comparison studies. Stat. Methods Med. Res. **8**(2), 135–160 (1999)
21. CLSI.: Estimation of total analytical error for clinical laboratory methods; approved guideline. CLSI document EP21-A. Wayne, PA: Clinical and Laboratory Standards Institute (2003)

Deep Learning for Plant Disease Identification from Disease Region Images

Aliyu Muhammad Abdu$^{(\boxtimes)}$, Musa Mohd Mokji, and Usman Ullah Sheikh

Universiti Teknologi Malaysia, 81310 Skudai, Johor, Malaysia
aliyu104@yahoo.com

Abstract. This paper proposes a deep learning (DL) plant disease identification approach at leaf surface level using image data of pathologically segmented disease region or region of interest (ROI). The DL model is an exceptional technique used in automatic plant disease identification that employs a series of convolutions for feature representation of the visible disease region, mainly characterized as the combination of the chlorotic, necrotic, and blurred (fuzzy) lesions. The majority of current DL model approaches apply whole leaf image data for which studies have shown its consequential tendencies of leading to irrelevant feature representations of the ROI. The effects of which are redundant feature learning and low classification performance. Consequently, some state-of-the-art deep learning methods practice using the segmented ROI image data, which does not necessarily follow the pathological disease inference. This study proposes an extended ROI (EROI) algorithm using pathological inference of the disease symptom to generate the segmented image data for improved feature representation in DL models. The segmentation algorithm is developed using soft computing techniques of color thresholding that follows an individual symptom color feature that resulted in the incorporation of all lesions. The results from three different pre-trained DL models AlexNet, ResNet, and VGG were used to ascertain the efficacy of the approach. The advantage of the proposed method is using EROI image data based on pathological disease analogy to implement state-of-the-art DL models to identify plant diseases. This work finds application in decision support systems for the automation of plant disease identification and other resource management practices in the field of precision agriculture.

Keywords: Deep learning · Plant disease identification · Disease region segmentation · Pathological inference

1 Introduction

Disease outbreaks are increasingly becoming rampant globally, especially since some are extremely difficult to control and can lead to famine [1, 2]. Notably, viral plant diseases attributed to diseases caused by pathogens, such as early and late blight, are known to reduce the overall yield of vegetable crops and are a great menace affecting both home gardeners and large productions [1]. With their large productions and adaptability comes high risk and high susceptibility to viral plant diseases. Plant disease

© Springer Nature Switzerland AG 2020
C. S. Chan et al. (Eds.): ICIRA 2020, LNAI 12595, pp. 65–75, 2020.
https://doi.org/10.1007/978-3-030-66645-3_6

identification is a crucial component in precision agriculture that primarily deals with observing the stages of diseases in plants [3]. With 60% to 70% of all visible observations first appearing on the leaves compared to the stem and fruits, plant diseases are most commonly observed on the leaves. Thus, early symptom detection is vital to the effect of disease diagnosis, control, and damage assessment. The traditional manual methods of plant disease identification, both direct and indirect, have been providing definitive solutions for centuries. However, despite using modern equipment, such methods are labor-intensive, exhaustive, and time-consuming in providing answers promptly. Also, most of these techniques require specialized tools and consumable chemical reagents, thus becoming inefficient and unsustainable [4]. As part of machine vision technology, machine learning (ML) systems can mimic the direct identification method through pattern recognition by providing accurate and timely identification.

The ML systems are generally classified into a conventional classifier (CC) and deep learning (DL) (or deep convolution neural network (D-CNN)) methods [5, 6]. The feature representation in a CC method involves explicitly extracting features as patterns with properties that sufficiently portray the quantifiable details of the disease symptom, the region of interest (ROI). Then, the complete automatic identification is achieved using a machine learning classifier built using the features. On the other hand, the DL method involves automated implicit feature representation. In this process, pixels in the entire image, a neighborhood, or a group, are considered characteristics for the feature learning. Hence, the typical feature representation becomes more of an implicit process than a stage and an embedded part of the model architecture. There are two methods of training a DL: training from scratch and transfer learning. The process of training a DL from "scratch" involves designing and building the network layer by layer. This process is often a complicated and time-consuming process due to the deep architecture design [6]. Transfer learning, as introduced by Bengio [7], involves applying an already established architecture that has been successful in other computer vision domain problems and can adapt to the problem under consideration, significantly reducing the complexity.

Throughout the literature, many works use the whole leaf image as training data with various degrees of precision. Sharada et al. [8] presented the first DL model to identify multiple plant diseases on a relatively comprehensive Plant Village (PV) image dataset. It constitutes over 54,000 images of 14 different crop species and 26 disease pairs (healthy and unhealthy), including early and late blight. The transfer learning pre-trained networks used were AlexNet and GoogLeNet, which were fine-tuned, trained on the training data, and validated on the testing data. Zhang et al. [9] also implemented AlexNet, GoogleNet, and ResNet pre-trained models on the same dataset to identify tomato diseases. Xu et al. proposed using a VGG-16 model trained with transfer learning [10]. Fuentes et al. also applied the DL architecture for real-time implementation on tomato crops [11]. Also, lightweight CNNs are proving useful in reducing some of the limitations associated with the approach. Geetharamani et al. and Durmus et al. focused on architecture simplification by reducing the number of deep layers [12, 13]. Durmus et al. proposed using squeezeNet, a compressed and lightweight version of the D-CNN with fewer layers, to identify the tomato diseases [13]. The tomato images used are part of the PV dataset. Dasgupta et al. also proposed a lightweight CNN to detect plant diseases designed for mobile applications [14]. Chen et al. proposed using a modified lightweight MobileNet-V2 model and a "step" transfer learning to optimize feature learning for multiple plant disease classification [15]. During training, the weights and parameters of the deeper layers are frozen, while those of the upper layers are updated. Such models have less computation and are cost-effective. However, the accuracy of using lightweight models is relatively lower than that achieved using regular models. Other recent D-CNN methods also employ manual data augmentation and use segmented ROI images as input data. Sharma *et al.* manually cropped the images to only include ROIs in the input data before training the D-CNN, instead of using the whole leaf image [16]. Barbedo also implemented ROI localization on the image data based on similar criteria of size and color of symptoms [17]. Sun et al. proposed using NN generated lesion images to augment existing image data [18]. However, despite the use of ROI image data, a significant number of DL methods still record low classification accuracies often attributed to the unavailability of sufficient data and the combination of diseases with similar symptoms without the basis of an inference rule [17].

A recent study by Lee et al. concluded that the DL feature representations learned from whole images do not necessarily focus on the ROIs [19]. The features are instead learned from areas with the most common distinctive characteristics, such as leaf venation. Toda and Akura also supported this claim, noting that a DL model to learn visual shape characteristics such as that of the profoundly grooved leaf edges of tomato crop instead of the visible symptom characteristics [20]. Regarding ROI image data, different levels of subjectivity arise during the segmentation, mainly due to loosely characterizing the segmentation without the necessary pathological inference. This problem influences the separation boundary limit during the segmentation resulting in the removal of the blurred region from the ROI. Whereas, earlier research studies indicated its prominence in improving the quality of learned features [21]. In regards to this, this paper proposes a DL plant disease identification method using an extended region of interest (EROI) segmented images as the training data. Instead of the typical ROI image data, the proposed method has the advantage of using segmented EROI data that is inclusive of the blurred region to enhance feature representation during training and improve classification accuracy.

2 Materials and Method

This study implements a DL with transfer learning to identify plant diseases using proposed segmented EROI image data. The ROI segmentation is typically practiced in conventional classifier methods, which involves isolating the visible disease lesions from the rest of the leaf. In this paper, the term identification refers to detection and classification. ROIs for the disease were identified and segmented using a proposed pathological segmentation algorithm. Three popular pre-trained networks, including AlexNet, ResNet-50, and VGG-16, have been used. The paper also considered the vegetable early blight and late blight diseases [22, 23], both of which cut across tomato, pepper, potato, and eggplant with similar symptoms [24]. These crops have planting areas ranging from small backyard plots to much larger field acreages and greenhouses, which makes them exceptional candidates for research in precision agriculture. Thus, even though only two diseases are considered in this study, the research impact is significant.

2.1 Image Dataset

The primary image data of 1,400 images of the potato plant leaf images used in this study was obtained from the comprehensive PV dataset [8]. In total, 500 images showed the symptoms caused by the EB, 500 showed the symptoms caused by the LB, and 400 are healthy leaves. Figure 1 shows samples of each disease symptom [8]. All images are in RGB format and of equal size 256 × 256 pixels.

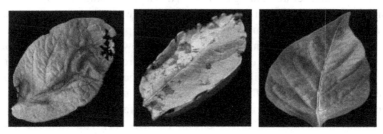

Fig. 1. Example of potato leaf image samples from the PV dataset. From right: healthy, early blight, and late blight.

2.2 Pathological Extended Region of Interest (EROI) Segmentation

As the infection in EB manifest, it forms concentric rings with a bulls-eye pattern emanating from a dark-brown focus and surrounded by a yellowish chlorosis zone [22]. In the case of LB, the infection has a small center of a dark lesion, after which it also manifests to dark-brown or black bordered by a water-soaked lesion with a pale whitish-green border that fades into the healthy tissue [23]. The dark foci (brown) areas are the necrotic regions, while the chlorosis boundary zones are the symptomatic regions.

Following the characteristics of two blight diseases, the symptoms typically show a significant color difference from the other surrounding tissue areas, which involves changing variation from light green to yellow, brown, or black. The proposed pathological segmentation method uses the proportion of each (RGB) color channel intensity for tissue pixels of a healthy leaf image. Typically, the pixels within a leaf image exhibiting higher intensity deviations towards the green hue than blue and red belong to healthy tissue [25, 26]. Mathematically, $G > R \gg B$. Hence, in order to establish the degree of certainty threshold, the proposed ROI segmentation starts with computing the percentage of the pixels' green color intensity in the original *RGB* image. Thus, an input leaf image $I(x, y)$ of size $M \times N$ is made up of several pixels $p_{i,j}(x, y)$ (*for* $i, j = 0, 1, 2, \ldots, M - 1, N - 1$). Each pixel has a color value which is the combination of the three *RGB* colors, and each channel color intensity ranges from 0 (no color) $-$ 255 (maximum intensity). The average percentage of each color value to the combined color values is computed using Eq. (1) – (4).

$$r_g = \sqrt{\frac{G_{i,j}}{R_{i,j} + G_{i,j} + B_{i,j}}} \times 100\% \tag{1}$$

$$r_r = \sqrt{\frac{R_{i,j}}{R_{i,j} + G_{i,j} + B_{i,j}}} \times 100\% \tag{2}$$

$$r_b = \sqrt{\frac{B_{i,j}}{R_{i,j} + G_{i,j} + B_{i,j}}} \times 100\% \tag{3}$$

$$r_{gr} = \sqrt{\frac{R_{i,j}}{R_{i,j} + G_{i,j}}} \times 100\% \tag{4}$$

From the results of Eqs. (1) – (4), pixel intensity values of healthy tissue (green color) with the highest degree of certainty are 42.4% for r_g, 34.5% for r_r, and 23.3% for r_b. Through further experimentations, it is found that lower percentages mean less green color (tone) pixels and vice versa. Hence, adjusting the values would change the segmentation boundary between the healthy and disease region tissues. Following this, four threshold values are proposed to generate four binary masks from $I(x, y)$ to incorporate the blurred region and allow invariancy against small intensity variations using Eqs. (5)–(8).

$$g_1 = \begin{cases} 0 & 38\% < r_g \leq 47\% \\ 1 & otherwise \end{cases} \tag{5}$$

$$g_2 = \begin{cases} 0 & 32\% \leq r_r < 37\% \\ 1 & otherwise \end{cases} \tag{6}$$

$$g_3 = \begin{cases} 0 & 18\% \leq r_b < 29\% \\ 1 & otherwise \end{cases} \tag{7}$$

$$g_4 = \begin{cases} 0 & 65\% < r_{gr} < 85\% \\ 1 & otherwise \end{cases} \tag{8}$$

The four masks are then combined to generate two binary segmentation masks; $m_1 = g_1 || g_2 || g_3$ and $m_2 = g_1 || g_4$. The first mask, m_1, succeeds in the segmentation of healthy tissue pixels incorporating the lighter green pixels. The second mask, m_2, segments the darker green pixels. Finally, the binary segmentation mask is given by Eq. (9).

$$S_{mask} = m_1 || m_2 \tag{9}$$

Some morphological post-processing operations are applied to clean-up the binary mask and remove isolated border pixels. A closing operation using a disk structural element of radius three (3) is applied, followed by a dilation operation using the same structuring element. Applying the completed S_{mask} to the original input image masks the healthy green tissue pixels, turning them to black (or zero). The mathematical expression is given in Eq. (10).

$$I_{EROI}(x, y) = \{(x, y) \in I(x, y) | S_{mask}\} \tag{10}$$

Algorithm 1 shows the pseudocode for the proposed EROI segmentation, and Fig. 2 shows a sample result.

Algorithm 1: Whole ROI Segmentation Algorithm

Input: I_{in} (whole leaf color image) of size $M \times N \times 3$
Output: I_{EROI} (EROI segmented image) of size $M \times N \times 3$

---

for $i = 1$ to M   do
    for $j = 1$ to N   do
Compute:
$r_g \leftarrow$ (Percentage of green color intensity) using equation (1) on I_{in} channels
$r_r \leftarrow$ (Percentage of red color intensity) using equation (2) on I_{in} channels
$r_b \leftarrow$ (Percentage of blue color intensity) using equation (3) on I_{in} channels
$r_{gr} \leftarrow$ (Percentage of red – green color intensity) using equation (4) on I_{in}
    end for
end for

Initialize g_k (for $k = 1,2,3,4$) to zeros of size M $\times$ N:

---

for $i = 1$ to M   do
    for $j = 1$ to N   do
        if conditions in equation (5) – (8) are satisfied, **then**
            $g_k \leftarrow 0$.
        else
            $g_k \leftarrow 1$.
        end if
    end for
end for

Compute:
$m_1 \leftarrow g_1||g_2||g_3$ (binary mask to segment green, light pixels)
$m_2 \leftarrow g_1||g_2$ (binary mask to segment green, dark pixels)
$S_{mask} \leftarrow m_1 \cup m_2$ (binary mask to segment healthy green pixels)

Perform the following morphological operations on S_{mask}:
    Closing
    Dilation
$S_{mask} \leftarrow$ new S_{mask}
$I_{EROI}(x, y) \leftarrow$ (EROI image using equation (10))

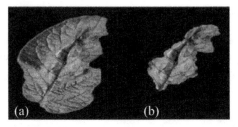

Fig. 2. EROI segmentation sample result showing input image (a) EROI segmented image (b).

2.3 Disease Identification with DL Classifiers

In this study, three transfer learning pre-trained CNN models, AlexNet [27], VGG-16 [28], and ResNet-50 [29], have been implemented using the segmented EROI images as input data.

The AlexNet model has a depth of eight (8) layers. To re-train the model, the last three fully-connected layers of the AlexNet are trimmed and replaced with new layers that will classify the three classes of EB, LB, and HL. This way, the features from the rest of the layers are kept, i.e., the transferred layer weights. However, the weights and biases in the new layers are increased by a factor of 10 to enable faster learning than in the transferred layers. Before training, the images were resized to 227×227 pixels, which is acceptable by the network and augmented to optimize training given minimal data. The augmentation includes flipping vertically and horizontally, scaling, and translating. The modified network is then re-trained with stochastic gradient descent (sgd) optimization with 1×10^{-4} as an initial learning rate, a mini-batch size of 10 for a maximum of 12 epochs.

The ResNet-50 model is 50 layers deep and requires input images of 224×224 pixels. Thus, for implementation, the images were resized to the network acceptable size. The layer with the learnable weights is the last fully-connected layer; similar to the process applied in AlexNet transfer learning, this layer, along with the output layer, is replaced by new ones with the number of outputs equal to the number of disease classes. However, in this case, while the weights of earlier layers are re-initialized, the weights of the first ten layers of the network are frozen by setting their learning rates to zero. This process speeds up the network training since the gradient in those layers will not update. Furthermore, it limits the risk of overfitting since the data is relatively small. The modified network is then trained with sgd optimization with 3×10^{-4} as an initial learning rate, a mini-batch size of 10 for a maximum of 12 epochs

The VGG-16 model features extremely homogeneous architecture that performs a 3×3 convolution with 1-stride 1-padding and 2×2 pooling with 2-strides from the starting to finishing layers. It is 16 layers deep, and, like in ResNet-50, the images must be resized to 224×224 pixels. For implementation, the images are resized to 224×224, and the same re-training and optimization hyper-parameters used in the ResNet model were applied.

2.4 Performance Measures

The standard performance used to compare the classifier performances are precision, recall, F_1 score, and overall accuracy. These are computed from the confusion matrixes using Eqs. 11–14.

$$Precision = \frac{TP}{TP + FP} \tag{11}$$

$$Recall(TPR) = \frac{TP}{TP + FN} \tag{12}$$

$$F_1 Score = \frac{2TP}{2TP + FP + FN} \tag{13}$$

$$Accuracy = \frac{TP + TN}{TP + FP + FN + TN} \tag{14}$$

3 Experimental Results and Discussion

The EROI image dataset was split into 80% training and 20% testing sets. For benchmarking, a separate data generated using the typical ROI segmentation approach described in [25] was used to train and test the models under the same setup. Table 1 summarizes the performance measures on the ROI image testing data for the three implemented DL models.

Table 1. Performance measures on ROI test data

Performance measures		F–M (%)	R (%)	P (%)	ACC (%)
Approach	Class				
AlexNet	EB	93.40	100	87.61	**93.86**
	HL	100	100	100	
	LB	92.39	100	100	
ResNet-50	EB	94.69	98.99	90.74	**95.18**
	HL	100	100	100	
	LB	94.18	89.90	98.89	
VGG-16	EB	93.40	100	87.61	**93.86**
	HL	100	100	100	
	LB	92.39	85.86	100	

From Tables 1 and 2, the ResNet-50 model achieved the highest performance measure results on both the ROI and EROI data, while VGG-16 recorded the lowest. With deeper layers, the ResNet-50 model harbors denser learned features but at the expense of simplicity in implementation as it uses a lot of memory and parameters. The classification results on the segmented image data generated using the proposed pathological segmentation method achieved higher performance measures across all three implemented DL models. Using the typical ROI data, AlexNet, ResNet-50, and VGG-16 have 93.86%, 95.18%, and 93.86% average accuracies, respectively. On the other hand, the accuracies improved by 1.75%, 2.19%, and 0.44% with EROI data, respectively.

Table 2. Performance measures on EROI test data

Performance measures		F–M (%)	R (%)	P (%)	ACC (%)
Approach	Class				
AlexNet	EB	95.19	100	90.83	**95.61**
	HL	100	100	100	
	LB	94.68	89.90	100	
ResNet-50	EB	97.03	98.99	95.15	**97.37**
	HL	100	100	100	
	LB	96.91	94.95	98.95	
VGG-16	EB	93.05	87.88	98.86	**94.30**
	HL	100	100	100	
	LB	93.78	98.99	89.09	

Furthermore, there is a significant improvement in the metric measures of EB and LB classes, which shows a better characterization of the two disease symptoms. From the results (Table 1 and 2), the change in the performance measure statistics is attributed to misclassifications relative to LB symptoms recognized as that of EB. Regardless, the improved results indicated improved feature representation for classification due to incorporating the extended blurred region. Hence, improved data quality leads to better feature learning, and there is better efficiency in performance since the classification accuracy has been improved given fewer data

4 Conclusion

In this work, a DL plant disease identification is actualized using segmented image data from a proposed pathological disease region segmentation algorithm. Instead of applying the typical disease region (ROI) image data, the proposed approach uses the advantage of pathological inference to incorporate extended region of interest (EROI), the fuzzy blurred region. Comparative results using state-of-the-art pre-trained DL models show the efficaciousness of the proposed approach in improving feature representation and classification performance.

Acknowledgements. The authors thank the Ministry of Education Malaysia and Universiti Teknologi Malaysia (UTM) for their support under the Flagship University Grant, grant number Q.J130000.2451.04G71.

References

1. Ojiambo, P.S., Yuen, J., van den Bosch, F., Madden, L.V.: Epidemiology: past, present, and future impacts on understanding disease dynamics and improving plant disease management—a summary of focus issue articles. Phytopathol. **107**(10), 1092–1094, 01 October 2017 (2017)
2. Kaur, S., Pandey, S., Goel, S.: Plants disease identification and classification through leaf images: a survey. Arch. Computat. Methods Eng. **26**(2), 507–530 (2018). https://doi.org/10.1007/s11831-018-9255-6
3. Tripathy, A.S., Sharma, D.K.: Image processing techniques aiding smart agriculture. In: Modern Techniques for Agricultural Disease Management and Crop Yield Prediction: IGI Global, pp. 23–48 (2020)
4. Fang, Y., Ramasamy, R.P.: Current and prospective methods for plant disease detection, (in eng). Biosensors **5**(3), 537–561 (2015)
5. Arnal Barbedo, J.G.: Digital image processing techniques for detecting, quantifying and classifying plant diseases. SpringerPlus **2**(1), 1–12 (2013). https://doi.org/10.1186/2193-1801-2-660
6. Singh, A.K., Ganapathysubramanian, B., Sarkar, S., Singh, A.: Deep learning for plant stress phenotyping: trends and future perspectives. Trends Plant Sci. **23**(10), 883–898 (2018)
7. Bengio, Y.: Deep learning of representations for unsupervised and transfer learning. In: Proceedings of ICML workshop on unsupervised and transfer learning, pp. 17–36 (2012)
8. Mohanty, S.P., Hughes, D.P., Salathé, M.: Using deep learning for image-based plant disease detection. Front. Plant Sci. **7**, 1419 (2016)
9. Zhang, K., Wu, Q., Liu, A., Meng, X.: Can deep learning identify tomato leaf disease? Adv. Multimedia, 2018, 1–10 (2018)
10. Xu, P., Wu, G., Guo, Y., Yang, H., Zhang, R.: Automatic wheat leaf rust detection and grading diagnosis via embedded image processing system. Procedia Comput. Sci. **107**, 836–841 (2017)
11. Fuentes, A., Yoon, S., Kim, S.C., Park, D.S.: A robust deep-learning-based detector for real-time tomato plant diseases and pests recognition. Sensors **17**(9), 2022 (2017)
12. Geetharamani, G., Pandian, A.J.: Identification of plant leaf diseases using a nine-layer deep convolutional neural network. Comput. Electr. Eng. **76**, 323–338, 01 June 2019 (2019)
13. Durmuş, H., Güneş, E.O., Kırcı, M.: Disease detection on the leaves of the tomato plants by using deep learning. In 2017 6th International Conference on Agro-Geoinformatics, pp. 1–5 (2017)
14. Dasgupta, S.R., Rakshit, S., Mondal, D., Kole, D.K.: Detection of diseases in potato leaves using transfer learning. In: Das, A.K., Nayak, J., Naik, B., Pati, S., Pelusi, D. (eds.) Computational Intelligence in Pattern Recognition. AISC, vol. 999, pp. 675–684. Springer, Singapore (2020). https://doi.org/10.1007/978-981-13-9042-5_58
15. Chen, J., Zhang, D., Nanehkaran, Y.A.: Identifying plant diseases using deep transfer learning and enhanced lightweight network. Multimed Tools Appl. **79**(41), 1–19 (2020). https://doi.org/10.1007/s11042-020-09669-w
16. Sharma, P., Berwal, Y.P.S., Ghai, W.: Performance analysis of deep learning CNN models for disease detection in plants using image segmentation. Inf. Process. Agric. (2019)

17. Barbedo, J.G.A.: Plant disease identification from individual lesions and spots using deep learning. Biosys. Eng. **180**, 96–107 (2019)

18. Sun, R., Zhang, M., Yang, K., Liu, J.: Data Enhancement for Plant Disease Classification Using Generated Lesions. Appl. Sci. **10**(2), 466 (2020)

19. Lee, S.H., Goëau, H. Bonnet, P., Joly, A.: New perspectives on plant disease characterization based on deep learning. Comput. Electron. Agric. **170**, p. 105220, 01 March 2020 (2020)

20. Toda, Y., Okura, F.: How convolutional neural networks diagnose plant disease. Plant Phenomics **2019**, p. 9237136 (2019)

21. Abdu, A.M., Mokji M., Sheikh, U.U.: An investigation into the effect of disease symptoms segmentation boundary limit on classifier performance in application of machine learning for plant disease detection. Int. J. Agric. For. Plantation (ISSN No: 2462–1757) **7**(6), 33–40, Art. no. IJAFP_39, December 2018

22. Rands, R.D.: Early blight of potato and related plants. Agricultural Experiment Station of the University of Wisconsin (1917)

23. Fry, W.: Phytophthora infestans: the plant (and R gene) destroyer. Mol. Plant Pathol. **9**(3), 385–402 (2008)

24. Morris, W.L., Taylor, M.A.: The solanaceous vegetable crops: potato, tomato, pepper, and eggplant A2 - thomas, brian. In: Murray, B.G., Murphy, D.J. (eds.) Encyclopedia of Applied Plant Sciences (Second Edition), pp. 55–58. Academic Press, Oxford (2017)

25. Barbedo, Jayme Garcia Arnal: A new automatic method for disease symptom segmentation in digital photographs of plant leaves. Eur. J. Plant Pathol. **147**(2), 349–364 (2016). https://doi.org/10.1007/s10658-016-1007-6

26. Khan, M.A., et al.: An optimized method for segmentation and classification of apple diseases based on strong correlation and genetic algorithm based feature selection. IEEE Access **7**, 46261–46277 (2019)

27. Krizhevsky, A., Sutskever, I., Hinton, G.E.: Imagenet classification with deep convolutional neural networks. In: Advances in Neural Information Processing Systems, pp. 1097–1105 (2012)

28. Simonyan, K., Zisserman, A.: Very deep convolutional networks for large-scale image recognition, *arXiv preprint* arXiv:1409.1556 (2014)

29. He, K., Zhang, X., Ren, S., Sun, J.: Deep residual learning for image recognition. In: Proceedings of the IEEE conference on computer vision and pattern recognition, pp. 770–778 (2016)

Task-Oriented Collision Avoidance in Fixed-Base Multi-manipulator Systems

Jia-Wei Luo, Jinyu Xu, Yongjin Hou, Hao Xu, Yue Wu, and Hai-Tao Zhang[✉]

School of Artificial Intelligence and Automation, Huazhong University of Science and Technology, Wuhan 430074, China
zht@mail.hust.edu.cn

Abstract. Collision avoidance implies that extra motion in joint space must be taken, which might exert unexpected influences on the execution of the desired end-effector tasks. In this paper, a novel framework for generating collision-free trajectories while respecting task priorities is proposed. Firstly, a data-driven approach is applied to learn an efficient representation of the distance decision function of the system. The function is then working as the collision avoidance constraints in the inverse kinematics (IK) solver, which avoids the collision between manipulators. To eliminate undesired influences of the extra motion for collision avoidance on the execution of tasks, task constraints are proposed to control the task priorities, offering the system with the ability to trade off between collision avoidance and task execution. Furthermore, the overall framework is formulated as a QP (quadratic programming), therein guarantees a real time performance. Numerical simulations are conducted to demonstrate the effectiveness and efficiency of the presented method.

Keywords: Task planning · Collision avoidance · Multi-robot systems

1 Introduction

Multi-robot systems have powered many aspects of modern society: from navigation and exploration in extreme environments to objects maneuver and manufacture and large-scale robotic construction [2,8,9,18], and they have attracted the attention of more and more researchers. In this paper, we mainly focus on a general but significant problem appearing in multi-manipulator systems, that is, how to avoid collision between manipulators, while respecting task priorities.

Many approaches have been proposed to generate a collision-free path in the configuration space, or C-space, of the robot. The artificial potential field approach [4,17] enables the robot to move from the starting point to the end without collision by setting a repulsive field around the obstacle, and a gravitational field at the target position. This method, however, requires a very regular shape of the obstacle, otherwise it will be too expensive to be implemented. A less computationally demanding method is to map the position and orientation of the obstacle into the C-space of the robot [14], where each dimension corresponds to one degree-of-freedom (DoF) of the robot. Therefore, the obstacles can be seen

C. S. Chan et al. (Eds.): ICIRA 2020, LNAI 12595, pp. 76–87, 2020.
https://doi.org/10.1007/978-3-030-66645-3_7

as some points and lines in the C-space. Though the spatial complexity grows exponentially with the increase of the robot DoF and a priori knowledge of the obstacle is needed [21], the C-space mapping method have motivated the study of many graph-searching algorithms, such as probabilistic road map [5], rapidly-exploring tree [11], A* algorithm [13], etc. Due to the unavoidable inefficient and nonconvex computation of the minimum distance between links/joints of manipulators [16], the methods mentioned above usually generate collision-free trajectories offline. To solve the problem online, [22] proposed a gradient-based optimization method to generate a smooth and collision-free path. Like other local optimization algorithms, however, this method can also be stuck in the local optima. In recent years, data-driven learning-based algorithms are increasingly used for collision avoidance. For example, [10] proposed that reinforcement learning can be used to guide a mobile robot to avoid collision. Unfortunately, although deep reinforcement learning is capable of learning end-to-end robotic control tasks, the accomplishments have been demonstrated primarily in simulation rather than on actual robots [19]. How to transfer the learned experience from simulation to practice still remains a challenge. In [3], the authors proposed an efficient framework for solving the multi-arm self-collision avoidance problem, but the contradiction between collision avoidance and task execution is not considered. In summary, the existed works either do not achieve a good enough real-time performance, and thus couldn't be extended to the multi-manipulator scenarios, or do not take the influence of collision avoidance on the task execution into consideration.

In this paper, a novel framework is proposed to solve the task-oriented collision avoidance problem in fixed-base multi-manipulator systems. By imposing kinematics constraints, collision avoidance constraints and task constraints on the inverse kinematics solver, feasible joint space collision-free trajectories with little influence on Cartesian space tasks can be generated in real time. The main contributions of this paper are three-fold:

– The collision between manipulators is avoided by introducing the collision avoidance constraints. Besides, sparse kernel support vector machines (SVM) [7] are applied to train a distance decision function on a system-specific collision avoidance dataset, which substantially accelerates the computation of collision avoidance constraints.
– Task constraints are proposed to eliminate the influence of the extra motion taken for collision avoidance on the desired Cartesian space task. Furthermore, by setting different weight matrices, different priorities can be attached to the desired tasks, making the motion generated more reasonable.
– The overall framework is formulated as a QP (quadratic programming), which can be solved efficiently by numerical QP solvers [15].

2 Preliminaries

In the following, without loss of generality, we consider a multi-manipulator system composed of N identical n-DoF manipulator robots, numbered from

1 to N. Throughout the paper, the superscripts of an variable represent the manipulators it belongs to, and subscripts denote its corresponding elements.

Suppose the desired task dimension of manipulator i is m, then the forward kinematics is given by

$$x^i = f(q^i), \qquad (1)$$

where $x^i \in \mathbb{R}^m$ is the desired end-effector motion, or say, the Cartesian space tasks of the manipulator, and $q^i \in \mathbb{R}^n$ is the joint space coordinates. The manipulator is said to be redundant if $n > m$, which is the case throughout the paper. Differentiating (1) with respect to time gives the inverse kinematics

$$\dot{x}^i = J^i \dot{q}^i, \qquad (2)$$

where $\dot{x}^i \in \mathbb{R}^m$ is the task space velocity vector, $\dot{q}^i \in \mathbb{R}^n$ is the joint angular velocity vector, and $J^i = \partial f / \partial q^i \in \mathbb{R}^{m \times n}$ is the Jacobian matrix.

The formulas above can be naturally extended to the multi-manipulator scenarios by coupling the variables, i.e.

$$x = f(q) \qquad (3)$$

and

$$\dot{x} = J\dot{q}, \qquad (4)$$

where $x = [x^1; x^2; \ldots; x^N] \in \mathbb{R}^{Nm}$, $q = [q^1; q^2; \ldots; q^N] \in \mathbb{R}^{Nn}$, and $J =$ blkdiag$(J^1, J^2, \ldots, J^N) \in \mathbb{R}^{Nm \times Nn}$, where the function blkdiag$(\cdot)$ generates a block-diagonal matrix from the given elements. $\dot{x}$ and $\dot{q}$ share the same shape with x and q, respectively.

3   Method

3.1   Overall Framework

Given a desired Cartesian space task of the end-effectors, $x_d(t)$, or equivalently, $\dot{x}_d(t)$, the corresponding joint space motion $\dot{q}_d(t)$, however, might lead to collision between the manipulators. Here $\dot{q}_d(t)$ refers to any possible joint space trajectory that generates $\dot{x}_d(t)$ in Cartesian space. To avoid collision, extra motion must be taken, i.e., the actual motion $\dot{q}(t) = \dot{q}_d(t) + \Delta\dot{q}(t)$. We refer to $\Delta\dot{q}(t)$ as the collision avoidance motion. Unfortunately, $\Delta\dot{q}(t)$ often leads to undesired changes in $\dot{x}_d(t)$, i.e., the actual end-effector motion $\dot{x}(t) = \dot{x}_d(t) + \Delta\dot{x}(t)$. Thus, the goal is to guarantee $\Delta\dot{x}(t) \to 0$, while executing the extra collision avoidance motion. Formulating this problem as an optimization problem can handle this dilemma naturally, if proper constraints are introduced.

Minimize the Cost of Motion. Generally, the objective is to minimize the cost of moving the robots, i.e.

$$\min_{\dot{q}} \quad \frac{1}{2}\dot{q}^\top W \dot{q}, \qquad (5)$$

where W is a positive semi-definite matrix. If $w_{ii} \gg w_{jj}$, the cost of moving joint i is much expensive than moving joint j. Thus, the solution will prefer moving joint j than moving joint i to reach some goal.

Kinematics Constraints. First and foremost, the inherent physical limits of the manipulators must be respected, i.e.

$$\dot{q}^- \leq \dot{q} \leq \dot{q}^+, \tag{6}$$

where $\dot{q}^-$ and $\dot{q}^+$ are the lower and upper bounds of the joint angular velocity, respectively.

Collision Avoidance Constraints. As in [21], we first claim that the configuration of the multi-manipulator system can only be in three states: *forbidden* states $\mathbb{Q}_f$, in which some parts of the manipulators would overlap, i.e., collision would occur; *safe* states $\mathbb{Q}_s$, in which no collision occurs; and *boundary* states $\mathbb{Q}_b$, in which the manipulators touch each other exactly on the surface.

Suppose there exists a continuous differentiable function $\mathcal{F}(q^{ij}) : \mathbb{R}^{2n} \mapsto \mathbb{R}$ to approximate the minimum distance between manipulators i and j, where $q^{ij} = [q^i; q^j]$ is the concatenated joint coordinates vector of manipulators i and j. Thus, the definition of the system states can be modified as follows:

$$\begin{aligned} \mathcal{F}(q^{ij}) < 0 &\iff q^{ij} \in \mathbb{Q}_f, \\ \mathcal{F}(q^{ij}) = 0 &\iff q^{ij} \in \mathbb{Q}_b, \\ \mathcal{F}(q^{ij}) > 0 &\iff q^{ij} \in \mathbb{Q}_s. \end{aligned} \tag{7}$$

Apparently, the trajectory $\dot{q}(t)$ is collision-free if and only if $\mathcal{F}(\cdot) > 0 \ \forall t$. To achieve this goal, we propose the collision avoidance constraint

$$-\nabla \mathcal{F}\left(q^{ij}\right)^\top \dot{q}^{ij} \leq \mathcal{F}\left(q^{ij}\right), \tag{8}$$

where $1 \leq i, j \leq N$ and $i \neq j$. On the one hand, once some $\mathcal{F}(q^{ij}) > 0$ is approaching zero, constraints (8) will be activated, forcing $\dot{q}^{ij}$ to keep in line with the gradient of $\mathcal{F}$, denoted as $\nabla \mathcal{F}$. This results in an increase in the value of $\mathcal{F}$, hence keeping the manipulators stay in safe states. On another, if $q^{ij}(t) \in \mathbb{Q}_s \ \forall t$, constraints (8) will be relaxed, which means no extra motion for collision avoidance is executed.

Task Constraints. The collision avoidance constraints try to move any joint possible to avoid collision. However, this might lead to unpredictable behaviors of the end-effectors, or say, $\Delta \dot{x}(t)$ might be huge. Hence, it is critical to introduce the task constraints. Hard task constraints like (4), however, might lead to extra problems in finding a feasible solution. To handle this, we propose that the task priorities can be taken into consideration by penalizing a non-negative slack variable $\xi \in \mathbb{R}^{Nm}$, i.e.

$$\begin{aligned} \min_{\xi} \quad & \frac{1}{2}\xi^\top V \xi \\ \text{s.t.} \quad & |\dot{x}_d - J\dot{q}| \leq \xi, \end{aligned} \tag{9}$$

where $|\cdot|$ takes the element-wise absolute value of the input vector, and $V \in \mathbb{R}^{Nm \times Nm}$ is a positive semi-definite matrix. In fact, (9) provides the system with the ability to decide what task can be sacrificed for collision avoidance. The larger the element of V, the more penalties are taken on the corresponding dimension of the end-effector motion. If $v_{ii} = 0$, the end-effector can move freely in the corresponding dimension. At one extreme, if $V = 0$, the task constraints are not working any more. At another, if all elements of V are large enough, the extra motion $\Delta \dot{q}(t)$ caused by collision avoidance will have little influence on $\dot{x}(t)$, i.e., $\Delta \dot{x}(t) \approx 0$.

To summarize, the task-oriented collision avoidance problem can be handled by solving the following optimization problem,

$$
\min_{\dot{q}, \xi} \quad \frac{1}{2} \left(\dot{q}^{\top} W \dot{q} + \xi^{\top} V \xi \right)
$$
$$
\text{s.t.} \quad \begin{cases} \dot{q}^{-} \leq \dot{q} \leq \dot{q}^{+} \\ -\nabla \mathcal{F} \left(q^{ij} \right)^{\top} \dot{q}^{ij} \leq \mathcal{F} \left(q^{ij} \right) \\ |\dot{x}_{\mathrm{d}} - J \dot{q}| \leq \xi. \end{cases}
\tag{10}
$$

Since the objective is quadratic and the constraints are linear, the optimization problem (10) is a QP, which can be solved efficiently by numerical QP solvers [15].

3.2   Dataset Construction

A dataset containing the system configurations and the corresponding minimum distance must be constructed to learn the distance decision function $\mathcal{F}(\cdot)$. In consideration of both accuracy and computation efficiency, we use capsule bodies to represent the manipulator links [12]. Let C_k^i denote the k-th capsule body of manipulator i with a radius of r_k^i, and P_k^i the starting point on the central axis of C_k^i, the minimum distance $d(C_k^i, C_l^j)$ between the two corresponding capsule bodies is then given by

$$
\min_{\mu, \nu} \quad \left\| P_{\mu}^i - P_{\nu}^j \right\|_2^2,
$$
$$
\text{s.t.} \quad P_{\mu}^i = (1 - \mu) P_k^i + \mu P_{k+1}^i,
$$
$$
P_{\nu}^j = (1 - \nu) P_l^j + \nu P_{l+1}^j,
\tag{11}
$$
$$
0 \leq \mu \leq 1, \quad 0 \leq \nu \leq 1,
$$

as shown in Fig. 1. Given some sampling resolution Δq, we can explore the whole configuration space and compute the corresponding minimum distance between the manipulators through (11). The Cartesian coordinates of P_k^i and P_l^j are given by the forward kinematics (3).

Instead of using the minimum distance as the label directly and formulating the learning problem as a regression problem, we treat it as a classification one. Slightly different from (7), we label the system states as two classes according to the minimum distance:

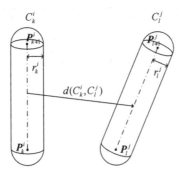

Fig. 1. The distance metrics between two capsule bodies. Here, P and r are the starting point and radius of the corresponding capsule body, respectively, and $d(\cdot, \cdot)$ represents the minimum distance.

$$y(\boldsymbol{q}^{ij}) = \begin{cases} -1 \text{ if } & \min_{k,l}[d(C_k^i, C_l^j)] < b^-, \\ +1 \text{ if } b^- \leq \min_{k,l}[d(C_k^i, C_l^j)] \leq b^+, \\ \emptyset \text{ if } & \min_{k,l}[d(C_k^i, C_l^j)] > b^+, \end{cases} \qquad (12)$$

where y is the label of configuration $\boldsymbol{q}^{ij}$, $b^- = r_k^i + r_l^j$ represents the critical distance at which collision occurs, and b^+ is used to balance the number of positive/negative samples, since in a normal way, the number of safe configurations is orders of magnitude larger than that of forbidden ones. The dataset construction algorithm for a 2-manipulator system is summarized in Algorithm 1.

3.3   Learning Details

We follow the idea proposed in [3,16] to learn an efficient representation of $\mathcal{F}(\cdot)$ from the collision avoidance dataset $\mathcal{D}$, which is in essence a binary classification problem. In practice, SVMs are especially well-performed in solving classification problems, and their performance and efficiency have been proved by a tremendous number of works [6,20]. By applying nonlinear kernels κ, SVMs are capable of learning very complicated functions, i.e.

$$g(\boldsymbol{x}) = \sum_{i=1}^{N_{sv}} \beta_i \kappa(\boldsymbol{x}, \boldsymbol{x}_i), \qquad (13)$$

where $\boldsymbol{x}_1, \ldots, \boldsymbol{x}_{N_{sv}}$ are called support vectors. In our case, we choose the radial basis function (Gaussian) kernel:

$$\kappa(\boldsymbol{x}, \boldsymbol{y}) = \exp\left(-\frac{1}{2\sigma^2}\|\boldsymbol{x} - \boldsymbol{y}\|_2^2\right). \qquad (14)$$

Notice that the euclidean distance $\|\cdot\|$ applied in the RBF kernel is possibly not an appropriate metrics for the configuration space features [12]. Thus, the

Algorithm 1. Collision Avoidance Dataset Construction for 2-Manipulator System

---

Input:

The manipulator DoF n, the exploration sampling resolution Δq, and the forward kinematics function f ;

Output:

The collision avoidance dataset $\mathcal{D}$;

1: **Initialize** q^{ij} ;
2: **repeat**
3:     **for** $k = 1 \to n$ **do**
4:         $P_k^i \leftarrow f([q_1^i, \ldots, q_k^i]^{\mathrm{T}})$;
5:         **for** $l = 1 \to n$ **do**
6:             $P_l^j \leftarrow f([q_1^j, \ldots, q_l^j]^{\mathrm{T}})$;
7:             Compute $d(C_k^i, C_l^j)$ by applying (11) ;
8:         **end for**
9:     **end for**
10:    $d_{\min} \leftarrow \min_{k,l}[d(C_k^i, C_l^j)]$;
11:    Compute $y(q^{ij})$ by applying (12) ;
12:    Record q^{ij} and $y(q^{ij})$;
13:    Update q^{ij} with Δq ;
14: **until** exploration terminated.

---

joint space features in dataset $\mathcal{D}$ are mapped to Cartesian space via forward kinematics (3) before training. The distance decision function is then given by

$$\mathcal{F}\left(f\left(q^{ij}\right)\right) = \sum_{k=1}^{N_{\mathrm{sv}}} \beta_k \kappa \left(f\left(q^{ij}\right), f\left(q_k^{ij}\right)\right). \qquad (15)$$

It can be verified that (15) is continuously differentiable. The gradient of (15) is given by

$$\nabla \mathcal{F}(q^{ij}) = \frac{\partial f(q^{ij})}{\partial q^{ij}} \cdot \frac{\partial \mathcal{F}(f(q^{ij}))}{\partial f(q^{ij})} \qquad (16)$$

where the first term is the Jacobian matrix, and the second term can be easily deduced from (15).

Notice that the evaluation time of (15) grows with the increase of N_{sv}, which implies that a sparser training method can boost the computation of (8). For example, the cutting-plane algorithm (CPSP) [7] allows to estimate the original sample space by a given number of support vectors, with slightly worsen classification performance. Thus, the computation complexity of (8) can be controlled.

4   Numerical Simulations

We first test the effectiveness of training the distance decision function on a system composed of two 1-dimension manipulators using traditional SVM [1]. Following Algorithm 1, a collision avoidance dataset composed of 732 positive samples and 899 negative samples is constructed. Results show a satisfactory classification performance of the learned function, as illustrated in Fig. 2. Furthermore, the relationship (7) is satisfied, hence the effectiveness of the collision avoidance constraint (8).

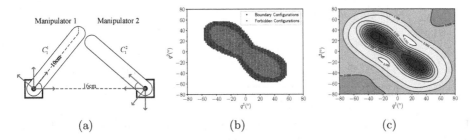

Fig. 2. (a) The illustration of two 1-dimension manipulators, where red arrows represent the initial frame and the blue arrows denote the present frame. (b) The corresponding collision avoidance dataset generated by Algorithm 1. (c) The contour of the learned function $\mathcal{F}(\cdot)$, which shows a satisfactory classification performance.

A system consisting of two 3-DoF manipulators is also considered for further investigation, as shown in Fig. 3 and Table 1. In this case, we have $N = 2$, $m = 2$ and $n = 3$. The forward kinematics (3) can be easily deduced as follows:

$$
x = \begin{bmatrix} x^1 \\ x^2 \end{bmatrix} = f(q^{12}) = \begin{bmatrix} L(c_1^1 + c_{12}^1 + c_{123}^1) \\ L(s_1^1 + s_{12}^1 + s_{123}^1) \\ L(c_1^2 + c_{12}^2 + c_{123}^2) + L_0 \\ L(s_1^2 + s_{12}^2 + s_{123}^2) \end{bmatrix}, \tag{17}
$$

where $c_{123}^i = \cos(q_1^i + q_2^i + q_3^i)$, $s_{123}^i = \sin(q_1^i + q_2^i + q_3^i)$, and the rest can be done in the same manner. The Jacobian matrix is given by $J = \mathrm{blkdiag}(J^1, J^2)$, where

$$
J^i = \begin{bmatrix} -L(s_1^i + s_{12}^i + s_{123}^i) & -L(s_{12}^i + s_{123}^i) & -Ls_{123}^i \\ L(c_1^i + c_{12}^i + c_{123}^i) & L(c_{12}^i + c_{123}^i) & Lc_{123}^i \end{bmatrix}.
$$

Following Algorithm 1, exploring the configuration space of each manipulator by $\Delta q = 15°$ yields 4693 samples, and hence ≈ 22 million samples for the dual-manipulator system. Among them, 580921 are labeled as positive and 675876 as negative with $b^+ = 6$ cm and $b^- = 2$ cm. Since this dataset is relatively large, the cutting-plane training method [7] is applied to guarantee a real-time performance. The performances of traditional SVM and the CPSP algorithm on different training set sizes are compared in Table 2. As can be seen, the cutting-plane training method reduces the evaluation time of $\mathcal{F}(\cdot)$ dramatically when the dataset is large, with a relatively small loss of classification performance.

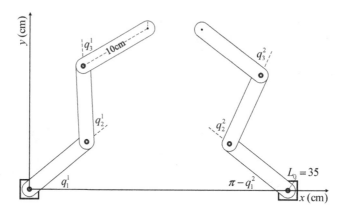

Fig. 3. A more complicated example composed of two identical 3-DoF manipulators, whose bases are 35 cm apart on the x-axis. The technical parameters of each link are specified in Table 1.

To analyze how the proposed constraints shape the trajectories of the manipulators, we give a task for the end-effector of manipulator 1 to move from initial position $x_i^1 = [0, 30]^\top$ to terminal position $x_t^1 = [20, 0]^\top$ through the upper right part of the ellipse $\frac{x^2}{20^2} + \frac{y^2}{30^2} = 1$. The initial configuration of manipulator 2 is $q_i^2 = [\frac{5}{6}\pi, 0, -\frac{\pi}{3}]^\top$ with no task attached. Clearly, if no collision avoidance methodology is applied, the manipulators will collide, see Fig. 4.

As claimed in Sect. 2, W controls the cost of moving each joint. In Fig. 5a, W is set as an identity matrix, which implies that the cost of moving any joint makes no difference. In comparison, when w_{44} is made larger, the changes in the corresponding joint q_1^2 (the first joint of manipulator 2) reduce dramatically (see Fig. 5b). The trajectories of all joints in both cases are illustrated in Fig. 4 for clarity. Besides, the values of $\mathcal{F}(\cdot)$ keep positive during the whole motion in either case (Fig. 4), which again demonstrates the effectiveness of the collision avoidance constraints.

Figure 5 shows how the weight matrix V assigns different priorities to the corresponding dimension of the task. In Fig. 5a and 5b, manipulator 1 is forced to execute the task while manipulator 2 can move freely, since v_{11}, v_{22} are large and $v_{33} = v_{44} = 0$. In other words, manipulator 1 executing the ellipse is given higher priority, thus manipulator 2 (with lower priority) must be "pushed away" to make

Table 1. Technical parameters of the 3-DoF manipulator.

Joint	q^- (rad)	q^+ (rad)	$\dot{q}^-$ (rad/s)	$\dot{q}^+$ (rad/s)	Length L (cm)	Radius r (cm)
1	0	π	-3	3	10	1
2	$-3\pi/4$	$3\pi/4$	-3	3	10	1
3	$-3\pi/4$	$3\pi/4$	-3	3	10	1

Table 2. Comparison between the traditional SVM [1] and the CPSP algorithm [7]. The evaluation time of CPSP is faster than traditional SVM for the ability of keeping N_{sv} fixed, at the cost of slightly worsen classification performance.

Method	Training set size	ACC (%)	F1 (%)	TPR (%)	FPR (%)	N_{sv}	Computation time of $\mathcal{F}(\cdot)$ and $\nabla \mathcal{F}$ (ms)
SVM	5K	85.27	84.09	84.22	13.83	1996	11.96 ± 1.66
	10K	87.41	86.38	86.40	11.72	3553	21.52 ± 2.86
	100K	93.34	92.77	92.53	5.97	20213	123.65 ± 7.36
CPSP	5K	83.07	81.91	82.93	16.81	2000	12.29 ± 1.66
	10K	85.16	84.04	84.56	14.32	2000	12.29 ± 1.66
	100K	90.28	89.97	90.24	8.65	2000	12.29 ± 1.66

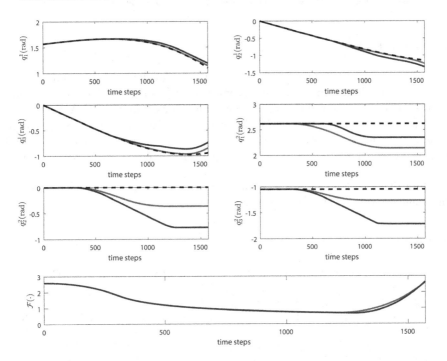

Fig. 4. The joint space trajectories of the dual-manipulator system with different W (solid red and blue), and the trajectory when the system is not controlled (dashed black) (Color figure online).

room for manipulator 1. As shown in Fig. 4, the trajectories of manipulator 1 are almost unaffected. In Fig. 5c, manipulator 1 still owns high priority, while the end-effector of manipulator 2 is forced to move along the x-axis. On the contrary, when manipulator 2 staying put is given higher priority, manipulator 1 is not able to execute the ellipse any more, hence the trajectory in Fig. 5d.

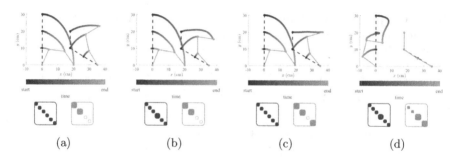

Fig. 5. (Top) The trajectories of each joint over time under different $W \in \mathbb{R}^{6 \times 6}$ and $V \in \mathbb{R}^{4 \times 4}$. Initial (dashed blue) and terminal (solid green) configurations of each manipulator are plotted as well. (Bottom) The relative sizes of the elements in W (blue) and V (green). Hollow squares represent zero (Color figure online).

5 Conclusion

In this work, a novel framework for solving the task-oriented collision avoidance problem in fixed-base multi-manipulator systems is proposed. The introduction of the kinematics constraints, collision avoidance constraints and task constraints guarantees the generated motion to be feasible, collision-free and task-oriented. Besides, the distance decision function is constructed sparsely and the overall framework is formulated as a QP, which guarantees the real-time performance.

Funding. This work is supported by National Natural Science Foundation (NNSF) of China under Grant U1713203, 51729501 and 61803168.

References

1. Chang, C.C., Lin, C.J.: Libsvm: a library for support vector machines. ACM Trans. Intell. Syst. Technol. (TIST) **2**(3), 27 (2011)
2. Dogar, M., Knepper, R.A., Spielberg, A., Choi, C., Christensen, H.I., Rus, D.: Multi-scale assembly with robot teams. Int. J. Robot. Res. **34**(13), 1645–1659 (2015)
3. Figueroa Fernandez, N.B., Mirrazavi Salehian, S.S., Billard, A.: Multi-arm self-collision avoidance: A sparse solution for a big data problem. In: In Proceedings of the Third Machine Learning in Planning and Control of Robot Motion (MLPC) Workshop., CONF (2018)
4. Ge, S.S., Cui, Y.J.: New potential functions for mobile robot path planning. IEEE Trans. Robot. Autom. **16**(5), 615–620 (2000)
5. Hsu, D., Kavraki, L.E., Latombe, J.C., Motwani, R., Sorkin, S., et al.: On finding narrow passages with probabilistic roadmap planners. In: Robotics: the Algorithmic Perspective: 1998 Workshop on the Algorithmic Foundations of Robotics, pp. 141–154 (1998)
6. Joachims, T.: Training linear svms in linear time. In: Proceedings of the 12th ACM SIGKDD International Conference on Knowledge Discovery and Data Mining, pp. 217–226. ACM (2006)

7. Joachims, T., Yu, C.N.J.: Sparse kernel SVMs via cutting-plane training. Mach. Learn. **76**(2–3), 179–193 (2009)
8. Keating, S.J., Leland, J.C., Cai, L., Oxman, N.: Toward site-specific and self-sufficient robotic fabrication on architectural scales. Sci. Robot. **2**(5), eaam8986 (2017)
9. Knepper, R.A., Layton, T., Romanishin, J., Rus, D.: Ikeabot: an autonomous multi-robot coordinated furniture assembly system. In: 2013 IEEE International Conference on Robotics and Automation, pp. 855–862. IEEE (2013)
10. Kröse, B.J., Van Dam, J.W.: Learning to avoid collisions: a reinforcement learning paradigm for mobile robot navigation. In: Artificial Intelligence in Real-Time Control 1992, pp. 317–321. Elsevier, Amsterdam (1993)
11. Lavalle, S.M.: Rapidly-exploring random trees: A new tool for path planning. Technical report, Citeseer (1998)
12. LaValle, S.M.: Planning Algorithms. Cambridge University Press, Cambridge (2006)
13. Likhachev, M., Ferguson, D.I., Gordon, G.J., Stentz, A., Thrun, S.: Anytime dynamic a*: an anytime, replanning algorithm. ICAPS **5**, 262–271 (2005)
14. Lozano-Perez, T.: Spatial planning: a configuration space approach. In: Autonomous Robot Vehicles, pp. 259–271. Springer, Berlin (1990)
15. Mattingley, J., Boyd, S.: Cvxgen: a code generator for embedded convex optimization. Optim. Eng. **13**(1), 1–27 (2012)
16. Mirrazavi Salehian, S.S., Figueroa, N., Billard, A.: A unified framework for coordinated multi-arm motion planning. Int. J. Robot. Res. **37**(10), 1205–1232 (2018)
17. Park, M.G., Jeon, J.H., Lee, M.C.: Obstacle avoidance for mobile robots using artificial potential field approach with simulated annealing. In: ISIE 2001. 2001 IEEE International Symposium on Industrial Electronics Proceedings, vol. 3, pp. 1530–1535. IEEE (2001)
18. Petersen, K.H., Napp, N., Stuart-Smith, R., Rus, D., Kovac, M.: A review of collective robotic construction. Sci. Robot. **4**(28), eaau8479 (2019)
19. Sünderhauf, N., Brock, O., Scheirer, W., et al.: The limits and potentials of deep learning for robotics. Int. J. Robot. Res. **37**(4–5), 405–420 (2018)
20. Vedaldi, A., Zisserman, A.: Efficient additive kernels via explicit feature maps. IEEE Trans. Pattern Anal. Mach. Intell. **34**(3), 480–492 (2012)
21. Wise, K.D., Bowyer, A.: A survey of global configuration-space mapping techniques for a single robot in a static environment. Int. J. Robot. Res. **19**(8), 762–779 (2000)
22. Zucker, M., et al.: Chomp: covariant hamiltonian optimization for motion planning. Int. J. Robot. Res. **32**(9–10), 1164–1193 (2013)

An Adaptive Seam-Tracking System with Posture Estimation for Welding

Zhi Yang[1], Shuangfei Yu[1], Yisheng Guan[1(✉)], Yufeng Yang[2], Chuangwu Cai[2], and Tao Zhang[1(✉)]

[1] Guangdong University of Technology, Guangzhou, Guangdong Province, China
{ysguan,tzhang}@gdut.edu.cn
[2] Foshan Biowin Robot and Automation Co., Ltd., Foshan, China

Abstract. In this paper, a robotic seam tracking system with welding posture estimation is proposed that can adapt to different conditions encountered in the field of welding, such as uncertain working positions and complex workpiece shapes. In the proposed system, the target coordinate system of the welding torch is established in real-time at each welding position, and the rotation angles are obtained to change the welding posture of the robot. Gaussian kernel correlation filter is used to track the weld feature in real-time that improves the accuracy and robustness of welding seam tracking. Compared with morphological methods, this method can quickly and accurately find the position of the weld from the noisy image. Finally, experimental results show that the method can be used to calculate the welding posture, which meet the welding requirements of a complex environment.

Keywords: Seam tracking · Gaussian kernel correlation filter · Welding posture

1 Introduction

The development of industrial technology towards automation and intelligence has placed, increasingly strict requirements for automatic welding technology [1,2]. Traditional welding robots use the teaching-reproduction method to realize automatic welding, but this method cannot be used in a multi-type small-batch production [3,4]. At present, the principle of laser triangulation [5] is widely used in weld inspection. The 3D coordinate of the weld can be obtained by the laser stripe in the image acquired by a camera [6]. During welding, in order to reduce the interference of arc light, the laser vision sensor is installed in parallel at a certain distance in front of the welding torch. However, a large front-view distance will cause tracking lag as well as reduce welding accuracy, and small front-view distance will make the

The work in this paper is partially supported by the Key Research and Development Program of Guangdong Province (Grant No. 2019B090915001) and the Program of Foshan Innovation Team of Science and Technology (Grant No. 2015IT100072).

C. S. Chan et al. (Eds.): ICIRA 2020, LNAI 12595, pp. 88–99, 2020.
https://doi.org/10.1007/978-3-030-66645-3_8

image vulnerable to splash and arc light interference [7,8]. Therefore, researchers have carried out a significant amount of research on the topic of accurately finding the position of the weld from a noisy image. In order to reduce the influence of noise on the positioning accuracy of weld feature point, Li et al. [9] used sequential gravity method (SGM) to improve the system robustness. Ge et al. [10] proposed a string descriptor to describe a contour. First, the contour is searched using Kalman filtering. Second, the least square method is used to fit a series of centers of mass obtained by scanning the columns in a rectangular window. Last, a string method is used to qualitatively describe the weld contour. In the Ref [11], a structured-based visual weld recognition method is proposed, which uses the information of the previous frame to process the current frame. This method is highly robust and can operate in the presence of strong arc light and splash interference.

However, in the existing work, the adaptive adjustment of welding posture during welding was not widely studied. The actual welding process can take place under many complicated working conditions. For example, in shipbuilding, the shape of the workpiece to be welded is complex. During the automatic welding, it is necessary to obtain the weld position, and simultaneously obtain the welding posture in real-rtime. Therefore, traditional welding seam tracking systems with fixed welding posture cannot meet these particular requirements. In this article, we propose a welding seam tracking method with welding posture estimation, The proposed methon can not only accurately finds the position of the welding seam from the noisy images, but also estimate the welding posture during the welding process. The laser stripe in a image is easily covered by the arc light, therefore it is difficult to accurately extract the weld feature point from the noise-contaminated image using morphological methods. In this article, the weld feature point is only extracted in the first frame when welding is not started. In the subsequent frames, Gaussian kernel correlation filter (KCF) [12] is used to track the target point in the first frame to determine the location of the weld feature point. The feature points of the weld in the image are mapped to the world coordinate system and saved in order to estimate the welding posture of the weld torch. The target coordinate system of the welding torch is calculated using the saved weld feature points, and the rotation angles R_x, R_y and R_z of the robot are calculated through the target coordinate system. Finally, the calculated rotation angles are transmitted to the robot for realizing real-time posture control.

2   Basic Principle of Laser Vision Sensor

Robotic welding needs to obtain the 3D information of the weld through the features of the 2D weld image, therefore the establishment of an accurate mapping model is very important for accurate weld tracking. The ideal imaging model is shown in Fig. 1, where planes A and B are the measurement and imaging planes respectively. The world coordinate system, robot tool coordinate system, pixel coordinate system and camera coordinate system are represented by $O_W X_W Y_W Z_W$, $O_T X_T Y_T Z_T$, $O_P X_P Y_P Z_P$, $O_C X_C Y_C Z_C$. The point on of intersection between the laser and the measurement object is given by P, and point

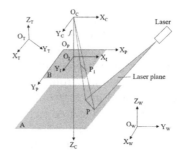

Fig. 1. Perspective projection model

P_1 is the mapping of point P in the pixel coordinate system. The mapping relationship between the pixel coordinates and camera coordinates based on the principle of camera imaging [13] is as follows:

$$\begin{bmatrix} X_C \\ Y_C \\ Z_C \end{bmatrix} = Z_C M^{-1} \begin{bmatrix} u \\ v \\ 1 \end{bmatrix} \qquad (1)$$

where M is the camera internal parameter matrix, that can be obtained by camera calibration. The pixel coordinate of P_1 is given by $\begin{bmatrix} u\ v\ 1 \end{bmatrix}^T$, and $\begin{bmatrix} X_C\ Y_C\ Z_C \end{bmatrix}^T$ is the coordinate of P in the camera coordinate system. It can be noted from Eq.(1), that there is no one-to-one correspondence between the points in the image coordinate and the points in the three-dimensional coordinate, therefore a constraint equation must be added. When $Z_C = 1$, Eq.(1) can be expressed as follows:

$$\begin{bmatrix} X'' \\ Y'' \\ Z'' \end{bmatrix} = M^{-1} \begin{bmatrix} u \\ v \\ 1 \end{bmatrix} \qquad (2)$$

We consider the equation of the laser plane in the camera coordinate system as:

$$AX_C + BY_C + CZ_C + D = 0 \qquad (3)$$

Substituting Eqs.(1) and (2) into (3), Z_C can be written as follows:

$$Z_C = -D/(AX'' + BY'' + CZ'') \qquad (4)$$

Substituting Eq.(4) into Eq.(1), the coordinates of P in the camera coordinate system can be obtained as follows:

$$\begin{bmatrix} X_C \\ Y_C \\ Z_C \end{bmatrix} = -D/(AX'' + BY'' + CZ'')M^{-1} \begin{bmatrix} u \\ v \\ 1 \end{bmatrix} \qquad (5)$$

Finally, the point P in the camera coordinate system is mapped to the world coordinate system as:

$$
\begin{bmatrix} X_W \\ Y_W \\ Z_W \\ 1 \end{bmatrix} = T_1 T_2 \begin{bmatrix} X_C \\ Y_C \\ Z_C \\ 1 \end{bmatrix} \tag{6}
$$

where T_1 and T_2 are the robot D-H matrix and hand-eye calibration matrix respectively . In this paper, the Halcon vision library calibration system is used to calibrate the internal parameters of the camera. At the same time, the laser plane parameters are obtained by using multiple non-coincident laser light bars to fit the laser plane equation.

3   Welding Seam Tracking Algorithm

The KCF tracking algorithm selects the tracked target in the first frame and trains the tracker with the target information position. The tracker is used to conduct sampling point response near the predicted target location of the next frame. The position with the strongest response point is chosen as the target location. There is no arc light interference in the image prior to welding, and the feature point of the weld is extracted by morphological methods. The $M \times N$ pixels area extracted from the image with the feature point as the center is the positive sample. The negative samples of training are obtained by the cyclic displacement of cyclic matrix. We assume that the training sample is (x_i, y_i) , and the linear regression function is as follows:

$$
f(x_i) = w^T x_i \tag{7}
$$

where w is the weight coefficient. The purpose of training is to find the appropriate weight to minimize the value of a cost function, where this cost function is given as follows:

$$
min \sum (f(x_i) - y_i)^2 + \lambda \|w\| \tag{8}
$$

In Eq.(8), λ is the penalty coefficient. We can find w as follows by letting the derivative of Eq.(8) equal to 0:

$$
w = (X^H X + \lambda I)^{-1} X^H y \tag{9}
$$

where $X = [x_1, x_2, \ldots, x_n]$, with each column of X representing the feature vector of a sample, and $X^H = (X^*)^T$, where X^* is the conjugate of X. The label of each samply is represented by a column vector y, and I is the identity matrix. To improve the classifier's performance, the ridge regression in the linear space is mapped to the nonlinear space through the Gaussian kernel function, namely $x \to \alpha(x)$. We define $k(x, x') = \alpha(x)\alpha(x')$ as the kernel function. The classifier weight can be expressed as a linear combination of training samples [14] as follows:

$$
w = \sum_{i=1}^{n} \sigma_i x_i = \sum_{i=1}^{n} \sigma_i \alpha(x_i) \tag{10}
$$

where σ_i is the weight of the training sample. Considering σ as a weight vector whose entries correspond to σ_i, the response value of the target model z in the classifier is given by:

$$f(z) = w^T z = \sigma^T \alpha(X)\alpha(z) \tag{11}$$

Substituting Eq.(11) into Eq.(8) and puting the derivative of the resulting equation equal to 0, the ridge regression based on the kernel function can be expressed as:

$$\sigma = (K + \lambda I)^{-1} y \tag{12}$$

where $K = \alpha(X)\alpha(z)$ is the kernel matrix. Therefore, the optimal problem for w is transformed into the optimal problem for σ. To speed up the calculation of Eq.(12), a cyclic matrix is introduced into the algorithm. All negative training samples of KCF are obtained by cyclic displacement of the cyclic matrix. All cyclic matrices have diagonal properties in the Fourier space [14], K can be expressed as follows:

$$K = C(k^{xz}) = F diag(\hat{k}^{xz}) F^H \tag{13}$$

where $\hat{}$ denotes the discrete Fourier transform. k^{xz} is the first row of the kernel matrix. Finally, Eq.(12) can be written as follows:

$$\hat{\sigma} = \frac{\hat{y}}{\hat{k}^{xz} + \lambda} \tag{14}$$

The response of the target model z in the current frame can be expressed as:

$$\hat{f}(z) = (\hat{k}^{xz})^* \hat{\sigma} \tag{15}$$

In a new frame, the coordinates of the feature point are given by:

$$(c_i.s_i) = max f(z) \tag{16}$$

where $f(z)$ is obtained by the inverse Fourier transform of $\hat{f}(z)$. In the next frame, we only need to update σ and training samples x. After the weld feature point detection in the previous frame is completed, a new training sample x' is generated and a new σ_i' is obtained through training. Finally, the parameters are updated as follows:

$$\begin{cases} \sigma_i = (1 - \beta)\sigma_{i-1} + \beta\sigma_i' \\ x_i = (1 - \beta)x_{i-1} + \beta x_i' \end{cases} \tag{17}$$

where σ_i and σ_{i-1} are the σ of the ith frame and the $(i-1)$th frame, and x_i and x_{i-1} are the training samples of the ith frame and the $(i-1)$th frame. The above process is repeated in the next frame tracking process to obtain a new weld feature point.

4   Welding Posture Calculation

In order to adapt to different weld tracks, the posture of welding torch should be adjusted in real-time to ensure that the torch and welding surface are perpendicular. We also need to simultaneously ensure that the weld feature point in the image is in the middle of the image. We calculate the welding posture by calculating the target posture of the robot's tool coordinate system in real-time. Each coordinate system of the welding system is shown in Fig. 2. The robot base coordinate system is represented by $O_B X_B Y_B Z_B$, which is also the world coordinate in this article. The tool coordinate system, the camera coordinate system and the target coordinate system of welding torch at welding point are represented by $O_T X_T Y_T Z_T$, $O_C X_C Y_C Z_C$ and $O_M X_M Y_M Z_M$.

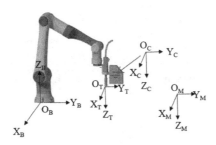

Fig. 2. Relative position of each coordinate system

4.1   Constructing Target Coordinate System of Welding Torch

Solving Unit Vector $\overrightarrow{Z_M}$. It is necessary to save the weld feature point detected by the sensor, as the laser vision sensor is fixed in front of the welding torch. During welding, the welding seam tracking algorithm described in Sect. 3 tracks the points p_1, p_2 and p_3, whose positions are shown in Fig. 3. The points in each frame are mapped into 3D coordinates by the method described in Sect. 2, and stored in the container in the form of point cloud. The points p_1, p_2 and p_3 are expressed as p_1', p_2' and p_3' respectively in robot base coordinate system. When the welding torch reaches the target welding point, we select from the container the point cloud data of the last 20 frames with respect to the current welding point. The cubic parameter equation is used to represent the spatial weld, and least squares method is utilized to fit the weld trajectory. The weld trajectory equation is as follows:

$$\begin{cases} X = X \\ Y = A_1 X^3 + B_1 X^2 + C_1 X + D_1 \\ Z = A_2 X^3 + B_2 X^2 + C_2 X + D_2 \end{cases} \tag{18}$$

Taking the derivative of Eq.(18), the tangent vector of the weld can be expressed as $\vec{V_1} = \left[1\ d(Y)/d(X)\ d(Z)/d(X) \right]^T$. During welding, we expect the welding torch to be always on the bisector of the groove. Let the vector $\vec{V_{v1}} = P_2' - P_1'$, $\vec{V_{v2}} = P_2' - P_3'$. The vectors $\vec{V_{v1}}$ and $\vec{V_{v2}}$ are unitized to obtain the unit vectors $\vec{V_{v1n}}$ and $\vec{V_{v2n}}$ respectively. The vector of the angle bisector of $\vec{V_{v1n}}$ and $\vec{V_{v2n}}$ can be expressed as follows:

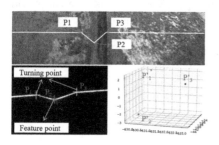

Fig. 3. The positional relationship of p_1, p_2 and p_3

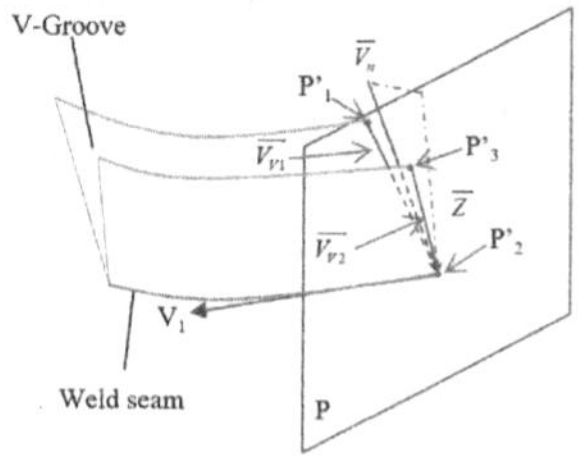

Fig. 4. The solution principle of vector $\vec{Z_M}$

$$\vec{V_n} = \vec{V_{v1n}} + \vec{V_{v2n}} \tag{19}$$

A plane P created at the weld feature point P_2' is shown in Fig. 4. The normal vector of the plane P is the tangent vector $\vec{V_1}$ of the space weld. Let the projection of the vector $\vec{V_n}$ on the plane P be $\vec{Z}$, i.e.

$$\vec{Z} = (\vec{V_1} \times \vec{V_n}) \times \vec{V_1} \tag{20}$$

where $\times$ is the outer product of the vector. The normal vector $\vec{Z}$ is unitized to obtain the unit vector $\vec{Z_M}$ for facilitate the calculation of the rotation matrix.

Solving Unit Vector $\vec{X_M}$. During the welding process, the weld feature point captured by the camera should always be kept in the middle of the image. This prevents the point from being lost when complex welds are tracked. Therefore, the tangent vector $\vec{V_1}$ of the spatial weld track cannot be used as $\vec{X_M}$. It is assumed that the current weld feature point is expressed as $P_f = (x_0, y_0, z_0)$, shown in Fig. 5. A plane P_0 is created at the weld feature point P_f, and $\vec{Z_M}$ is the normal vector of P_0. The equation of plane P_0 in the robot base coordinate system can be expressed as follows:

$$A(x - x_0) + B(y - y_0) + C(z - z_0) = 0 \tag{21}$$

where $\vec{Z_M} = \begin{bmatrix} A\ B\ C \end{bmatrix}$. It is assumed that the extension line of the welding seam and the optical center O_C of the camera intersects the plane P_0 at a point P_C.

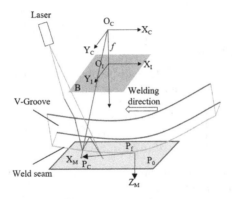

Fig. 5. The solution principle of vector $\vec{X_M}$

The unit vector connecting the points P_C and P_f is the posture vector $\vec{X_M}$. In order to obtain the coordinates of the point P_C in the robot base coordinate system, the 3D coordinates of the point P_C should be solved in the camera coordinate system. Consider a point A on the vector $\vec{Z_M}$. The points A and P_f are mapped from the world coordinate system to the camera coordinate system by the hand-eye calibration matrix T_1 and the robot D-H matrix T_2, respectively. Points A and P_f can be expressed as A_C and P_{fc} respectively, in the camera coordinate system, where $P_{fc} = \begin{bmatrix} x_c & y_c & z_c \end{bmatrix}$. The vector $\vec{Z_M}C$ can be expressed in the camera coordinate system as:

$$\vec{Z_{MC}} = A_C - P_{fc} = \begin{bmatrix} A_c & B_c & C_c \end{bmatrix} \tag{22}$$

and the plane P_0 can be expressed in the camera coordinate system as:

$$\begin{cases} A_c x + B_c y + C_c z + D_c = 0 \\ D_c = -(A_c x_c + B_c y_c + C_c z_c) \end{cases} \tag{23}$$

Combining Eq.(2), Eq.(4), Eq.(5) and Eq.(23), the coordinate of P_C in the camera coordinate system can be obtained. Finally the coordinate of the point P_C in the world coordinate system can be obtained using Eq.(6). Let $\vec{X} = P_C - P_f$, vector $\vec{X_M}$ is the unit vector of $\vec{X}$.

Solving Unit Vector $\vec{Z_M}$. The unit vectors $\vec{Y_M}$, $\vec{X_M}$ and $\vec{Z_M}$ are orthogonal to one another, then:

$$\vec{Y_M} = \vec{Z_M} \times \vec{X_M} \tag{24}$$

4.2   Rotation Angles Calculation

The unit vectors $\vec{X_M}$, $\vec{Y_M}$ and $\vec{Z_M}$ of the tool coordinate system are projected to the robot base coordinate system $\vec{X_B}$, $\vec{Y_B}$ and $\vec{Z_B}$ through the transformation matrix. The rotation matrix is calculated as follows:

$$R = \begin{bmatrix} r_{11} & r_{12} & r_{13} \\ r_{21} & r_{22} & r_{23} \\ r_{31} & r_{32} & r_{33} \end{bmatrix} = \begin{bmatrix} \vec{X_M} & \vec{Y_M} & \vec{Z_M} \end{bmatrix} \begin{bmatrix} \vec{X_B} & \vec{Y_B} & \vec{Z_B} \end{bmatrix}^{-1} \quad (25)$$

The rotation angles R_x, R_y and R_z can be obtained by calculating the inverse of the rotation matrix. The calculations can be expressed as follows:

$$\begin{cases} R_x = atan2(r_{32}, r_{33}) \\ R_y = atan2(-r_{31}, \sqrt{r_{31}^2, r_{32}^2}) \\ R_z = atan2(r_{21}, r_{11}) \end{cases} \quad (26)$$

Finally, the rotation angles of the robot are transmitted to the controller for real-time control of the robot.

5   Experiments and Results

To verify the validity of the method in this paper, we perform offline seam tracking test and welding posture estimation test, and the experimental results are evaluated and analyzed.

5.1   Offline Weld Tracking Test

This experiment uses the image data from field welding to verify the proposed algorithm. The vertical height of the camera and workbench is 16 cm, and the laser is installed 50 mm in front of the camera. When recording image data, the welding current and speed are 218 A and 6 mm/s, respectively. The weld feature point in the image is manually marked, and the marked label point is taken as the actual position of the weld feature point in the image. The tracking error in the experiment is measured by the Euclidean distance between the target point of label and the target point of tracker. calculated as follows:

$$E = \sqrt{(x' - x)^2 + (y' - y)^2} \quad (27)$$

where (x', y') is the position of the detected weld feature point in the pixel coordinate system, and (x, y) is the manually marked weld feature position. In Fig. 6(a), the noise in the image is low as the welding is not turned on. The morphological method is used to determine the initial position of the weld, with the initial position as the center and the extracted size 80×20 pixels area as the   tracking target. Then the algorithm described in Sect. 3 of this paper is used to continuously track the weld feature area located in the first frame. We randomly selected three pictures (Figs. 6(b)(d)) during the welding process. It can be observed from Fig. 6 that regardless of the level of interference in the image, the tracking algorithm can accurately track the weld.

In order to evaluate the algorithm proposed in this paper more rigorously, Eq.(27) is used to calculate the tracking error, which is shown in Fig. 7. As can be seen from the figure, the tracking error in welding is less than 5 pixels. The overall performance is good, and the experiment verifies that the algorithm can carry out weld detection in actual welding conditions.

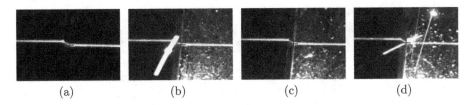

Fig. 6. Offline weld tracking test. (a)Initial frame and morphological method were used to extract the weld feature point, (b), (c), (d) are the weld feature tracking results of th KCF algorithm

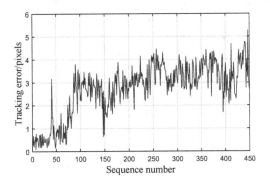

Fig. 7. Weld feature tracking error

5.2 Welding Posture Estimation

A curved weld is used for the experimental verification of posture estimation and verify the effectiveness of the pose estimation method proposed in this article. One end of the workpiece is lifted so that its dip angle relative to the horizontal plane is four degrees. During the experiment, the welding torch moves along the negative direction of the y-axis of the robot base frame. The actual posture of the weld is obtained through teaching prior to tracking. The experimental platform is shown in Fig. 8.

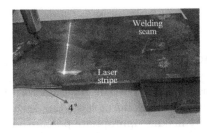

Fig. 8. Experimental platform for welding posture estimation

Table 1. Pose errors of welding experiments

The rotation angle around the axes	Mean error(°)	Max error(o)
Rx	1.12	3.45
Ry	1.42	4.56
Rz	1.36	4.54

The posture estimation result is shown in Fig. 9, and Table 1 shows the analysis of the trajectory error. The broken blue line is the desired welding angle obtained through teaching, and the red broken line is the rotation angle calculated using the method proposed in Sect. 4. It can be observed from Table 1 that the maximum error of attitude estimation is about 4.6°. The main reason for this error is that our workpiece surface is not smooth due to processing. Although the estimated attitude still has some error, but is within an acceptable range. Overall, the estimated and actual welding angles are approximately similar, which proves the feasibility of this method for posture estimation.

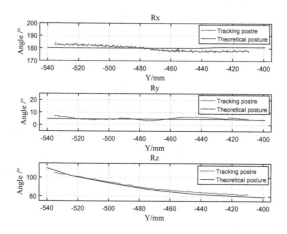

Fig. 9. Experimental results of welding posture estimation (Color online figure)

6   Conclusion and Future Work

In this paper, we proposed a seam tracking method with welding posture estimation. This method avoided the failure of welding seam tracking under complex working conditions by making appropriate posture adjustments. During the welding process, the Gaussian kernel correlation filter was used to track the feature point in real-time. Subsequently, we established the 3D point cloud data of the feature point, constructed the target coordinate system of the welding torch in real-time, and obtained the rotation angles. We verified the method through offline seam tracking and posture estimation experiments. The results of offline seam tracking experiment showed that the tracking error is less than 5 pixels, and the seam tracking algorithm was highly accurate. The posture estimation experiment showed that the method can realize the welding posture estimation in real-time. Our future research will focus on adapting posture estimation for various welding types.

References

1. Ding, H., Li, B.: Scientific problems originated from key techniques of advanced manufacturing equipments. Mach. Build. Autom. **040**(1), 1–5 (2011)
2. Zhao, W., Li, S., Zhang, B.: Present situation and prospect of intelligent technology for welding robot. Dev. Appl. Mater. **3**, 108–114 (2016)
3. Zou, Y., Wang, Y., Zhou, W.: Line laser visual servo control system for seam tracking. Opt. Precis. Eng. (11), 10 (2016)
4. Chen, H., Yu Huang, H., Lin, T., Jun Zhang, H.: Situation and development of interlligentized techonology for arc welding robot. Electric Weld. Mach. **43**(4), 8–15 (2013)
5. Gao, X., You, D., Katayama, S.: Seam tracking monitoring based on adaptive Kalman filter embedded Elman neural network during high-power fiber laser welding. IEEE Trans. Ind. Electron. **59**(11), 4315–4325 (2012)
6. Li, L., Lin, B., Zou, Y.: Study on seam tracking system based on stripe type laser sensor and welding robot. Chinese J. Lasers **04**, 8–15 (2013)
7. Zou, Y., Zhou, W., Chen, X.: Research of laser vision seam detection and tracking system based on depth hierarchical feature. Chinese J. Lasers **4**, 95–106 (2017)
8. xiao Qiao, D., Zheng, J., luan Pan, J.: Dual structure laser vision sensor and its character. Electric Weld. Mach. (11), 31–33+85 (2010)
9. Li, X., Li, X., Khyam, M.O., Ge, S.S.: Robust welding seam tracking and recognition. IEEE Sens. J. **17**(17), 5609–5617 (2017)
10. Li, X., Li, X., Ge, S.S., Khyam, M.O., Luo, C.: Automatic welding seam tracking and identification. IEEE Trans. Ind. Electron. **64**(9), 7261–7271 (2017)
11. Wang, N., Zhong, K., Shi, X., Zhang, X.: A robust weld seam recognition method under heavy noise based on structured-light vision. Robot. Comput. Integr. Manufact. **61**, 101821 (2020)
12. Henriques, J.F., Caseiro, R., Martins, P., Batista, J.: High-speed tracking with kernelized correlation filters. IEEE Trans. Pattern Anal. Mach. Intell. **37**(3), 583–596 (2015)
13. Zhang, R., Shu, Z., Nan, G.: Calibration method for line-structured light. Laser Optoelectron. Progress **56**(22) (2019)
14. Schölkopf, Bernhard: Learning with Kernels : Support Vector Machines, Regularization, Optimization, and Beyond. MIT Press, Cambridge (2002)

A Miniature Robot with Changeable Multiple Locomotion Modes

Wenju Ye, Jingheng Chen, Yisheng Guan$^{(\boxtimes)}$, and Haifei Zhu

Biomimetic and Intelligent Robotics Lab (BIRL),
School of Electro-mechanical Engineering, Guangdong University of Technology,
Guangzhou, Guangdong Province, China
ysguan@gdut.edu.cn

Abstract. High mobility is always a very important feature of mobile robots, highly depending on locomotion modes. Integrating multiple locomotion modes will provide robot with better mobility and higher adaptability to a variety of terrain. Therefore, how to integrate multiple locomotion modes in one robot has been an interesting and important issue. In this paper, a miniature robot with at least three locomotion modes: wriggling like a caterpillar, winding like a snake and biped walking, is presented. The study describes how such a non-wheeled mobile robot can possess different locomotion modes, and moreover, how the transition between them can be implemented. The feasibility and effectiveness of the different modes, as well as transition among them are illustrated with some simple experiments.

Keywords: Locomotion modes · Locomotion transition · Miniature robot · Mobile robot · Modular robot

1 Introduction

The rapid development of robotic science and technology has led to the growing employment of mobile robots in various fields. As one of the most important features of mobile robots, mobility depends largely on locomotion modes. Integrating multiple locomotion modes will undoubtedly enhance mobility of robots and higher adaptability to a variety of environments or terrains. However, a conventional mobile robot usually has only one locomotion mode. How to integrate multiple locomotion modes in one robot has been an interesting and important issue.

According to locomotion modes, mobile robots on land can be divided into certain categories: wheeled robots, tracked robots, legged robots, creeping robots and hybrid robots. Regarding the first three categories, there are extensive studies, such as the work done on legged robots DRC-Hubo [13], NOROS [16] and

The work in this paper is partially supported by the Key Research and Development Program of Guangdong Province (Grant No. 2019B090915001) and the Program of Foshan Innovation Team of Science and Technology (Grant No. 2015IT100072).

© Springer Nature Switzerland AG 2020
C. S. Chan et al. (Eds.): ICIRA 2020, LNAI 12595, pp. 100–111, 2020.
https://doi.org/10.1007/978-3-030-66645-3_9

NOROS-II [2], demonstrating their unique advantage on discrete supporting points or non-smooth terrains. However, their mobility or adaptability is limited, due to their one locomotion mode.

Rolling is the most common locomotion mode in wheeled and wheel-track hybrid mobile robots, while walking is the essential mode of legged mobile robots. Rolling with wheels, crawling with tracks and/or walking with legs are sometimes integrated into robotic systems to increase their mobility. For example, HRP-2 [7], Gorilla Robot [4], MOBIT [3] would show higher mobility and better adaptability to complex terrains or environments, if they were integrated with multiple locomotion modes.

On the other hand, creeping mode, commonly found in crawling animals, where their whole body touches on the ground, such as snakes, lizards, earthworms and caterpillars, is used in many bio-inspired robots. Many snake-like robots [8,11,12] and caterpillar-like robots [14] have been developed, while their locomotion modes have been studied [10,15,17,18]. This kind of bio-inspired robots usually consists of numerous joints, connected one by one in serial mode, offering flexibility with multiple degrees of freedom. Though their locomotion efficiency is much lower than that of wheeled robots, they can be more adaptive to complex environments. Nevertheless, how to enhance their mobility is a challenging issue of great importance.

A mobile robot usually has only one locomotion mode. However, by novel design and appropriate control strategy, it is possible to implement multiple locomotion modes into one robot. In this paper, a non-wheeled mobile robot with several locomotion modes is presented. The robot is equipped with three locomotion modes, including wriggling like a caterpillar, winding like a snake and biped walking. Moreover, these different modes are interchangeable. The system design, locomotion features, and locomotion transition are presented in detail. Finally, the effectiveness of the locomotion modes and the transition among them have been verified with experiments.

2   The Robot System

2.1   Design Methods

The creeping locomotion, widely found in many animals such as snake, caterpillar, earthworm and so on, inspired numerous snake-like robots and caterpillar-like robots with one locomotion mode, either snake-like winding or caterpillar wriggling, respectively.

On the other hand, inchworm is very skillful in climbing on a variety of object surfaces. Biped climbing, similar to biped walking, is also an important locomotion mode when discrete supporting means are involved. Naturally, in the past decades, inchworm-like biped climbing robots have also been developed [9,19]. Especially, Climbot [5] and its miniature version miniClimbot [1] have shown strong mobility capabilities, involving three climbing gaits.

The aforementioned robots, excluding the miniClimbot, have only one locomotion mode each. The issue in question is whether it is possible to combine these locomotion modes into one robot. Since the animals that gave inspiration, like snake, caterpillar, earthworm and inchworm have similar bodies, and the corresponding bio-inspired robots have similar configuration and mechanical structure, the answer to the above question would be positive, provided appropriate design.

Furthermore, as bodies of the aforementioned animals are slender and flexible, with hyper redundant degrees of freedom, the corresponding bio-inspired robots consist of numerous joint modules. Based on this configuration, the presented system is developed according to a modularity approach. Compared to conventional robots, modular robots have a lot of advantages, such as reconfigurability, scalability, low cost and high fault tolerance.

2.2   Hardware System

According to the modularity design approach, three types of basic modules have been developed, among which, two are joint modules and the third is the end-effector module, as shown in Fig. 1. Each of the joint modules has only one degree of freedom, rotation. The two types of joint modules include I-type and T-type, where the rotation axes and the module axes are in parallel and perpendicular to each other, respectively. It is noted that, in the T-type module, two tiny wheels on one shaft are attached to the one side, perpendicular to the joint axis (and they can be detached easily), which serve only for snake-like locomotion. The end-effector module, called G-type module, may be used as gripper for manipulator, attaching device for the biped climbing robot mode, foot for the biped walking robot formation and as required auxiliary apparatus for transition between the two locomotion modes.   Each module is driven by

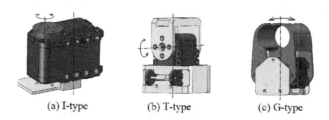

(a) I-type   (b) T-type   (c) G-type

Fig. 1. Three basic modules

a servo motor, AX-12A from Dynamixel, which is a small compact devices, an integrated motor for driving, gear set for transmission and PCB for control. The respective specifications and parameters are listed in Table 1.

Using one I-type module, four T-type modules and two G-type modules, seven modules in total, a miniature robot is composed, as shown in Fig. 2, where I-type module is in the middle, connecting two T-type modules on each side, and each end of the robot is a G-type module. The kinematic chain or the

Table 1. Parameters of the moduels

Module type	Size (mm^3)	Weight (g)	Output
I-type	$72 \times 37 \times 53$	96	1.62 (Nm)
T-type	$60 \times 44 \times 59$	106	1.62 (Nm)
G-type	$53 \times 54 \times 44$	128	8 (N)

configuration of the robot can be described in Fig. 2(b), where, G, T and I means G-type, T-type and I-type modules, respectively, $\perp$ and $\parallel$ symbols stand for the perpendicular and parallel relationship between the rotation axes of two modules, and $-$ symbol represents a fixed connection.

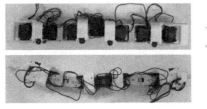

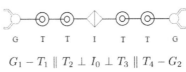

$$G_1 - T_1 \parallel T_2 \perp I_0 \perp T_3 \parallel T_4 - G_2$$

(a) Prototype: side view (above) and top view (below)   (b) Kinematic configuration

Fig. 2. The prototype and configuration of the snake-like robot

The robot possesses multiple locomotion modes, including snake-like winding, caterpillar-like wriggling and biped walking on the ground. It can also climb on poles like an inchworm, and the different locomotion modes can be interchanged.

The T-type modules play a key role in all locomotion modes. The G-type modules are employed as the two feet in biped walking, and the grasping devices in pole-climbing. In addition, the modules also play an important role in locomotion transition. The I-type module is necessary for mode transition between snake-like winding and caterpillar-like wriggling.

2.3   Control System

The control system includes communication and software. The communication framework of the robotic system, illustrated in Fig. 3(a), contains three layers, namely the application layer, transport layer and execution layer. The application layer is mainly the control program running on the PC, whose functions include human-computer interaction, motion sequence generation and the transport layer control. The transport layer includes a USB-to-TTL module and a semi-duplex serial communication bus. As the driving unit of each module is AX-12 type, the PC can send instructions to the robot through a USB-to-TTL module. The execution layer includes the bottom-level units to be controlled, i.e., the modules of the robot. They can rotate and change velocities, according to the instructions they receive. If the execution layer receives an instruction to read current status, a corresponding status package will be transmitted back to the application layer.

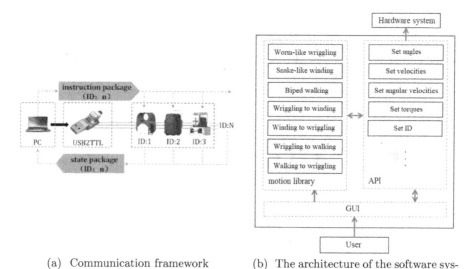

(a) Communication framework   (b) The architecture of the software system

Fig. 3. Control system

The architecture of the software system, as shown in Fig. 3(b), includes three parts: Graphic User Interface (GUI), motion library and Application Programming Interface (API). The GUI contains several push-buttons, which trigger different locomotion modes. The motion library is a class of locomotion modes and the transition among them, and the motion sequences of locomotion are also stored there. The application interface functions are provided by the company ROBOTIS and can be used to control the AX-12A. At the moment where a motion sequence is issued, a torque or velocity value is inputted as the function argument of the API, and an instruction package is sent to the hardware system by the API.

3   Four Locomotion Modes

As stated previously, the robot has several locomotion modes, including wriggling like a caterpillar, winding like a snake, biped walking and biped climbing like an inchworm. Following, these locomotion modes are described in detail.

← Forward direction

Fig. 4. Wriggling of the caterpillar-like robot

3.1 Caterpillar-Like Wriggling

Figure 4 illustrates the wriggling mode. The robot lays on the ground with the joint axes of the T-type modules parallel to the ground surface. The motion is like a caterpillar moving, however, unlike the caterpillar having legs and claws to attach to the object surfaces, the robot has to rely on the friction between its body and the ground. To this end, the robot bends part of its body, using some of T-type joint modules, and the two G-type modules are open or close accordingly, to adjust the contact areas and friction forces between the front and rear parts of the robot and the ground.

The procedure is as follows: (1) The G-type module at the front end is opened, so that one of its fingers makes contact with the ground to generate friction, while the one at the rear end is closed, to avoid contact with the ground. (2) The robot shrinks its body and achieves forward motion, due to the unbalanced frictional forces between its two sections and the ground. (3) The front G-type module "closes" and the rear one "opens", so they lose or gain the contact with the ground respectively. (4) The robot stretches and moves forward, owing to the higher frictional force on the rear part. (5) Re-iterating the above steps, the robot can move forward continuously.

Note that the procedure can be reversed, so that the robot may move backwards. This locomotion shows low efficiency, but it is suitable for narrow spaces, such as pipe, slit and so on.

3.2 Snake-Like Winding

Winding is the regular locomotion mode of a snake, whose motion path is similar to a sine function. The robot lays on the ground with the tiny wheels of the T-type modules contacting the ground and supporting the robot, and the joint axes of the T-type modules are perpendicular to the ground. The tiny wheels prevent side sliding and guide the motion perpendicular to the wheel axis, under different friction in these two directions [6].

The swinging angle of each T-type module is a sine function described as:

$$\theta_i = \alpha_0 + A sin\{[(i + 0.5)b + 2\omega(t_1 - t_2) \pm \omega t]\pi/180\} + C \qquad (1)$$

where, α_0 is the swinging angle of the middle module, A the maximum swinging angle of modules, θ_i the swinging angle of the i-th module, i ($i = 0, 1 \cdots, N$-1) is the serial number of the modules, $\pm\omega$ the shift frequency of swinging angle (+ for forward motion and - for backwards), b the parameters of motion waveform, t

the motion time, t_1 and t_2 are phase compensation variables in switching forward and backwards.

When the swing angle of each T-type module satisfies this function, the robot moves in a snake-like winding gait. I-type module in this locomotion can be considered as a T-type, when calculating the swing angle, and it does not affects the motion. This locomotion is highly efficient, and suitable to smooth terrain.

Forward direction ⟶

Fig. 5. Winding of the snake-like robot

3.3 Biped Walking

The biped walking presented here is a special locomotion mode. One end of the robot flips or turns over, while another one offers support on the ground. This walking gait was originally proposed for a biped robot, overcoming obstacles [5].

The biped walking process is shown in Fig. 6. First, the two G-type modules at the ends open to support the main body as two feet. Next, the G-type module of the swinging leg closes and moves toward to the supporting leg. Following, supported on one leg, the robot raises the other end and flips the upper body over to the other side. Meantime, to maintain stability, I-type module stays on the other side, opposite to the swinging end. Then, the swinging leg is put down onto the ground. Finally, the swinging G-type module is opened to be a foot again, supporting the robot. Re-iterating the above steps, the robot can "walk" continuously.

During the walking process, the robot moves at a low speed, and it may be regarded as a quasi-static system, whose stability may be guaranteed by COM (Center-of-Mass) method. This locomotion has low efficiency and it is suitable for striding across tall and thin barriers, such as threshold.

Forward direction ⟶

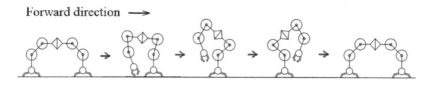

Fig. 6. Walking of the biped robot

3.4   Biped Climbing Like an Inchworm

With two G-type modules on the two ends, the robot can climb poles like an inchworm. The climbing gait is the same as the biped walking mode, as described before. If two I-type modules are mounted on the two ends of the robot, then the robot may climb with three basic climbing gaits or their combination, as the miniClimbot [1]. Since locomotion on the ground is the main consideration and biped climbing has already been extensively studied [1,5,19], their detailed description is not included in this paper.

4   Transition Among Locomotion Modes

As stressed previously, the different locomotion modes of the robot may be interchanged. Caterpillar-like wriggling is the intermediate mode among the three locomotion modes, as discussed in the previous section. The other two locomotion modes can be transited via the intermediate mode. In this section, the mode transition implementation is discussed in detail.

4.1   Transition Between Wriggling and Winding

In caterpillar-like wriggling, the rotation axes of all T-type joint modules should be parallel to the ground, while during the snake-like wriggling, these should be perpendicular to the ground and the robot should be supported by the tiny wheels under the T-type modules. Therefore, the transition between these two locomotion modes is about changing the rotation axes of all the T-type modules with respect to the ground.

The procedure to switch from the caterpillar-like wriggling mode to the snake-like winding mode is as follows: (1) Reset all the T-type modules so that the robot fully stretches on the ground. (2) One of the G-type modules sufficiently opens to contact the ground, and the main body of the robot is hence lifted up from the ground, enabling the I-type module to rotate. (3) The I-type joint module rotates by 90° and forces the half part of the robot to rotate with the opened G-type module, so that the side with the tiny wheels faces the ground and the orientation of the joint axes changes from parallel to perpendicular to the ground. (4) Finally, the I-type module continues to rotate by 90°, since the rotated part of the robot meets resistance by the opened G-type module, the other half of the robot turns over, so that the whole body of the robot rotates by 90°.

The transition from winding to wriggling is reverse to the one just described.

4.2   Transition Between Wriggling and Walking

Biped walking requires that the robot is supported by the two opened G-type modules, acting as two feet, while the caterpillar-like wriggling requires that the robot lays on the ground. So, the transition between them is mainly the process to move the robot from the lying position to the standing one, or reverse.

From wriggling to walking

From walking to wriggling

Fig. 7. Transition between wriggling and walking modes

The process of the transition from caterpillar-like wriggling to biped walking is as follows. At the beginning, the robot extents and lays on the ground. The robot bends and retracts one part, to raise its middle body until it can be supported by the near G-type module, as one foot. At this point, the robot looks like two edges of a triangle laying on the ground. Then, the other G-type module moves slowly toward the supporting leg and opens. Finally, the robot reshapes into a biped one in the standing state. The transition is complete and reversing the steps can transforms the biped walking robots into a wriggling one. The steps are simply illustrated in Fig. 7.

5   Experiments

The locomotion modes and their transition process have been implemented with an experimental robot, in order to verify their effectiveness. In the experiments, the robot starts from caterpillar-like wriggling to snake-like winding, followed by biped walking formation without interruption.

(a) Wriggling like a caterpillar   (b) Winding like a snake   (c) Biped walking with the robot

Fig. 8. Three locomotion modes

Figure 8(a) illustrates the process of locomotion in wriggling mode, using some snapshots. It can be seen that two G-type modules at the two ends open and close in turn, modifying the contact area and friction between robot and the ground. T-type modules make the body of the robot to stretch and contracts at the same time.

The process of locomotion with snake-like winding is demonstrated in the snapshots in Fig. 8(b). The path of this mode is similar to a sine function.

In biped walking, the G-type module on the swinging leg closes and moves toward the supporting leg. Then, the swinging leg flips over to the other side. During this phase, the body of the robot should change and retract so that the center of gravity remains within the supporting area, serving the balance and standing stability. Finally, the swinging G-type module moves down and open to contact the ground as another supporting foot. The process is illustrated in Fig. 8(c).

(a) From wriggling to winding (b) From winding to wriggling (c) From wriggling to walking (d) From walking to wriggling

Fig. 9. Locomotion transition between three locomotion modes

Figure 9(a) and Fig. 9(b) show the process of the transition between caterpillar-like wriggling and snake-like winding. The G-type module at one end opens to make the robot body floating in the air. The I-type module rotates so that the half body with the opened G-type module turns over by 90°. The I-type module continues to actuate and then the half of the robot rotates by 90°, while the other half part keeps stationary due to the resistance of the opened G-type module. The whole robot is finally turning over for snake-like winding. The reverse transition is implemented trivially.

The experiments of transition between the caterpillar-like wriggling and biped walking are also carried out successfully. The process has been described clearly in the previous section, while it is supplemented by the snapshots in Fig. 9(c) and Fig. 9(d).

Table 2. Speed of three locomotion modes

Locomotion mode	Wriggling	Winding	Biped walking
Speed (mm/s)	7.1	23.8	13.3

Table 2 shows that, winding is the fastest locomotion mode on smooth terrain. Biped walking is slower than winding, but it is suitable for striding across barriers. Wriggling moves in a slow speed, but it is suitable for narrow space. These three locomotion modes all have their advantages and each better suits different terrains.

6   Conclusion

Compared to a robot with single locomotion mode, a robot with multiple locomotion modes shows higher mobility and strong adaptability to complex terrains. Therefore, development of robots capable of multiple locomotion modes and the study of these modes is significant both in theory and in practice.

In this paper, a miniature non-wheeled mobile robot with several locomotion modes is proposed. Based on novel design, using bio-inspired and modularity methods, the described robot is endowed with four locomotion modes, including caterpillar-like wriggling, snake-like winding, biped walking and biped climbing. Specifically, different locomotion modes may be interchanged, applying appropriate control strategy. The design methods, basic modules, configuration and the system of the robot have been introduced in this paper, while the locomotion modes and their transition have also been analyzed. The effectiveness of the presented locomotion modes and their transition have been verified by the conducted experiments, where the high mobility, in terms of locomotion modes, has also been illustrated. Such robots may be used for inspection or simple and slight manipulation, in complicated environments, in disaster scenarios, or in trusses and trees or on high poles.

This kind of robot has a wide application range, because of its multiple locomotion modes. When assigned to complete a task on a complex terrain, this robot can flexibly and quickly adjust into a suitable locomotion mode. All is required is to control the swing angle of each module, and then the robot can well finish the task, showing high adaptability in complex environment.

The efficiency of each locomotion mode can be enhanced, and the robot can improve, advancing its reconfigurability by complete modules with unified connection interfaces. In addition, energy-supplying modules are also in development. Self-contained energy can enable the robot to move in larger ranges. External sensors will also be integrated to the robot, to enhance the sensing function. Autonomous locomotion, inspection and manipulation are the ultimate goals of the development of such robots.

References

1. Cai, C., et al.: A biologically inspired miniature biped climbing robot. In: IEEE International Conference on Mechatronics and Automation, pp. 2653–2658 (2009)
2. Ding, X., Li, K., Xu, K.: Dynamics and wheel's slip ratio of a wheel-legged robot in wheeled motion considering the change of height. Chinese J. Mech. Eng. (in Chinese) **25**(5), 1060–1067 (2012)
3. Duan, X., et al.: Kinematic modeling of a small mobile robot with multi-locomotion modes. In: IEEE/RSJ International Conference on Intelligent Robots and Systems, pp. 5582–5587 (2006)

4. Fukuda, T., et al.: Multilocomotion robot: novel concept, mechanism, and control of bio-inspired robot. In: Artificial Life Models in Hardware. Springer, London (2009). https://doi.org/10.1007/978-1-84882-530-7_4

5. Guan, Y., et al.: Climbot: a bio-inspired modular biped climbing robot c system development, climbing gaits and experiments. ASME J. Mech. Robot. 8:021026-1-17 (2016)

6. Yamada, H., Hirose, S.: Snake-like robots machine design of biologically inspired robots. IEEE Robot. Autom. Mag. **16**(1), 88–98 (2009)

7. Kanehiro, F., et al.: Locomotion planning of humanoid robots to pass through narrow spaces. In: IEEE International Conference on Robotics and Automation, vol. 1, pp. 604–609 (2004)

8. Klaassen, B., Paap, K.L.: Gmd-snake2: a snake-like robot driven by wheels and a method for motion control. In: IEEE International Conference on Robotics and Automation, vol. 4, pp. 3014–3019 (1999)

9. Lam, T., Xu, Y.: Climbing strategy for a flexible tree climbing robot treebot. IEEE Trans. Robot. **27**(16), 1107–1117 (2011)

10. Ma, S.: Analysis of creeping locomotion of a snake-like robot. Adv. Robot. **15**(2), 205–224 (2001)

11. Nilsson, M.: Snake robot-free climbing. IEEE Control Syst. Mag. **18**(1), 21–26 (1998)

12. Poi, G., Scarabeo, C., Allotta, B.: Traveling wave locomotion hyper-redundant mobile robot. In: IEEE International Conference on Robotics and Automation, vol. 1, pp. 418–423 (1998)

13. Wang, H., et al.: Drc-hubo walking on rough terrains. In: IEEE International Conference on Technologies for Practical Robot Applications, pp. 1–6 (2014)

14. Wang, W., Wu, S.: A caterpillar climbing robot with spine claws and compliant structural modules. Robotica **34**(7), 1553–1565 (2016)

15. Wang, W., Zhang, H., Zhang, J.: Crawling locomotion of modular climbing caterpillar robot with changing kinematic chain. In: IEEE/RSJ International Conference on Intelligent Robots and Systems, pp. 5021–5026 (2009)

16. Wang, Z., et al.: Conceptual design of a novel robotics system for planetary exploration. In: IEEE 6th World Congress on Intelligent Control and Automation, pp. 8962–8965 (2006)

17. X. Wu and S. Ma. Adaptive creeping locomotion of a CPG-controlled snake-like robot to environment change. Autonomous Robots, **28**(3): 283 C294 (2010)

18. Ye, C., Ma, S., Li, B., Wang, Y.: Turning and side motion of snake-like robot. In: IEEE International Conference on Robotics and Automation, pp. 5075–5080 (2004)

19. Yoon, Y., Rus, D.: Shady3D: a robot that climbs 3D trusses. In: IEEE International Conference on Robotics and Automation, pp. 4071–4076 (2007)

Problem of Robotic Precision Cutting of the Geometrically Complex Shape from an Irregular Honeycomb Grid

M. V. Kubrikov[1], M. V. Saramud[1,2](✉) ⓘ, and M. V. Karaseva[1,2]

[1] Reshetnev Siberian State University of Science and Technology,
Krasnoyarsky Rabochy Av. 31, Krasnoyarsk 660037, Russian Federation
msaramud@gmail.com
[2] Siberian Federal University, 79 Svobodny Avenue, Krasnoyarsk 660041, Russian Federation

Abstract. The article considers solving the problem of precision cutting of honeycomb blocks. The urgency of using arbitrary shapes application cutting from honeycomb blocks made of modern composite materials is substantiated. The problem is to obtain a cut of the given shape from honeycomb blocks. The complexity of this problem is in the irregular pattern of honeycomb blocks and the presence of double edges, which forces an operator to scan each block before cutting. It is necessary to take into account such restrictions as the place and angle of the cut and size of the knife, its angle when cutting and the geometry of cells. For this problem solving, a robotic complex has been developed. It includes a device for scanning the geometry of a honeycomb block, software for cutting automation and a cutting device itself. The software takes into account all restrictions on the choice of the location and angle of the operating mechanism. It helps to obtain the highest quality cut and a cut shape with the best strength characteristics. An actuating device has been developed and implemented for both scanning and cutting of honeycomb blocks directly. The necessary tests were carried out on real aluminum honeycomb blocks. Some technical solutions are used in the cutting device to improve the quality of cutting honeycomb blocks. The tests have shown the effectiveness of the proposed complex. Robotic planar cutting made it possible to obtain precise cutting with a high degree of repeatability.

Keywords: Modeling · Robotics technology · Mobile robots · Honeycomb blocks

1 Introduction

The development of industry, especially aviation, rocket and space technologies, is closely related to the development of new design and technological solutions based on modern materials, including high-strength fibrous polymer composite materials [1]. A great attention is paid to the efficiency of modern spacecraft's; it leads to the necessity to search for new technological solutions [2]. The application of three-layer structures is the main and most important field of research.

© Springer Nature Switzerland AG 2020
C. S. Chan et al. (Eds.): ICIRA 2020, LNAI 12595, pp. 112–120, 2020.
https://doi.org/10.1007/978-3-030-66645-3_10

Three-layer structures are two load-bearing faces and lightweight core materials located between them. This lightweight core material is mainly honeycomb [3]. Load-bearing faces perceive longitudinal loads (tension, compression, and shear) in their plane and transverse bending moments [4]. The lightweight core material absorbs the shear forces at bending of the three-layer structure and ensures the joint work and stability of the load-bearing faces. The honeycomb core is designed to produce lightweight, rigid and heat-insulating panels [5]. Due to its characteristics, this type of structure has found active use in spacecraft designs, since it provides a significant reduction in the mass of their structures, and, as a consequence, an increase in the mass efficiency of spacecraft's in general. In this regard, space platforms of various classes have been developed and they are actively used on the basis of structures made of honeycomb panels [6, 7].

The cost of the main structural lightweight core materials is very high, and the production of curved panels or panels of variable thickness on their basis is a complex technological problem. The application of multilayer structures is increasing. Scientific and practical problems associated with them are of a great importance [8].

The article is devoted to solving the urgent problem of producing structures using honeycomb blocks, namely, to the solution of automated precision cutting of honeycomb blocks from various materials.

2 Problem of Cutting Honeycomb Blocks

The application of a honeycomb block of aluminum, aramid or low-thickness carbon fiber as the main material does not allow using classical methods of material processing, such as milling. A shear line and nearby cells are compressed when milling is a thin jumper of a honeycomb block. The main problem when cutting such material is passing the cutting tool along a path that does not take into account the honeycomb structure of the block. Passing the cutting tool through the points connection of the cells often leads to the cells deformation. Also it creates an unreinforced cell wall. A wall without reinforcement with other cells is not able to withstand the applied loads. As a result, an edge of the sandwich panel will have less strength in comparison with the main surface.

The technology for obtaining a high-quality cut of using a special knife is investigated. It makes cutting a honeycomb block of complex geometry, along a trajectory with a minimum radius of curvature equal to the cell size possible. Figure 1 presents cutting a wall far away from the place of gluing. This cutting leads to its strong deformation. This effect can be reduced by moving the cut closer to the junction. Figure 1 shows a manufacture cut along the outer contour of the block and the cut, presented by authors of the article, i.e., it is located one cell from an edge.

We are dealing with an aluminum honeycomb block with a height of 40 mm and a cell edge length of 10 mm. A photo of a honeycomb block is given in Fig. 1.

The method of cutting with a special knife was chosen as the most suitable for the problem of cutting honeycomb blocks solving.

Four main problems when cutting edges of honeycomb blocks arise. They are a choice of the correct angle of the knife blade relative to the cut edge, a choice of the cutting place on the edge relative to the nodal points of intersection of the edges, a choice of single edges and the location of the knife in space to exclude unwanted contact with

Fig. 1. Aluminum honeycomb block

adjacent edges. The cutting edges process is shown schematically in Fig. 2; glued edges with double thickness are shown in blue. The honeycomb structure is obtained by gluing and further extension of aluminum strips. Thus, edges of double thickness are formed.

It is required to automate the determination of points and cutting angles to eliminate the described problems. It is necessary to obtain a "map" of a real honeycomb block to solve this problem. For this, a honeycomb block must be fixed on the surface where the cutting will be performed. A scanning device should be installed and synchronized with the software. And then a honeycomb block must be scanned.

The algorithm is required; it will optimally locate a cut shape on the surface of the honeycomb block, taking into account the described requirements. The algorithm selects the optimal location of the knife at each point of intersection of the shape and the edges of the honeycomb block. As a result, one gets an array of cutting points with an indication of the angle of the knife. All these points meet our requirements, i.e., a cut is no closer than 0.5 mm from the intersection points of the edges, a knife is at the optimal angle to the cut edge and it does not touch adjacent edges. If it is possible, a cut passes along single edges, as close as possible to the nodal points. This cutting will help to get the most reliable structure.

After that, the software converts the resulting data array into a G-code that can be understood by the equipment.

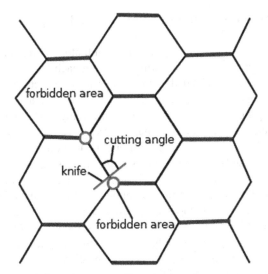

Fig. 2. Representation of the edge cutting in the honeycomb block.

3   Honeycomb Block Scanning

A honeycomb block located on the working surface of the robotic complex is used as an initial object. Due to the high flexibility of the honeycomb structure of the block, resulting from the small thickness of the partitions, the cells have different geometries. The use of technical vision is a necessary measure to obtain cutting only along the edges of the honeycomb, without destroying the nodal joints.

A scanning process begins with the sequential passage of the actuator with the installed video camera over the surface of the honeycomb block. The resulting video sequence is formed into a panoramic image of the entire surface of the honeycomb block.

Nodal points (intersection of several edges) and honeycomb edges are formed from a panoramic image with the help of technical vision. Technical vision helps to bind the coordinates of the nodal points to the coordinates of the working surface of the robotic complex. Thereby it is possible to obtain a digital copy of a real honeycomb block. Also, double and single edges are marked on the map. The process is simplified by their arrangement, i.e., double edges are formed as a result of gluing aluminum strips and they are located on one straight line.

The OpenCV library is used to select the required objects [9]; it is used in the languages C++, Python, Java. A neural network is used to detect the required objects. The TensorFlow [10] and Keras [11] libraries are applied to train it. The Python programming language is applied for its implementation.

As a result of this operation, it is possible to get a "map" of our honeycomb block, with the real dimensions and positions of all the node intersection points of the faces, marked with double faces.

4   Software Implementation of the Proposed Approach

So, there exists the scanned geometry of a honeycomb block in real dimensional units, node points of intersection of faces, information about each face, i.e., single or glued. The form that needs to be cut is loaded into the program. First, a check for sufficient size is carried out. If a form cannot be positioned on the block even not taking into account the restrictions, the program generates an error, informing a user about it. If the size of the block is sufficient for the selected shape, selection of the optimal location of the shape on it begins. Then, a shape is aligned along the longest straight line, if any, and it is located parallel to the block so that this straight line passes along the single edges at the optimal distance from the intersection points. The original location of the form is in the lower left corner for the most rational application of the material. The program also specifies whether an existing block edge can be applied. It will greatly simplify the process if the existing block edge is cut cleanly and smoothly, and one of the sides of the shape is straight. In this case, the edge of the form is aligned with the edge of the block and further movement is carried out in the same plane.

Further, all points of intersection of the shape and honeycomb structure are checked. Each point is first checked for coincidence with the unwanted area, an area with a radius of 0.5 mm around the anchor points of intersection of edges or double edges if there are such intersections. The point is placed as a problematic one and a shape is moved. If there are several problem points, the search for suitable places for cutting within a radius of 5 mm from these points is carried out.

When the location of the shape is found without problem points, the further verification is performed. At each point, the contour of the knife is located at the optimal angle to the cut face and it is checked whether it touches adjacent edges. If there exist intersections, a knife angle changes within acceptable values to ensure a correct cutting. If changing the angle eliminates the intersection problem, it is stored by the program, if not, a point is marked as problematic and the shape changes its location again.

These iterations continue until a location is found where there are no problem points. As a result, we get an array of cutting points, i.e., their coordinates and knife angles for each of them. The software generates the G-code required to control a operating mechanism.

The proposed algorithm is implemented in the form of a software environment for the automation of cutting honeycomb blocks. The interface of the software system is shown in Fig. 3.

Having fixed a honeycomb block on the cutting surface, it moves a operating mechanism to the zero point (Home). It helps to apply operating mechanism coordinates for binding the geometry of the honeycomb cell structure.

After that, use a button Load to upload the required pattern. The pattern file contains an array of vectors. The program searches for closed contours and visualizes the beginning of the cutting and its direction.

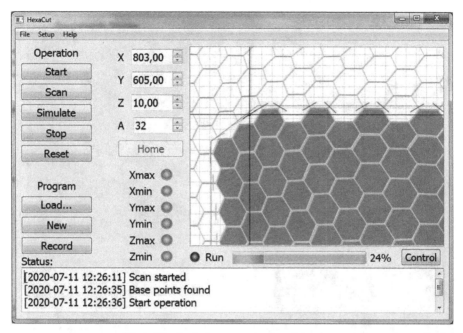

Fig. 3. Software system interface

By clicking on a button Scan, a cell scanning process starts. It is necessary to take into account the parameters of the honeycomb block in order to control the height of the machine vision camera. Linear scanning produces a series of images to obtain an overall picture of the entire honeycomb blocks. The complete image is sent for honeycomb geometry recognition and vectorized image acquisition.

The recognized vectors are transferred back to the software. The automated algorithm seeks to locate optimally the pattern on an irregular honeycomb structure, while the pattern should not go beyond the surface of the honeycomb block. The next step in the algorithm realization is to find the cutting points, while several conditions are met. The main is that a cutting point must be outside the closed contour and at least 0.5 mm from the nodal point of the cell. Also it is necessary to take into account all the other restrictions described above. At this stage, the selected points and the location of the knife are visualized. Clicking a button Simulate the G-code is generated and a simulation of the operating mechanism is displayed in the graphics window. A button Start is activated upon completion of the simulation. So, one can start the fulfillment of the operating mechanism for the actual operation.

5   Operating Mechanism for Cutting

An actuating unit, an operating mechanism, has been developed for cutting edges of honeycomb blocks and machine vision for the practical application of the described approach. A photograph of such a operating mechanism is given in Fig. 4.

Fig. 4. Operating mechanism for cutting

The operating mechanism consists of several autonomous systems on a rotary mechanism. A rotary mechanism is made on radial axial bearings with a central hollow rod. It is driven by a stepper motor with the ability to rotate with an accuracy of 0.1°. The main mounting plate is attached to the rotate mechanism. A machine vision camera is fixed on its one side. A mechanism for vertical movement of the knife is on the other side. The knife speed reaches 1500 mm/s. The cutting knife is made of tool steel; its thickness is only 0.3 mm. The knife's width is 5 mm. A photo of the knife profile is given in Fig. 5.

A polyurethane backing is installed on the working surface. Its use allows reducing knife wear and increasing the influence of knife positioning errors.

The operating mechanism has a spring-loaded platform, ensuring that the honeycomb block is pressed to the surface. Due to the high flexibility of the material, pressing is necessary to ensure that the honeycomb block remains in position at the time the knife is removed.

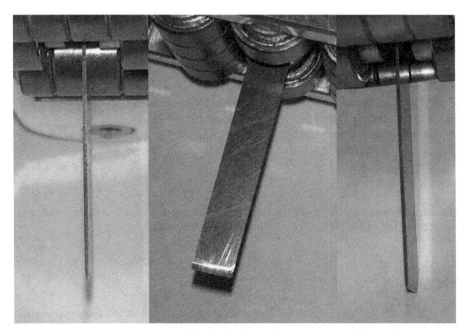

Fig. 5. Cutting knife

6   Result of the System Operation

The developed robotic complex was tested on a real aluminum honeycomb block. Seven different types of shapes were cut, including sections of different curvature and sections with a change in the generatrix at right angles. There are also holes for embedded fasteners with a diameter of 40 mm. The cell edge of the honeycomb block is 10 mm and its thickness is 100 μm.

There were five runs for each form. The accuracy of determining the geometry of the cells was 0.1 mm as a result of 35 experiments. The positioning accuracy of the knife relative to the cellular structure of the block was no more than 0.1 mm. Meanwhile, the scanning speed was 1000 mm^2/s, and the cutting speed was 5 mm/s. The average area of the object is 0.17 m^2.

The maximum indentation while maintaining the integrity of the cellular structure from the boundary of the object is 5 mm. This spread is easily compensated for in the finished product due to the high elasticity of the honeycomb block.

Thus, the obtained edge of the honeycomb block as a part of the finished sandwich panel has stable transverse strength. It has a positive effect when several panels are joined into a finished product.

7   Conclusion

The investigation in the field of the process of spatial precision cutting of honeycomb blocks made of composite materials, guarantees the development of high-quality lightweight structures in the aerospace and related industries.

The technical problem posed for robotic precision cutting of geometrically complex shapes from an irregular honeycomb grid was solved. As a result, a cutting algorithm was obtained. It provides a final shape with the best strength characteristics. The resulting algorithm is implemented in the form of a software and hardware complex. This complex includes a vision module, cutting automation, generation of a control code for the operating mechanism with an operating mechanism and the operating mechanism itself. It allows both scanning and cutting of honeycomb blocks.

A series of experiments made it possible to find the optimal solution for the development a special knife.

The robotic spatial cutting made it possible to obtain precise cutting with a high degree of repeatability.

Acknowledgements. This work was supported by the Ministry of Science and Higher Education of the Russian Federation (State Contract No. FEFE-2020-0017)

References

1. Bayraktar, E.: Section 12 composites materials and technologies (2016). https://doi.org/10.1016/b978-0-12-803581-8.04108-4
2. Crupi, V., Epasto, G., Guglielmino, E.: Comparison of aluminium sandwiches for lightweight ship structures: honeycomb vs. foam. Mar. Struct. **30**, 74–96 (2013)
3. Jin, X., Li, G., Gao, S., Gong, J.: Optimal design and modeling of variable-density triangular honeycomb structures. In: 2017 8th International Conference on Mechanical and Intelligent Manufacturing Technologies (ICMIMT), Cape Town, pp. 138–143 (2017)
4. Crupi, V., Epasto, G., Guglielmino, E.: Collapse modes in aluminium honeycomb sandwich panels under bending and impact loading. Int. J. Impact Eng. **43**, 6–15 (2012). https://doi.org/10.1016/j.ijimpeng.2011.12.002
5. Hu, Z., et al.: Design of ultra-lightweight and high-strength cellular structural composites inspired by biomimetics. Compos. B Eng. (2017). https://doi.org/10.1016/j.compositesb.2017.03.033
6. Yan, L., Xu, H., Deng, Y.: 3D digital mockup for honeycomb sandwich panels of satellites. In: 2018 IEEE 4th Information Technology and Mechatronics Engineering Conference (ITOEC), Chongqing, China, pp. 1956–1959 (2018)
7. Wenjian, J., Lanlan, Z., Feng, W., Jinwen, S., Yun, L.: Structural design and realization of a mechanical reconfigurable antenna. In: 2018 International Conference on Electronics Technology (ICET), Chengdu, pp. 349–353 (2018)
8. Regassa, Y., Lemu, H., Sirabizuh, B.: Trends of using polymer composite materials in additive manufacturing. In: IOP Conference Series: Materials Science and Engineering, vol. 659, p. 012021 (2019). https://doi.org/10.1088/1757-899x/659/1/012021
9. OpenCV Homepage. https://opencv.org. Accessed 30 July 2020
10. TensorFlow Homepage. https://www.tensorflow.org. Accessed 30 July 2020
11. Keras Homepage. https://keras.io. Accessed 30 July 2020

Research on Key Technology of Logistics Sorting Robot

Hongwei Xiang[1], Yong Wang[2], Yang Gao[2], Zhe Liu[1], and Diansheng Chen[1,3(✉)]

[1] School of Mechanical Engineering and Automation, Beihang University, Beijing, China
hhhtxhw@163.com, {liuzhe18,chends}@buaa.edu.cn
[2] Zhejiang Cainiao Supply Chain Management Co., Ltd., Hangzhou, Zhejiang, China
{richard.wangy,liancang.gy}@cainiao.com
[3] Beijing Advanced Innovation Center for Biomedical Engineering,
Beihang University, Beijing 100191, People's Republic of China

Abstract. It is an important direction to reduce the cost and improve the efficiency in the field of logistics to use automatic equipment such as mechanical arm to complete logistics sorting efficiently and accurately. This paper introduces a kind of mechanical arm sorting system. It improves two technologies—instance segmentation and poses estimation by using instance segmentation technology based on deep learning and calculating the point cloud vector of the depth camera. By using the Rapid-exploration Random Tree-Connect method (RRT-Connect) and Probabilistic Roadmap Method (PRM) algorithm, it can complete motion planning and select sucker to absorb and hold objects according to commodity size and weight. It is proved that the model recognition rate of the training system is high, the working efficiency of the system is 500 pieces/h, and the success rate of grasping is more than 99%.

Keywords: Logistics sorting · Manipulator · Motion planning · Instance segmentation

1 Introduction

In recent years, the rapid development of e-commerce, online shopping has been very popular in China. Logistics speed has become one of the most important factors to let consumers have a better online shopping experience. Generally speaking, the time required for the logistics sorting link accounts for more than 40% of the operation time of the logistics center, and the required cost accounts for 15–20% of the total cost [1]. However, the logistics sorting link in China is basically in the stage of manual sorting, with low accuracy, high cost, and low efficiency. Therefore, in the evolution process of the whole popular industry from traditional to automatic and intelligent, the use of mechanical arm and other automation equipment instead of labor is an important direction to reduce cost and improve efficiency in the field of logistics.

The use of mechanical arms for logistics sorting requires the use of computer vision technology for commodity identification, segmentation, pose estimation, planning the

© Springer Nature Switzerland AG 2020
C. S. Chan et al. (Eds.): ICIRA 2020, LNAI 12595, pp. 121–132, 2020.
https://doi.org/10.1007/978-3-030-66645-3_11

sorting trajectory, flexible and stable grasp of goods. FCN, FasterRCNN, and so on are used to solve the segmentation of dense goods in the bin. FCN can use a deep convolution neural network to segment the input image of any size end-to-end pixel by pixel semantic segmentation, [2] but it needs to obtain the target by post-processing. FasterRCNN proposes the RPN network, which can detect the target faster, but the output result has a deviation error [3]. The traditional path planning algorithm is used Methods include artificial potential field method, A* algorithm, etc., but they can not guarantee to search the optimal path and are not suitable for the complex environment; [4] in logistics sorting scenario, the gripper needs to stably grasp the goods with different sizes and weights in a relatively narrow space, which also has higher requirements for the gripper.

To complete the logistics sorting task efficiently and reliably, we design a kind of mechanical arm sorting system, which is very suitable for the logistics sorting scene.

2   The Overall Scheme of the System

The sorting system is composed of camera, computer, manipulator and gripper. Firstly, the camera captures the 2D & 3D image of the object. After the image signal is transmitted to the computer, the machine vision technology is used to identify the object, grasp planning and path planning. According to the end pose, the inverse solution of the manipulator is solved, and then the motion command is sent to the manipulator controller. The manipulator cooperates with the gripper to complete the sorting task.

3   End Effectors

3.1   Structural Design

The sorting scene has the characteristics of a complex sorting environment, narrow working area, the irregular shape of goods, and large size span. Therefore, we design an end effector which covers multi-size goods.

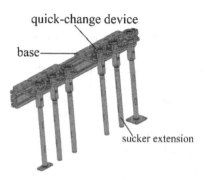

Fig. 1. An end effector for covering multi-size goods

Fig. 2. The quick-change device

As shown in Fig. 1, The end effector is composed of six different sizes of suction cups, sucker extension parts, and quick-change devices (Fig. 2). The extension part of the suction cup is rod-shaped, and one end is connected with the suction cup, which can extend the operation range of the mechanical arm so that the suction cup can be extended into the material box; the quick-change device is used to connect the mechanical arm and the suction cup extension piece; the base can temporarily place the end actuator.

The designed end effector has six suckers of different sizes, including four 1 × 1 sucker, 12 × 1 sucker, and 12 × 2 suckers. The diameters of four 1 × 1 suckers are 1 cm, 2 cm, 3 cm, and 4.5 cm, corresponding to the products of 1 cm–2.5 cm, 2.2 cm–4 cm, 3.2–8 cm, 5 cm–20 cm, respectively; one 2 × 1 suction cup sucks heavy goods between 10–20 cm; one 2 × 2 sucker absorbs more than 15 cm goods. There is a certain overlap in the range of products absorbed by the six suckers, which can ensure that 2102 kinds of goods can be covered in the business scenario (the specific number is shown in Table 1).

The manipulator we used is m-10iA/7L of FANUC company. It has six axes, the maximum load is 7 kg, the working radius is 1632 mm, the repositioning accuracy is 0.03 mm, and the working space is shown in the Fig. 3.

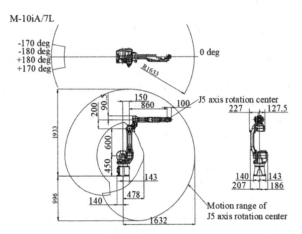

Fig. 3. The working space of M-10iA/7L

Table 1. Product and sucker mapping table

Suckers categoriestt	1 cm 1 × 1	2 cm 1 × 1	3 cm 1 × 1	4 cm 1 × 1	2 × 1	2 × 2
Number of commodity types	108	335	482	1305	179	668

3.2 The Perception and Control Strategy in Grasping

Before the system goes online, all goods are marked, and the mapping table can be automatically generated according to the product size in the database and the product size that the suction cup can absorb, to establish the mapping relationship between the commodity and the suction cup. But in the process of grabbing, there will be the phenomenon that the size of the product to be absorbed meets the requirements of the suction cup, and the weight exceeds the maximum suction range that the suction cup can bear. Therefore, we add an online learning strategy to the system. When the system counts that it has failed to absorb a certain commodity for more than 5 times, it will automatically switch to the suction device with a larger suction force following the size.

In the process of system scheduling, we develop a set of intelligent perception and control strategies. Because the local server system has the mapping table of goods and suckers, when the corresponding goods arrive at the picking position, the system will send the corresponding suction cup instructions to the upper computer in advance, trigger the pre-defined fast switch control strategy of the upper computer, and switch the suckers according to the mapping table of the goods and suckers.

4 Object Segmentation and Pose Estimation

4.1 Object Recognition, Segmentation and Pose Estimation

To recognize the object and perceive its pose, hand-eye calibration is needed first, which is to determine the pose relationship between camera and robot. Our sorting system adopts the "eye in hand" mode, that is, the camera and robot end are fixed together for calibration [5]. The related principle is shown in Fig. 4.

Instance segmentation is to segment the object and background at pixel level while detecting the object [6]. Although the depth camera obtains the depth information of the image while obtaining the color image, and can quickly obtain the color scene three-dimensional point cloud image after hand-eye calibration [7]. However, in the sorting scene, the goods in the bin are closely linked together, and it is usually impossible to distinguish each commodity by using the depth camera. Therefore, we use multi-sensor data fusion based on a structured light depth camera and RGB industrial camera to obtain the depth and texture information of goods in the material box and use the case segmentation algorithm based on deep learning to solve the problem of commodity identification and segmentation in the material box and focus on the recognition of tightly fitting and non-textured goods.

The algorithm we use benefits from the Mask-RCNN algorithm by Kaiming He [8]. The Mask-RCNN algorithm combines Faster R-CNN and FCNs, adopts ResNet-FPN

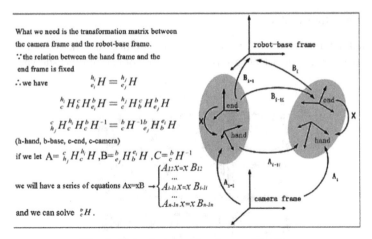

Fig. 4. The principle of hand-eye calibration

architecture in feature extraction, and uses ROI align to replace ROI pooling in the connection part, which both enhance the robustness of the algorithm for small target detection. The algorithm also adds a mask prediction branch, which can predict a more accurate target contour and distinguish each instance [9].

Besides, due to the problem of packaging materials, there will be some reflection, which will lead to holes in the point cloud, which will have an impact on the calculation and grasp position and posture of the manipulator. If there is a more accurate commodity segmentation mask, the point cloud can be repaired by using the mask. Therefore, the algorithm features of Mask-RCNN are very suitable for our business scenarios.

To avoid the capture failure caused by grasping the edges and corners of the product when the manipulator grabs, we take the surface of each commodity as an example when defining the target instance, and mark the surface of each commodity that can be seen by each commodity in the view of the camera, that is, to mark the outline of each commodity surface. When determining the category and category, since the goods in the business scene are generally box-shaped, bottle-shaped, cylindrical, bag-shaped, and sheet-shaped, and are stacked in the material box in a square, incomplete or circular shape, the categories are defined as the above three categories. We have carried out experiments to divide the three categories into one category, two categories, and three categories respectively. We find that the effect of combining the three categories into one category is the best and the generalization ability is the strongest. In essence, merging the three categories into one category is to let the algorithm learn the outline of goods.

After the recognition and segmentation of the case, we use 2D/3D bit technology based on point cloud normal vector to realize the pose estimation of the goods in the material box. That is, through the 3D point cloud data collected by the depth camera, the point cloud vector is calculated to determine the position and posture of the object in the manipulator coordinate system through the previously calibrated hand-eye coordinate conversion parameters.

4.2 Experimental Results

We collected pictures of nearly 500 kinds of commodities, and the trained model can effectively detect more than 1000 kinds of untrained commodities as shown in Figs. 5, 6, 7 and 8.

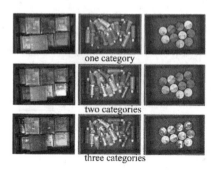

Fig. 5. Training classification results

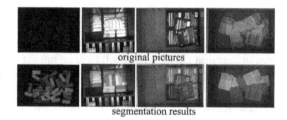

Fig. 6. Segmentation results

Fig. 7. Product segmentation results in production operation

The image data we collected are placed in the bin, which effectively increases the background semantic information, so that the algorithm can detect a certain area in the bin as a commodity (the virtual image of the bin wall mirror), avoiding the manipulator to grab the bin. We have also done a lot of data augmentation, including contrast, brightness, up and down flip, left and right flip, color channel transformation, salt and pepper noise, image dropout, etc., which has greatly improved the robustness of the model.

Fig. 8. Commodity pose estimation

5 Real-Time Planning and Control of Mechanical Arm Sorting

5.1 The Overall Technical Scheme

Given the kinematics and dynamics description, environment model, initial and target states, kinematics, and dynamics constraints of the manipulator, the path that meets the constraint conditions can be obtained by solving the motion planning of the manipulator, to realize the flexible and intelligent operation of the manipulator. This process (Fig. 9) can be roughly divided into three parts: environmental information storage and perception, path segmentation and search, and real-time motion planning.

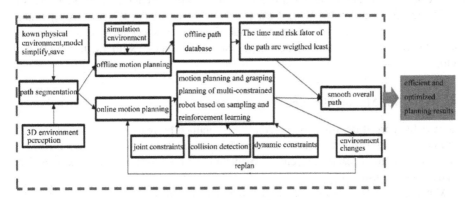

Fig. 9. The overall technical scheme

5.2 Environmental Information Storage and Perception

The environment of manipulator motion is divided into the known environment and the unknown environment. The known environment needs simplification, modeling and pre-storage, while the unknown environment needs perception.

For general crawling, most environments are known. To simplify, model and store it in advance, many triangles can be used to approximate the appearance details of the object as much as possible, store the normal vector and coordinate information of all the triangles, or only retain the basic contour, and use the basic geometry to represent the known environment as much as possible, to improve the detection speed. Each link of manipulators can also be represented by simple basic geometry, and the information of manipulators is stored in URDF files.

A small part of the environment is unknown, such as the material box to be caught, the items in the material box, and so on. The multi-sensor data fusion of depth camera based on structured light and RGB industrial camera can be used to obtain the depth and texture information of goods in the material box, and the collision detection model can be formed by storing in octrees.

5.3 Path Segmentation and Search

Based on the position, posture, shape, weight and other characteristics of the object in the box, the path and trajectory of the manipulator grasping the object to the halfway point will change constantly, while the trajectory from the halfway point to the destination box will not change. To reduce the space calculation range of real-time motion planning, the fixed trajectory is optimized offline.

The robot arm is composed of joints, so the trajectory formed by the acceleration and deceleration of the joint is the trajectory of the manipulator. The optimal trajectories can be determined by weighting the risk factors and the shortest time-consuming. The risk factors can be determined in many ways, including whether the trajectory can achieve high acceleration, the possibility of collision with the environment, and can also be obtained from the reinforcement experience of other similar types of manipulator movement. We can set the threshold of risk factor, and then select the trajectory with the shortest time, or set the threshold of time to select the trajectory with the minimum risk factor.

5.4 Real-Time Motion Planning

The variable part of the trajectory needs to be obtained by real-time motion planning. Motion planning is to get the path that meets the constraint conditions by given the kinematics and dynamics description, environment model, initial and target states, kinematics and dynamics constraints.

Mathematical description of motion planning is as follows:

1. World W is R^3 or R^2;
2. Closure o $\in$ W;
3. Robot a consists of m links:
4. Configuration space C is the set of all possible transformations of the robot. Based on this, Cobs and C_{free} were produced;
5. The configuration q_1 is the initial configuration;
6. The configuration q_G is the target configuration;
7. Provide a continuous path τ: [0, 1] $\rightarrow$ C_{free} such that $\tau(0) = q_1$, $\tau(1) = q_G$, or correctly report that such a path does not exist.

We use a 6-DOF manipulator to plan the motion in the joint space, which only needs to solve the forward kinematics.

Feasibility Test of Grab Point. The target point of motion planning is the output of the visual part of the environment perception, and there are many points to be grasped in the visual output. Therefore, it is necessary to detect the feasibility of grasping points, including judging whether there is an inverse solution (i.e. judging whether the grasping pose is reachable) and whether there is a collision.

Inverse Solution Detection. Since the axes of the three joints of the end of the industrial manipulator intersect at one point, there is an analytical inverse solution [10]. However, in the process of solving, singular problems may appear. For serial manipulators with degrees of freedom less than or equal to 6, the singular configuration will bring about infinite solutions and sudden stop. So we should avoid the appearance of singular value.

After the inverse solution of the manipulator is completed, the joint angle needs to be mapped to the actual joint angle of the manipulator. The joint angle of the manipulator may be greater than $[0, 2\pi]$, so the actual joint angle can be determined according to the principle of minimum weighted joint angle, and the inverse solution beyond the range of joint angle should be discarded.

Collision Detection. Motion planning also avoids collisions. We build the environment model through known environment modeling and environment perception, and collision detection can be carried out through the environment model.

In the sorting system, we use AABB algorithm for collision detection [11]. In this method, objects in space are surrounded by a cube bounding box parallel to the coordinate axis, which is called AABB bounding box. Then, from the eight vertices of the bounding box of object a and the eight vertices of the bounding box of object B, two maximum and minimum vertices are selected for comparison. If the coordinates of these two vertices intersect, then the two bounding boxes must intersect and collide, otherwise, they will not. This algorithm is fast and widely used, but it is rough.

Sample-Based Motion Planning. There are two kinds of motion planning algorithms based on sampling: PRM and RRT.

Probabilistic Roadmap Method (PRM) is to select n nodes randomly in the planning space, and then connect the nodes, and remove the connection line with the obstacles, to get a random road map. In this method, if the number of nodes is too small, the global path may not be obtained; if the number of nodes is too large, the quality of the global path obtained is difficult to guarantee [12].

Rapid-exploration random tree method (RRT) extends a tree structure outward from the starting point, and the expansion direction of the tree structure is determined by random sampling points in the planning space. The shortcomings of this method are obvious:① the convergence speed is slow; ② the obtained path is unstable and has a large deviation from the optimal path [13]. ③ the sampling process is random, and the best path point may not be sampled [14]. If two fast-expanding random trees are grown from the initial state point and the target state point simultaneously to search the state space, and the greedy strategy is added to the growth mode, the useless search in the blank area can be reduced and the search speed in the space can be significantly improved. This algorithm is called RRT connect algorithm.

RRT-connect algorithm has a longer step size and faster tree growth; besides, two trees continue to expand toward each other alternately instead of random expansion. This heuristic extension makes tree expansion more greedy and explicit, so it is more effective [15].

Our sorting system uses RRT connect and PRM algorithm to complete the motion planning. Because there is more than one target point to capture, the planning is completed once it is connected with the start tree for each target, which speeds up the speed of sampling planning; for repeated planning in a similar environment, the PRM algorithm is adopted, one thread samples and checks the effectiveness, and the other thread searches. When planning for many times, the previous sampling points are retained, and the existing map is used next time to construct the path from the starting point to the endpoint. Once the construction is successful, the validity of the previous points used in the path is detected. If the nodes are invalid, they will be removed and reconnected.

5.5 Actual Planning Process

Firstly, we simplify the modeling, store the known environment, and perceive the unknown environment; then, we optimize the fixed path through offline motion planning and optimize the variable path through online motion planning, that is, sampling in the joint space allowed by the manipulator, and the sampling results construct the overall posture of the manipulator through the forward kinematics of the manipulator, and then complete the online integration in consideration of joint constraints, dynamic constraints and collision detection sports planning. Combined with offline and online motion planning, the overall smooth path can be obtained. After the manipulator has been running for a while, the off-line learning is carried out periodically to improve the path planning strategy of the manipulator and realize the strategy update iteration.

6 System Integration and Verification

Our logistics sorting system is currently used in the sorting task in the Central China INTPLOG beauty warehouse (Fig. 10). After the sorting task is sent to the robot control system, the robot control system coordinates all hardware to perform relevant actions: first, trigger the 2D & 3D camera to take a photo, and obtain the 2D color image and 3D of the goods in the material box Point cloud image information, 2D & 3D images are transmitted to the image industrial computer through TCP/IP, and the corresponding visual algorithm is used to process the conversion relationship between the manipulator and the commodity in a Cartesian coordinate system. The conversion relationship is transmitted to the RCS through TCP/IP, and the RCS sends motion instructions to the manipulator controller, and then the manipulator performs a sorting task. The sorting tasks are repeated according to the execution order of the algorithm.

The orders processed by the manipulator accounted for 50%–80% of the total orders of the roadway, and the theoretical coverage rate of the whole warehouse SKU (more than 2400 kinds) reached 88.7%. The actual working verification shows that the actual picking and placing time of the manipulator is about 4.5 s, the grasping success rate is more than 99%, and the working efficiency of the manipulator workstation is 500 pieces/h.

Fig. 10. The working scene

7 Conclusion

We designed a kind of mechanical arm sorting system applied in logistics sorting scenes. The system is composed of manipulators, 2D & 3D camera, image industrial control computer, end effector, vibration table (used to adjust the position and posture of goods in the bin) and hood, etc. When the system receives the sorting task, the 2D & 3D camera takes pictures of the goods, and then obtains the transformation relationship between the manipulator and the goods in a Cartesian coordinate system through the corresponding visual algorithm, including the use of case segmentation technology based on deep learning to complete the identification and segmentation of goods in the bin, especially the identification of tightly fitting and non-textured goods, and the use of the point-based method. The 2D/3D position technology including the cloud normal vector is used to realize the pose estimation of the goods in the bin. Then, the trajectory of the manipulator is planned based on the RRT-Connect algorithm and the PRM algorithm. The manipulator is grasped by a specially designed end gripper with six specifications of suckers and can learn online. Finally, the sorting task is completed efficiently and accurately.

Acknowledgement. This research was supported by the National Key R&D Program of China 2018YFB1309300. The authors would like to personally thank all the team members.

References

1. Zhongtai, Z.: Design and key technology research of multi-AGV logistics sorting system. South China University of Technology (2018)
2. Gang, Z.: Research on key technologies of remote sensing image semantic segmentation based on deep learning. University of the Chinese Academy of Sciences (Institute of Optoelectronic Technology, Chinese Academy of Sciences) (2020)
3. Xiuliang, X.: Target recognition and location of sorting robot based on deep learning. Anhui Engineering University (2019)
4. Shangfei, Z.: Path planning method based on deep reinforcement learning and its application. Shandong normal University (2020)
5. Jiexuan, R., Xu, Z., Shaoli, L., Zhi, W., Tianyi, W.: A high precision hand-eye calibration method for robot. Modern Manufact. Eng. **2020**(04), 44–51 (2020)

6. Guodong, L.: Depth estimation, case segmentation and reconstruction based on visual semantics. University of the Chinese Academy of Sciences (Shenzhen Institute of Advanced Technology, Chinese Academy of Sciences) (2020)

7. Zhikun, Y., Bingwei, H., Shuiyou, C., Yaling, Z.: Research on surgical navigation registration algorithm based on depth camera. Mach. Manufact. Autom. **49**(04), 36-million (2020)

8. He, K., Gkioxari, G., Dollár, P., Girshick, R.: Mask R-CNN. In: 2017 IEEE International Conference on Computer Vision (ICCV), Venice, pp. 2980–2988 (2017). https://doi.org/10. 1109/iccv.2017.322

9. Zhihong, Y., Shizhong, H., Wei, F., Qiuqiu, L., Weichu, H.: Intelligent identification and application of wear particles based on mask R-CNN network. J. Tribol. 1–16, 14 Aug 2020. http://kns.cnki.net/kcms/detail/62.1095.O4.20200720.1442.002.html

10. Qiang, Z.: Robotics. Tsinghua University Press, Beijing (2019)

11. Zhifeng, G.: Research on collision avoidance technology of two manipulators based on AABB bounding box. Sci. Technol. Horiz. **17**, 95–98 (2020)

12. Jinfeng, L.V., Jianwei, M., Xiaojing, L.: Ship route planning based on improved random path graph and harmony algorithm. Control Theor. Appl. 1–9, 14 Aug 2020. http://kns.cnki.net/ kcms/detail/44.1240.TP.20200804.1654.008.html

13. Yuwei, Z., Zuo Yunbo, W., Xiaoli, G.X.: Research on path planning based on improved Informed-RRT algorithm. Modular Mach. Tool Autom. Mach. Technol. **07**, 21–25 (2020)

14. Ting, L.: Research on path planning algorithm based on reinforcement learning. Jilin University (in press, 2020)

15. Jiupeng, F., Guohui, Z., Bo, H., Zhijun, F.: Narrow path planning based on RRT-Connect. Comput. Appl. 1–6, 14 Aug 2020. http://kns.cnki.net/kcms/detail/51.1307.TP.20190610.1618. 006.html

Towards Safe and Socially Compliant Map-Less Navigation by Leveraging Prior Demonstrations

Shiqing Wei, Xuelei Chen, Xiaoyuan Zhang, and Chenkun Qi[✉]

Shanghai Jiao Tong University, 800 Dongchuan Rd., Shanghai, China
{weishiqing,chenkqi}@sjtu.edu.cn

Abstract. This paper presents a learning-based approach for safe and socially compliant map-less navigation in dynamic environments. Our approach maps directly 2D-laser range findings and other measurements to motion commands, and a combination of imitation learning and reinforcement learning is deployed. We show that, by leveraging prior demonstrations, the training time for RL can be reduced by 60% and its performance is greatly improved. We use Constrained Policy Optimization (CPO) and specially designed rewards so that a safe and socially compliant behavior is achieved. Experiment results prove that the obtained navigation policy is capable of generalizing to unseen dynamic scenarios.

Keywords: Mobile robot · Navigation · Deep reinforcement learning · End-to-end motion planning

1 Introduction

One of the main challenges in the field of ground robot navigation is to enable robots to move around autonomously and intelligently. In environments where global knowledge of the map is known, navigation algorithms are now well studied [1], and optimization goals such as shortest travel path or minimal travel time can also be applied [2]. However, exploration of unknown environments remains a common problem for tasks such as search, rescue, mining, etc. Particularly, in rapidly changing environments, e.g., social scenarios with the presence of people, it can be very difficult to get a reliable information of the global environment and robots have to navigate merely based on their local perception of the environment. Thus, map-less navigation strategies are required.

Learning-based methods prove to be very suitable for this kind of real-time end-to-end navigation problems, and can be divided into two broad categories. The imitation learning (IL) based ones are trained on expert demonstrations and try to mimic the behavior of the expert, while the reinforcement learning (RL) based ones employ the trial-and-error strategy to learn their navigation policy

This work was funded by the National Key Research and Development Plan of China (2017YFE0112200).

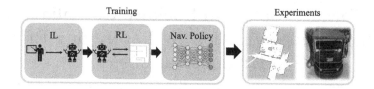

Fig. 1. A safe and socially compliant navigation policy is obtained by combining imitation learning and reinforcement learning, and then tested in simulations.

with the help of reward signals. IL is sample efficient, but usually has weak generalization abilities and a tendency of overfitting, and the learned navigation policy is limited by the quality of expert demonstrations. On the contrary, RL is theoretically more robust to unseen scenarios, since agents learn to act by trial and error, gradually improving their performance at the task as learning progresses. Still, weaknesses of RL include sample inefficiency and high time consumption, as training data are forward simulated. However, RL offers the possibility to encode desired behaviors through reward or constraint design, which makes it a conceptually better alternative to end-to-end motion planning.

In this work, we focus on the problem of navigating an unknown environment to a goal position in a *safe* and *socially compliant* way. Safety is of crucial significance for mobile robot navigation, as collision could do damage to both pedestrians and robots. In the meanwhile, in pedestrian-rich environments, robots are additionally required to understand and comply with mutually accepted rules, i.e., behave in socially compliant manners to better interact with people [3]. To this end, we adopt an approach that combines the advantages of IL and RL: the navigation policy is pre-trained by a supervised IL and then subsequently trained by RL. We use Constrained Policy Optimization (CPO) [4] for RL because of its ability to incorporate constraints during training.

In summary, the main contributions of this paper are:

- an effective deep RL-based approach for safe and socially compliant navigation through raw laser range findings and other measurements
- a case study for combining IL and RL for map-less navigation
- deployment and tests on a simulated robotic platform in unknown dynamic environments.

2   Related Work

Imitation Learning Based Methods: IL, also known as behavior cloning (in the narrow sense), takes expert demonstrations as training data and directly learn the navigation policy. Tai *et al.* [5] achieve a model-less obstacle avoidance behavior by using a convolutional neural network (CNN) and taking raw depth images as input, but their approach can only generate discrete steering commands, limiting their application to a continuous state problem. Pfeiffer *et al.* [6] and

Sergeant *et al.* [7] adopt an end-to-end approach mapping 2D laser range findings to control commands. Other methods using inverse reinforcement learning (IRL) ([3,8]) and generative adversarial imitation learning (GAIL) ([9,10]) have also occurred in recent years.

Reinforcement Learning Based Methods: Zhu *et al.* [11] apply a deep siamese actor-critic model based on deep RL to target-driven visual navigation. Li *et al.* [12] propose an approach incorporating a neural network to learn an exploration strategy to extend a continuously updated map. Zhele *et al.* [13] augment the normal external reward for deep RL algorithms with an additional term in function of curiosity. However, these approaches mentioned above remain basically in the range of static environments and do not deal with dynamic obstacles. Long *et al.* [14] use Proximal Policy Optimization (PPO) and obtain a decentralized collision avoidance policy in a multi-robot situation. Lütjens *et al.* [15] embed MC-Dropout and Bootstrapping in a RL framework to achieve uncertainty estimates and uncertainty-aware navigation around pedestrians. These methods are primarily collision avoidance strategies. Although they address motion planning in dynamic environments, their navigation capabilities are limited.

Socially Compliant Navigation: Simplistic approaches that treat pedestrians as dynamic obstacles with simple kinematics often generate unnatural robot behaviors [3], while predictive approaches that reason about nearby pedestrians' hidden intents can lead to the freezing robot problem once the environment surpasses a certain level of dynamic complexity [16]. One possible solution to this problem is to anticipate the impact of the robot's motion on nearby pedestrians. Existing work on socially compliant navigation can be split into model-based ([17,18]) and learning-based ([19,20]) approaches.

In this work, we introduce a deep RL based method that uses IL as pretraining. The expert demonstrations are generated by Timed-Elastic-Bands (TEB) [2] and a navigation policy mapping the raw measurements to motion commands is learned. This temporary navigation policy is rudimentary, and will be improved by the subsequent deep RL. What's more, we design specific rewards to encourage human-like navigation through cooperative collision avoidance, thus addressing the problem of socially compliant navigation.

3   Approach

3.1   Problem Formulation

Socially compliant navigation can be formulated as a Markov Decision Process (MDP). Let s_t, a_t denote an agent's states and action at time t. Each agent has a current position $\mathbf{p} = [p_x, p_y]$, a current velocity $\mathbf{v} = [v_x, v_y]$ and a collision radius r, and they form the agent's external information $\mathbf{y}^{ext}$. Furthermore, each

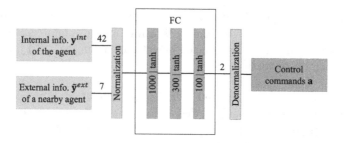

Fig. 2. Network structure of the policy π_θ.

agent has its goal position $\mathbf{p}_g = [p_{gx}, p_{gy}]$, orientation ψ (impossible to observe since agents are seen as discs) and laser range findings $\mathbf{r}_{laser}$, and they form the agent's internal information $\mathbf{y}^{int}$. To reduce redundancy, we reparameterize the internal information of a given agent $\mathbf{y}^{int}$ and external information of another agent $\tilde{\mathbf{y}}^{ext}$ in the local frame of the given agent, with the x-axis pointing to the front and y-axis to the left:

$$\mathbf{y}^{int} = [d_g, \psi_g, v_x, v_y, \psi, r, \mathbf{r}_{laser}], \tag{1}$$

$$\tilde{\mathbf{y}}^{ext} = [\tilde{d}_a, \tilde{p}_x, \tilde{p}_y, \tilde{v}_x, \tilde{v}_y, \tilde{r}, \tilde{\phi}], \tag{2}$$

where d_g is the agent's Euclidean distance to goal, ψ_g is the agent's relative orientation to goal, $\tilde{d}_a$ is the Euclidean distance between the two agents and $\tilde{\phi} = \arctan(\tilde{v}_y/\tilde{v}_x)$ is the nearby agent's heading direction. Thus, the states of the MDP are $\mathbf{s} = [\mathbf{y}^{int}, \tilde{\mathbf{y}}^{ext}]$[1]. We want to find a navigation policy π_θ parameterized by $\boldsymbol{\theta}$, which maps $\mathbf{s}$ to a motion control command $\mathbf{a}$:

$$\mathbf{a} = \pi_\theta(\mathbf{s}). \tag{3}$$

The control command $\mathbf{a}$ is composed of the translational and rotational velocities. Since the mapping from the states and the desired control commands can be really complicated, a neural network representation of the navigation policy shows great potential.

3.2 Neural Network Model and Pre-training

In this work, we combine IL and RL to obtain a robust navigation. The neural network representation of π_θ is shown in Fig. 2. Compared with [5] and [6], where a convolutional neural network is deployed to extract environmental features, our model is simplified but still adequate. Unlike image-based methods which often have to deal with virtual-to-real problems, our method uses 2D laser range findings and can be easily implemented on a real-world robotic platform.

---

[1] Here we only consider a two-agent collision avoidance problem, but the resulting two-agent navigation policy is parallelizable in a multiagent scenario, because it consists of a large number of independent queries of the trained neural network.

We use minimum pooling to downsample the laser range data and reduce input dimensions. In our case, $\mathbf{r}_{laser} \in \mathbb{R}^{36}$, i.e., we keep the minima out of 36 equally divided intervals of laser range data. The inputs $\mathbf{s}$ are normalized to $[-1, 1]$, and the outputs of the neural network are denormalized to obtain actual control commands.

In the hope of improving RL performance and reducing training time, we use supervised IL, also called behavior cloning, to pre-train the navigation policy. This is similarly done as in [5] and [6], and the resulting policy of IL will continue to be ameliorated in the subsequent RL.

3.3   Reinforcement Learning

Background: A Markov Decision Process (MDP) can be defined as a tuple (S, A, R, P, μ), where S is the set of states, A is the set of actions, $R : S \times A \times S \to \mathbb{R}$ is the reward function, $P : S \times A \times S \to [0, 1]$ is the transition probability function and $\mu : S \to [0, 1]$ is the starting state distribution. A stationary policy $\pi : S \to \mathcal{P}(A)$ is a map from states to probability distribution over actions. In RL, we aim to find a policy π_{θ} parameterized by θ which maximizes the expected discounted return of reward $J(\theta)$, i.e.,

$$J(\boldsymbol{\theta}) = \mathop{E}_{\tau \sim \pi_{\theta}} [\sum_{t=0}^{T} \gamma^t R(s_t, a_t, s_{t+1})], \tag{4}$$

where T is the time horizon of an episode, $\gamma \in [0, 1)$ is the discount factor, τ denotes a trajectory ($\tau = (s_0, a_0, s_1, ...)$), and $\tau \sim \pi_{\theta}$ indicates that the distribution over trajectories depends on π_{θ}.

Trust Region Policy Optimization (TRPO) has recently gained much attention among policy optimization algorithms because of its ability to avoid the problem of high gradient variance commonly seen in policy gradient methods and guarantee monotonic improvement. However, when a system involves physical interaction with or around humans, it is indispensable to define and satisfy safety constraints. Therefore, we introduce Constrained Policy Optimization (CPO), a constrained version of TRPO, and applies it to obtain a safe and socially compliant navigation policy.

We augment the MDP with a an auxiliary cost function $C : S \times A \times S \to \mathbb{R}$ and a limit d. This augmented version of MDP is called Constrained Markov Decision Process (CMDP). Let $J_C(\boldsymbol{\theta})$ denote the expected discounted return of policy π_{θ} with respect to cost function C:

$$J_C(\boldsymbol{\theta}) = \mathop{E}_{\tau \sim \pi_{\theta}} [\sum_{t=0}^{T} \gamma^t C(s_t, a_t, s_{t+1})]. \tag{5}$$

CPO finds the optimal policy $\pi_{\theta^\star}$, i.e.,

$$\theta^\star = \arg\max \quad J(\boldsymbol{\theta}) \tag{6}$$

$$\text{s.t.} \quad J_C(\boldsymbol{\theta}) \leq d \tag{7}$$

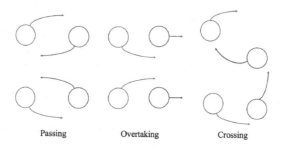

Passing   Overtaking   Crossing

Fig. 3. Left-handed (top row) and right-handed (bottom row) rules for the blue agent to pass, overtake and cross the red agent. (Color figure online)

Reward and Cost Design: The agent aims to reach the target position while learning to confirm to social norms and avoiding collisions with the environment or other agents. The design of the reward is crucial, since the reward function gives feedback to the agent so that desired behaviors can be learned. In our case, the overall reward function consists of two parts: the navigation-inducing and the norm-inducing rewards, i.e.,

$$R = R_{nav} + R_{norm} \tag{8}$$

To avoid the problem of sparse reward, we define the navigation-inducing reward as

$$R_{nav}(\mathbf{s}_t) = \begin{cases} 10, & \text{if goal reached} \\ -\alpha(d_{g,t} - d_{g,t-1}), & \text{otherwise,} \end{cases} \tag{9}$$

where $d_{g,t}$ and $d_{g,t-1}$ are lengths of the agent's shortest path to goal position at timestamps t and $t-1$, and α is a positive scaling parameter ($\alpha = 1$). In this way, we guarantee that the agent receives a feedback at all states, and that there is a significant increase in total reward when the target is reached.

As shown in Fig. 3, there are symmetries in a collision avoidance scenario. To induce particular social norms, an additional reward can be introduced in the learning process so that one set of social norms is preferred over the other. In this work, we choose to encourage the right-handed rules, and the reward function R_{norm} is specified as follows,

$$R_{norm}(\mathbf{s}) = \beta I(\mathbf{s} \in \mathcal{S}_{norm}) \tag{10}$$

$$\text{s.t.} \quad \mathcal{S}_{norm} = \mathcal{S}_{pass} \cup \mathcal{S}_{overtake} \cup \mathcal{S}_{cross} \tag{11}$$

$$\mathcal{S}_{pass} = \{\mathbf{s} \mid d_g > 3,\ 1 < \tilde{p}_x < 4, 0 < \tilde{p}_y < 2,\ |\tilde{\phi} - \psi| > 3\pi/4\} \tag{12}$$

$$\mathcal{S}_{overtake} = \{\mathbf{s} \mid d_g > 3,\ 0 < \tilde{p}_x < 3,\ |\mathbf{v}| > |\tilde{\mathbf{v}}|, -1 < \tilde{p}_y < 0,\ |\tilde{\phi} - \psi| < \pi/4\} \tag{13}$$

$$\mathcal{S}_{cross} = \{\mathbf{s} \mid d_g > 3,\ \tilde{d}_a < 2,\ 0 < \tilde{p}_y < 2,\ \pi/2 < |\tilde{\phi} - \psi| < \pi\}, \tag{14}$$

where β is a positive parameter ($\beta = 0.5$), $I(\cdot)$ is the indicator function, and $\tilde{\phi} - \psi$ is wrapped in $[-\pi, \pi]$.

Algorithm 1. CPO with Two Agents

1: Initialize the policy network π_θ and value network V randomly from a normal distribution or use pre-trained weights from IL.
2: **for** epoch $= 1, 2, ...$ **do**
3:     // *Collect data in parallel*
4:     **for** agent $i = 1, 2$ **do**
5:         Run policy π_θ for T_i timesteps, and collect $\{s_{i,t}, R_{i,t}, a_{i,t}, C(s_{i,t})\}$, where $t \in [0, T_i]$
6:         Compute advantages and safety cost using GAE [21]
7:     **end for**
8:     Update policy π_θ using CPO
9:     Update value network V using Conjugate Gradient Decent
10: **end for**
11: **return** π_θ, V

Enforcing negative rewards to discourage collisions has been a common practice for RL-based navigation methods as in [14]. However, this penalty-oriented approach could lead to undesirable trade-offs between policy exploration and policy performance in the training process. Luckily, in CMDPs, we can impose a constraint on the average number of collisions per episode. Let $S_{collision} \in S$ denote the set of all collision states, a cost function C can be defined as

$$C(\mathbf{s}) = I(\mathbf{s} \in S_{collision}). \tag{15}$$

We set the constraint value d to 0.1, so that we can expect a very safe navigation policy after the learning process.

Training: A value network V is trained at the same time as the navigation policy π_θ. Its network structure is similar to that in Fig. 2 except that its output is a value estimation for the input states. We adopt the *centralized learning, decentralized execution* paradigm in the training process. As summarized in Algorithm 1, each agent execute the policy in parallel and generate trajectories from the shared policy π_θ, and the policy π_θ and value network V are updated at the end of each epoch. This parallel algorithm can be easily scaled to multi-agent systems ($n > 2$) and reduce the time of sample collection.

4   Experiments

This section contains training process of four different models and experiments conducted in simulations. We aim to investigate the effect of pre-training by IL and the use of safety constraint. This work is not intended to outperform a model-based planner with global knowledge of the map. We would like to show the effectiveness of RL-based method in achieving safe and socially compliant behaviors without the use of a map and its generalization ability to unseen dynamic environments.

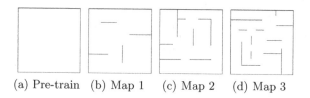

(a) Pre-train   (b) Map 1   (c) Map 2   (d) Map 3

Fig. 4. Training maps for IL and RL.

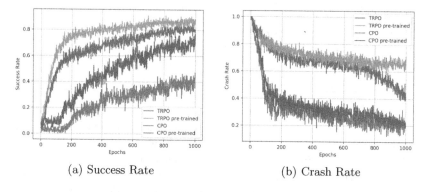

(a) Success Rate   (b) Crash Rate

Fig. 5. Training results of four different models.

4.1 Model Training

We apply two different procedures to train the model[2]: (i) pure RL and (ii) a combination of IL and RL. As shown in Fig. 4, the map 4(a) is used for IL, and the maps 4(b), 4(c) and 4(d) for RL. The three training maps for RL vary in complexity in an increasing order from left to right.

In pure RL, at the beginning of each episode, a map is randomly selected among the three maps (4(b), 4(c) and 4(d)), and then a start position and a target position are also randomly generated for each agent. The motion commands are published at a frequency 5 Hz, and a maximum of 300 motion commands is allowed per agent in each episode. The training lasts 1,000 epochs and one epoch consists of 60,000 episodes.

In the other procedure, we pre-train the navigation policy in 4(a). We put two agents in the map with one wandering randomly and the other receiving control commands from the TEB algorithm [2]. Ten trajectories including approximately 2,000 state-action pairs are sampled and fed into a supervised IL. The pre-trained navigation policy is then used to initialize RL as described above.

---

[2] The models are trained on a computer equipped with an Intel i7-8700 processor and an NVIDIA GeForce GTX 1660 GPU, running Ubuntu 16.04 and ROS Kinetic. The training is conducted in the accelerated Stage simulator with a differential drive Kobuki TurtleBot2 platform.

Figure 5(a) and 5(b) show respectively the evolution of the percentage of successful navigation (goal reached) and that of navigation with collision(s) among one epoch of four different models, and two points can be made:

Influence of the Pre-training: As shown in Fig. 5(a) and 5(b), for both TRPO and CPO based methods, the trainings that start from a pre-trained result show a better performance and a greater learning rate with a similar crash rate to their not pre-trained version. Even though the pre-training is conducted on a rather simplistic map (Fig. 4(a)), this rudimentary pre-trained navigation policy helps reduce the RL training time by 60% (if we compare the number of epochs used to reach the same success rates, e.g.., 0.8 for TRPO-based methods and 0.4 for CPO-based methods), and the overall performance is improved.

Influence of Imposing a Safety Constraint: For TRPO-based methods, a fixed penalty of -0.2 is used for each collision during the training process. We can see that although the two TRPO-based methods reach a high success rate at the end of the training process, they also have a very high crash rate compared with CPO-based methods. This blind pursue of augmenting expected discounted rewards could compromise the safety of both RL training and its real-world application. On the contrary, methods using a safety constraint, the CPO-based methods, have lower crash rates. In particular, the curve *CPO Pre-trained* reaches a similarly high success rate but with a much lower crash rate.

4.2   Experiments on Navigation Ability

In this part, we conduct several experiments in a pre-recorded map to evaluate the navigation ability of the four models trained in Sect. 4.1. The experiment map (see Fig. 6) consists of a corridor and two connected labs, and the edge length of the grid is 1 m. During the experiment, each model is given 20 runs, and in each run, the agent starts from point A, and is allowed 300 s to navigate the points B, C and D in order.

As summarized in Table 1, we note the number of successful runs (no collisions or time-outs), the number of runs with collisions, the average time of successful runs $\bar{t}_s$ and the average time till the first collision $\bar{t}_c$. The experiment results correspond well with the training curves in Fig. 5. The pre-trained models show higher success rate, and the CPO-based models, which are trained using a safety constraint, have lower crash rate. In particular, the model *TRPO Pre-trained* has both shortest $\bar{t}_s$ and $\bar{t}_c$, which is due to its risky decision making. On the contrary, the model *CPO Pre-trained* shows better safety with an acceptable trade-off with its navigation time. The model *CPO* has zero crash rate as the result of many time-outs.

By observing the sampled trajectories from the model *CPO Pre-trained* and TEB in Fig. 6, we can see that our method generates nearly the same trajectories but with more oscillations. As said at the beginning of Sect. 4, this work is not intended to outperform a model-based planner with global knowledge of the

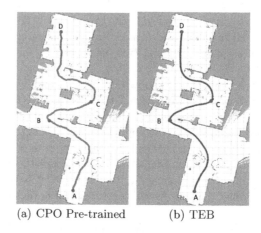

(a) CPO Pre-trained   (b) TEB

Fig. 6. Two trajectories sampled from our method CPO Pre-trained and TEB [2].

map. This experiment proves that our method is effective in addressing the problem of map-less navigation and generalizes well to unseen environments although trained in only four maps (see Fig. 4).

Table 1. Experiment results in a Pre-recorded Map

Model name	Success (%)	$\bar{t}_s$ (s)	Crash (%)	$\bar{t}_c$ (s)
TRPO	40	152.8	55	65.6
TRPO pre-trained	55	**93.3**	45	25.5
CPO	20	180.1	-	-
CPO pre-trained	**75**	150.3	**20**	**101.3**

4.3 Experiments on Social Compliance

As shown in Fig. 7, four different experiments are conducted to test if the trained navigation policy (*CPO Pre-trained*) has a good collision avoidance ability while respecting the introduced right-handed rules (see Fig. 3). Timestamps (in seconds) are marked around the trajectories. In experiments 7(a), 7(b) and 7(c), the blue agent is set to pass, overtake and cross the red agent. In experiment 7(d), each agent is set to switch its position with the agent on the opposite side. We can see in all four experiments, the agents successfully reach their target position in a safe and socially compliant way. What's more, although our method uses a two-agent navigation policy for multiagent scenarios, more complex interaction patterns have occurred.

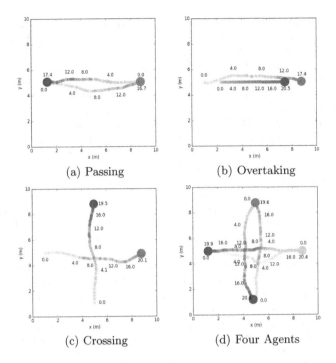

Fig. 7. Tests on social compliance and collision avoidance. (Color figure online)

5   Conclusion

This work developed a learning-based approach for safe and socially compliant map-less navigation in dynamic environments. Our method uses a neural network representation and combines imitation learning and reinforcement learning. Experiments show that by leveraging prior demonstrations, the training time for RL can be reduced by around 60% and the overall performance is improved. Different tests in simulations prove that our method generalizes well to unseen environments, and a safe and socially compliant behavior is achieved. Future work will consider alternatives to better handle situations with a high number of nearby agents.

References

1. LaValle, S.M.: Planning Algorithms. Cambridge University Press, Cambridge (2006)
2. Rosmann, C., Hoffmann, F., Bertram, T.: Timed-elastic-bands for time-optimal point-to-point nonlinear model predictive control. In: 2015 European Control Conference (ECC), Linz, Austria, pp. 3352–3357. IEEE, July 2015
3. Kretzschmar, H., Spies, M., Sprunk, C., Burgard, W.: Socially compliant mobile robot navigation via inverse reinforcement learning. Int. J. Robot. Res. **35**(11), 1289–1307 (2016)

4. Achiam, J., Held, D., Tamar, A., Abbeel, P.: Constrained policy optimization. In: International Conference on Machine Learning, May 2017. arXiv: 1705.10528
5. Tai, L., Li, S., Liu, M.: A deep-network solution towards model-less obstacle avoidance. In: 2016 IEEE/RSJ International Conference on Intelligent Robots and Systems (IROS), Daejeon, South Korea, pp. 2759–2764. IEEE, October 2016
6. Pfeiffer, M., Schaeuble, M., Nieto, J., Siegwart, R., Cadena, C.: From perception to decision: a data-driven approach to end-to-end motion planning for autonomous ground robots. In: 2017 IEEE International Conference on Robotics and Automation (ICRA), Singapore, Singapore, pp. 1527–1533. IEEE, May 2017
7. Sergeant, J., Suenderhauf, N., Milford, M., Upcroft, B.: Multimodal deep autoencoders for control of a mobile robot. In: Li, H., Kim, J. (eds.) Proceedings of the Australasian Conference on Robotics and Automation 2015, Australia, pp. 1–10. Australian Robotics and Automation Association (2015)
8. Wulfmeier, M., Wang, D.Z., Posner, I.: Watch this: scalable cost-function learning for path planning in urban environments. In: 2016 IEEE/RSJ International Conference on Intelligent Robots and Systems (IROS), Daejeon, South Korea, pp. 2089–2095. IEEE, October 2016
9. Li, Y., Song, J., Ermon, S.: InfoGAIL: Interpretable Imitation Learning from Visual Demonstrations. NIPS 2017, November 2017. arXiv: 1703.08840
10. Kuefler, A., Morton, J., Wheeler, T., Kochenderfer, M.: Imitating driver behavior with generative adversarial networks. In: 2017 IEEE Intelligent Vehicles Symposium (IV), Los Angeles, CA, USA, pp. 204–211. IEEE, June 2017
11. Zhu, Y., et al.: Target-driven visual navigation in indoor scenes using deep reinforcement learning. In: 2017 IEEE International Conference on Robotics and Automation (ICRA), Singapore, Singapore, pp. 3357–3364. IEEE, May 2017
12. Li, H., Zhang, Q., Zhao, D.: Deep reinforcement learning-based automatic exploration for navigation in unknown environment. IEEE Trans. Neural Networks Learn. Syst. **31**(6), 2064–2076 (2020). https://doi.org/10.1109/TNNLS.2019.2927869
13. Zhelo, O., Zhang, J., Tai, L., Liu, M., Burgard, W.: Curiosity-driven exploration for mapless navigation with deep reinforcement learning. In: ICRA 2018 Workshop in Machine Learning in the Planning and Control of Robot Motion, At Brisbane, May 2018. arXiv: 1804.00456
14. Long, P., Fanl, T., Liao, X., Liu, W., Zhang, H., Pan, J.: Towards optimally decentralized multi-robot collision avoidance via deep reinforcement learning. In: 2018 IEEE International Conference on Robotics and Automation (ICRA), Brisbane, QLD, pp. 6252–6259. IEEE, May 2018
15. Lutjens, B., Everett, M., How, J.P.: Safe reinforcement learning with model uncertainty estimates. In: 2019 International Conference on Robotics and Automation (ICRA), Montreal, QC, Canada, pp. 8662–8668. IEEE, May 2019
16. Trautman, P., Ma, J., Murray, R.M., Krause, A.: Robot navigation in dense human crowds: statistical models and experimental studies of human-robot cooperation. Int. J. Robot. Res. **34**(3), 335–356 (2015)
17. Ferrer, G., Garrell, A., Sanfeliu, A.: Robot companion: a social-force based approach with human awareness-navigation in crowded environments. In: 2013 IEEE/RSJ International Conference on Intelligent Robots and Systems, Tokyo, pp. 1688–1694. IEEE, November 2013
18. Mehta, D., Ferrer, G., Olson, E.: Autonomous navigation in dynamic social environments using Multi-Policy Decision Making. In: 2016 IEEE/RSJ International Conference on Intelligent Robots and Systems (IROS). pp. 1190–1197. IEEE, Daejeon, South Korea (Oct 2016)

19. Kim, B., Pineau, J.: Socially adaptive path planning in human environments using inverse reinforcement learning. Int. J. Soc. Robot. **8**(1), 51–66 (2016)
20. Kuderer, M., Kretzschmar, H., Sprunk, C., Burgard, W.: Feature-based prediction of trajectories for socially compliant navigation. Robot. Sci. Syst. **2012**, 8 (2012)
21. Schulman, J., Moritz, P., Levine, S., Jordan, M., Abbeel, P.: High-Dimensional Continuous Control Using Generalized Advantage Estimation. ICLR 2016, October 2018. arXiv: 1506.02438

Human-Robot Interaction

Variational Augmented the Heuristic Funnel-Transitions Model for Dexterous Robot Manipulation

Jiancong Huang[1,2], Yijiong Lin[1], Hongmin Wu[2], and Yisheng Guan[1(✉)]

[1] BIRL, Guangdong University of Technology, Guangzhou,
Guangdong, People's Republic of China
jiancong.huang@mail2.gdut.edu.cn, ysguan@gdut.edu.cn
[2] IIM, Guangdong Academy of Sciences, Guangzhou,
Guangdong, People's Republic of China

Abstract. Learning from demonstrations is a heuristic technique that can only obtain the intentional dynamics of robot manipulation, which may fail to the task with unexpected anomalies. In this paper, we present a method for enhancing the diversity of multimodal signals collected from few-shot demonstrations using Variational Auto-encoders (VAEs), which can provide sufficient observations for clustering many funnel representations of the complex and multi-step task with anomalies. Then a funnel-base reinforcement learning is applied to obtain the policy from the synthetic funnel-transition model. Experimental verifications are based on an open-source force/torque dataset and our previous kitting experiment setup that equips with a well-constructed framework for multimodal signal collection, anomaly detector, and classifier. The baseline is used traditional funnel policy learning (without use augmented signals), the result shows significant improvement on the success rate from 70% to 90% on performing the kitting experiments after combined with the VAEs augmented signals to compute the funnel-transitions model. To the best of our knowledge, our scheme is the first attempt for improving robot manipulation by few demonstrations, which not only can respond to the normal manipulation but also can well adapt to the unexpected abnormal out of the demonstration. Our method can be extended to the environment that not only difficult to collect sufficient transitions online but having unpredictable anomaly. For example learning long-horizon household skills.

Keywords: Robotic · Learning from demonstration · Multimodal data augmentation · High-level deep reinforcement learning

1 Introduction

Learning a task from demonstration [18,35] instead of daunting undertaking from scratch is fitting in real robot manipulation because motion on real robot

This work is partially supported by the Key R&D Programmes of Guangdong Province (Grant No. 2019B090915001), the Frontier and Key Technology Innovation Special Funds of Guangdong Province (Grant No. 2017B050506008), and the National Natural Science Foundation of China (Grant No. 51975126, 51905105).

C. S. Chan et al. (Eds.): ICIRA 2020, LNAI 12595, pp. 149–160, 2020.
https://doi.org/10.1007/978-3-030-66645-3_13

is slow. Robot learning from demonstration enables robots to perform skills without much analytically decomposing and manually programming the desired skill faster and safer [26]. However, learning from demonstration is limited when a failure occurs in normal execution; the end-user needs to provide more demonstrations like recovery from the abnormal state [35]. For example, the robot will not grasp a cup again when it collides with the table. End-user needs to demonstrate a novel skill to recover or re-enact from this abnormal state. Moreover, it must be an impossible work to demonstrate every possible abnormal state.

As we all know, reinforcement learning [31] is powerful with the trial-and-error formulation, which indicates that using reinforcement learning could explore the environment whether the execution is normal or abnormal. This leads to our research question: can we use reinforcement learning from the demonstration to improve the robustness of the policy model instead of following a human-defined demonstration? Despite the significant performance of reinforcement learning, a common shortcoming of reinforcement learning is rely on many abundant transitions that will result in time-consuming when collecting real robot data. Kalashnikov et al. [16] present the first attempt to train reinforcement learning for object grasping on real robots directly, the data collected with several robots in about 800 robot hours and consequently obtained the useful and robust grasping policy model.

To address the problem of needing a massive amount of data for reinforcement learning, most of works generated synthetic signals from a few signals to augment the transitions. The first is to use simulator to randomize the feature in observation [32] and combine it with the real world transitions to learn a robustness policy. But the limitation is still existing some gaps between the simulator and the real world. The second direction is to transfer the trajectory from the simulator, which calculates with the obtained policy model and demonstrates that trajectory on a real robot [2,21]. But less efficient because it needs to transfer the trajectory from simulator to real robot and real-world initialize states from real-world to simulator every time.

Instead, we use data augment technique to due with the real robot multimodal signals directly. But how to augment the multimodal signals become a challenge. Because it still needs abundant work to label all the multimodal signals for augment. To overcome the challenge, an unsupervised generative model name VAEs [9] is used to augment the signals without any label work. As illustrated in Fig. 1, every complete trajectory is seperated from demonstration as different skills, and use the multimodal signals including: pose and velocity of end-effector, tactile signals on fingertips and force/torque signals on the wrench (See Fig. 3 for more detail about our experiment platform) of this skill as a sample to update the VAEs model. Then, we use the decoder from the VAEs to generate skill to augment the dataset. Finally, we cluster the synthetic signals as different funnel representation as mention in [34] to reinforcement learning the kitting experiment policy.

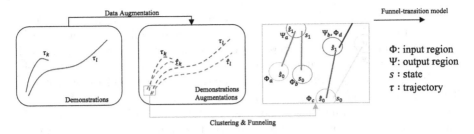

Fig. 1. Illustration of our framework of obtaining the funnel-transitions model for reinforcement learning with the data augmented manipulation trajectory. The red trajectories indicate human demonstration. The blue trajectories are the augmentation obtained with VAEs. The different circles in the third box indicate different regions. The different lines in the third box indicate different actions to different regions. Different actions to the same region will as a funnel chain. In this way, the agent will base on this funnel transition to search the optimal policy with reinforcement learning. (Color figure online)

The contribution is three-fold. 1) a sequential multimodal augmentation method proposes using the VAEs; 2) funnel-based transition model of robot complex task; 3) an autonomous recovery policy is learned based on Funnel RL.

2   Related Work

Robotic learning from demonstration (LfD) can quickly and easily learn skills in kitting experiments [3,35], and visual-based manipulations [28]. However, pure LfD is challenging to adapt to a novel situation. For example, robot arm can not recover from a novel collision position which nearby the demonstrated place.

Actually, we can improve the adaptability of LfD with reinforcement learning. Reinforcement learning has succeeded in solving more complex and undesirable tasks like games [4,30], recommender system [8], and dexterous manipulation [25]. In robotic task, reinforcement learning can be used to planning end-to-end [12,20,24] and can finished difficult task like grasping thousand of different shape objects [16], opening the door with different user-specify angles [27], and finishing multi-tasks [37]. However, traditional reinforcement learning still needs a bunch of time to online exploration [16,27], which is not useful in real-robot setup until we have a hardware robot can move as faster as in the simulator to collect transition. Because a robust policy model should rely on sufficient transitions to cover all the possible states from the environment.

Instead of learning without any human-knowledge, Hester et al. [13] use demonstration to pre-train the Q value neural network without using a massive amount of data before they reach reasonable performance; Vecerk et al. [33] use demonstration instead of manually shaping the reward function to faster learning the challenging motor insertion tasks specified by sparse rewards on real robot. Finn et al. [11] use meta-imitation to enable a robot to learn how to learn more efficiently, allowing it to acquire new skills from just a single demonstration.

However, reinforcement learning from demonstration is a challenge because without sufficient transitions to obtain a robustness policy. To improve the robustness, Wang et al. [34] propose to combine the high-level planning method base on different funnel states to obtain the recover skill from the undesirable anomaly which unknown in the prior demonstration. In funnel representation, different state may cluster into the same region to as same funnel state to optimize the policy, meaning that funnel-base reinforcement learning can very suit to get the representation of the close signals data points with rare data collected environment and get robuster performance than the traditional reinforcement learning.

In this paper, we base on funnel reinforcement learning, further increasing the diversity of the funnel-transitions through data augmentation, and improve the recovery capability from anomaly finally.

Data augmentation is widely using in image domain [22] There are many data augmentation techniques like Noise injection [14], Kernel filters [17] and GAN-based augmentation [10]. However, the above methods assembling enormous datasets can be a very daunting task due to the manual effort to collect and label data. Instead, VAEs is an unsupervised learning method that can use to represent [15] or generate [6] data without any label. VAEs can be used in the augment different domain signals like Spoken Language Understanding [36], Historical Documents [5], EEG-based Emotion [23].

3   Preliminaries

3.1   Data Augmentation with VAEs

VAEs [9,15] is an unsupervised method that can be used to compress and generate data with the help of a loss function to compare the reconstruction and the original input, which formulated as:

$$\mathcal{L}_{\psi,\phi} = \mathbb{E}_{p(s)} \left[\beta \mathrm{KL}(p_\phi(z|s) || \mathcal{N}(z)) - \mathbb{E}_{p_\phi(z|s)} \log p_\psi(s|z) \right]. \tag{1}$$

As described in Eq. 1, the above loss function is combined with three components, including the encoder $p_\phi(z|s)$, the decoder $p_\psi(s|z)$, and the prior $\mathcal{N}(z)$, which represents the negative of the normalized evidence lower bound ($-$ELBO) on the marginal likelihood of observations [1]. Particularly, the term $\beta \mathrm{KL}(p_\phi(z|s) || \mathcal{N}(z))$ is the Kullback–Leibler divergence [19] between distributions $p_\phi(z|s)$ and $\mathcal{N}(z)$ with a penalty value $\beta > 1$; $\mathbb{E}_{p_\phi(z|s)} \log p_\psi(s|z)$ is the reconstruction error. In this paper, the reconstruction error is the L2 distance between input and the outout of VAEs. The overall minimization of the loss function is to learn a latent encoding Z that is maximally expressive about the reconstruction while maximally compressing the samples in an adversarial way. In this paper, we use the decoder from VAEs to generated signals. Therefore, motivated by the continued success of VAEs to produce realistic-looking data, we propose a method to generate normal signals or abnormal signals from the obtained latent distribution to enrich the datasets.

3.2   Funnel Q-Learning

High-level reinforcement learning is more efficient and stable than traditional low-level reinforcement learning. In particular, we need to ensure safety in the real robot manipulation learning phase. The main different between funnel reinforcement learning with traditional reinforcement learning is the state transition model $P_{\Psi,\Phi}(s'|s,a)$. The funnel-transition model is considered with the input region Φ and output region Ψ, which are clustered from the collected data. The different transition (s,a,s') may cluster to within the same region due to the hyperparameters of the clustering method. With the transition model, we then can model the reward function, and the Q function can be written as:

$$Q(s,a) = r(s,a) + \gamma \int_{\mathbb{S}} P_{\Psi,\Phi}(s'|s,a) \max_{a'} Q(s',a')ds' = \frac{\sum_i q_i \Phi_i(s)\delta(a_i,a)}{\sum_j \Phi_j(s)\delta(a_j,a)}. \quad (2)$$

For the cluster method, we use DBSCAN (Density-Based Spatial Clustering of Applications with Noise) [29].

4   Method

In this section, we first introduce how we pre-process different multimodal signals as the input of VAEs and introduce how to use all the synthetic signals to cluster as different funnels for Q-learning.

4.1   Pre-process the Multimodal Signals

We first introduce how to use VAEs to augment the sequence multimodal signals. The dexterous robot manipulation signals were collected as a trajectory labeled by the skill name. However, every skill trajectory having different steps. To align them for having the same size and then take as input to the VAEs, we extend the shorter trajectory with the last step signal. For example, the standard trajectory length has 50 steps, a shorter trajectory with 35 steps will be appended with 15 steps begin from the end, which has the same multimodal signals with the last step (step 35). And then use the one-hot encoder to encode the skill name also as a conditional input of the VAEs. See Fig. 2 for more detail of our VAEs structure.

4.2   Funnelling the Trajectory as Transition Model

In the collected data executed from the demonstration, there are many trajectories τ_j which include many state-skill-state transitions: (s,a,s'), where $s_t = \{p_{xyz}, o_{xyzw}, \omega_{xyz}, v_{xyz}, f_{xyz}, m_{xyz}, t_{lr}\}$ and a is the correspond skill illustrated in Fig. 4. We then augmented the τ_j as many different $\hat{\tau}_j$ with VAEs as introduced in Sect. 4.1. We first cluster the synthetic datasets $\{s_t^j, s_t^j\}$ as different sets Ψ with normal distance function of DBSCAN $D_s(s_i, s_j) < \varepsilon_s$. And then

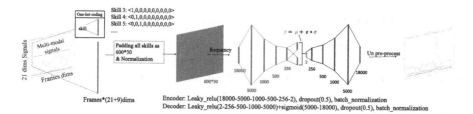

Fig. 2. The pre-process method and neural network structure of our VAEs augmentation model for Kitting experiment dataset.

merge the region if Bhattacharyya distance $D_\Psi(\Psi_k, \Psi_l) < \varepsilon_\psi$. Finally abtained the distribution function with the state: $\Psi_k(s_i)$ to model the probability of state s_i in region Ψ_k. Φ is the input region and ψ is the output region. Finally, we get the funnel-transition model is as below:

$$P(s'|s, a) = \frac{\sum_i \Psi_i(s')\Phi_i(s)\delta(a_i, a)}{\sum_j \Phi_j(s)\delta(a_j, a)} \tag{3}$$

We then base on the funnel-transition model to update the policy with Q-learning as used in [34].

5 Experiments

In this section, we first verify our data augmented method with the open-source force/torque dataset, which is to insert multi-shape peg [7]. And them verify on the kitting experiment with anomaly datesets [35] collect by ourself. And then compare the diversity of the clustered regions from the synthetic data which combining the generated data. Finally, we evaluate the policy learned from funnel Q-learning on the Baxter Kitting experiment. We aim to answer the following question through our experiments:

1. Dose data augmentation can enrich the multimodal signals?
2. The use of synthetic data for clustering can get more diverse regions?
3. The augmented funnel-transitions model can guarantee a more robust performance than use the traditional funnel-transitions model?

5.1 Experiments Setup

Force/Torque dataset is collected with the task by manipulate a Universal Robots arm to insert peg in the hole. There are five different shapes of pegs and corresponding holes. For every different peg-hole pair, there are 144 peg positions × 11 peg orientations × 1000 time-steps. This will have different signal structures with successful inserting or not, including the XYZ force, XYZ moment, the peg position, and the angle of the arm.

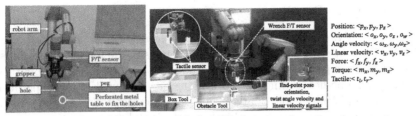

(a) The multi-shape insertion environment

(b) Our Kitting experiment environment

Fig. 3. Illustrate the two different environments for collect multimodal signals

Our kitting experiment dataset is collect on Rethink Baxter robot. We use the right arm camera to detect the object position and use our Sense-Plan-Act-Introspect-Recover (SPAIR) system [35] to do the human-robot collaborative kitting task. The kitting trajectory is finishing with five normal skills and two recover skills, including *move to pick, pre-pick to pick, pick to pre-pick, place, pre-place with an empty hand, re-enactment from anomaly* and *adaptation from the anomaly*. Moreover, there are also having seven types of anomalies in different skills including *human collisions, tool collisions, object slips, human collision with object, wall collision, no object,* and *other*. A successful experiment is if the arm finished the skill 9 (pre-place with empty hand) and will start again when the object is detected. Ten different skills are collected within 600 experiments based on the demonstration rule as showing in Fig. 4.

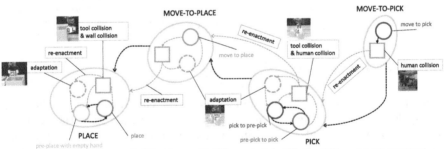

Skill3: move to pick, **skill4:** pre-pick to pick, **skill5:** pick to pre-pick, **skill7:** move to place, **skill8:** place, **skill9:** pre-place with empty hand
Skill1000: re-enactment, **skill1001~1003:** adaptation.

Fig. 4. All different skills from our demonstration along with different anomalies. We start the kitting experiment from skill three to skill nine. (At this figure is from right side to left side. In this figure, different colored circles represent a normal skill, squares represent abnormal occur, and dotted circles represent recover skills). (Color figure online)

5.2   Experiment Results

The input size of the encoder and output size of the decoder are different according to the different dimensions of multimodal signals in the force/torque dataset and our Kitting experiment dataset. For the force/torque dataset, we without combining the one-hot coding of the skill name like our Kitting experiment dataset. We empirically choices latent representation size $z = 2$ and $\beta = 1$. The VAEs augmented signals are showed in Fig. 5, the valid reconstruction error (on average) between the reconstruction and input sample is 0.03925 ± 0.002 for force/torque dataset and 0.04043 ± 0.003 for Kitting multimodal dataset.

After augmented the trajectory, we test the clustered result using synthetic data, the DBSCAN hyperparametes are: $\varepsilon = 0.8, minPts = 5$ (Other parameters performance in Table 1). As shown in Fig. 6, more regions obtained after cluster with the synthetic data.

Finally, we evaluate the performance of the policy learned from the augmented funnel-transition model. Compared with not use the augmentation, the success ratevof our method is increased from 70% to 90% average on totally 20 trials[1] (See Fig. 7 for one of the experiment performance of using our method).

Table 1. We compare our method with the baselines (not use augmented signals) in Baxter Kiting Environment. The parameters set: $\beta = 1, z = 2, \varepsilon = 0.8, minPts = 5$ can achieve nontrivial performance than others.

Methods	Baselines	Ours	Ours	Ours	Ours	Ours	Ours
VAE parameters	None	$\beta = 1, z = 1$	$\beta = 1, z = 2$	$\beta = 1, z = 4$	$\beta = 10, z = 2$	$\beta = 1, z = 2$	$\beta = 1, z = 2$
DBSCAN parameters	$\varepsilon = 0.8,$ $mniPts = 5$	$\varepsilon = 0.8,$ $mniPts = 5$	$\varepsilon = 0.8,$ $mniPts = 5$	$\varepsilon = 0.8,$ $mniPts = 5$	$\varepsilon = 0.8,$ $mniPts = 5$	$\varepsilon = 0.8,$ $mniPts = 5$	$\varepsilon = 0.8,$ $mniPts = 5$
Reconstruction Error	None	0.04 ± 0.002	$\mathbf{0.03 \pm 0.002}$	0.04 ± 0.003	0.12 ± 0.002	0.03 ± 0.002	0.05 ± 0.002
Success Rate	70 %	80%	**90%**	80%	70%	70%	70%

6   Discussion

In this paper, we successfully use funnel reinforcement learning with the augmented demonstration, due to can learn the recovery skill from the complex abnormal state which out of planning in demonstrations, e.g., if an abnormal state encountered that the demonstration did not reach, the robot will then recover to the normal state and continue kitting. Our method only needs to collect 600 episodes of multi-step tasks and finally perform a robust performance on robot dexterous manipulation.

---

[1] An experiment is success if the agent can finish a kitting experiment from skill 3 to skill 9.

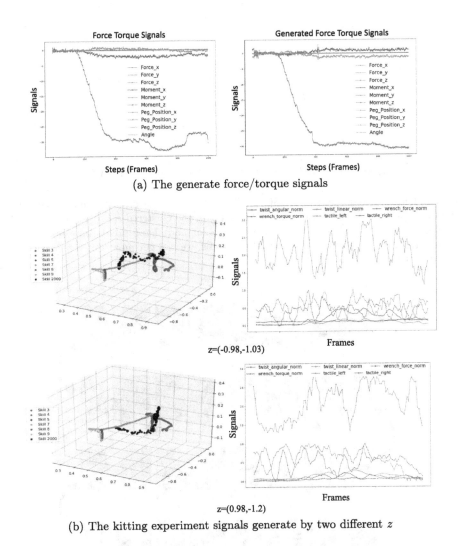

(a) The generate force/torque signals

$z=(-0.98,-1.03)$

$z=(0.98,-1.2)$

(b) The kitting experiment signals generate by two different z

Fig. 5. In (a), the true signal (left) and generated signal (right) of the force/torque dataset. In (b), are the generated signals of our Kitting experiment dataset. In the 3D charts, the black curve is our generated trajectory. The other color curves are the normal real trajectories. And the 2D charts are the other modal signals calculate as the norm. To visualize the linear velocity (v_x, v_y, v_z), angular velocity $(\omega_x, \omega_y, \omega_z)$, force (f_x, f_y, f_z) and torque (m_x, m_y, m_z) on two dimensional chart, we calculate the norm of each dimension, for example norm of linear velocity $n_v = \sqrt{v_x^2, v_y^2, v_z^2}$ of different dimensions are showing on the 2D charts. We can see that the generated signals are similar to the real one. In this case, we can base the synthetic signals to create more different regions to obtain a more robust funnel-transitions model than the traditional funnel-transitions model.

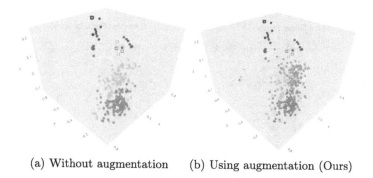

(a) Without augmentation   (b) Using augmentation (Ours)

Fig. 6. The clustered result by DBSCAN. The left side figure is the original signal. The right side plot is the synthetic signal (We just plot the clustered result with the initial positions and the final positions). The same color and shape point representing the same set. After combining our augmented signals, we can get more different clustered set. (Color figure online)

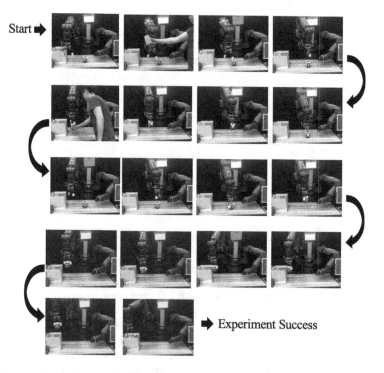

Fig. 7. Example of finish a kitting experiment with random anomalies successfully. We testing the policy learn from the augmented funnel-transitions model. Compare to without using the augment data. The funnel policy can achieve 20% success rate more than learn from pure funnel-transitions model to finish full kitting with anomalies.

References

1. Alemi, A.A., Poole, B., Fischer, I., Dillon, J.V., Saurous, R.A., Murphy, K.: Fixing a broken elbo. arXiv preprint arXiv:1711.00464 (2017)
2. Andrychowicz, M., et al.: Hindsight experience replay. In: Advances in Neural Information Processing Systems (2017)
3. Argall, B., Chernova, S., Veloso, M.M., Browning, B.: A survey of robot learning from demonstration. Robot. Auton. Syst. **57**, 469–483 (2009)
4. Burda, Y., Edwards, H.A., Storkey, A.J., Klimov, O.: Exploration by random network distillation. arXiv:1810.12894 (2019)
5. Cao, G., Kamata, S.-I.: Data augmentation for historical documents via cascade variational auto-encoder. In: 2019 IEEE International Conference on Signal and Image Processing Applications (ICSIPA), pp. 340–345 (2019)
6. Chen, T.Q., Li, X., Grosse, R.B., Duvenaud, D.K.: Isolating sources of disentanglement in variational autoencoders. In: NeurIPS (2018)
7. De Magistris, G., Munawar, A., Pham, T.-H., Inoue, T., Vinayavekhin, P., Tachibana, R.: Experimental force-torque dataset for robot learning of multi-shape insertion. arXiv preprint arXiv:1807.06749 (2018)
8. Deng, Z.-H., Huang, L., Wang, C.-D., Lai, J.-H., Yu, P.S.: Deepcf: A unified framework of representation learning and matching function learning in recommender system. arXiv:1901.04704 (2019)
9. Diederik, P.K., et al.: Auto-encoding variational bayes (2014)
10. Donahue, J., Krähenbühl, P., Darrell, T.: Adversarial feature learning. arXiv:1605.09782 (2017)
11. Finn, C., Yu, T., Zhang, T., Abbeel, P., Levine, S.: One-shot visual imitation learning via meta-learning. In: CoRL (2017)
12. Gupta, A., Eysenbach, B., Finn, C., Levine, S.: Unsupervised meta-learning for reinforcement learning. arXiv:1806.04640 (2018)
13. Hester, T., et al.: Deep q-learning from demonstrations. In: AAAI (2018)
14. Hettich, S., Blake, C., Merz, C.J.: UCI machine learning repository (1998)
15. Higgins, I.: Beta-vae: learning basic visual concepts with a constrained variational framework. In: ICLR (2017)
16. Kalashnikov, D., et al.: Qt-opt: Scalable deep reinforcement learning for vision-based robotic manipulation. arXiv:1806.10293 (2018)
17. Kang, G., Dong, X., Zheng, L., Yang, Y.: Patchshuffle regularization. arXiv:1707.07103 (2017)
18. Kim, B., Massoud Farahmand, A., Pineau, J., Precup, D.: Learning from limited demonstrations. In: NIPS (2013)
19. Kullback, S.: Information theory and statistics. Courier Corporation (1997)
20. Lee, M.: Making sense of vision and touch: self-supervised learning of multi-modal representations for contact-rich tasks. In: 2019 International Conference on Robotics and Automation (ICRA), pp. 8943–8950 (2019)
21. Lin, Y., Huang, J., Zimmer, M., Rojas, J., Weng, P.: Invariant transform experience replay. arXiv preprint arXiv:1909.10707 (2019)
22. Litjens, G.J.S., et al.: A survey on deep learning in medical image analysis. Med. Image Anal. **42** (2017)
23. Luo, Y., Zhu, L., Wan, Z., Lu, B.-L.: Data augmentation for enhancing EEG-based emotion recognition with deep generative models. arXiv:2006.05331 (2020)
24. Nair, A., Bahl, S., Khazatsky, A., Pong, V.H., Berseth, G., Levine, S.: Contextual imagined goals for self-supervised robotic learning. In: CoRL (2019)

25. OpenAI, et al.: Solving rubik's cube with a robot hand. arXiv:1910.07113 (2019)
26. Osa, T., Esfahani, A.M.G., Stolkin, R., Lioutikov, R., Peters, J., Neumann, G.: Guiding trajectory optimization by demonstrated distributions. IEEE Robot. Autom. Lett. **2**, 819–826 (2017)
27. Pong, V.H., Dalal, M., Lin, S., Nair, A., Bahl, S., Levine, S.: Skew-fit: State-covering self-supervised reinforcement learning. arXiv:1903.03698 (2019)
28. Rahmatizadeh, R., Abolghasemi, P., Bölöni, L., Levine, S.: Vision-based multi-task manipulation for inexpensive robots using end-to-end learning from demonstration. In: 2018 IEEE International Conference on Robotics and Automation (ICRA), pp. 3758–3765 (2018)
29. Schubert, E., Sander, J., Ester, M., Kriegel, H.P., Xu, X.: Dbscan revisited, revisited: why and how you should (still) use dbscan. ACM Trans. Database Syst. (TODS) **42**(3), 1–21 (2017)
30. Silver, D., et al.: Mastering the game of go without human knowledge. Nature **550**, 354–359 (2017)
31. Sutton, R.S., Barto, A.G.: Reinforcement learning: an introduction. IEEE Trans. Neural Networks **16**, 285–286 (1998)
32. Tobin, J., Fong, R.H., Ray, A., Schneider, J., Zaremba, W., Abbeel, P.: Domain randomization for transferring deep neural networks from simulation to the real world. In: 2017 IEEE/RSJ International Conference on Intelligent Robots and Systems (IROS), pp. 23–30 (2017)
33. Vecerík, M., et al.: Leveraging demonstrations for deep reinforcement learning on robotics problems with sparse rewards. arXiv:1707.08817 (2017)
34. Wang, A.S., Kroemer, O.: Learning robust manipulation strategies with multi-modal state transition models and recovery heuristics. In: 2019 International Conference on Robotics and Automation (ICRA), pp. 1309–1315 (2019)
35. Wu, H., Guan, Y., Rojas, J.: A latent state-based multimodal execution monitor with anomaly detection and classification for robot introspection. Appl. Sci. **9**, 1072 (2019)
36. Yoo, K.M., Shin, Y., Goo Lee, S.: Data augmentation for spoken language understanding via joint variational generation. In: AAAI (2019)
37. Yu, T., et al.: Meta-world: a benchmark and evaluation for multi-task and meta reinforcement learning. In: CoRL (2019)

A Guided Evaluation Method for Robot Dynamic Manipulation

Chuzhen Feng[1], Xuguang Lan[1(✉)], Lipeng Wan[1(✉)], Zhuo Liang[2(✉)], and Haoyu Wang[1(✉)]

[1] School of Artificail Intelligence, Institute of Artificial Intelligence and Robotics, Xi'an Jiaotong University, No. 28, Xianning West Road, Beilin District, Xi'an, Shannxi, China
fengchuzhen@163.com, xglan@mail.xjtu.edu.cn, {wanlipeng3,haoyuwang}@stu.xjtu.edu.cn
[2] China Academy of Launch Vehicle Technology, Beijing, China
liangzhuo_nust@163.com

Abstract. It is challenging for reinforcement learning (RL) to solve the dynamic goal tasks of robot in sparse reward setting. Dynamic Hindsight Experience Replay (DHER) is a method to solve such problems. However, the learned policy DHER is easy to degrade, and the success rate is low, especially in complex environment. In order to help agents learn purposefully in dynamic goal tasks, avoid blind exploration, and improve the stability and robustness of policy, we propose a guided evaluation method named GEDHER, which assists the agent to learn under the guidance of evaluated expert demonstrations based on the DHER. In addition, We add the Gaussian noise in action sampling to balance the exploration and exploitation, preventing from falling into local optimal policy. Experiment results show that our method outperforms original DHER method in terms of both stability and success rate.

Keywords: Reinforcement learning · Robot · Dynamic task · Demonstration · Evaluation mechanism

1 Introduction

Reinforcement learning with neural network has achieved great success in different fields, such as the Atari games, chess games and so on. Nowadays, more and more attention has been paid to robotics. Robot control problems have large state space and continuous action space, which makes it difficult to learn effective control policy [17]. One of the most serious challenge is the low sample efficiency caused by sparse reward. Previous methods try to set continuous reward function manually which is reliable in simple tasks. Reshaping reward function requires a comprehensive knowledge of reinforcement learning, robotics and kinematics. These information are difficult to obtain in complex robot tasks, and even with these knowledge, it is not easy to design a reasonable reward function. For the problems above, Hindsight Experience Replay (HER) transforms failed samples into successful ones by reconstructing goals of episodic tasks [14].

© Springer Nature Switzerland AG 2020
C. S. Chan et al. (Eds.): ICIRA 2020, LNAI 12595, pp. 161–170, 2020.
https://doi.org/10.1007/978-3-030-66645-3_14

However, HER algorithm only works well in fixed goals task. If the goal is dynamic, it is no longer effective. Because in the fixed goal environment, the target position at each time step in an episode is the same, the hindsight goal selected by HER for the target position can be applied to every time step in the whole episode. While in dynamic goal tasks, HER only selects the corresponding hindsight goals according to the goal position at the initial time, which is obviously unreasonable. Based on HER, DHER, which is effective in dynamic goal tasks, proposes to aggregate the successful trajectories according to the connection between two failed trajectories.

There are two main problems in DHER: 1) policy degradation; 2) poor performance in complex dynamic goal tasks. Therefore, we propose a guided evaluation method for dynamic goal tasks. We sample dynamic demonstrations to guide agent training [3,4]. At the same time, we use the evaluation to select the expert sample set to eliminate the demonstrations that deviate from the target trajectory, improving the quality of the demonstration set, and make the policy learning faster and more effective. Finally, in order to prevent the agent from overfitting to the demonstrations [2], we add additional loss to the policy loss function instead of directly restricting the policy learning within the scope of the demonstrations. Gaussian noise is added to help agent to explore the environment.

The contributions of our work are as follow:

1) In the simulation environment, the trajectories of the robot arm are generated according to the dynamic goal's trajectories, which are sampled as the expert demonstrations;
2) Establish an evaluation approach to evaluate the performance of the collected demonstrations and select satisfied samples;
3) We propose a novel reinforcement learning method for dynamic goal tasks in 7DOF robot to improve the stability and success rate of the robot control policy.

The rest of this paper is organized as follow: The background is introduced in Sect. 2. The problem and the proposed method are presented in details in Sect. 3. Finally, our experiments are illustrated in Sect. 4 and the conclusions are discussed in Sect. 5.

2   Background

2.1   Reinforcement Learning

Reinforcement learning is usually modeled as a standard Markov Decision Process(MDP) [7]. The MDP is defined by a tuple $<S, A, R, P, \gamma>$ which consist a set of states S, a set of action A, a reward function $R(s, a)$, a transition function $P(s'|s, a)$, and a discounted factor $\gamma \in [0, 1]$. In each state $s \in S$, the agent takes an action $a \in A$. After taking the action, the agent receives a reward $R(a, s)$ and reaches a new state s' according to the transition probability $P(s'|s, a)$. The goal of the agent is to learn a deterministic and stationary policy π mapping states to actions $\pi : S \rightarrow A$ that maximizes the expected discounted total reward which is formalized by the action value function: $Q^\pi(s_t, a_t) = E[\sum_{t=0}^{\infty} \gamma^t R(s_t, a_t)]$.

2.2   DDPG

Here, we focus on the Deep Deterministic Policy Gradient (DDPG) [1] to solve the robot control problems. The method is a off-policy model-free reinforcement learning algorithm for continuous action space [18].

DDPG, which combines the advantages of value-based and policy-based method, is an actor-critic algorithm [9] consisting of a policy network and a critic network parametrized by θ^μ and θ^Q respectively. DDPG learns an action-value function (critic) by minimize the error between the target and the estimated, while simultaneously learns a policy (actor) by directly maximizing the estimated action-value function with respect to the parameters of the policy. The experienced transitions collected from the interaction between agent and environment are stored in the replay buffer in the form of $< s_t, a_t, s_{t+1}, r_t >$. Actions are obtained by behavior policy adding noise for exploration: $a_t = \mu(s_t|\theta^\mu) + N$, where N is usually a Gaussian noise process. At each training time step, agent selects n transitions from the replay buffer to construct a minibatch to update the policy and critic network. The algorithm minimizes the following loss to update the critic:

$$L = \frac{1}{N} \sum_i (y_i - Q(s_i, a_i|\theta^Q)) \tag{1}$$

where the $y_i = r_i + \gamma Q(s_{i+1}, \pi(s_{i+1}))$ is the target Q value, and the i in the function represents time step. And the parameter of policy network θ^μ is updated by policy gradient:

$$\nabla_{\theta^\mu} J = \frac{1}{N} \sum_i \nabla_a Q(s, a|\theta^Q)|_{s=s_i, a=\pi(s)} \nabla_{\theta^\mu} \pi(s|\theta^\mu)|_{s_i} \tag{2}$$

The $Q(s, a|\theta^Q)$ and $\pi(s|\theta^\mu)$ represent the action-value function and the policy respectively.

In order to improve the stability of the trained policy and accelerate convergence, actor network and critic network set up the current network and target network respectively. The current network parameters are copied to the target network every certain steps in the training process, which is called soft update.

Since DDPG is an off-policy method, it is suitable to use off-line samples improving training, such as expert demonstrations. And we can also make use of the Q function to select better demonstrations.

2.3   Dynamic Hindsight Experience Replay

In robot control tasks, we usually use sparse reward function. Compared with shaped reward, sparse reward function is simple in form and does not need to be designed for specific tasks, so it has better generalization. However, low sample efficiency is a fatal problem of it. Here, we adopt the Hindsight Experience Replay (HER) algorithm to solve it.

HER takes state and target as input, modifies the transition reward by sampling hindsight goals [15], transforming the failed samples with reward of negative to the successful samples with reward of zero, and replays the modified transitions in the buffer. It has shown an great performance in fixed goal tasks, while the performance degrades when the goal is changing over time because the hindsight goals selected by HER contain almost no information of the dynamic target.

To solve the problem mentioned above, Dynamic Hindsight Experience Replay (DHER) [5] has made one step further on the basis of HER: the agent not only selects hindsight goals to modify the corresponding reward in the transition, but also finds the collection between the different episodes to aggregate successful experiences from two relevant failed trajectories. In the algorithm of DHER, E_i and E_j represent two different trajectories. $E_{i,p}^{ac}$ is the position of the i-th trajectory at time step p, and $E_{j,q}^{de}$ is the desired position of the j-th trajectory at time step q. If $E_{i,p}^{ac} = E_{j,q}^{de}$, then the desired goal of the j-th trajectory at each time after time q is regarded as the subgoal of the corresponding time step of the i-th trajectory after time p, cloning a goal trajectory $\{g_0', ..., g_m'\}_{m=min\{p,q\}}$ where $g_t' = g_{j,q-m+t}^{de}$ from E_j for E_i.

3 Method

Although DHER shows good performance in some dynamic tasks, it has some problems such as policy degradation and low success rate of some tasks. In this section, on the basis of DHER, we propose to add expert demonstrations to guide training to make the policy more stable. What's more, we establish the evaluation mechanism for demonstrations and filter the demonstrations with poor performance, so that agents can learn more from demonstrations to accelerate learning and improve success rate. And we also take some measures to encourage agent to explore randomly in the environment. The details of our method are introduced as follow.

3.1 Expert Guiding

Before training the policy, we first set the motion of robot controller to complete the corresponding tasks. These samples are regarded as the expert demonstrations [6,16]. Different from the previous method of storing the demonstrations and the experience obtained from the interaction between the agent and the environment together to form a mixed buffer [13], here we set up another buffer R_D named demo buffer in the same format as replay buffer to store the demonstrations. Each episode, we sample N_D demo samples from demo buffer R_D and N_R samples from replay buffer R to construct minibatch. Compared with the former methods, our algorithm can guarantee the proportion of demonstrations in minibatch, which is helpful to guide training.

3.2 Behavior Cloning Auxiliary Loss

We hope that after adding the demonstrations, the agent can learn a stable policy from the expert action faster, which is similar to the behavior clone in *Imitation Learning* [10]. The standard loss function of *Imitation Learning* is to calculate the mean square error between demonstration and sampled action:

$$L_{bc} = \sum_{i=1}^{N_D} ||\pi(s_i|\theta^\mu) - a_i||^2 \tag{3}$$

The π and a_i represent the learned policy and the demonstration behavior respectively. The loss function aims to let the agent learn more about the behavior of demonstration. However, if the demonstration is not good enough, it is prone to converge to suboptimal. Therefore, agent needs to explore further while learning from demonstrations. In order to balance exploration and exploitation, we take the loss function as an auxiliary loss of the actor loss. The gradient of the policy parameter θ^μ changes to the following form:

$$\lambda_1 \nabla_{\theta^\mu} J - \lambda_2 \nabla_{\theta^\mu} L_{bc} \tag{4}$$

λ_1 and λ_2 are the weight of two terms, and λ_1 is larger than λ_2.

3.3 Evaluation Mechanism

Using demonstration to guide agent learning policy has strict requirements for the performance of demonstration. However, in the common cases, we do not know the quality of demonstrations [8,12]. Therefore, we need to evaluate demonstrations and select high quality demonstrations from demo buffer to guide agent training.

In DDPG algorithm, critic network is used to calculate the action-value function to evaluate the current policy. The larger action-value function, the better the policy. Inspired by this, we consider using critic network to calculate the action-value function of demonstration $Q(s_i, a_i)$ at the same time, in which a_i refers to the demonstration action, and compare the action-value function of the current policy $Q(s_i, \pi(s_i))$. If $Q(s_i, a_i) > Q(s_i, \pi(s_i))$, it means that the action of demonstration is better than that of current policy. The sample is stored in the demo buffer and used to calculate the auxiliary loss function mentioned in the previous part. If $Q(s_i, a_i) < Q(s_i, \pi(s_i))$, the kind of demonstrations is filtered out and not used to train policy. The loss function becomes:

$$L_{bc} = \sum_{i=1}^{N_D} ||\pi(s_i|\theta^\mu) - a_i||^2 \mathbb{I}_{Q(s_i,a_i)>Q(s_i,\pi(s_i))} \tag{5}$$

The $\mathbb{I}$ is the indicator function.

In this way, we can filter out the unsatisfactory demonstrations and avoid the agent learn the poor demonstrations so that deviating from target trajectory.

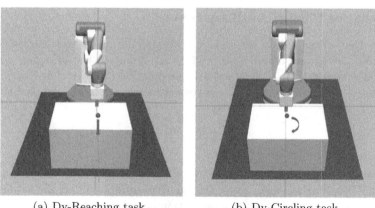

(a) Dy-Reaching task   (b) Dy-Circling task

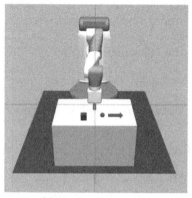

(c) Dy-Pushing task

Fig. 1. These are the dynamic environments, and the arrow indicates the velocity direction of the target object.

4   Experiment

In this section, we evaluate our method on three dynamic tasks: Dy-Reaching, Dy-Circling, Dy-Pushing, which are usually needed for testing. Our method is also compared with the classic method in terms of policy stability and success rate.

4.1   Environments

In all experiments, we use a 7-DOF Fetch Robot in the *MuJoCo* physics engine [11]. The robot is trained to follow the goal in a straight line (Dy-Reaching) and in a circle (Dy-Circling), push a block to follow the goal in a straight line (Dy-Pushing) as shown in Fig. 1. The target speed is set to 0.011 m/s. The agent gets the position, gesture and velocity of the object and the robot gripper as observation. The action space is continuous with 4 dimension. We adopt a sparse reward function: $r(s_t, a_t, g_t) = -\mathbb{I}_{condition}(|s_{t'}^{obj} - g_{t'}| \geq \varepsilon)$, where ε indicates the distance tolerance, $s_{t'}^{obj}$ and $g_{t'}$ respectively refer to the state and dynamic goal position of next time step. We use data generation script to generate 100 demonstrations before training in the dynamic environments with our method.

4.2   Experiment Details

The actor and critic networks contain three hidden layers, each of which has 256 neurons with the *tanh* as the activation function. Different from policy network, the input of critic network includes not only normalized observation and goal, but also the output action of policy network. Normalized parameters, $norm_{eps}$ and $norm_{clip}$, are set to 0.01 and 5 respectively. To ensure the policy training more stable and effective, we use *Adam* as the optimizer with learning rate 10^{-3} for both networks. We set the size of replay buffer and demo buffer to 10^6, and randomly sample 128 transitions from both buffers to form minibatch for each episode.

In the training process, two approaches are raised to encourage exploration. First, agent can select actions randomly with a certain probability of 0.3. Second, we use Gaussian process with mean value 0 and standard deviation 0.2 to produce noise which is added to help the agent explore during training. The evaluated success rate is defined as the average success rate of all episodes per epoch.

4.3   Results

We compare our method with DDPG + HER, DDPG + DHER, DDPG + DHER + Demonstration. In each task, we set several random seeds to train and take a average as the results. The Fig. 2 shows the success rate curve of the four methods in the Dy-Reaching, Dy-Circling and Dy-Pushing task. As shown in Fig. 2 (a), in the Dy-Reaching task, DDPG + HER has the lowest success rate, which is about 20%. Compared with HER algorithm, DHER has a great improvement in the success rate, but there will be serious policy degradation. While DDPG + DHER + Demo can alleviate the policy degradation to some degree. Our method outperforms the others, not only improving the success rate compared with HER, but also completely solving the problem of policy degradation of DHER algorithm.

Figure 2 (b) shows the result of Dy-Circling task. HER algorithm has a success rate about 60%, while the other methods can reach 100%. Compared with the Dy-Reaching task, the target object moves in a circle in Dy-Circling. When the motion radius and linear velocity are fixed, the position change of the target in the same time interval is smaller than that of linear movement, so it is easier for agent to learn the policy.

In Dy-Pushing task, the agent first needs to find the location of the block, and then push the block to follow the moving goal. From Fig. 2 (c), we can see that HER shows weakness in the Dy-Pushing task with almost zero performance. Compared with DHER, DHER+demo works better on the whole, but

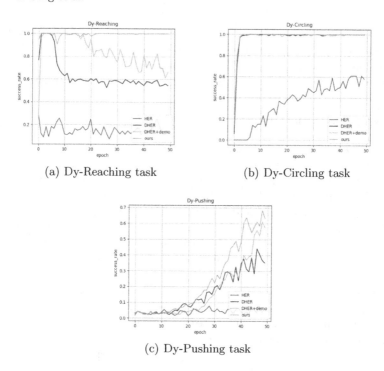

(a) Dy-Reaching task   (b) Dy-Circling task

(c) Dy-Pushing task

Fig. 2. Test results on three dynamic tasks

its deficiency lies in the slow learning before epoch = 30. This is due to the existence of demonstrations that do not meet the conditions in the collected demo buffer, which guide the agent learn the action far away from the target trajectory. Our method filters such samples, which makes agent learn policies faster and improves the success rate further. In fact, Dy-Pushing task can be divided into two subtasks: locating and following. In comparison with Dy-Reaching, the agent needs more iterations to learn robust policy. It can be learned form the experiment results that Dy-Pushing task is more difficult.

To illustrate the performance of our method in more complex task, we study the four methods in Dy-Reaching task with higher velocity of 0.016, as shown in Fig. 3. Our method still performs best in this setting.

Table 1 shows the average success rate of the four methods in the Dy-Reaching, Dy-Circling and Dy-Pushing task. HER algorithm performs the worst in there dynamic tasks because it does not consider that the desired goal at each time in the dynamic tasks is constantly changing when selecting hindsight goals, so it only obtain rarely trajectory information about the dynamic goal. Our method has a success rate of 100% in Dy-Reaching task and Dy-Circling task, and greatly improves performance in Dy-Pushing than original DHER.

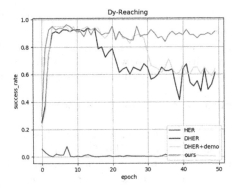

Fig. 3. The result of Dy-Reaching task with 0.016 velocity

Table 1. This is the success rates of the four algorithms in there dynamic tasks.

Success rate	Dy-Reaching	Dy-Circling	Dy-Pushing
HER	0.25	0.6	0.05
DHER	1	1	0.47
DHER + demo	1	1	0.63
OURS	1	1	**0.73**

5 Conclusion

In this paper, we present a method to utilize the evaluated demonstrations to solve reaching and pushing task in dynamic environment on 7-DOF robot. The advantage of adding demonstrations is to guide the agent to learn from expert policy. However, we can not ensure the quality of original demonstrations. Some samples deviate from the target trajectory, which will mislead the agent and cause learning slower or even failure. The evaluation can eliminate such samples to ensure that the demonstrations in the expert sample set can effectively guide the agent learning. In addition, random action and Gaussian noise are added to balance exploration and exploitation. Experiment results demonstrate that our method is the best in terms of policy stability and success rate.

Funding Information. This work was supported in part by Trico-Robot plan of NSFC under grant No.91748208, National Major Project under grant No. 2018ZX01028-101, Shaanxi Project under grant No.2018ZDCXLGY0607, NSFC No.61973246, and the program of the Ministry of Education.

References

1. Lillicrap, T.P., et al.: Continuous control with deep reinforcement learning. arXiv preprint arXiv:1509.02971 (2015)

2. Nair, A., et al.: Overcoming exploration in reinforcement learning with demonstrations. In: ICRA (2018)
3. Vecerik, M., et al.: Leveraging Demonstrations for Deep Reinforcement Learning on Robotics Problems with Sparse Rewards. arXiv:1707.08817 (2017)
4. Wang, Y., et al.: An experienced-based policy gradient method for smooth manipulation. In: IEEE-CYBER (2019)
5. Fang, M., et al.: DHER: hindsight experience replay for dynamic goals. In: ICLR (2019)
6. Mnih, V., Kavukcuoglu, K., Silver, D., et al.: Human-level control through deep reinforce-ment learning. Nature **518**(7540), 529 (2015)
7. This, Paul R. Markov decision processes. Comap, Incorporated, (1983) (MDP)
8. Yang, G., et al.: Reinforcement learning form imperfect demonstrations. In: International Conference on Machine Learning, Stockholm, Sweden, PMLR, vol. 80 (2018)
9. Mnih, V., et al.: Asynchronous methods for deep reinforcement learning. In: ICML (2016)
10. Ratliff, N., Bagnell, J.A., Srinivasa, S.S.: Imitation learning for locomotion and manipulation. In: 2007 7th IEEE-RAS International Conference on Humanoid Robots (2007)
11. Todorov, E., Erez, T., Tassa, Y.: "MuJoCo": a physics engine for model-based control. In: The IEEE/RSJ International Conference on Intelligent Robots and Systems (2012)
12. Popov, I., et al.: Data-efficient Deep Reinforcement Learning for Dexterous Manipulation. arXiv preprint arXiv:1704.03073 (2017)
13. Haarnoja, T., et al.: Deep Reinforcement Learning for Robotic Manipulation with Asynchronous Off-Policy Updates. arXiv preprint arXiv:1610.00633 (2016)
14. Andrychowicz, M., et al.: Hindsight experience replay. In: Advances in Neural Information Processing Systems, pp. 5048–5058 (2017)
15. Bakker, B., Schmidhuber, J.: Hierarchical reinforcement learning based on subgoal discovery and subpolicy specialization. In: Proceedings of the 8-th Conference on Intelligent Autonomous Systems, pp. 438–445
16. Hester, T., et al.: Learning from Demonstrations for Real World Reinforcement Learning. arXiv preprint arxiv:1704.03732 (2017)
17. Xu, K., Liu, H., Shen, H., Yang, T.: Structure design and kinematic analysis of a partially-decoupled 3T1R parallel manipulator. In: Yu, H., Liu, J., Liu, L., Ju, Z., Liu, Y., Zhou, D. (eds.) ICIRA 2019. LNCS (LNAI), vol. 11742, pp. 415–424. Springer, Cham (2019). https://doi.org/10.1007/978-3-030-27535-8_37
18. Heess, N., et al.: Learning continuous control policies by stochastic value gradients. In: Proceedings of the International Conference on Neural Information Processing Systems, pp. 2944–2952 (2015)

A Learning Approach for Optimizing Robot Behavior Selection Algorithm

Basile Tousside[(✉)], Janis Mohr, Marco Schmidt, and Jörg Frochte

Bochum University of Applied Science, 42579 Heiligenhaus, Germany
{basile.tousside,janis.mohr,marco.schmidt,joerg.frochte}@hs-bochum.de

Abstract. Algorithms are the heart of each robotics system. A specific class of algorithm embedded in robotics systems is the so-called behavior − or action − selection algorithm. These algorithms select an action a robot should take, when performing a specific task. The action selection is determined by the parameters of the algorithm. However, manually choosing a proper configuration within the high-dimensional parameter space of the behavior selection algorithm is a non-trivial and demanding task. In this paper, we show how this problem can be addressed with supervised learning techniques. Our method starts by learning the algorithm behavior from the parameter space according to environment features, then bootstrap itself into a more robust framework capable of self-adjusting robot parameters in real-time. We demonstrate our concept on a set of examples, including simulations and real world experiments.

Keywords: Robot learning · Behavior selection · Parameter optimization

1 Introduction

Modern robots are nowadays empowered with more and more behavior selection algorithms (BSA) [6,15]. A behavior is an action a robot can take in response to objectives or sensor inputs. These behaviors might be either the algorithm of interest or a component of an algorithm with a larger scope. A behavior selection algorithm therefore selects appropriate behaviors or actions for a robot to perform. It is used in various applications to allow robots to autonomously perform diverse tasks like vision, navigation or grasping [8,13].

Robot behavior selection algorithms often involve many parameters that have a significant impact on the algorithms efficiency. In this setting, a key challenge concerns the tuning of those parameters. Consider for example a parametric path planning algorithm operating in a dynamic environment. A parameter configuration that avoids a moving human is not guaranteed (without further tuning) to be effective at avoiding other dynamic obstacles like a random moving cat for example. The parameters tuning of behavior selection algorithms is sometimes delegated to a human and accomplished through a process of trial-and-error [1].

© Springer Nature Switzerland AG 2020
C. S. Chan et al. (Eds.): ICIRA 2020, LNAI 12595, pp. 171–183, 2020.
https://doi.org/10.1007/978-3-030-66645-3_15

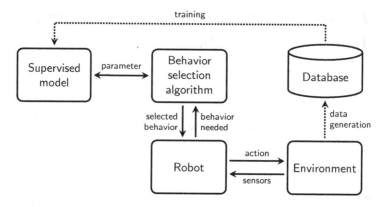

Fig. 1. Our strategy for implementing a learnable module, which gives a robot the ability to self-adjust its parameters while performing a specific task. The dashed lines represent connections taking place exclusively at the first stage of our method, which focus on learning the parameters of the BSA w.r.t environment features.

Unfortunately, manual tuning of robot parameters is not only time consuming, it also tends to make the algorithms vulnerable. As a result, the question of how to optimally select parameters of robot behavior selection algorithms is still an open problem.

In this paper, we demonstrate that these parameters can be learned from observation data via supervised learning techniques. This automates the choice of parameters therefore providing an appealing alternative to the complex manual tuning. Another significant benefit is that the learning solution will approximate an optimum in contrast to manual solutions which are often not of the best quality. Our approach is to build a supervised learner, which learns the parameters of a behavior selection algorithm – therefore learning the behavior of the behavior selection algorithm – from observation data. This supervised learner is then later used to optimize the parameters of the behavior selection algorithm for the specifics of the environment. This is schematically illustrated in Fig. 1. Other authors have already approached the problem this way. However, the learning framework they propose almost always uses specialized techniques such as salient landmarks [12], language measure [14] or stochastic Petri nets [10]. Our framework in contrast uses supervised learning technique making it general and transferable to any other robotics parameter optimization problem. It starts with a data generation process and bootstraps itself towards a more general framework capable of self-adjusting robot parameters while the robot is performing a given task. One challenge when using supervised learning technique is the need of labelled data. Our framework takes this into account by incorporating a data generation phase, which will be presented in Sect. 3.6.

In this paper, we focus on the path planning behavior selection problem, which is a crucial component of each robot navigation framework. When navigating in a dynamic environment, a path planning algorithm – which is a behavior selection algorithm – might be confronted with the issue of which behavior to adopt with respect to the environment, that is, in order to accurately avoid a specific obstacle for example. The behavior the path planning algorithm adopts is determined by its parameters. The problem can then be rephrased as follows: Which parameters should be chosen to avoid a certain type of obstacle. Later in this paper, we will demonstrate how our method is able to solve this problem by learning from observation data the behavior of the path planning algorithm when avoiding obstacles.

One problem that can pose serious difficulties to the learning algorithm is that for some behavior selection algorithm like path planning, searching for the optimal parameters belongs to the class of the so-called ill-posed problems since there is not a unique solution. In fact, different paths produced by various configurations in the parameter space might not significantly differ in terms of quality. To overcome the ill-posedness of parameter search, the training data needs to be filtered and cleaned-up, which is done in a preprocessing step.

Another challenge when applying parameter learning to behavior selection algorithm is the fact that robotic algorithms are mostly interconnected. Consequently, integrating external factors such as learned parameters in one algorithm without compromising the functionality of some others is not trivial. This difficulty is well illustrated in our path planning use case. In fact, robot navigation can mostly be achieved by successfully connecting several key components like map building, localizers, path planner or range sensors among others. However, as the number of components increases, their relationships become difficult to manage. As a side effect, if we learn the behavior of a path planning algorithm, the problem arises of how to integrate the learned behavior into the navigation framework without comprising the functionality of the overall framework. To solve this problem, a framework, which formalizes the navigation system and its components was required. This will be revisited in Sect. 3.4. The main contributions of this paper are as follows:

- Strategy that generates labelled data for a robot behavior selection algorithm from environment observations.
- Supervised-learning approach for learning parameter (and thus the behavior) of robot behavior selection algorithms.
- Framework that exploits a supervised learned model to optimally adjusting robot parameter in real-time.

2   Methodology

We aim at demonstrating how to empower robots with the ability to automatically tweak their parameters, therefore reacting to environment changes. In our setting, the actions – or behaviors – of the robot are controlled by a behavior selection algorithm. Therefore, by controlling the behavior selection algorithm we can indirectly control the robot behavior (see Fig. 1). Since behavior selection algorithms are parameter-dependent, they can be controlled via their parameters. Our method builds on this and can be subdivided into two steps, first, it

---

Algorithm 1. Our method for optimizing parameters of BSA

Input: Deterministic behavior selection algorithm $\mathcal{A}$ with parameters P e.g.,
   parametric path planning algorithm (RRT, APF)
Result: Machine learning model, which improve in real-time the robot behavior
   with respect to a specific environment
1 parameterize the environment with some features $\mathcal{X}$;
2 generate observation data with features $\mathcal{X}$ and response P ;
3 train a supervised-learning model $\mathcal{M}$ to learn P w.r.t $\mathcal{X}$;
4 integrate $\mathcal{M}$ into the robot framework to which $\mathcal{A}$ belong e.g.,
   navigation framework ;
5 robot use $\mathcal{M}$ to self-adjust its behavior by optimizing P according to
   environmental conditions ;

---

focus on learning the parameters of the behavior selection algorithm with respect to some environment features. Doing so, it somehow learns the behavior of the behavior selection algorithm according to environmental conditions. In a second step, the learned model is exploited to optimize in real-time the parameters of the behavior selection algorithms for the specifics of the environment. Let us illustrate this with a concrete use-case. In a navigation setup, path planning is a typical behavior selection algorithm. It selects the appropriate behaviors or actions for a robot to take in response to environment input (such as free space, obstacles, etc.) gained by robot sensors. Consider an environment including a static chair and two dynamic obstacles, namely, a deterministic moving human and a random moving cat. A robot equipped with a parametric path planner – such as Rapidly-exploring random tree (RRT) [11] or the artificial potential field (APF) [9] method – is required to navigate that environment while avoiding obstacles. Thus, when facing an obstacle, the robot – in fact the path planning algorithm – should decide which behavior (or action) to select. More concretely, the path planning algorithm, which is here the behavior selection algorithm should decide which parameter configuration to use. It is difficult to manually find near-optimal choice of parameter configuration to avoid all three types of obstacles. In fact, it can be expected that a skilled human will often be able to do an as good job as machine learning at finding parameters that avoid each obstacle individually, especially for the chair and the human. But for avoiding all the obstacles sequentially, machine learning will be of great help for novice as well as expert users.

Our method for using machine learning to solve such a behavior selection problem consists of first (i) learning the parameters of the path planning algorithm according to some environment feature such as the obstacles's behavior. To do so, we build a labelled data-set, where features are environment characteristics (obstacle type, obstacle speed, etc.) and responses are the parameters to be learned. This is carried out in a bootstrap process involving an intelligent grid-search and a compact local search. This is revisited in Sect. 3.6. Once the data are collected and preprocessed, regression techniques can be used to learn

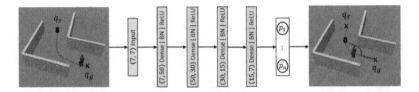

Fig. 2. Our learning model: Predicted parameters are fed to the robot via the control system. Input and output dimensions of each layer are shown on the connections.

path planning parameters corresponding to environment features. However, as pointed out earlier, parameter learning is often an ill-posed problem, since different parameter configurations can lead to an equally good result. This is the case for our path planning parameter learning use-case as well as for many other robotics problems like the inverse kinematic [2]. To tackle such problems using standard regression technique like a multilayer Perceptron (MLP) or a k-nearest neighbors (KNN), the labelled data needs to be filtered and cleaned up as will be presented in the experiments section, otherwise those techniques will fail at finding good solutions as shown in [4]. Once the data is generated and preprocessed, they can be fed to an appropriate regression model, which learns to predict the parameter values the robotics algorithm should use in real-time.

The second step of our method for solving the path planning behavior selection problem is to (ii) deploy the learned model in the navigation framework, in order to optimize the path planning parameter - therefore controlling and improving the path planning behavior - regarding environment features. Our general framework for solving the BSA parameters optimization problem is summarized in Algorithm 1. This framework applied to the path planning behavior selection problem as described above has shown positive results, which will be presented in Sect. 4.

We now turn our attention to the regression model implemented to learn the parameters setting in our path planning use-case. It is a multilayer Perceptron with parameters θ, comprising $L = 6$ layers (see Fig. 2). Each hidden layer implements a non-linear transformation $\mathcal{H}_l(.)$, where l indexes the layer. We define the transformation $\mathcal{H}_l(.)$ as a composite function of two consecutive operations: batch normalization (BN) [7], followed by a rectified linear unit (ReLU) [5].

3 Experimental Setup

Our concept as described above is experimented at the path planning behavior selection problem, which is a fundamental challenge in robotics. We simulate an environment in Gazebo including as obstacle: a static chair, a human moving in a deterministic manner and a cat moving randomly. In this environment, a robot is given the task to reach a goal position without hitting obstacles. The rest of this section is organized as follows: it starts by formalizing the path planning problem we aim to solve, this is followed by a description of the environment

Fig. 3. Setup for experiments in simulation.

in which the robot is navigating. Furthermore, our navigation framework and navigation evaluation measure is presented. Finally, attention is paid to data generation and supervised model training methodology.

3.1 Problem Formulation

Consider a mobile robot navigating in an unknown environment, we denote by $\mathcal{C} \subset \mathbb{R}^q$ the complete configuration space of the robot, where $q \in \mathbb{N}$ is the dimension of the configuration space. $\mathcal{C}_{\text{free}}$ and $\mathcal{C}_{\text{obs}} = \mathcal{C} \backslash \mathcal{C}_{\text{free}}$ denote respectively the valid configuration in the planning space and the obstacle space consisting of robot state obstructed by collisions with either the obstacles or the environment. The robot navigates from a start state $\mathbf{q}_s \in \mathcal{C}_{\text{free}}$ with the task of reaching a goal state $\mathbf{q}_g \in \mathcal{C}_{\text{free}}$. To execute this task, the robot plans a path $\xi_{\mathbf{q}_s \to \mathbf{q}_g}$ using a path planning algorithm – a behavior selection algorithm –, which can be represented by a function $\mathcal{F}$, that solves the following motion planning problem:

$$\xi_{\mathbf{q}_s \to \mathbf{q}_g} = \mathcal{F}(\mathbf{q}_s, \mathbf{q}_g, \mathcal{C})$$
$$\text{s.t. } \xi_{\mathbf{q}_s \to \mathbf{q}_g} = \{\mathbf{q}_1, \ldots, \mathbf{q}_m\} \tag{1}$$
$$\text{and } \forall \ \mathbf{q}_i \in \xi_{\mathbf{q}_s \to \mathbf{q}_g}, \mathbf{q}_i \in \mathcal{C}_{\text{free}}$$

The output of the path planning is therefore a sequence of states $\mathbf{q}_i \in \mathcal{C}_{\text{free}}$, which allow the robot to navigate without hitting obstacles. A path $\xi_{\mathbf{q}_s \to \mathbf{q}_g}$ is said to be successful if $\xi_{\mathbf{q}_s \to \mathbf{q}_g} \subset \mathcal{C}_{\text{free}}$ from the beginning to the end of the driving task. As already mentioned, path planning is an ill-posed problem, meaning that there is not a unique solution to avoid a chair for example. Our evaluation measure for comparing path solutions will be presented in Sect. 3.5. Furthermore, we denote by Υ_ξ the execution cost of a path $\xi_{\mathbf{q}_s \to \mathbf{q}_g}$. For a straight line between state $\mathbf{q}_a$ and $\mathbf{q}_b$ the cost of the path $\xi_{\mathbf{q}_a \to \mathbf{q}_b}$ is given by the distance: $\Upsilon_\xi(\mathbf{q}_a, \mathbf{q}_b) = \| \mathbf{q}_a - \mathbf{q}_b \|$. Since a path is the summation of consecutive small straight movements, the overall execution cost of a path $\xi_{\mathbf{q}_s \to \mathbf{q}_g} = \{\mathbf{q}_1, \ldots, \mathbf{q}_m\}$ is then formalized as:

$$\Upsilon(\xi_{\mathbf{q}_s \to \mathbf{q}_g}) = \sum_{i=1}^{m-1} \Upsilon_\xi(\mathbf{q}_i, \mathbf{q}_{i+1}) \tag{2}$$

The path planning algorithm $\mathcal{F}$ can be configured by a set of parameter $P = \{p_1, \ldots, p_n\} \in \mathcal{P}$, where $\mathcal{P}$ is the parameter space and $p_i \in \mathbb{R}$. Let J denote the cost metrics defined for $\mathcal{F}$ when using a parameter set P in an environment configuration $\mathcal{C}$. Our supervised learning model aims to find in a dynamic environment, for a given robot state and configuration space, a parameter set P^* that drives the robot to the goal without collision while optimizing J.

$$P^* = \arg \min_{p \in \mathcal{P}} J(\mathcal{F}(\mathcal{P}), \mathcal{C}_{\text{obs}}) \tag{3}$$

Whenever an obstacle (detected by the depth camera mounted on the robot) is encountered on a currently followed path $\xi_{\mathbf{q}_s \to \mathbf{q}_g}$, our method solves (3) finding the optimal parameter set P^* for the new configuration space that moves the robot collision-free towards the goal state $\mathbf{q}_g$.

3.2 Environment Representation

The robot is required to navigate to randomly selected target positions on an 12m x 12m, 2D map of the environment. In simulation, the environment consists of a corridor with variable width and curvature (see Fig. 3). The 2D map of the environment is updated for each simulation as the corridor dimension and curvature change. The following subsections address each of the environment component individually.

Corridor. The navigation environment consists of a corridor of width w and length l. The corridor is made of 4 walls, which are pairwise parallel as shown in Fig. 3. Wall 1 and 2 (the 2 top walls) have an angle α with the vertical axis, whereas walls 3 and 4 (the bottom walls) have an angle β with the same (vertical) axis. Note that the feature w, α and β are variable parameters. This allows the corridor to be modeled as: $\zeta = f(w, \alpha, \beta)$.

Obstacle. Inside the corridor, we modeled three obstacle types $O_t = \{O_1, O_2, O_3\}$ consisting of a static chair, a human character moving in a deterministic manner and a cat moving randomly (see Fig. 3). During the simulation, each obstacle (except the chair, which is static) move at a constant speed O_s, where O_s is a variable parameter. Note that there is only one obstacle per simulation, that is, the cat and the human for example never appear together.

World. Joining the two later sections together, we can formally define a simulated world W as: $W = f(w, \alpha, \beta, O_t, O_s)$.

3.3 Robotic Platform

A kobuki based Turtlebot is used as robotic platform in simulation as well as in real world. It is equipped with a depth camera sensor mounted on it, which collects depth images in a 5 m range, therefore allowing to sense local obstacles. The computing center of our real Turtlebot is an Intel NUC5PPYH with 8 Gb RAM running Ubuntu 16.04 and the Robot Operating System (ROS).

3.4   Navigation Framework

Our navigation framework builds on the ROS navigation stack, which provides a two level motion planner consisting of a global planner and a local planner. At a time t, the global planner produces a global path (around obstacles) from the current state $\mathbf{q}_c$ to the goal state $\mathbf{q}_g$ based on the current state configuration $\mathcal{C}_{\text{free}}$ and $\mathcal{C}_{\text{obs}}$. The local planner acts as a controller with the role of following the global path as close as possible. The ROS navigation stack is modified to use our implementation of a path planning behavior selection algorithm (RRT) as global planner. The dynamic window approach (DWA) [3] is used as default local planner in the ROS navigation stack, we do not modify this behavior. Our navigation framework therefore consists of a behavior selection algorithm (BSA) – with parameter $p_1, \ldots, p_n$ –, which plans a global path around obstacles and a local planner, which follows the planned path. If the local planner encounters an obstacle (detected by the depth camera) on the global path it is following, it aborts the process and asks the BSA for a new behavior to adopt, that is a new global path to follow. The BSA then delivers a new path (if one exists) regarding the new configuration space $\mathcal{C}$. This process is illustrated in Fig. 4.

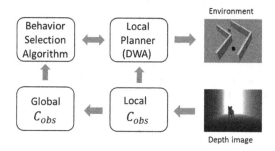

Fig. 4. Our navigation framework.

3.5   Evaluation Metric

Consider two parameter sets P_1 and P_2 of the BSA both successfully avoiding an obstacle in a navigation scenario. An important question is which of P_1 and P_2 produces the best planning strategy. To tackle this question we define the *robot collision* as metric, meaning that we expect the robot to reach the goal without colliding with obstacles. More concretely, during an entire simulation run, we compute the smallest distance $d_{\min}$ between the robot and the obstacle. A too small $d_{\min}$ implies that the avoidance maneuver was close to fail. A big $d_{\min}$ means that the robot has avoided the obstacle with a safer margin. On the other side, another requirement is that the avoidance maneuver should not move the robot too far from the shortest path leading to the goal, this is the straight line connecting start state $\mathbf{q}_s$ and goal state $\mathbf{q}_g$ assuming all obstacles are ignored. This later requirement implies that a big $d_{\min}$ is also not a good choice since it deviates the robot trajectory too far from the shortest path. Denoting the length

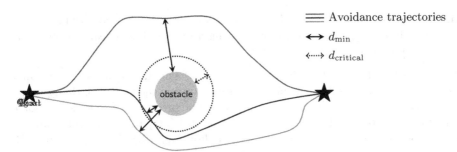

Fig. 5. Illustration of our metric. The red path avoids the obstacle with a safer margin but deviate too far from the shortest path between robot and goal states. The blue path produces the smallest d_{min} but lies inside the critical region. The green path is the one which exhibits a best trade-off between l_p and d_{min}. It produces a d_{min} not deviating too much from the shortest path while yielding at the same time a safe margin to the obstacle.

of the path produced by the avoidance maneuver as l_p, the best trajectory is therefore the one producing a path that minimizes l_p while maximizing d_{min}. This trade-off between l_p and d_{min} is solved by introducing a critical distance $d_{critical}$, which acts as a threshold distance between the obstacle and the robot. Figure 5 illustrates the robot collision metric as described above. The red, blue and green paths show 3 different avoidance trajectories for the same environment and configuration space $\mathcal{C}_{free}$, $\mathcal{C}_{obs}$. The n best avoidance trajectories, are the one with the smallest $d_{min} \in \mathbb{D}$ in ascending order, such that

$$\mathbb{D} = \{d_{min}^x \geq d_{critical} \geq 0 \ , \ x \in W\} \tag{4}$$

3.6   Data Generation

We aim at generating labelled data for the path planning BSA. The data-set consists of environment features and BSA parameters, which are the labels. To generate the data, we ran a lot of simulations, which were distributed on a computer cluster with 5 nodes. Each node consisted of a PC with Intel $i7-2600$ CPU processor clocked at 2.4 GHz, which is able to run the simulation in real time. In each simulation run, the robot is required to navigate from a start position $\mathbf{q}_s$ to a goal position $\mathbf{q}_g$ whereby the Euclidean distance between start and goal is typically 8 m. Each simulation run takes around 40 s and returns a Boolean value signifying whether or not the simulation was successful. That is if the robot reached the goal state within the 40 s and the smallest distance between the robot and the obstacle satisfies (4). Our data generation strategy consists of three consecutive steps, which will be addressed individually in the following subsections:

Initial Guess. The first step is searching for potentially good parameter settings for the path planning BSA by running a quick random search using wide ranges of parameter values.

Selective Grid Search. The second step performs a search using a smaller range of values centered on the best ones found during the first step. Doing so, we zoom in on a good set of planner parameters. The process is as follows: For some random generated world W, we try a grid of 90 BSA parameter configurations and choose the 3 best as the one with the smallest d_{min}, whereby d_{min} satisfies (4).

Local Search. The third step is a local regressor, which allows to automatically generate far more data out of the previously created database. First, we train a k-nearest neighbors multi-output regressor on the data resulting from the previous step. In a second stage, the trained KNN is used to predict the parameters setting P_i that might work well for a previously unseen world W_i. Once a parameter set producing a path satisfying (4) is found, the pair (W_i, P_i) is added to a database.

3.7 Supervised Model Training

We learn the parameters of the path planning BSA using the network presented in Fig. 2. The training objective consists of two parts. The first part minimizes the mean-squared-error (MSE) loss between predicted parameters $\hat{P}_i$ and true parameter P_i, where i indexes the training instance. The second part accounts at each training instance for the cost $\Upsilon_i(\xi_{\mathbf{q}_s \to \mathbf{q}_g})$ of the path $\xi_{\mathbf{q}_s \to \mathbf{q}_g}$ produced by the parameters $\hat{P}_i$. This cost has been defined in (2) and is weighted in our training objective by a parameter λ, which acts as a regularization hyper-parameter. The overall loss function of the network is then computed as:

$$L(\theta) = \frac{1}{n \cdot m} \sum_{j=1}^{n} \sum_{i=1}^{m} (P_i^j - \hat{P}_i^j)^2 + \frac{\lambda}{m} \sum_{i=1}^{m} \left(1 - \frac{\Upsilon_i(\xi_{\mathbf{q}_s \to \mathbf{q}_g})}{\bar{\Upsilon}}\right)^2 \qquad (5)$$

where, m is the batch size, n the number of neurons in the output layer – that is the number of parameters to predict –, and $\bar{\Upsilon}$ the length of the straight line connecting start and goal state.

4   Experimental Results

We evaluate our methodology at two levels. In a first level, we want to rate the impact of using supervised learning to adjust in real-time the parameters of a robot behavior selection algorithm. For this purpose, a human solution is used as baseline. We compare the solution of a human expert to the supervised learning solution at finding parameters of the path planning BSA able to navigate the robot in the environment described in 3.2. Remember that this environment embeds three obstacle types that should be avoided: A static chair, a human moving in a deterministic manner and a random moving cat. We ran 3 type of simulations, each consisting at avoiding one of the three obstacles. Table 1 shows that for static obstacles (here the chair), an expert human is almost as good as a supervised learning solution at providing the BSA with parameters that avoid the obstacle. For dynamic obstacles, however, the supervised learning solution dramatically improves the robots ability to safely navigate that environment. This is due to the fact that the supervised learning solution has learned to adapt the robot behavior - via the BSA parameters - to environment features such as the obstacle type, speed or trajectory.

Table 1. Baseline expert human vs supervised learning at avoiding obstacles

Obstacles type	% of success out of 1000 simulations	
	Human expert	Supervised learning
Chair	98%	100%
Human	56%	96%
Cat	44%	92%

The second level of our methodology evaluation consists of comparing our multilayer Perceptron (MLP) learning approaches to two standard regression techniques, namely, a k-nearest neighbors with $k = 5$ and a random forest with 800 decision tree estimators. We choose those two regression techniques as they have little hyperparameters and are known to perform well on structured data and multi-regression problems. Table 2 shows that our MLP works well and is stable compared to standard regression techniques.

Table 2. Result of comparing our MLP to two supervised learning methods, when predicting parameters of BSA for the same environment configuration.

Regressor	% of success out of 1000 simulations		
	Chair	Human	Cat
Our MLP	100%	96%	92%
kNN (k = 5)	99%	95%	79%
Random forest (800 estimators)	100%	93%	91%

5   Conclusion

We have presented a method that can learn and optimize the behavior of a robot behavior selection algorithm via its parameters. The key idea was to build a supervised learner, which is trained to learn the parameters with respect to some environment features. The application of this approach to a path planning problem showed that the method is able to adjust in real-time the parameters of the algorithm therefore efficiently controlling the robot behavior while performing the navigation task. Simulations showed great results, however our method could not be extensively tested in reality due to a lack of real-world training data. In a prospective future work, we wish to empower our multilayer Perceptron with transfer learning techniques in order to forward simulation insight to reality without the need to generate a large amount of real-world data.

Acknowledgments. This work was funded by the federal state of North Rhine-Westphalia and the *European Regional Development Fund* FKZ: ERFE-040021.

References

1. Bajcsy, A., Losey, D. P., O'Malley, M. K., Dragan, A.D.: Learning from physical human corrections, one feature at a time. In: Proceedings of the 2018 ACM/IEEE International Conference on Human-Robot Interaction, pp. 141–149 (2018)
2. DeMers, D., Kreutz-Delgado, K.: Learning global direct inverse kinematics. In: Advances in Neural Information Processing Systems, pp. 589–595 (1992)
3. Fox, D., Burgard, W., Thrun, S.: The dynamic window approach to collision avoidance. IEEE Robot. Autom. Mag. **4**(1), 23–33 (1997)
4. Frochte, J., Marsland, S.: A learning approach for Ill-posed optimization problems. In: Le, T.D., et al. (eds.) AusDM 2019. CCIS, vol. 1127, pp. 16–27. Springer, Singapore (2019). https://doi.org/10.1007/978-981-15-1699-3_2
5. Glorot, X., Bordes, A., Bengio, Y.: Deep sparse rectifier neural networks. In: Proceedings of the Fourteenth International Conference on Artificial Intelligence and Statistics, pp. 315–323 (2011)
6. Huang, Z., Chen, Y.: An improved artificial fish swarm algorithm based on hybrid behavior selection. Int. J. Control Autom. **6**(5), 103–116 (2013)
7. Ioffe, S., Szegedy, C.: Batch normalization: Accelerating deep network training by reducing internal covariate shift. arXiv preprint arXiv:1502.03167 (2015)
8. Izumi, K., Habib, M.K., Watanabe, K., Sato, R.: Behavior selection based navigation and obstacle avoidance approach using visual and ultrasonic sensory information for quadruped robots. Int. J. Adv. Robot. Syst. **5**(4), 41 (2008)
9. Khatib, O.: Real-time obstacle avoidance for manipulators and mobile robots. In: Autonomous Robot Vehicles, pp. 396–404. Springer (1986). https://doi.org/10.1007/978-1-4613-8997-2_29
10. Kim, G., Chung, W.: Navigation behavior selection using generalized stochastic petri nets for a service robot. IEEE Trans. Syst. Man Cybern. Part C (Appl. Rev.) **37**(4), 494–503 (2007)
11. LaValle, S.M.: Rapidly-exploring random trees: A new tool for path planning
12. Liu, D., Cong, M., Du, Y., Gao, S.: Robot behavior selection using salient landmarks and object-based attention. In: 2013 IEEE International Conference on Robotics and Biomimetics (ROBIO), pp. 1101–1106. IEEE (2013)
13. Murphy, T.G., Lyons, D.M., Hendriks, A.J.: Stable grasping with a multi-fingered robot hand: a behavior-based approach. In: Proceedings of 1993 IEEE/RSJ International Conference on Intelligent Robots and Systems (IROS 1993), vol. 2, pp. 867–874. IEEE (1993)

14. Wang, X., Ray, A., Lee, P., Fu, J.: Optimal control of robot behavior using language measure. In: Quantitative Measure for Discrete Event Supervisory Control, pp. 157–181. Springer (2005). https://doi.org/10.1007/0-387-23903-0_6

15. Wang, Y., Li, S., Chen, Q., Hu, W.: Biology inspired robot behavior selection mechanism: using genetic algorithm. In: Li, K., Fei, M., Irwin, G.W., Ma, S. (eds.) LSMS 2007. LNCS, vol. 4688, pp. 777–786. Springer, Heidelberg (2007). https://doi.org/10.1007/978-3-540-74769-7_82

sEMG Feature Optimization Strategy for Finger Grip Force Estimation

Changcheng Wu[1,3]($\boxtimes$), Qingqing Cao[2], Fei Fei[1], Dehua Yang[1], Baoguo Xu[3], Hong Zeng[3], and Aiguo Song[3]

[1] College of Automation Engineering, Nanjing University of Aeronautics and Astronautics, Nanjing, China
changchengwu@nuaa.edu.cn
[2] School of Aviation Engineering, Nanjing Vocational University of Industry Technology, Nanjing, China
[3] School of Instrument Science and Engineering, Southeast University, Nanjing, China

Abstract. Finger Grip force estimation based on sEMG plays an important role in dexterous control of a prosthetic hand. In order to obtain higher estimation accuracy, one of the commonly used methods is to extract more features from sEMG and input them into the regression model. This practice results in a large amount of computation and is not suitable for practical use in low cost commercial prosthetic hand. In this paper, a sEMG feature optimization strategy for thumb-index finger grip force estimation is proposed with the purpose that using less features to achieve higher estimation accuracy. Four time-domain features are extracted from raw sEMG signals which captured from four muscle surfaces of the subject's forearm. GRNN is employed to realize the estimation of the finger grip force. RMS and MAE are adopted to validate the performance of estimation. The effects of different feature sets on the estimation performances are evaluated by ANOVA. The results show that sEMG features have a significant influence on the grip force estimation results. The optimal feature combination is VZ, which provides an accuracy of 1.13N RMS and 0.85N MAE.

Keywords: Finger grip force · sEMG · GRNN · ANOVA

1 Introduction

Prosthetic hands have irreplaceable significance for the upper limb amputees. It can not only decorate their missing limbs in appearance, but also restore body functions to a certain extent. Nowadays, there are several surface electromyography(sEMG) based pattern recognition systems investigated for prosthetic hands control [1–4]. These systems capture sEMG from amputees' residual arm, recognize the movement patterns and control the movements of the prosthetic hands. However, in the dexterous control of a prosthetic hand, not only gestures but also finger grip force is critical.

In numerous studies, the process of handgrip force estimation based on sEMG mainly involves signal collection, signal pre-processing and regression. Yokoyama et al. conducted the research of hand grip force estimation based on the neural networks with

© Springer Nature Switzerland AG 2020
C. S. Chan et al. (Eds.): ICIRA 2020, LNAI 12595, pp. 184–194, 2020.
https://doi.org/10.1007/978-3-030-66645-3_16

different number of hidden layers and the four channels of sEMG captured from dorsal interosseous muscles. In their experiment, subjects were asked to grip a dynamometer with six different force levels as the targets. And the root mean square of the sEMG was extracted and fed into the neural network to realize the force prediction. The correlations between predicted data and observed data were 0.840, 0.770, and 0.789 in the intrasession, intrasubject, and intersubject evaluations, respectively [5]. Baldacchino et al. proposed a simultaneous force regression strategy based on unified bayesian framework. The strategy was tested by using data from the publicly released NinaPro database. And four features were extracted from the sEMG and applied to the fore prediction respectively [6]. Cao et al. investigated handgrip force prediction based on sEMG of forearm muscles and extreme learning machine (ELM). The results of ELM were compared with support vector machine (SVM) and multiple nonlinear regression. Their results demonstrate that ELM possessed a relatively good accuracy and consumed short time, although SVM was effective for handgrip force estimation in terms of accuracy [7]. Yang et al. used three methods, local weighted projection regression, artificial neural network (ANN) and SVM, to find the best regression relationship between sEMG and grasp force. And their experiment results indicated that SVM performed best of the three methods [8, 9]. Huang et al. used 128 channels of sEMG and nonnegative matrix factorization algorithm to realize the estimation of isometric muscle force. And their method provided a way to find proper electrode placement for force estimation [10]. Kamavuako et al. investigated simultaneous and proportional estimation of force in two degree of freedoms from intramuscular electromyography. ANN was used to find the relationship between five EMG features and force. The research results show that correlation coefficients between the actual force and the estimated force was 0.88 ± 0.05 with post processing [11]. Yang et al. investigated handgrip force predictions by using gene expression programming (GEP) based on sEMG. 10 features of the sEMG time domain were extracted and used as the input of the GEP based model. The prediction results of the GEP model were compared with the back propagation neural network (BPNN) and SVM. And the comparative results show that the GEP model provided the highest accuracy and generalization capability among the studied models [12].

The studies mentioned above used different regression methods and sEMG features to realize the hand force prediction. However, most of them did not consider the influence of combination of features and number of features on the force prediction results. These factors are related to the computational burden of the prediction algorithm. For real-time systems and embedded systems with limited processing capacity, these factors cannot be ignored.

In this paper, grip force of thumb-index finger is estimated based on sEMG and general regression neural network (GRNN). The root mean square (RMS), mean of absolute error (MAE) are employed to evaluate the performance of the grip force estimation. The influences of the number of features and feature combination on the results are analyzed. Meanwhile we optimize the combination of feature sets by using analysis of variance (ANOVA) and Tukey HSD testing.

The rest paper is as follows. Section 2 introduces the materials and methods. Section 3 describes and analyses the results in detail. The conclusions are given in the Sect. 4.

2   Materials and Methods

2.1   System Structure and Experiment Platform

Figure 1 shows the structure of the finger grip force estimation based on sEMG and neural network. In the procedure of model training, finger grip force and sEMG are captured simultaneously. Then, several features are extracted from the sEMG. Finally, the captured finger grip force and sEMG features are feed into the neural network for the purpose of network training. In the procedure of grip force estimation, only features extracted from the raw sEMG are feed into the trained neural network. And the neural network will output the estimated finger grip force.

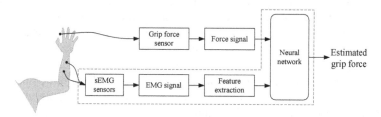

Fig. 1. The structure of the grip force estimation

The experiment platform set up in this paper consists of a four-channel sEMG sensor, a finger grip force sensor, a data collector and a C# based computer software. The output signals of the sEMG sensor and the finger grip force sensor are digitalized by a data collector (USB5936). The resolution of USB5936 is 12 bit. In this paper, the sample rate is set to 1 kHz per-channel. USB5936 has a USB interface through which data can be exchanged between the data collector and computer. Figure 2 shows the four-channel sEMG sensor and the finger grip force sensor. The pass-band and voltage gain of the sEMG sensor are 10–500Hz and 1000, respectively. The resolution of the grip force sensor which consists of a thin film pressure sensor (FSR400) and two convex platforms is 0.1N.

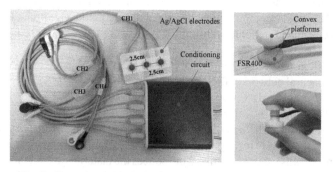

Fig. 2. Four-channel sEMG sensor and finger grip force sensor

2.2 Data Acquisition

Six healthy subjects, four males and two females (mean age = 23.2 years, S.D. = 2.8 years), took part voluntarily in the experiments.

Figure 3(a) shows the experiment scene and the measurement positions of the sEMG. The measurement positions of four sEMG sensors are brachioradialis, extensor digitorum, flexor ulnar carpal and thumb extensor, respectively. The paste diagram of sEMG sensors is shown in Fig. 3(b). In the experiment, subject quickly applies force to the grip force sensor through thumb and index finger and then relaxes fingers. During this process, sEMG and grip force are recorded synchronously. Figure 4 shows the sEMG signals and the finger grip force when a subject grasps the grip sensor five times continuously.

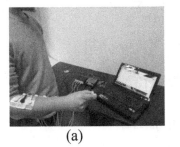 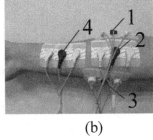

(a)   (b)

Fig. 3. Experimental scene and distribution of the measurement positions. (a) Experimental scene; (b) Measurement positions of the sEMG signals.

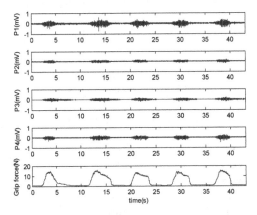

Fig. 4. Four channels of sEMG and finger grip force.

2.3 Feature Extraction

As shown in Table 1, four time domain sEMG features: variance, zero crossing, Willison amplitude and slope sign changes are extracted from four channels of raw sEMG signals. In the table, x_i represents the current data of the sEMG signal, x_{i-n} represents the previous nth data of the sEMG signal, N is the sampling number in the sliding window. As shown in Table 2, combining the four sEMG features, we can get 15 different feature sets which can be used for grip force estimation.

Table 1. The EMG features in time domain and its mathematical equation

Feature	Mathematical equation
Variance of sEMG (VAR)	$VAR_i = \frac{1}{N} \sum_{j=i-N+1}^{i} x_j^2$
Zero crossing (ZC)	$ZC_i = \sum_{j=i-N+1}^{i} \text{sgn}(x_j x_{j-1})$ $\text{sgn}(x) = \begin{cases} 1, x < 0 \\ 0, x \geq 0 \end{cases}$
Willison amplitude (WAMP)	$WAMP_i = \sum_{j=i-N+1}^{i} f(x_j - x_{j-1})$ $f(x) = \begin{cases} 1, |x| > threshold \\ 0, \quad otherwise \end{cases}$
Slope Sign Changes (SSC)	$SSC_i = \sum_{j=i-N+1}^{i-1} g((x_i - x_{i-1})(x_i - x_{i+1}))$ $g(x) = \begin{cases} 1, \quad x \geq threshold \\ 0, \quad\quad otherwise \end{cases}$

Table 2. sEMG feature sets

Label of the feature set	Combination of sEMG features	Label of the feature set	Combination of sEMG features
V	VAR	ZS	ZC + SSC
Z	ZC	WS	WAMP + SSC
W	WAMP	VZW	VAR + ZC + WAMP
S	SSC	VZS	VAR + ZC + SSC
VZ	VAR + ZC	VWS	VAR + WAMP + SSC
VW	VAR + WAMP	ZWS	ZC + WAMP + SSC
VS	VAR + SSC	VZWS	VAR + ZC + WAMP + SSC
ZW	ZC + WAMP		

2.4   Regression Method

In this paper, GRNN which with good nonlinear mapping ability is employed to estimate the grip force. As shown in Fig. 5, the network consists of input layer, pattern layer, sum layer and output layer. The input dimension of the network is determined by the number of the features in the selected feature set. In the procedure of the network training, 80% of the data are randomly selected for network training, and the rest 20% are used for network testing. The data of each feature set are used to train the GRNN for 15 times.

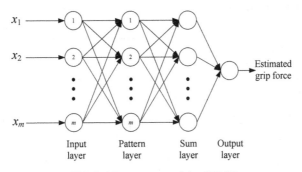

Fig. 5. The structure of the GRNN

2.5   Performance Evaluation

RMS and MAE are employed to evaluate the performance of the grip force estimation. A greater MAE indicates that the estimated result is poorer. And a greater RMS indicates that the estimated result has a larger fluctuation.

$$RMS = \sqrt{\frac{\sum_{i=1}^{N}(\tilde{f}_i - f_i)^2}{N-1}}, \tag{1}$$

$$MAE = \frac{1}{N}\sum_{i=1}^{N}\left|\tilde{f}_i - f_i\right|, \tag{2}$$

where, $\tilde{f}_i$ is the estimate of force; f_i is the force meassured by the grip force sensor; N is the length of the data.

2.6   Feature Optimization Strategy

One-way ANOVA is employed to analyze the results which estimated from 15 different feature sets. The obviousness of different combinations on grip force estimated results can be obtained by p-value of the ANOVA. Feature sets with a similar effect on estimation accuracy are grouped into one subset by performing Tukey HSD testing. The optimal feature set is the one that contains the least number of features in the subset in which the highest estimation accuracy is obtained. In this paper, ANOVA and Tukey HSD testing are conducted in IBM SPSS 19.

3   Experiment Results

Figure 6 shows the grip force estimation results when different feature sets are used. And Fig. 7 shows the grip force estimation results when different number of features is used. The estimation results vary from one feature set to another. The average estimation accuracy increases as the number of features increases. And the fluctuation of the results decreases with the increase of the number of features. In some case, using feature set which with less features, such as VZ, may get an equivalent result which obtained from feature set with more features.

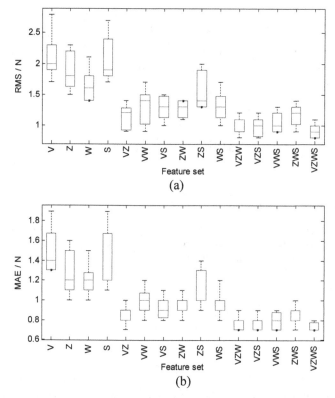

Fig. 6. Estimated results by using different feature sets. (a) RMS of the estimation error; (b) MAE of the estimation error.

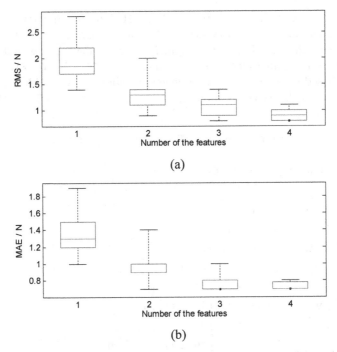

Fig. 7. Estimated results by using different number of features. (a) RMS of the estimation error; (b) MAE of the estimation error.

As shown in Table 3, the results of ANOVA (p-value < 0.05) indicate that in the both evaluation factors, RMS and MAE, the sEMG feature set has a significant effect on grip force estimation results.

Table 3. The results of One-way ANOVA

		Sum of squares	df	Mean square	F	Sig.
RMS	Between groups	31.667	14	2.262	43.659	0.000
	Within groups	10.880	210	0.052		
	Total	42.547	224			
MAE	Between groups	11.711	14	0.836	41.017	0.000
	Within groups	4.283	210	0.020		
	Total	15.994	224			

Table 4 and Table 5 show the results of Tukey HSD testing. In the result of RMS, the 15 feature sets are divided into 9 subsets. And the 15 feature sets are divided into 8 subsets in the result of MAE. For both evaluation factors, the subset with the highest estimation accuracy contains the same six feature sets (VZWS, VZS, VZW, VWS, VZ and ZWS). Based on the optimization principle mentioned above, VZ is the optimal feature set, which exhibits a estiamtion accuracy of 1.13N RMS and 0.85N MAE.

Table 4. The results of Tukey HSD testing for RMS

Festure set	Subset (alpha = 0.05)								
	1	2	3	4	5	6	7	8	9
VZWS	0.91								
VZS	1.00	1.00							
VZW	1.03	1.03	1.03						
VWS	1.05	1.05	1.05	1.05					
VZ	1.13	1.13	1.13	1.13	1.13				
ZWS	1.17	1.17	1.17	1.17	1.17				
ZW		1.27	1.27	1.27	1.27	1.27			
VS		1.29	1.29	1.29	1.29				
WS			1.33	1.33	1.33				
VW				1.33	1.33				
ZS						1.55	1.55		
W							1.64	1.64	
Z								1.87	1.87
S									2.09
V									2.12
Sig.	.144	.076	.118	.060	.513	.060	.999	.251	.175

Table 5. The results of Tukey HSD testing for MAE

Festure set	Subset (alpha = 0.05)							
	1	2	3	4	5	6	7	8
VZWS	0.7267							
VZS	0.7533	0.7533						
VWS	0.7800	0.7800	.7800					
VZW	0.7800	.7800	.7800					
ZWS	0.8467	.8467	.8467	.8467				

(*continued*)

Table 5. (*continued*)

Festure set	Subset (alpha = 0.05)							
	1	2	3	4	5	6	7	8
VZ	0.8533	.8533	.8533	.8533				
ZW		.9200	.9200	.9200	.9200			
VS		.9267	.9267	.9267	.9267			
WS			.9467	.9467	.9467			
VW				.9867	.9867			
ZS					1.0867	1.0867		
W						1.1933		
Z						1.2533	1.2533	
S							1.3867	1.3867
V								1.4800
Sig.	.496	.069	.098	.322	.098	.098	.405	.900

4 Conclusion

In this paper, a strategy based on ANOVA and GRNN for optimizing sEMG features is proposed to deal with the problem that the increase in the number features adds computational burden to grip force estimation based on sEMG and GRNN. Four time domain features are extracted from the four channels of sEMG signals. Fifteen different feature sets are used for grip force estimation respectively. And RMS, MAE are employed to evaluate the performance of grip force estimation. ANOVA and Tukey HSD testing are conducted to analysis the results. Our experimental results indicate that sEMG features have a significant influence on the estimation results. The optimal feature set for grip force estimation is VZ.

For future work, we will investigate the optimization of sEMG measurement positions for hand output force estimation.

Acknowledgement. This paper is supported by the National Natural Science Foundation of China under Grant No. 61803201, 91648206. Jiangsu Natural Science Foundation under Grant No.BK20170803. The China Postdoctoral Science Foundation under Grant No.2019M661686.

References

1. Khezri, M., Jahed, M.: Real-time intelligent pattern recognition algorithm for surface EMG signals. Biomed. Eng. Online **6**(1), 45 (2007)
2. Masood-ur-Rehman, A.K.K., Kasi, J.K., Bokhari, M., et al.: Design and development of sEMG prosthetics for recovering amputation of the human hand. Pure Appl. Biol. (PAB) **8**(3), 1935–1942 (2019)

3. Wu, C., Song, A., Ling, Y., et al.: A control strategy with tactile perception feedback for EMG prosthetic hand. J. Sens. **2015**, 869175 (2015)
4. Parajuli, N., Sreenivasan, N., Bifulco, P., et al.: Real-time EMG based pattern recognition control for hand prostheses: a review on existing methods, challenges and future implementation. Sensors **19**(20), 4596 (2019)
5. Yokoyama, M., Koyama, R., Yanagisawa, M.: An evaluation of hand-force prediction using artificial neural-network regression models of surface EMG signals for handwear devices. J. Sens. **2017**, 3980906 (2017)
6. Baldacchino, T., Jacobs, W.R., Anderson, S.R., et al.: Simultaneous force regression and movement classification of fingers via surface EMG within a unified Bayesian framework. Front. Bioeng. Biotechnol. **6**, 13 (2018)
7. Cao, H., Sun, S., Zhang, K.: Modified EMG-based handgrip force prediction using extreme learning machine. Soft. Comput. **21**(2), 491–500 (2017)
8. Yang, D., Zhao, J., Gu, Y., et al.: EMG pattern recognition and grasping force estimation: Improvement to the myocontrol of multi-DOF prosthetic hands. In: 2009 IEEE/RSJ International Conference on Intelligent Robots and Systems, pp. 516–521. IEEE (2009)
9. Yang, D., Zhao, J., Gu, Y., et al.: Estimation of hand grasp force based on forearm surface EMG. In: 2009 International Conference on Mechatronics and Automation, pp. 1795–1799. IEEE (2009)
10. Huang, C., Chen, X., Cao, S., et al.: An isometric muscle force estimation framework based on a high-density surface EMG array and an NMF algorithm. J. Neural Eng. **14**(4), 046005 (2017)
11. Kamavuako, E.N., Englehart, K.B., Jensen, W., et al.: Simultaneous and proportional force estimation in multiple degrees of freedom from intramuscular EMG[J]. IEEE Trans. Biomed. Eng. **59**(7), 1804–1807 (2012)
12. Yang, Z., Chen, Y., Tang, Z., et al.: Surface EMG based handgrip force predictions using gene expression programming. Neurocomputing **207**, 568–579 (2016)

Master-Slave Control of a Bio-Inspired Biped Climbing Robot

Ting Lei[1,2], Haibin Wei[3], Yu Zhong[3], Liqiang Zhong[1,2], Xiaoye Zhang[1,2], and Yisheng Guan[3(✉)]

[1] Electric Power Research Institute of Guangdong Power Grid Co., Ltd, Guangzhou, China
[2] Guangdong Diankeyuan Energy Technology Co., Ltd, Guangzhou, China
[3] School of Electro-Mechanical Engineering, Guangdong University of Technology, Guangzhou, China
ysguan@gdut.edu.cn

Abstract. Biped climbing robot is able not only to climb a variety of media, but also to grasp and manipulate objects. It is just like a "mobile" manipulator, and it has great application prospects in high-rise tasks in agriculture, forestry, and architecture fields. Motivated by these potential applications, design of a climbing robot (Climbot) was proposed in this study, and a modular master robot, isomorphic to the Climbot was designed. Then, a master–slave robot system with joint-to-joint mapping strategy was developed, which was com-bined with the Climbot and master robot. In this system, the master robot could control the slave robot intuitively, and experiments on climbing poles were conducted to verify the feasibility and efficiency of the proposed master–slave robot system.

Keywords: Master-slave control · Climbing robot · Tele-operation · Master robot

1 Introduction

Robots have wide range of potential application prospects in numerous fields such as industry, military, medicine, education, entertainment, home, and social services. With the continuous development of robotics, robots are expected to carry out high-rise works (such as picking up fruits, trimming or cutting branches, truss detection and maintenance, high altitude reconnaissance, etc.) in agriculture, forestry, and architecture fields, for human being. To this end, many climbing robots (Climbots) are being developed at the global scale. A prototype of "Treebot" with high maneuverability was presented. It is composed of a pair of omni-directional tree grippers for holding the robot on a tree surface and a novel three degrees of freedom (3-DOF) continuum manipulator for maneuvering [1]. Moreover, in order to inspect the pipes for defects or damage in a nuclear power plant, a climbing robot has been developed, which has a five-degree-of-freedom manipulator and two grippers, it can climb up and down to cross over such pipes,

Y. Guan–The work in this paper is supported in part by the Science and Technology Project of China Southern Power Grid Co., Ltd. (Grant No. GDKJXM20173031).

and able to overcome obsta-cles such as valves, pipe flanges, and T-shaped branches [2]. Moreover, inspired by the climbing motion of inchworms, a biped wall-climbing robot— W-Climbot was developed, which consists of five joint modules connected in series and two suction modules mounted at the two ends [3]. Furthermore, there are many other climbing robots, such as a bio-inspired climbing robot with flexible pads and claws [4], a climbing robot attached to the wall based on the principle of vacuum [5], and a wall-climbing robot that uses magnetic adsorption [6], etc.

Most of the researches on climbing robot are focused on the mechanical structure design and gait analysis, and many excellent robot prototypes have been created, but in terms of efficient control of climbing robots has few attention. In recent years, much attention also has been paid to global path planning, single-step motion planning, autonomous grasping or adsorption of climbing robots. A novel global path rapid determination approach is proposed by Zhu in 2018, this scheme is efficient for finding feasible routes with respect to the overall structural environment [7]. Yang studied a stepping path planning approach for a climbing robot in 2016 [8]. In 2018, Kim presented a vision-based scheme for grasping a cylindrical pipe semi-autonomously [2]. Qiang Zhou proposed a localization scheme based on convolutional neural network. The light local wheel detector can quickly and accurately detect the four wheel points of a remote wall-climbing robot, and use the detected wheel points to calculate the position and direction angle of the robot [9]. The purpose of the above researchers is to achieve autonomous climbing of climbing robot. But limited by the factors of stability, reliability, accuracy and efficiency, it is still a challenging work to achieve autonomous climbing of climbing robot in the unstructured environment.

The human brain is a perfect combination of perception, analysis, and decision-making, it is still a good choice for the human to operate the robot to complete climbing and operation through teleoperation technology at present. In the previous research, we used a joystick (Saitek Cyborg X) to control the climbing robot [10]; however, the operation was not intuitive. To this end, this study aims to develop a modular, isomorphic master–slave robot system, to improve the operation intuitively, the efficiency of robot movement, and environmental adaptability.

This paper is organized as follows: An introduction to the Climbot is presented in Sect. 2. Design of a master robot, which is used to control the Climbot, is presented in Sect. 3. The prototype of the master–slave robot is described in Sect. 4, while the experiments are summarized in Sect. 5. Finally, conclusions are given in Sect. 6.

2 The Biped Climbing Robot: Climbot

The climbot, assembled with three basic modules, was developed via a modular approach. Figure 1 exhibits that the basic modules include T-type and I-type joint modules and a gripper module. The revolute axis of the T-type joint module is perpendicular to the link axes; and the rotational axis of the I-type joint is collin-ear with the link axes. Moreover, the gripper module is used in Climbot for both supporting the robot when it climbs and for grasping objects when it manipulates.

Further, the Climbot is composed of three T-type joint modules, two I-type joint modules and two grippers, as shown in Fig. 2. All modules are connected in series.

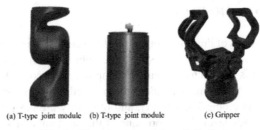

(a) T-type joint module   (b) T-type joint module   (c) Gripper

Fig. 1. Three basic modules of climbot

Three T-type modules are in the middle with their axes parallel, two I-type modules are at the two ends, and two grippers are then connected to each of the I-type modules, with the following kinematic chain,

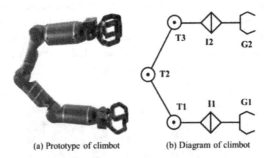

(a) Prototype of climbot   (b) Diagram of climbot

Fig. 2. Climbing robot-climbot

$$G_1 - I_1 \perp T_1 \parallel T_2 \parallel T_3 \perp I_2 - G_1 \tag{1}$$

where G1 and G2 represent the two gripper modules, I1 and I2 stand for the two I-type joint modules, T1, T2 and T3 for the three T-type joint modules, $\parallel$ and $\perp$ mean the parallel and perpendicular relationship between the joint module axes, respectively. More information about this robot system can be found in the literature study [1].

Compared to other robotic systems, the biggest advantage of this robot is that it can climb to different positions in the environment to perform different tasks, in other words, it integrates functions of both climbing and manipulation. When one of the gripper firmly grasps an object to support the robot (this end is fixed as the robot base), the other can be used to grasp objects in manipulation tasks, or the other end of the robot moves with the swinging gripper grasps to a target position, then the former gripper is released and moves to a next position so that the movement of this Climbot is realized by two grippers interchanging their roles. Moreover, the robot has three basic climbing gaits, including the inchworm gait, swinging-around gait, and flipping-over gait, more details can be found in literature study [2]. The perfect climbing ability is accomplished by two grippers interchanging their roles; however, the problem that needs to be solved is to use flexible posture to avoid obstacles in the path and make corresponding actions accurately when climbing.

3 Design of Master Robot

The potential applications of this Climbot include high-rise tasks in agriculture, forestry, or building industry, requiring robots possessing climbing function. The working environment of this robot is unstructured, thus it is impossible to control the movement of this Climbot through the program. The ideal control mode is the autonomous motion of this robot; however, its efficiency, stability, and reliability are not guaranteed because of the current technology level. Therefore, one of the most suitable control methods is master–slave teleoperation, the slave robot (Climbot) only needs to receive the control signals from the master robot to perform the motion and action.

In order to develop a master–slave robot system, a master robot is needed which is used as a human interface device to control the slave robot. In terms of structure, the master robot has two types: isomorphic and non-isomorphic compared to the slave robot. The isomorphic master–slave robot system has the advantages of intuitive operation and simple mapping; therefore, the isomorphic master robot is designed by modular method as well. Figure 3 demonstrates the design of two types (Imas module and Tmas module) of basic 1-DoF master robot joint modules corresponding to those of slave joint modules, and the isomorphic master robot can be easily assembled according to the specific slave robot using the above mentioned modules (Fig. 4).

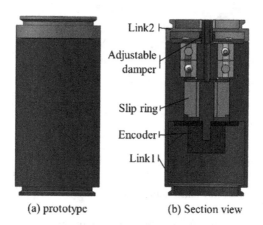

(a) prototype   (b) Section view

Fig. 3. Imas joint module of master robot

In general, the two types of master joint modules have similar structures. First, one encoder of 18 M series (1024P/R, $\varphi 18 \times 15$ mm) by NEMICON was employed in each master joint module, which was used to record the motion of master robot joint module itself driven by a human operator. Besides, a special adjustable damping structure was designed. Its working principle is presented in Fig. 5, the magnitude of damping between the two links can be adjusted by adjusting the pressure of the gasket for preloading in the bolt connection. Therefore, the neutral balance of the master robot can be achieved by the adjustable damping, which can reduce the difficulty of operation, reduce misoperation, and improve human-computer friendliness. In particular, a slip ring part is also used in the Imas module, which enables the module to rotate infinitely.

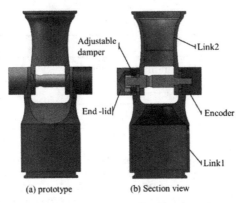

Fig. 4. Tmas joint module of master robot

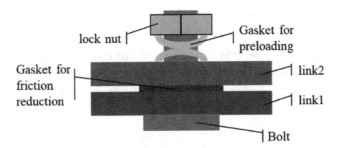

Fig. 5. Sketch of adjustable damping structure

The diameters of the master joint modules are about 40 mm. Their lengths are proportional to those of the corresponding slave modules with a ratio of 1:2.14, which is important for isomorphism. The main parts of the modules are made of aluminum alloy. The weights of the Tmas and Imas master joint modules are about 170 and 135 g, respectively. The rotation of the Imas joint has no limits, and the rotational range of the Tmas joint is up to 120, which are the same as the ranges of slave joints.

4 Development of Master-Slave System

In the master–slave control system, the operator controls the master robot, and the posture information of the robot is processed on the PC side, and then trans-mitted to the slave robot. Figure 5 shows the logic of the master–slave control system

A literature report [7] presents a more detailed analysis of the kinematics at the position level of Climbot, and when the method of master–slave control is used to control the climbing of the robot, the relationship between the spatial position of the master robot and the control feasibility needs to be analyzed.

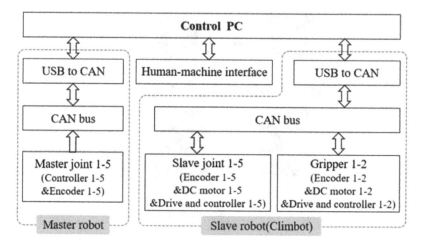

Fig. 6. The hardware architecture

There are two types, one is to fix the base of one end of the master robot on the platform, the platform can rotate around the X-axis, Y-axis, and Z-axis of the coordinate system; and the second is to arrange multiple base fixing brackets in three-dimensional space. The main robot can flexibly adjust its position in space. The specific structure layout is shown in the figure.

In this system, the master and slave robots are connected to the control PC as two relatively independent subsystems. The motion information of the main robot is read by the encoder and preprocess by the controller in the joint module, then transmitted it to the control PC through the CAN bus as the control signal of the slave robot. In the subsystems of slave robot, joint module and gripper read/write data and signals from/to control PC through CAN bus. The human-machine interface is used to visualize the data from the master robot and slave robot, it also has some control functions, such as controlling the action of the gripper. The hardware architecture shown in Fig. 6.

5   Climbing Analysis and Mapping Strategy

Compared to other robots, the biggest advantage and characteristic of this Climbot is that it can be moved in the working environment by gripping the pole with the two grippers. The flexibility of the Climbot is improved significantly; however, it is more difficult to control the robot. When the Climbot climbs or performs some tasks, one of the grippers must hold the pole firmly as the base of the Climbot all the time. In particular, the body of the Climbot is in the same plane at any time, as shown in Fig. 7(a), which is beneficial to the intuition of master–slave operation. Therefore, a simple mapping strategy with one-to-one mapping can be used in the isomorphic master–slave robot system. In other words, the motion of the corresponding joint of the slave robot is directly controlled by the joint of the master robot. The corresponding relationship between the joints of the master and the slave robot can be described as follows:

$$M : \left[\theta_1', \theta_2', \theta_3', \theta_4', \theta_5'\right] \rightarrow S : [\theta_1, \theta_2, \theta_3, \theta_4, \theta_5] \tag{2}$$

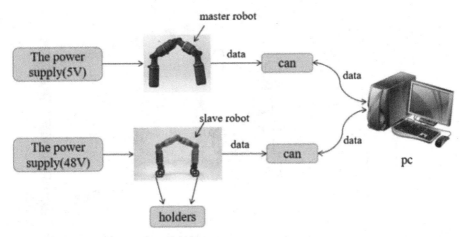

Fig. 7. Prototype of master-slave robot system

where θ'_x and θ_x represent the joint position of the master and slave robots respectively. In this mapping strategy, the slave robot will always follow the motion and pose of the master robot. A mapping pose of the master and slave robots at a certain time is shown in Fig. 8.

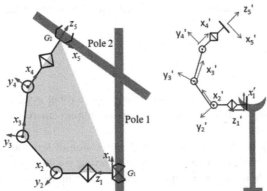

(a) The diagram of the slave robot   (b) The diagram of the master robot

Fig. 8. A mapping pose of the master and slave robots

The two claws of the climbing robot are used alternately; therefore, if one end of the master robot is fastened to the base, the mapping of master–slave control becomes a problem worth considering. In order to make the master–slave control more intuitive and feasible, the end base of the master robot can be temporarily fixed on the fixture. When the climbing robot changes the gripper, the base of the main robot is also changed. This ensures the spatial freedom of the master robot and the intuitiveness of master–slave control.

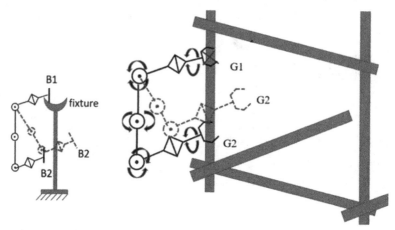

(a)The control process of master robot   (b)Climbing process of the slave robot

Fig. 9. The climbing analysis of the master-slave robot system

In Fig. 8, the dashed lines indicate the posture changes of the master robot and the slave robot. When the gripper G2 grabs the rod 2, the gripper G1 is released. Correspondingly, the B2 of the master robot is placed at the fixture, and B1 is released, and the climbing action can continue to be completed with the isomor-phic master–slave mapping.

6   Experiments

To verify the feasibility and efficiency of the master-slave robot system, we carry out experiments with prototype. To this end, we set up a climbing environment using steel pipes of 48 mm in diameter.

Before the experiment, the initialization and calibration of the master-slave robot system are completed. First, the Climbot is in the starting position with two grippers grasp the vertical pole, then the upper gripper opens to climb up with the inchworm gait, as shown in Fig. 9. Next, transit to a horizontal pole with the swinging-around gait, and finally climb along the horizontal pole with the flipping-over gait, as shown in Figs. 10 and 11 respectively. In the whole experiment, the movement of the Climbot is controlled by the master robot, and the action of the gripper is controlled by the human-machine interface on the control PC (Fig. 12).

In the experiment, the Climbot climbs from the vertical pole to the horizontal pole in three different gaits, and the whole experiment can be completed within 9 min, therefore the master-slave robot system has good control efficiency. The designed master robot, mapping strategy and control method in this paper was also be verified to be reasonable and feasible in the experiment.

Fig. 10. Climbot climbing experiment with the inchworn gait

Fig. 11. Climbot climbing experiment with the swinging-around gait

Fig. 12. Climbot climbing experiment with the flipping-over gait

7 Conclusions

In this study, a modular and isomorphic master-slave robot system is presented, it is mainly composed of climbing robot, master robot and control PC. The climbing robot is able not only to climb a variety of media, but also has grippers to grasp and manipulate objects, it just like a "mobile" manipulator. The modular master robot which is isomorphic to the Climbot, it is used to control the slave robot. One end of the master robot is fixed on the desktop to facilitate the operation of operator, but it can high effectively and intuitively control the climbing robot which has two exchangeable base (either of the two gripper which end of the Climbot can grasps an object firmly to support itself as base). Control PC is a data receiving and processing, control planning and signal sending unit, and it has a human-machine interface to visualize the received data and realize the action control of the gripper. In this master-slave robot system with joint-to-joint mapping strategy, numerous experiments have been conducted. Experimental results reveal that it is a good way to control the Climbot to complete climbing movement and operation at present.

References

1. Lam, T., Xu, Y.: A flexible tree climbing robot: Treebot-design and implementation. In: 2011 IEEE International Conference on Robotics and Automation, pp. 5849–5854 (2011)
2. Guan, Y., et al.: Climbot: A modular bio-inspired biped climbing robot. In: 2011 IEEE/RSJ International Conference on Intelligent Robots and Systems, pp. 1473–1478 (2011)
3. Zhu, H., et al.: The superior mobility and function of W-Climbot—A bio-inspired modular biped wall-climbing robot. In: 2011 IEEE International Conference on Robotics and Biomimetics, pp. 509–514 (2011)
4. Ji, A., Zhao, Z., Manoonpong, P., Wang, W., Chen, G., Dai, Z.: A bio-inspired climbing robot with flexible pads and claws. J. Bionic Eng. **15**(2), 368–378 (2018)
5. Chen, X., Jin, Z., Liao, S., He, K.: Design of a vacuum adsorption wall climbing robot. DEStech Transactions on Engineering and Technology Research, amee (2019)
6. Zhang, Q., Xue, K., Luo, Y.: Structure design and research of magnetic attraction wall climbing robot. Intern. Core J. Eng. **6**(4), 218–224 (2020)
7. Zhu, H., Gu, S., He, L., Guan, Y., Zhang, H.: Transition analysis and its application to global path determination for a biped climbing robot. Appl. Sci. **8**(1), 122 (2018)
8. Yang, C., Paul, G., Ward, P., Liu, D.: A path planning approach via task-objective pose selection with application to an inchworm-inspired climbing robot. In: 2016 IEEE International Conference on Advanced Intelligent Mechatronics, AIM, pp. 401–406 (2016)
9. Zhou, Q., Li, X.: Visual positioning of distant wall-climbing robots using convolutional neural networks. J. Intell. Rob. Syst. **98**(3), 603–613 (2020)
10. Cai, C., Guan, Y., Zhou, X., et al.: Joystick control of two-handed claw bionic climbing robot. Robotics **34**(003), 363–368 (2012)

Mobile Robots and Intelligent Autonomous System

Leveraging Blockchain for Spoof-Resilient Robot Networks

Tauhidul Alam[1($\boxtimes$)], Jarrett Taylor[1], Jonathan Taylor[1], Shahriar Badsha[2], Abdur Rahman Bin Shahid[3], and A.S.M. Kayes[4]

[1] Department of Computer Science, Louisiana State University Shreveport, Shreveport, LA 71115, USA
{talam,taylorj48,taylorj6975}@lsus.edu
[2] Department of Computer Science and Engineering, University of Nevada, Reno, NV 89557, USA
sbadsha@unr.edu
[3] Department of Computer Science, Concord University, Athens, WV 24712, USA
ashahid@concord.edu
[4] Department of Computer Science and Information Technology, La Trobe University, Bundoora, Victoria 3086, Australia
a.kayes@latrobe.edu.au

Abstract. Autonomous robots, such as unmanned aerial or ground robots, are vulnerable to cyber attacks since they use sensor data heavily for their path planning and control. Furthermore, consensus is critical for resilient coordination and communication of robots in multi-robot networks against a specific adversarial attack called the spoofing attack, where robots can be compromised by an adversary. Therefore, we leverage Blockchain in a network of robots to coordinate their path planning and present a consensus method utilizing their transferred Blockchain data to detect compromised robots. Our simulation results corroborate the fact that the proposed method enhances the resilience of a robot network by detecting its spoofed client robots or compromised server at a significant rate during the spoofing attack.

Keywords: Robot networks · Spoofing attack · Blockchain · Communication

1 Introduction

Robot networks have made a notable impact in several applications such as drone delivery, infrastructure inspections, disaster information gathering, agriculture precision, border and area surveillance, and search and rescue operations. An example application of robot networks is Wing's drones [1] certified by the Federal Aviation Administration (FAA) in the U.S. for the first time that deliver small packages, including food, medicine, and household items, directly to homes in minutes following flight paths. In practice, these robots in a network utilize wireless communication sensors for their path planning, coordination, control,

© Springer Nature Switzerland AG 2020
C. S. Chan et al. (Eds.): ICIRA 2020, LNAI 12595, pp. 207–216, 2020.
https://doi.org/10.1007/978-3-030-66645-3_18

and collision avoidance. However, malicious agents can jam or intercept these wireless sensors to gain access to the network that makes the robots vulnerable to cyber attacks and malicious traffic. In particular, a robot network can be disrupted by a spoofing attack which is also known as a Sybil attack [4]. An adversary can forge multiple spurious identities or impersonate several existing client robots in the network during the spoofing attack [13] after having complete control over GPS [16] or optical flow sensors [3].

Our work is motivated by the problem of defending a robot network against the spoofing attack, e.g.., a set of client robots is drawn away by an adversary from their service robot as illustrated in Fig. 1. Specifically, we consider a robot network in which a group of aerial delivery vehicles delivers packages launching from a central fulfillment or distribution station to designated customer (goal) locations and a server robot controls the operations of delivery vehicles. However, an adversary attempts to gain control of multiple delivery vehicles by spoofing their customer locations or compromising the server robot. This spoofing attack is easy to carry out but difficult to prevent in multi-robot settings. Consequently, our problem of safeguarding the network by detecting this attack is challenging.

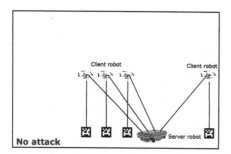 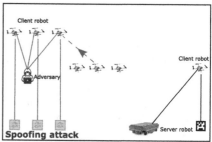

Fig. 1. Spoofing attack. A server robot controls pre-computed flight paths to client robots for their intended customer locations when no attack is present. In a spoofing attack, an adversary spoofs many client robots by drawing them away from their original paths and directing them toward pseudo-customer locations.

Our work is closely related to spoof-resilient solutions to multi-robot networks using information extracted from Wi-Fi communication signals for detecting spoofed client robots [7] and providing a consensus algorithm with bounded performance guarantees [6]. Consensus methods in mobile and distributed networks are also considered using transmitted values [14,15] to remove adversarial agents from the consensus and using exchanged keys [8] or tokens [9] to secure networks with different initial topologies [20]. However, these consensus methods are prone to failures when robots in a network fail or communicate incorrect messages. Additionally, a malicious agent can generate a number of false identities in a robot network that utilizes wireless signals for security instead of using a trusted system within the network. Unlike prior work, this method utilizes trusted Blockchain technology for the resilient coordination and communication of other robots in the network in the presence of malfunctioning and malicious robots (Byzantine robots).

The potential benefits of using Blockchain technology for addressing security issues in swarm robotic systems are discussed in [5]. The Nakamoto's white paper [10] first introduced Blockchain technology as a trusted database of encrypted and linked data transactions with timestamps stored by participating agents of a peer-to-peer network. There are very few approaches that exploit Blockchain technology for robotic security systems. A Blockchain-based collective decision making approach [17,19] for managing Byzantine robots in homogeneous robot swarms is presented. Blockchain has been utilized in heterogeneous robot swarms as well for collaboration in [12]. Existing consensus algorithms are compared with the Blockchain consensus algorithm in [18]. The pivotal advantages of using Blockchain are the immutability of transactions, decentralized consensus, fault tolerance, and so on. As such, this work leverages a permissioned or private Blockchain, where a centralized entity has control over its participants, to detect spoofed client robots or the compromised server through the validation by a devised committee of robots in the network.

Contributions: This paper makes the following contributions.

- A consensus method with a subset of robots in the committee making use of transferred data transactions on Blockchain for detecting compromised robots in the network.
- A simulation study that validates the performance of our method for different types of compromised robots in dealing with the spoofing attack.

The remainder of this paper is laid out as follows. First, we define several notations required for our robot network setting and formulate our problem of interest in Sect. 2. Then, we describe our method to detect the spoofing attack in the network in Sect. 3. The results from the implementation of our method appear next in Sect. 4. Finally, we summarize our paper along with future directions in Sect. 5.

2 Preliminaries

We examine a robot network setting in which m client delivery robots (drones) $\mathcal{D} = \{D_1, \ldots, D_m\}$ obtain computed flight paths $\mathcal{T} = \{\tau_1, \ldots, \tau_m\}$ for their product delivery from a server robot S located at a central distribution center or service station. These client delivery robots are identified by a set of identification keys which are denoted as $\mathcal{I} = \{i_1, \ldots, i_m\}$. They communicate their path information (location, velocity, time, distance, etc.) with the server robot through the identification keys $\mathcal{I}$. Let $P = \{p_1, \ldots, p_m\}$ denote the client delivery robots' locations in $\mathbb{R}^3$. Let $V = \{v_1, \ldots, v_m\}$ denote the client delivery robots' velocities in $\mathbb{R}$. Let $G = \{g_1, \ldots, g_m\}$ denote the client delivery robots' goal or customer locations in $\mathbb{R}^2$. Let p_s be the location of S in $\mathbb{R}^2$. Let a flight path of a client delivery robot D_k, where $k \in \{1, \ldots, m\}$, be $\tau_k : [0, t] \to \mathbb{R}^3$ such that $\tau_k(0) = p_s$ and $\tau_k(t) = g_k$ for a finite time interval $[0, t]$. We consider in this setting that either a subset of client delivery robots denoted by $\mathcal{A}$, where $\mathcal{A} \subset \mathcal{D}$, can be spoofed, or the server S is compromised by adversaries. It is assumed that an adversary sends various messages over the network with identification keys to make client delivery robots spoofed by taking control over their GPS sensors and that the knowledge of which client delivery robots are spoofed is unknown [6]. Furthermore, all the client delivery robots are considered to be spoofed when the server robot is compromised. In this context, we formulate the following problem of interest.

Problem 1 (*Detecting spoofed robots*): *Given m client delivery robots, the server robot S in a network, and their pre-computed paths $\mathcal{T}$, detect $\mathcal{A}$ spoofed client delivery robots or the compromised server S in the case of a spoofing attack.*

3   Methods

This section details our Blockchain leveraged consensus method for detecting the spoofing attack in a robot network.

In our method, we first construct a server robot, and then it establishes a network with m client delivery robots. The server robot provides the identification keys $\mathcal{I}$ to client delivery robots for communication and sends their pre-computed paths $\mathcal{T}$ toward their goal locations G.

In our next step, we employ a private Blockchain on the server robot in order to keep track of transferred data over the network. Client delivery robots communicate data related to their locations, velocities, distances, and time periodically with the server robot, and the server robot incorporates them into the transferred Blockchain data. The transferred Blockchain data is defined as B. We assume that client delivery robots act honestly in communicating their data.

Afterward, we develop a consensus Algorithm 1 in the robot network for detecting the spoofing attack. To achieve this, we devise a verification committee with the server robot and a subset of random client delivery robots. The server robot alone can also detect spoofed client delivery robots but cannot detect itself being compromised. Let $\mathcal{C} \subset \mathcal{D} \setminus \mathcal{A}$ be the subset of client delivery robots in the verification committee. It is considered that n client committee members, where $|\mathcal{C}| = n$ and $n < m$, can access the transferred Blockchain data B. We also consider that both the server and the client committee members are not compromised at the same time.

The verification committee members can vote in a weighted manner for detecting spoofed client delivery robots or a compromised server. Let w_c be the weight for each committee member, including the server robot. Let w_s be the weight for the server robot's vote and w_{cc} be the weight for each client committee robot's vote. The client committee robot's weight for voting is calculated as $w_{cc} = (1 - w_s)/n$. The weights for all committee members can be represented as a $(n + 1)$-dimensional vector as follows.

$$w = (w_1, \ldots, w_{n+1}) = (w_s, w_1, \ldots, w_n).$$

The elements of w should satisfy two conditions: 1) $w_c > 0$ for all $c \in \{1, \ldots, n+1\}$ and 2)

$$\sum_{c=1}^{n+1} w_c = w_s + \sum_{cc \in \mathcal{C}} w_{cc} = 1.$$

Let Q be the consensus trigger by the first committee member to start the consensus process. Let $\mathcal{L} = \{l_1, \ldots, l_m\}$ be the set of each committee member's votes for client delivery robots. Let π be the threshold value for the path deviation of a client delivery robot. In Algorithm 1, we calculate the set of votes for client delivery robots to determine one or more spoofed client delivery robots. We apply the consensus Algorithm 1 for each verification committee member, including the server robot. For each client delivery robot with an identification key i_k, we check its path deviation from the provided path τ_k by the server robot. Since we consider that a client delivery robot's GPS is spoofed, we make use of its locations data that are stored in Blockchain to find the path deviation. Thus, the PATHDEVIATION function takes the input of Blockchain data B, identification key i_k, and path τ_k with the location of a client delivery robot's original destination. The function iterates through B to compare the client delivery robot's current location to its goal location to determine if the client delivery robot is getting closer to its intended destination. Once determined, the function returns a value t between 0 and 1. If it is less than the threshold value π, the algorithm determines that the client delivery robot is deviating from its intended path. Then, we add $-w_c$ to the set of votes for that client delivery robot. Otherwise, we add w_c to the set for the same client delivery robot.

Algorithm 1: CONSENSUS $(Q, \mathcal{T}, \mathcal{I}, B, w_c)$

Input: $Q, \mathcal{T}, \mathcal{I}, B, w_c$ – Consensus trigger, Client delivery robots' paths, Set of identification keys for client delivery robots, Blockchain data, Weight of each committee member

Output: $\mathcal{L}$ – Set of votes for client delivery robots

1   **for** $k = 1$ **to** $|\mathcal{I}|$ **do**
2      $t \leftarrow$ PATHDEVIATION(B, i_k, τ_k)
3   **if** $t < \pi$ **then**
4         $\mathcal{L} \leftarrow \mathcal{L} \cup \{-w_c\}$
5   **else**
6         $\mathcal{L} \leftarrow \mathcal{L} \cup \{w_c\}$

7   **return** $\mathcal{L}$

In our final step, we investigate two spoofing attack scenarios: 1) client delivery robots are spoofed and 2) the server is compromised. The verification committee validates these attack scenarios utilizing the transferred Blockchain data

B by detecting the compromised server or spoofed client delivery robots. In both scenarios, one of the client committee members begins the consensus by voting for each client delivery robot and alerting the other committee members, including the server, to initiate their voting. Therefore, the rest of the committee members also provide their votes for each client delivery robot. Once a consensus is reached by all the committee members, the total votes are combined to a tally for each client delivery robot. If the vote tally for a client delivery robot less than zero, the associated client delivery robot is detected as spoofed.

For the first scenario, a set of client delivery robots D is launched with n client committee members (robots) by the server. A subset of client delivery robots A is spoofed and changes directions mid-way to their intended destinations. The committee members detect a spoofed client delivery robot through their consensus algorithm after it moves away from its intended destination. It is important to mention for this scenario that the server robot itself can also detect a spoofed client delivery robot without the verification of other client committee members as both the server and client committee members can utilize the transferred Blockchain data B for verification.

For the second scenario, the server is no longer considered as a committee member. The n client committee members are relied upon to complete the consensus using their accessible communicated Blockchain data B. Once the consensus is initiated and completed, the votes are tallied and checked. All the client delivery robots will be detected as spoofed because the compromised server automatically spoofs non-committee client members. Once all non-committee client members are detected as spoofed, the server will be detected as compromised.

4   Experimental Results

In this section, we present the results from the implementation of our method.

Initially, we implemented a robot network through server-client socket programming in Python 3 with the simulation of port numbers and identification keys for robots. The server robot provided flight paths to client delivery robots (drones). Then, we simulated our own private Blockchain in this network setting using Python. Client delivery drones followed the provided flight paths which were simulated taking advantage of a Python Robotics tool [2]. While client delivery drones following the simulated paths, they communicated their locations, velocities, covered distances, and time with the server robot. These communicated data were transferred over the network through our implemented Blockchain.

We also implemented our consensus Algorithm 1 in simulation. In our implementation, we accounted for $n = 2$ random client delivery drones and the server robot for devising the verification committee. These committee members employed transferred Blockchain data for validation of non-committee client delivery drones' path deviation using their locations along their flight paths. For the voting process of the verification committee members, the weights we utilized for the server robot and the client delivery drone are $w_s = 0.4$ and $w_{cc} = 0.3$

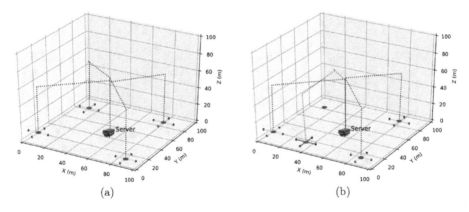

(a)   (b)

Fig. 2. Client delivery drone spoofing scenario. Flight paths of two client committee drones (depicted in cyan) toward their desired goal locations (green circles) and one non-committee drones (depicted in green) toward its desired goal location and another (depicted in red) toward its unintended goal location (red circle) starting from the server robot's location. Between two non-committee drones, the red non-committee drone is spoofed and detected. (Color figure online)

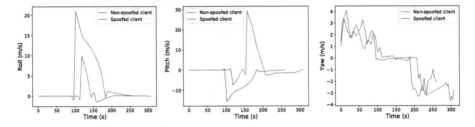

Fig. 3. Comparison of velocities and time for a non-spoofed client delivery drone and its spoofed counterpart after detection.

respectively. The threshold value $\pi = 0.9$ was used for finding the path deviation of a client delivery drone. Finally, the verification committee members recorded their votes for detecting spoofed client delivery drones.

Figure 2(a) delineates a drone delivery network setting, where $m = 4$ client delivery drones were launched from the server robot's location toward their green goal locations with $n = 2$ client committee drones marked in cyan and the remaining client delivery drones marked in green. Figure 2(b) presents the first spoofing attack scenario, where one of the non-committee client drones was attacked during its flight path execution and moved away from its intended goal location. In our implementation, the verification committee members detected the client delivery drone as spoofed based on their votes and turned it red, including its path and spoofed goal location. Figure 3 illustrates the variations of distinct velocities (roll, yaw, pitch) with respect to time for a non-spoofed client delivery drone and its detected spoofed counterpart. These results indi-

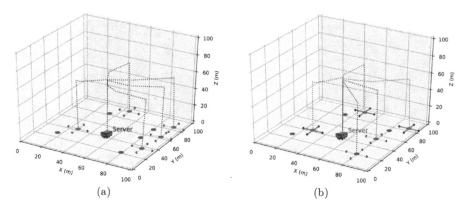

Fig. 4. Server robot compromising scenario. (a) Flight paths of five client delivery drones (depicted in green) starting from the server robot's location toward their spoofed goal locations (red circles) when the server robot is compromised and undetected. (b) Flight paths of two client committee drones (depicted in cyan) toward their desired goal locations (green circles) and three non-committee drones (depicted in red) toward their spoofed goal locations (red circles) starting from the server robot's location when it is detected that the server robot is compromised. (Color figure online)

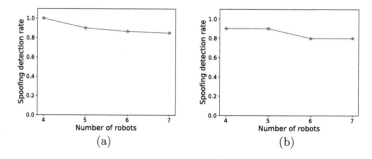

Fig. 5. Spoofing detection rate of our method for different numbers of robots in a network for the first scenario (a) and the second scenario (b).

cate the inconsistencies in velocities between an actual flight path and a deviated flight path of a client delivery drone when it is attacked.

Figure 4(a) shows a spoofing attack scenario where $m = 5$ drones were launched; however, the server was compromised and could not run its own consensus nor could any of the five drones launch their own consensus. Since the server was compromised, it redirected all non-committee drones to spoofed goal locations. Since no committee drones were present, this resulted in no drones being detected as spoofed. Figure 4(b) demonstrates the results of the second spoofing attack scenario, where five client delivery drones were launched but later the server was compromised. In this case, we converted $n = 2$ client delivery drones into committee members. As a result, the committee members were able to detect the remaining client delivery drones as spoofed. Since all non-committee drones were spoofed, it detected that the server was compromised which was depicted by coloring its name to red.

We computed the spoofing detection rate of our method for different numbers of robots, including the server robot, in a network for both scenarios which is illustrated in Fig. 5. The spoofing detection rate was computed from the average of 10 runs of our implementation for each number of robots for both scenarios. This result shows that our detection rate is significant but decreases slightly with the increase in the number of robots. The reason for this small detection rate decline is that the client committee members sometimes complete their flights or do not even start their flights while some non-committee client members are spoofed. This problem can be overcome by dynamically assigning client committee members that are on their flights to the verification committee.

5   Conclusion and Future Directions

In this paper, we presented a consensus method with a committee of robots in a network for detecting its spoofed client robots or compromised server utilizing transferred Blockchain data. Our simulation results demonstrate that our method makes a robot network resilient against the spoofing attack. We believe that we have just scratched the surface in leveraging Blockchain for detecting a cyber attack within a robot network. This effort paves the way for several interesting future research directions as detailed below.

In the future efforts of this stream of research, we will evaluate the vulnerabilities of our method by learning the characteristics of compromised robots by different attacks on a secure network using machine learning methods, and present solutions to these vulnerabilities to make our method more attack resilient. We also plan to test our method with a set of programmable drones as client robots and a ground vehicle as the server robot.

One potential problem of our method lies in storing transferred Blockchain data from a group of robots in a network due to the increase of its storage while the network keeps running with a large number of robots. To alleviate this problem, we will investigate an approach to reduce Blockchain data by transferring them on-demand or storing only hash values for these data [11].

Acknowledgements. This work is supported in part by the Louisiana Board of Regents Contract Number LEQSF(2020-21)-RD-A-14.

References

1. Wing - A commercial drone delivery service. https://wing.com. Accessed 30 June 2020
2. Python Robotics. https://github.com/AtsushiSakai/PythonRobotics. Accessed 5 April 2020
3. Davidson, D., Wu, H., Jellinek, R., Singh, V., Ristenpart, T.: Controlling UAVs with sensor input spoofing attacks. In: Proceedings of the 10th USENIX Workshop on Offensive Technologies (2016)

4. Douceur, J.R.: The sybil attack. In: Druschel, P., Kaashoek, F., Rowstron, A. (eds.) IPTPS 2002. LNCS, vol. 2429, pp. 251–260. Springer, Heidelberg (2002). https://doi.org/10.1007/3-540-45748-8_24

5. Castelló Ferrer, E.: The blockchain: a new framework for robotic swarm systems. In: Arai, K., Bhatia, R., Kapoor, S. (eds.) FTC 2018. AISC, vol. 881, pp. 1037–1058. Springer, Cham (2019). https://doi.org/10.1007/978-3-030-02683-7_77

6. Gil, S., Baykal, C., Rus, D.: Resilient multi-agent consensus using Wi-Fi signals. IEEE Control Syst. Lett. 3(1), 126–131 (2018)

7. Gil, S., Kumar, S., Mazumder, M., Katabi, D., Rus, D.: Guaranteeing spoof-resilient multi-robot networks. Auton. Robots 41(6), 1383–1400 (2017). https://doi.org/10.1007/s10514-017-9621-5

8. LeBlanc, H.J., Zhang, H., Koutsoukos, X., Sundaram, S.: Resilient asymptotic consensus in robust networks. IEEE J. Sel. Areas Commun. 31(4), 766–781 (2013)

9. Ma, H., Hönig, W., Kumar, T.S., Ayanian, N., Koenig, S.: Lifelong path planning with kinematic constraints for multi-agent pickup and delivery. In: Proceedings of the AAAI Conference on Artificial Intelligence, vol. 33, pp. 7651–7658 (2019)

10. Nakamoto, S.: Bitcoin: a peer-to-peer electronic cash system. Technical Report (2019)

11. Nishida, Y., Kaneko, K., Sharma, S., Sakurai, K.: Suppressing chain size of blockchain-based information sharing for swarm robotic systems. In: Proceedings of International Symposium on Computing and Networking Workshops, pp. 524–528 (2018)

12. Queralta, J.P., Westerlund, T.: Blockchain-powered collaboration in heterogeneous swarms of robots. Front. Robot. AI (2020)

13. Renganathan, V., Summers, T.: Spoof resilient coordination for distributed multi-robot systems. In: Proceedings of the International Symposium on Multi-Robot and Multi-Agent Systems, pp. 135–141 (2017)

14. Sargeant, I., Tomlinson, A.: Modelling malicious entities in a robotic swarm. In: Proceedings of IEEE/AIAA Digital Avionics Systems Conference, pp. 7B1-1–7B1-12 (2013)

15. Saulnier, K., Saldana, D., Prorok, A., Pappas, G.J., Kumar, V.: Resilient flocking for mobile robot teams. IEEE Robot. Autom. lett. 2(2), 1039–1046 (2017)

16. Shepard, D.P., Bhatti, J.A., Humphreys, T.E., Fansler, A.A.: Evaluation of smart grid and civilian UAV vulnerability to GPS spoofing attacks. In: Proceedings of Radionavigation Laboratory Conference (2012)

17. Strobel, V., Castelló Ferrer, E., Dorigo, M.: Managing byzantine robots via blockchain technology in a swarm robotics collective decision making scenario. In: Proceedings of the International Conference on Autonomous Agents and MultiAgent Systems, pp. 541–549 (2018)

18. Strobel, V., Castelló Ferrer, E., Dorigo, M.: Blockchain technology secures robot swarms: a comparison of consensus protocols and their resilience to byzantine robots. Front. Robot. AI 7, 54 (2020)

19. Strobel, V., Dorigo, M.: Blockchain technology for robot swarms: a shared knowledge and reputation management system for collective estimation. In: Proceedings of International Conference on Swarm Intelligence, pp. 425–426 (2018)

20. Wheeler, T., Bharathi, E., Gil, S.: Switching topology for resilient consensus using Wi-Fi signals. In: Proceedings of the International Conference on Robotics and Automation, pp. 2018–2024 (2019)

ImPL-VIO: An Improved Monocular Visual-Inertial Odometry Using Point and Line Features

Haoqi Cheng[1], Hong Wang[1($\boxtimes$)], Zhongxue Gan[2], and Jinxiang Deng[2]

[1] School of Mechanical Engineering and Automation, Harbin Institute of Technology, Shenzhen, China
hongwang@hit.edu.cn

[2] Engineering Research Center for Intelligent Robotics, Ji Hua Laboratory, Guangdong, China

Abstract. Most of the visual-inertial navigation systems (VINS) that use only point features usually work well in regular environment, but decay in low-texture scenes. Meanwhile, those systems rarely construct environmental map with structural information. In this paper, an improved tightly-coupled monocular visual-inertial odometry (ImPL-VIO) is developed. The whole system is composed of point and line feature tracking, inertial measurements processing, pose estimator and loop closure detection. For the better use of monocular line observations in the sliding window based pose estimator, an improved line triangulation algorithm is proposed after a detailed analysis of error sources. In addition, we, for the first time, employ the closest point (CP) representation for spatial lines to optimization-based VINS system, and derive the corresponding Jacobians analytically. Finally, simulation and real-world experiments are conducted to validate the proposed system.

Keywords: Visual-inertial odometry · Visual simultaneous localization and mapping · Line segment features

1 Introduction

Visual-inertial odometry techniques [1] have become an active area of research in robotic and computer vision communities due to urgent needs for mobile robot navigation, unmanned aerial vehicle (UAV) and augmented reality (AR). By fusing measurements of a camera and an inertial measurement unit (IMU), a visual-inertial odometry that using only point features usually can estimate the current 6DOF poses of moving platform robustly. Nevertheless, there are many factors that may decrease the accuracy of estimation, such as lack of robust features, illumination changing, platform shaking, etc. For example, when

Supported by the Program "Research on Basic and Key Technologies of Intelligent Robots" (No. X190021TB190), Ji Hua Laboratory, Guangdong, China.

C. S. Chan et al. (Eds.): ICIRA 2020, LNAI 12595, pp. 217–229, 2020.
https://doi.org/10.1007/978-3-030-66645-3_19

moving in a hotel corridor, the camera tends to barely capture enough point features, resulting in frame tracking failure.

There are two key questions when utilizing point and line features in a tightly-coupled visual-inertial odometry: how to fuse a large number of measurements from camera and IMU sensors with efficiency, how to describe a spatial line for triangulation and optimization. For the first question, quite a few state-of-art works have been proposed, and they are classified into two categories: the filter-based approaches, and the optimization-based approaches. Mourikis et al. [2] proposed a multi-state constraint Kalman filter (MSCKF). IMU measurements between two consecutive frames are integrated to predict the state, and the latest state is updated with visual measurements. The MSCKF maintains a fixed-size sliding window to achieve constant time optimization by marginalizing out the unobserved visual measurements, while sacrificing the accuracy of estimation. VINS-Mono [3], to some extent, is similar to the MSCKF. They both use the sliding window algorithm [4] and marginalize out the unobserved landmarks. However, the MSCKF only processed one measurement of the update step, while the pose estimator in VINS-Mono linearized the loss function multiple times for the batch optimization step.

For the second question, an amount of research is devoted to employing line features to visual simultaneous localization and mapping (SLAM). In the early exploration, most of these works [5,6] focused on mapping in pure visual SLAM that using only line features, which are parameterized and optimized with the algorithms proposed by Adrien et al. [7]. Zuo et al. [8] proposed a new visual SLAM system using point and line features based on the ORB-SLAM2 framework. They represented spatial lines with the Plücker coordinates for triangulation and proposed an orthonormal representation as the minimal model of line features for optimization. Then He et al. [9] introduced the Plücker coordinates and the orthonormal representation to point-line visual-inertial odometry (PL-VIO) based on the VINS-Mono framework. Gomez-Ojeda et al. [10] and Zhao et al. [11] developed pure visual SLAM by improving the loop closure detection with point and line features. Although the accuracy of localization is improved, the system has to suffer high computation cost. Yang et al. [12] summarized several commonly used line representations and triangulation algorithms in visual-inertial odometry, and conducted the experiments based on the MSCKF algorithm. However, the implementation details of the system are not introduced.

In this paper, we proposed the ImPL-VIO system, an improved VINS system combining point and line features. By following the pipeline of point feature tracker in VINS-Mono, we launched a new thread for line features tracking. In the pose estimator, line features are represented by the Plücker coordinates for proposed sliding window based triangulation algorithm. The CP representation is introduced to describe spatial lines for the sliding window based nonlinear optimization, and the related Jacobians are derived. The proposed ImPL-VIO system is implemented based on VINS-Mono framework, and a simulation is designed to verify the advantage of the proposed line triangulation algorithm.

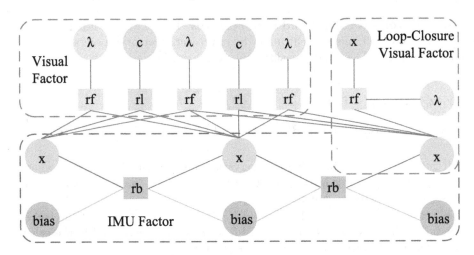

Fig. 1. Tightly-coupled factor graph framework of optimization. x is IMU pose in the world frame, $bias$ is bias of IMU. λ is inverse depth of feature in the world frame, c is line state vector represented by the proposed closest point representation. rb, rf and rl are the measurement residuals related to point, line and IMU measurements.

Since the proposed ImPL-VIO system is similar to VINS-Mono and PL-VIO, real-world experiments are conducted with the EuRoC MAV dataset [13] by comparing with those two state-of-art systems.

2   Methodology

2.1   Graph Optimization

The proposed ImPL-VIO system utilizes graph optimization technique and the sliding window algorithm to carry out batch nonlinear optimization. As shown in Fig. 1, the factor graph framework contains visual factor with point and line features, IMU factor, and loop-closure visual factor. The corresponding state vector in a sliding window at a certain time step is defined as:

$$\mathcal{X} = \left[\mathbf{x}_n, \mathbf{x}_{n+1}, \ldots, \mathbf{x}_{n+N}, \lambda_m, \lambda_{m+1}, \ldots, \lambda_{m+M}, \mathbf{c}_k, \mathbf{c}_{k+1}, \ldots, \mathbf{c}_{k+K}\right]^\top$$
$$\mathbf{x}_i = \left[\mathbf{p}_{b_i}^w, \mathbf{q}_{b_i}^w, \mathbf{v}_i^w, \mathbf{b}_a^{b_i}, \mathbf{b}_g^{b_i}\right]^\top, \; i \in [n, n+N] \tag{1}$$

where $\mathbf{x}$ is IMU body state, including the position $\mathbf{p}_b^w$, orientation $\mathbf{q}_b^w$, and velocity $\mathbf{v}^w$ in the world frame. $\mathbf{b}_a^{b_i}$ and $\mathbf{b}_g^{b_i}$ are bias of accelerometer and gyroscope respectively. λ is the inverse depth of point, and $\mathbf{c}$ is the line state described by the CP representation. The loss function of graph optimization is:

$$\mathcal{X}^* = \min_\mathcal{X} \|\mathbf{r}_p\|_{\Sigma_p}^2 + \sum \|\mathbf{r}_b\|_{\Sigma_b}^2 + \sum \|\mathbf{r}_f\|_{\Sigma_f}^2 + \sum \|\mathbf{r}_l\|_{\Sigma_l}^2 \tag{2}$$

with $\mathbf{r}_f$, $\mathbf{r}_l$, $\mathbf{r}_b$, and $\mathbf{r}_p$ indicating visual factor with point and line features, IMU factor and marginalization factor, respectively.

Visual Factor with Point and Line Features. In this work, Pinhole camera is modeled on each image to calculate the re-projection error of points and lines in camera frame. The point measurement model can be derived as:

$$
\begin{aligned}
\mathbf{r}_f(\mathbf{z}_{\mathbf{f}_k}^{c_j}, \mathcal{X}) &= \hat{\mathbf{f}}_k^{c_j} - \mathbf{f}, \ \hat{\mathbf{f}}_k^{c_j} = \left[\hat{x}_k^{c_j}/\hat{z}_k^{c_j} \ \hat{y}_k^{c_j}/\hat{z}_k^{c_j} \right]^\top \\
\mathbf{f}_k^{c_j} &= \mathbf{R}_b^c(\mathbf{R}_w^{b_j}(\mathbf{R}_{b_i}^w((\mathbf{R}_c^b \frac{1}{\lambda_k} \left[x_k^{c_i} \ y_k^{c_i} \ 1 \right]^\top + \mathbf{p}_c^b) + \mathbf{p}_{b_i}^w) - \mathbf{p}_{b_j}^w) - \mathbf{p}_c^b)
\end{aligned} \tag{3}
$$

with $[x_k^{c_i} \ y_k^{c_i} \ 1]^\top$ indicating the first observation of kth feature in camera c_i where λ_k is the corresponding inverse depth. $[\hat{x}_k^{c_j}/\hat{z}_k^{c_j} \ \hat{y}_k^{c_j}/\hat{z}_k^{c_j}]^\top$ is the observation of the same feature in the camera frame c_j.

Given a line observation $\hat{\mathbf{l}} = [l_1 \ l_2 \ l_3]^\top$ in the image plane, and the corresponding endpoints, $\mathbf{f}_s = [u_s \ v_s \ 1]^\top$ and $\mathbf{f}_e = [u_e \ v_e \ 1]^\top$, extracted in the other image, the line re-projection error can be written as:

$$
\mathbf{r}_l(\mathbf{z}_{\mathcal{L}_k}^{c_j}, \mathcal{X}) = \left[\frac{\mathbf{f}_s^\top \hat{\mathbf{l}}}{\sqrt{l_1^2 + l_2^2}} \quad \frac{\mathbf{f}_e^\top \hat{\mathbf{l}}}{\sqrt{l_1^2 + l_2^2}} \right]^\top \tag{4}
$$

where $\hat{\mathbf{l}}$ can be projected from k^{th} spatial line $\mathcal{L}$ in the world frame. More detailed contents are introduced in later sections, including the proposed line triangulation algorithm, the CP representation for line optimization.

Pre-Integrated IMU Factor. The local motion can be inferred from the IMU measurements, which include linear acceleration $\mathbf{a}$ and angular velocity $\boldsymbol{\omega}$. While considering noises and biases, those readings can be expressed as:

$$
\begin{aligned}
\hat{\boldsymbol{\omega}}^b &= \boldsymbol{\omega}^b + \mathbf{b}_g^b + \mathbf{n}_g^b \\
\hat{\mathbf{a}}^b &= \mathbf{q}_w^b (\mathbf{a}^w + \mathbf{g}^w) + \mathbf{b}_a^b + \mathbf{n}_a^b
\end{aligned} \tag{5}
$$

where $\mathbf{n}_g$ and $\mathbf{n}_a$ are assumed to be zero-mean Gaussian noise. To avoid high computation cost and re-propagation, IMU pre-integration approach is introduced to construct IMU factor between two key frames b_i and b_j. The IMU measurement model can be written as:

$$
\mathbf{r}_b\left(\mathbf{z}_{b_i b_j}, \mathcal{X}\right) = \begin{bmatrix} \mathbf{r}_p \\ \mathbf{r}_\theta \\ \mathbf{r}_v \\ \mathbf{r}_{ba} \\ \mathbf{r}_{bg} \end{bmatrix} = \begin{bmatrix} \mathbf{q}_w^{b_i}\left(\mathbf{p}_{b_j}^w - \mathbf{p}_{b_i}^w - \mathbf{v}_i^w \Delta t + \frac{1}{2}\mathbf{g}^w \Delta t^2\right) - \hat{\boldsymbol{\alpha}}_{b_i b_j} \\ 2\left[\hat{\mathbf{q}}_{b_i}^{b_j} \otimes \left(\mathbf{q}_w^{b_i} \otimes \mathbf{q}_{b_j}^w\right)\right]_{xyz} \\ \mathbf{q}_w^{b_i}\left(\mathbf{v}_j^w - \mathbf{v}_i^w + \mathbf{g}^w \Delta t\right) - \hat{\boldsymbol{\beta}}_{b_j}^{b_i} \\ \mathbf{b}_a^{b_j} - \mathbf{b}_a^{b_i} \\ \mathbf{b}_g^{b_j} - \mathbf{b}_g^{b_i} \end{bmatrix} \tag{6}
$$

where Δt is the duration between two key frames, $[\mathbf{q}]_{xyz}$ are the imaginary part of unit quaternion $\mathbf{q}$. $\hat{\boldsymbol{\alpha}}_{b_j}^{b_i}$, $\hat{\boldsymbol{\beta}}_{b_j}^{b_i}$ and $\hat{\mathbf{q}}_{b_j}^{b_i}$ are pre-integration measurements as follows:

$$\hat{\boldsymbol{\alpha}}_{b_{k+1}}^{b_i} = \hat{\boldsymbol{\alpha}}_{b_k}^{b_i} + \hat{\boldsymbol{\beta}}_{b_k}^{b_i}\delta t + \frac{1}{2}\hat{\mathbf{a}}\delta t^2, \quad \hat{\boldsymbol{\beta}}_{b_{k+1}}^{b_i} = \hat{\boldsymbol{\beta}}_{b_k}^{b_i} + \hat{\mathbf{a}}\delta t, \quad \hat{\mathbf{q}}_{b_{k+1}}^{b_i} = \hat{\mathbf{q}}_{b_k}^{b_i} \otimes \begin{bmatrix} 1 \\ \frac{1}{2}\hat{\boldsymbol{\omega}}\delta t \end{bmatrix} \quad (7)$$

with δt indicating the duration between two IMU measurements. In practice, We adopt mid-point integration approach to propagate the kth mean of pre-integration measurements from the zero initial assumption. Additionally, the linearized error state dynamics can be derived as:

$$
\begin{bmatrix} \delta\boldsymbol{\alpha}_{b_{k+1}} \\ \delta\boldsymbol{\theta}_{b_{k+1}} \\ \delta\boldsymbol{\beta}_{b_{k+1}} \\ \delta\mathbf{b}_a^{k+1} \\ \delta\mathbf{b}_g^{k+1} \end{bmatrix} =
\begin{bmatrix}
\mathbf{I} & \mathbf{f}_{12} & \mathbf{I}\delta t & -\frac{1}{4}(\mathbf{q}_{b_k}^{b_i} + \mathbf{q}_{b_{k+1}}^{b_i})\delta t^2 & \mathbf{f}_{15} \\
0 & \mathbf{I} - [\boldsymbol{\omega}]_\times\delta t & 0 & 0 & -\mathbf{I}\delta t \\
0 & \mathbf{f}_{32} & \mathbf{I} & -\frac{1}{2}(\mathbf{q}_{b_k}^{b_i} + \mathbf{q}_{b_{k+1}}^{b_i})\delta t & \mathbf{f}_{35} \\
0 & 0 & 0 & \mathbf{I} & 0 \\
0 & 0 & 0 & 0 & \mathbf{I}
\end{bmatrix}
\begin{bmatrix} \delta\boldsymbol{\alpha}_{b_k} \\ \delta\boldsymbol{\theta}_{b_k} \\ \delta\boldsymbol{\beta}_{b_k} \\ \delta\mathbf{b}_a^k \\ \delta\mathbf{b}_g^k \end{bmatrix}
$$
$$
+ \begin{bmatrix}
\frac{1}{4}\mathbf{q}_{b_k}^{b_i}\delta t^2 & \mathbf{g}_{12} & \frac{1}{4}\mathbf{q}_{b_{k+1}}^{b_i}\delta t^2 & \mathbf{g}_{14} & 0 & 0 \\
0 & \frac{1}{2}\mathbf{I}\delta t & 0 & \frac{1}{2}\mathbf{I}\delta t & 0 & 0 \\
\frac{1}{2}\mathbf{q}_{b_k}^{b_i}\delta t & \mathbf{g}_{32} & \frac{1}{2}\mathbf{q}_{b_{k+1}}^{b_i}\delta t & \mathbf{g}_{34} & 0 & 0 \\
0 & 0 & 0 & 0 & \mathbf{I}\delta t & 0 \\
0 & 0 & 0 & 0 & 0 & \mathbf{I}\delta t
\end{bmatrix}
\begin{bmatrix} \mathbf{n}_a^k \\ \mathbf{n}_g^k \\ \mathbf{n}_a^{k+1} \\ \mathbf{n}_g^{k+1} \\ \mathbf{n}_{b_a^k} \\ \mathbf{n}_{b_g^k} \end{bmatrix}
\quad (8)
$$

where $[\cdot]_\times$ indicates the skew-symmetric matrix, and

$$
\begin{aligned}
\mathbf{f}_{12} &= -\frac{1}{4}(\mathbf{q}_{b_k}^{b_i}\left[\mathbf{a}^{b_k} - \mathbf{b}_a^k\right]_\times \delta t^2 + \mathbf{q}_{b_{k+1}}^{b_i}\left[(\mathbf{a}^{b_{k+1}} - \mathbf{b}_a^k)\right]_\times (\mathbf{I} - [\boldsymbol{\omega}]_\times\delta t)\delta t^2) \\
\mathbf{f}_{32} &= -\frac{1}{2}(\mathbf{q}_{b_k}^{b_i}\left[\mathbf{a}^{b_k} - \mathbf{b}_a^k\right]_\times \delta t + \mathbf{q}_{b_{k+1}}^{b_i}\left[(\mathbf{a}^{b_{k+1}} - \mathbf{b}_a^k)\right]_\times (\mathbf{I} - [\boldsymbol{\omega}]_\times\delta t)\delta t) \\
\mathbf{f}_{15} &= -\frac{1}{4}(\mathbf{q}_{b_{k+1}}^{b_i}\left[(\mathbf{a}^{b_{k+1}} - \mathbf{b}_a^k)\right]_\times \delta t^2)(-\delta t) \\
\mathbf{f}_{35} &= -\frac{1}{2}(\mathbf{q}_{b_{k+1}}^{b_i}\left[(\mathbf{a}^{b_{k+1}} - \mathbf{b}_a^k)\right]_\times \delta t)(-\delta t) \\
\mathbf{g}_{12} &= \mathbf{g}_{14} = -\frac{1}{4}(\mathbf{q}_{b_{k+1}}^{b_i}\left[(\mathbf{a}^{b_{k+1}} - \mathbf{b}_a^k)\right]_\times \delta t^2)(\frac{1}{2}\delta t) \\
\mathbf{g}_{32} &= \mathbf{g}_{34} = -\frac{1}{2}(\mathbf{q}_{b_{k+1}}^{b_i}\left[(\mathbf{a}^{b_{k+1}} - \mathbf{b}_a^k)\right]_\times \delta t)(\frac{1}{2}\delta t)
\end{aligned}
\quad (9)
$$

Marginalization Factor. There are two criteria to marginalize frames to maintain the constraints with the previous measurements and keep the fixed-size sliding window. When the latest frame reaches, the average parallax between this frame and the last frame are calculated. If the value is greater than threshold, factors related to the oldest frame will be marginalized out. Otherwise, we will marginalize out the visual factors of this frame and propagate corresponding IMU measurements to maintain IMU factors. All the operations of marginalization are carried out at the time processing the next frame.

2.2   Line Triangulation

Before graph optimization, line features in the sliding window are triangulated to 3D position in the world frame. We leverage the Plücker coordinates $(\mathbf{n}, \mathbf{v})$

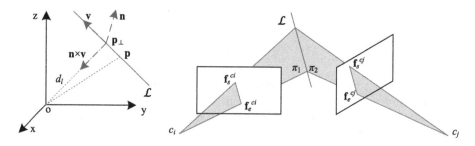

Fig. 2. Left: Plücker line coordinates. Right: Line triangulation between two frames.

to represent a line $\mathcal{L}$ in triangulation. In Fig. 2a, $\mathbf{v}$ is the vector representing line direction, and $\mathbf{n}$ is the normal vector of the plane determined by the line and origin. The norm $\|\mathbf{n}\|/\|\mathbf{v}\|$ gives the distance d_l from the origin to the line, achieved at the closest point $\mathbf{p}_\perp$ of line perpendicular to the origin.

Many works empirically select two key frames to triangulate the observed line $\mathcal{L}$. Given the corresponding line segment measurements, $(\mathbf{f}_s^{c_i}, \mathbf{f}_e^{c_i})$ and $(\mathbf{f}_s^{c_j}, \mathbf{f}_e^{c_j})$, which is obtained by line feature tracking, the intersecting planes, $\boldsymbol{\pi}_i$ and $\boldsymbol{\pi}_j$, can be derived as:

$$\boldsymbol{\pi}_i = \begin{bmatrix} \mathbf{f}_s^{c_i} \times \mathbf{f}_e^{c_i} \\ 0 \end{bmatrix}, \ \boldsymbol{\pi}_j = \begin{bmatrix} \mathbf{f}_s^{c_j} \times \mathbf{f}_e^{c_j} \\ 0 \end{bmatrix} \tag{10}$$

Once the relative pose between the selected key frames is obtained, plane $\boldsymbol{\pi}_j$ is converted to the camera frame c_i. Based on the planes $\boldsymbol{\pi}_i$ and $\boldsymbol{\pi}_j$ in the same frame, the Plücker matrix $\mathcal{L}^*$ can be computed as:

$$\mathcal{L}^* = \boldsymbol{\pi}_i \boldsymbol{\pi}_j^\top - \boldsymbol{\pi}_j \boldsymbol{\pi}_i^\top = \begin{bmatrix} [\mathbf{v}^{c_i}]_\times & \mathbf{n}^{c_i} \\ -\mathbf{n}^{c_i \top} & 0 \end{bmatrix} \tag{11}$$

with $(\mathbf{n}^{c_i}, \mathbf{v}^{c_i})$ indicating the Plücker coordinates of line $\mathcal{L}$ in camera frames c_i. According to the pipeline of line triangulation algorithm elaborated above, it can be analyzed that there are three drawbacks when applied the algorithm to optimization-based visual-inertial odometry:

1. When the accuracy of line feature tracking and pose estimation decreases, the algorithm's performance decreases as well due to both line measurements and the relative pose are required in calculation of the Plücker matrix.
2. A line feature is triangulated by only two selected frames, while in the sliding window, it usually can be observed by much more key frames. The observations are not fully used to construct the visual factor for graph optimization.
3. The algorithm is deployed on the monocular camera, which means more possibilities to failure in some type of motions, such as pure rotation, moving toward the line, moving along the line direction, etc.

In this section, we utilize as many line observations as possible in the sliding window and propose a new triangulation algorithm. For convenience, the line $\mathcal{L}$ obtained by Eq. 11 is normalized to $(d_l \mathbf{n}_e, \mathbf{v}_e)$. Those parameters are written as:

$$d_l^{c_i} = \sum_{j=2}^{M} \frac{d_l^{c_i}(j)}{M-1}, \quad \mathbf{n}_e = \sum_{j=2}^{M} \frac{\mathbf{n}_e^{c_i}(j)}{M-1}, \quad \mathbf{v}_e = \sum_{j=2}^{M} \frac{\mathbf{v}_e^{c_i}(j)}{M-1} \tag{12}$$

where c_i is the first key frame where the line $\mathcal{L}$ is observed in the sliding window, M is the number of the required key frames. Specifically, not every key frame by which the line $\mathcal{L}$ is observed will be accepted. When the latest key frame reaches, its parallax angle and translation related to key frame c_i will be calculated. If one of these two values is lower than threshold, the key frame is excluded from the algorithm in case of poor visual tracking and degenerate motions. In addition, a line feature is considered to be stable only if M is within the setting range. When M is less than the minimum, it won't be triangulated. When M is greater than the maximum, it is regarded as a new line feature in database.

2.3   Line Representation and Jacobians

Since spatial lines only have four degrees of freedom (DoFs), the Plücker coordinates with the redundant DoFs is inappropriate for graph optimization. In this section, the CP representation is introduced as the minimal representation for spatial lines. Meanwhile, the measurement Jacobians are derived.

In Fig. 2a, the DoFs of a spatial line $\mathcal{L}$ encapsulate the line orientation $\mathbf{R}_l$ between the camera frame and the line frame, and the vertical distance d_l from the origin to this line. Specifically, the closest point $\mathbf{p}_\perp$ is the origin of the line frame, $\mathbf{n}$, $\mathbf{v}$ and $\mathbf{n} \times \mathbf{v}$ are three axes. Once the Plücker coordinates of this line is calculated, the line orientation $\mathbf{R}_l$ and the distance d_l can be derived as:

$$\mathbf{R}_l(\mathbf{q}_l) = \begin{bmatrix} \frac{\mathbf{n}}{\|\mathbf{n}\|} & \frac{\mathbf{v}}{\|\mathbf{v}\|} & \frac{\mathbf{n}\times\mathbf{v}}{\|\mathbf{n}\times\mathbf{v}\|} \end{bmatrix}, \quad d_l = \|\mathbf{n}\|/\|\mathbf{v}\| \tag{13}$$

where $\mathbf{q}_l$ is the unit quaternion associated with the rotation matrix $\mathbf{R}_l$. Hence, the vector $[\mathbf{q}_l \; d_l]^\mathsf{T}$ can be used to describe a spatial line in graph optimization, where $\mathbf{q}_l$ and d_l is updated by quaternion multiplication and scalar addition respectively. To unify the update operator, the CP representation is introduced:

$$\mathbf{c}_l = d_l\mathbf{q}_l = d_l \left[[\mathbf{q}_l]_{xyz}^\mathsf{T} \quad q_l \right]^\mathsf{T} \tag{14}$$

then the line state $\mathbf{c}_l$ can be easily updated with increment $\delta\mathbf{c}_l$ when solving the nonlinear optimization problem. The update formula can be written as:

$$\begin{aligned} \mathbf{c}_l' &= \mathbf{c}_l + \delta\mathbf{c}_l \\ d_l'\mathbf{q}_l' &= (d_l + \delta d_l)\delta\mathbf{q}_l \otimes \mathbf{q}_l = (d_l + \delta d_l)\begin{bmatrix} \frac{1}{2}\delta\boldsymbol{\theta}_l \\ 1 \end{bmatrix} \otimes \mathbf{q}_l \end{aligned} \tag{15}$$

Therefore, the Jacobians for line measurements are derived using line measurement model Eq. 4. In more detail, line observation $\mathbf{l}$ can be written as:

$$\mathbf{l} = [\mathcal{K} \quad \mathbf{0}_3]\mathcal{L}^c, \quad \mathcal{L}^c = \begin{bmatrix} d_l^c\mathbf{n}_e^c \\ \mathbf{v}_e^c \end{bmatrix} = T_{wc}^{-1}\mathcal{L}^w = \begin{bmatrix} \mathbf{R}_w^c & [\mathbf{p}_w^c]_\times \mathbf{R}_w^c \\ \mathbf{0}_3 & \mathbf{R}_w^c \end{bmatrix}\mathcal{L}^w \tag{16}$$

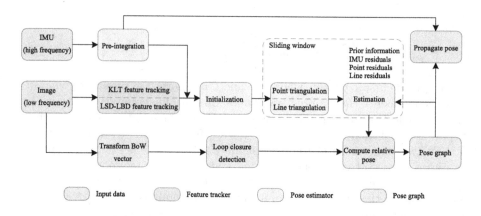

Fig. 3. Architecture of the proposed visual-inertial odometry.

where $\mathcal{K}$ is line projection matrix. Finally, the Jacobians can be written as:

$$\mathbf{J}_l = \frac{\partial \mathbf{r}_l}{\partial \mathbf{l}^c} \frac{\partial \mathbf{l}^c}{\partial \mathcal{L}^c} \left[\frac{\partial \mathcal{L}^c}{\partial \mathcal{L}^w} \frac{\partial \mathcal{L}^w}{\partial \delta \mathbf{x}} \quad \frac{\partial \mathcal{L}^c}{\partial \mathcal{L}^w} \frac{\partial \mathcal{L}^w}{\partial [\delta \boldsymbol{\theta}_l^\top \ \delta d_l]^\top} \frac{\partial [\delta \boldsymbol{\theta}_l^\top \ \delta d_l]^\top}{\partial \delta \mathbf{c}_l} \right] \tag{17}$$

with

$$\frac{\partial \mathbf{r}_l}{\partial \mathbf{l}^c} = \frac{1}{\sqrt{l_1^2 + l_2^2}} \begin{bmatrix} u_s - \frac{l_1 \mathbf{f}_s^\top \hat{\mathbf{i}}}{l_1^2 + l_2^2} & v_s - \frac{l_2 \mathbf{f}_s^\top \hat{\mathbf{i}}}{l_1^2 + l_2^2} \\ u_e - \frac{l_1 \mathbf{f}_e^\top \hat{\mathbf{i}}}{l_1^2 + l_2^2} & v_e - \frac{l_2 \mathbf{f}_e^\top \hat{\mathbf{i}}}{l_1^2 + l_2^2} \end{bmatrix}$$

$$\frac{\partial \mathbf{l}^c}{\partial \mathcal{L}^c} = \begin{bmatrix} \mathcal{K} \ \mathbf{0}_3 \end{bmatrix}, \ \mathcal{K} = \begin{bmatrix} f_v & 0 & 0 \\ 0 & f_u & 0 \\ -f_v c_u & -f_u c_v & f_u f_v \end{bmatrix}$$

$$\frac{\partial \mathcal{L}^c}{\partial \mathcal{L}^w} \frac{\partial \mathcal{L}^w}{\partial \delta \mathbf{x}} = \mathcal{T}_{bc}^{-1} \begin{bmatrix} \mathbf{R}_w^b \left[\mathbf{v}_e^w \right]_\times \\ \mathbf{0}_{3\times 3} \end{bmatrix} \begin{bmatrix} \left[\mathbf{R}_w^b \left(d_l^w \mathbf{n}_e^w + [\mathbf{v}_e^w]_\times \mathbf{P}_{wb} \right) \right]_\times \\ \left[\mathbf{R}_w^b \mathbf{v}_e^w \right]_\times \end{bmatrix} \mathbf{0} \ \mathbf{0} \ \mathbf{0} \end{bmatrix} \tag{18}$$

$$\frac{\partial \mathcal{L}^c}{\partial \mathcal{L}^w} \frac{\partial \mathcal{L}^w}{\partial [\delta \boldsymbol{\theta}_l^\top \ \delta d_l]^\top} = \mathcal{T}_{wc}^{-1} \begin{bmatrix} d_l^w [\mathbf{R}_l \mathbf{e}_1]_\times & \mathbf{R}_l \mathbf{e}_1 \\ [\mathbf{R}_l \mathbf{e}_2]_\times & \mathbf{0}_{3\times 1} \end{bmatrix}, \ \mathbf{e}_1 = [1 \ 0 \ 0]^\top, \ \mathbf{e}_2 = [0 \ 1 \ 0]^\top$$

$$\frac{\partial [\delta \boldsymbol{\theta}_l^\top \ \delta d_l]^\top}{\partial \delta \mathbf{c}_l} = \begin{bmatrix} \frac{2}{d_l^w} \left(q_l \mathbf{I}_3 - \left[[\mathbf{q}_l]_{xyz} \right]_\times \right) & -\frac{2}{d_l^w} [\mathbf{q}_l]_{xyz} \\ [\mathbf{q}_l]_{xyz}^\top & q_l \end{bmatrix}$$

3 Experimental Results

3.1 System Implementation

The proposed ImPL-VIO system (see Fig. 3) is implemented based on the VINS-Mono framework. The system has three main modules: Feature tracker, Pose estimator and Pose graph. In this section, we briefly describe each module while focusing on our own proposal.

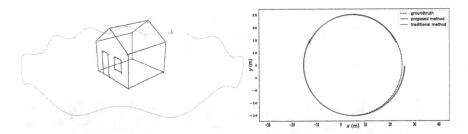

Fig. 4. Left: Example of scene reconstruction in simulation. Right: Trajectory on simulation data using our proposed triangulation method and traditional triangulation method in PL-VIO.

Feature Tracker. In our system, monocular image sequence is input to feature tracker module. When a new frame reaches, point and line features are tracked by two threads in parallel mode. The KLT optical flow algorithm [14] is applied for point feature tracking, while outliers are removed by the RANSAC algorithm with Epipolar geometry constraints. Line features are detected by the LSD method [15], and described by the LBD descriptor [16]. We sort the detected line features according to their response, and remove the line that length is lower than the threshold. In line feature matching, we filter out line matches according to geometric constraints, where the distance and the angle between two matched lines are considered to be criteria. Finally, the coordinates of the point features and the line endpoints are output to pose estimator module.

Pose Estimator. The input IMU measurements are first pre-integrated to predict the camera motion, then the point and line features are triangulated after initialization. The state vector in Eq. 1 are estimated with all the observations in the sliding window. Before graph optimization, the average parallax is calculated for key frame marginalization. Each key frame with the corresponding optimal state are finally sent to pose graph module for loop closure detection.

Pose Graph. Considering the computation cost, only point features are applied to pose graph module. For every input key frame, point features are detected by the FAST algorithm [17] and described as the BRIEF descriptors. Those descriptors are converted to a visual word in a bag-of-words vector and used to query the visual database. After temporal and geometrical consistency check, the relative pose between this key frame and loop closure is computed and added into the pose graph optimization to correct the deviation of trajectory.

3.2 Results

A simulation is conducted to verify the advantage of the proposed line triangulation algorithm. The proposed ImPL-VIO system is validated through real-world experiments on EuRoC MAV dataset. Both simulation and real-world experiments are analyzed with Evo evaluation tool[1]. All the experiments run on an Intel Core i5-4570 CPU @ 3.20GHz and 16GB RAM in ROS Kinetic without GPU parallelization.

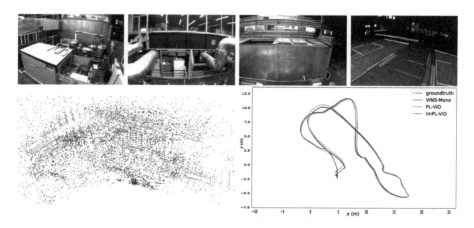

Fig. 5. Experiments on EuRoC datasets. Top: Example line detection of different MH_04_difficult scenes. Bottom Left: Example 3D map reconstruction of MH_04_difficult scene. Bottom Right: Example trajectory estimated in VINS-Mono, PL-VIO, and ImPL-VIO.

Simulation. We construct a scene that contains a house with 23 lines and line endpoints. A monocular camera moves around the house in a sine-like circle motion (20 m radius) and generates endpoint measurements. Along with this trajectory, IMU readings are generated with biases and Gaussian noise based on the model Eq. 5. We use three different covariance values (0.005, 0.01, and 0.05 m) of Gaussian noise to corrupt the line endpoints and conduct the Monte Carlo experiments 30 times. It is verified that the proposed method can successfully restore spatial lines in such motion, an example of scene reconstruction is presented in the left figure in Fig. 4. The average root mean square error (RMSE) of the absolute pose error (APE) is shown in Table 1.

Table 1. APE of the triangulation algorithms based on different noise levels.

Algorithm	$\sigma = 0.005\,m$		$\sigma = 0.01\,m$		$\sigma = 0.05\,m$	
	Trans.	Rot.	Trans.	Rot.	Trans.	Rot.
Proposed method	**0.399890**	**0.367473**	**0.307429**	**0.382442**	**0.379758**	**0.358920**
Traditional method	0.501712	0.398589	0.525720	0.409541	0.514036	0.414472

---

[1] https://github.com/MichaelGrupp/evo.git.

With the increase of noise levels, the performance of line triangulation will decrease due to the decay of line tracking. Table 1 shows that our proposed method has slightly higher accuracy of estimation than the traditional method in PL-VIO. The proposed method is verified to make up for the lack of line tracking. It also can be intuitively seen from the right figure of Fig. 4 that the estimated trajectory using our proposed method has lower deviation from the ground truth than using the traditional method.

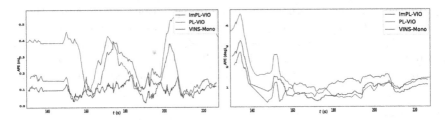

Fig. 6. APE of trajectories estimated by VINS-Mono, PL-VIO, and ImPL-VIO on EuRoC datasets. Left: Example of translation error in MH_04_difficult scene. Right: Example of rotation error in MH_04_difficult scene.

Experiments on EuRoC Datasets. The EuRoC MAV Dataset [13] records two indoor scenes, containing stereo images from a global shutter camera 20 Hz, and synchronized IMU readings 200 Hz. Our experiments are conducted on Machine Hall scenes(see the top figure in Fig. 5), which are recorded with rich line features. Our proposed ImPL-VIO system is validated by comparing with VINS-Mono, and PL-VIO, where the RMSE of APE is used as the evaluation metric. The 3D map reconstruction of the hall is shown in the bottom left figure in Fig. 5 to intuitively demonstrate the performance of ImPL-VIO.

Table 2. APE of VINS-Mono, PL-VIO, and ImPL-VIO in different scenes.

Seq.	VINS-Mono		PL-VIO		ImPL-VIO	
	Trans.	Rot.	Trans.	Rot.	Trans.	Rot.
MH_02_easy	0.178364	2.310627	0.142323	1.746199	**0.138953**	**1.722063**
MH_03_medium	0.194802	1.626538	0.263199	1.702571	**0.226694**	**1.598345**
MH_04_difficult	0.346282	1.497640	0.361471	1.641813	**0.309522**	**1.439429**
MH_05_difficult	0.302626	**0.716123**	0.275127	1.065700	**0.258255**	1.031266

Table 2 shows that ImPL-VIO is capable of different complex scenes and improves the accuracy of estimation. Notice that in most of the scenes, the absolute translation and rotation error of ImPL-VIO is lower than VINS-Mono and PL-VIO. For example, in MH_04_difficult scene(see the bottom right figure in Fig. 5), the estimated trajectory of ImPL-VIO fits the ground truth better, while the estimated trajectory of PL-VIO has a large drift. Furthermore, it can be seen from Fig. 6 in more details.

4   Conclusion

In this paper, we proposed ImPL-VIO system using point and line features. To improve the accuracy of estimation, a line triangulation algorithm and the CP representation is introduced to the sliding window based graph optimization. The Jacobians for line measurements are derived as well. In the future, We intend to explore semantic features to robustly describe the environment, and utilize the wheel odometer to implement a real-time robust VINS system for mobile robots.

References

1. Huang, G.: Visual-inertial navigation: a concise review. In: 2019 International Conference on Robotics and Automation (ICRA), pp. 9572–9582. IEEE (2019)
2. Mourikis, A.I., Roumeliotis, S.I.: A multi-state constraint kalman filter for vision-aided inertial navigation. In: Proceedings 2007 IEEE International Conference on Robotics and Automation, pp. 3565–3572. IEEE (2007)
3. Qin, T., Li, P., Shen, S.: Vins-mono: a robust and versatile monocular visual-inertial state estimator. IEEE Trans. Robot. **34**(4), 1004–1020 (2018)
4. Sibley, G., Matthies, L., Sukhatme, G.: Sliding window filter with application to planetary landing. J. Field Robot. **27**, 587–608 (2010)
5. Zhang, G., Lee, J.H., Lim, J., Suh, I.H.: Building a 3-d line-based map using stereo slam. IEEE Trans. Robot. **31**(6), 1364–1377 (2015)
6. Zhang, G., Suh, I.H.: Building a partial 3D line-based map using a monocular slam. In: 2011 IEEE International Conference on Robotics and Automation, pp. 1497–1502. IEEE (2011)
7. Bartoli, A., Sturm, P.: Structure-from-motion using lines: representation, triangulation, and bundle adjustment. Comput. Vis. Image Understanding **100**(3), 416–441 (2005)
8. Zuo, X., Xie, X., Liu, Y., Huang, G.: Robust visual slam with point and line features. In: 2017 IEEE/RSJ International Conference on Intelligent Robots and Systems (IROS), pp. 1775–1782. IEEE (2017)
9. He, Y., Zhao, J., Guo, Y., He, W., Yuan, K.: Pl-vio: tightly-coupled monocular visual-inertial odometry using point and line features. Sensors **18**(4), 1159 (2018)
10. Gomez-Ojeda, R., Moreno, F.A., Zuñiga-Noël, D., Scaramuzza, D., Gonzalez-Jimenez, J.: Pl-slam: a stereo slam system through the combination of points and line segments. IEEE Trans. Robot. **35**(3), 734–746 (2019)
11. Zhao, Wei., Qian, Kun., Ma, Zhewen., Ma, Xudong, Yu, Hai: Stereo visual SLAM using bag of point and line word Pairs. In: Yu, Haibin, Liu, Jinguo, Liu, Lianqing, Ju, Zhaojie, Liu, Yuwang, Zhou, Dalin (eds.) ICIRA 2019. LNCS (LNAI), vol. 11743, pp. 651–661. Springer, Cham (2019). https://doi.org/10.1007/978-3-030-27538-9_56
12. Yang, Y., Geneva, P., Eckenhoff, K., Huang, G.: Visual-inertial odometry with point and line features. Macau, China, November 2019

13. Burri, M., Nikolic, J., Gohl, P., Schneider, T., Rehder, J., Omari, S., Achtelik, M.W., Siegwart, R.: The euroc micro aerial vehicle datasets. Int. J. Robot. Res. **35**(10), 1157–1163 (2016)
14. Lucas, B.D., et al.: An iterative image registration technique with an application to stereo vision (1981)
15. Von Gioi, R.G., Jakubowicz, J., Morel, J.M., Randall, G.: LSD: a fast line segment detector with a false detection control. IEEE Trans. Pattern Anal. Mach. Intell. **32**(4), 722–732 (2008)
16. Zhang, L., Koch, R.: An efficient and robust line segment matching approach based on LBD descriptor and pairwise geometric consistency. J. Visual Commun. Image Representation **24**(7), 794–805 (2013)
17. Rosten, E., Porter, R., Drummond, T.: Faster and better: a machine learning approach to corner detection. IEEE Trans. Pattern Anal. Mach. Intell. **32**(1), 105–119 (2008)

Recent Trends in Computational Intelligence

Progressive Attentional Learning
for Underwater Image Super-Resolution

Xuelei Chen, Shiqing Wei, Chao Yi, Lingwei Quan, and Cunyue Lu$^{(\boxtimes)}$

Shanghai Jiao Tong University, Shanghai 200240, China
{chenxuelei,weishiqing,ucnpliterme,caleenkwon,lucunyue}@sjtu.edu.cn

Abstract. Visual perception plays an important role when underwater robots carry out missions under the sea. However, the quality of images captured by visual sensors is often affected by underwater environment conditions. Image super-resolution is an effective way to enhance the resolution of underwater images. In this paper, we propose a novel method for underwater image super-resolution. The proposed method uses CNNs with channel-wise attention to learn a mapping from low-resolution images to high-resolution images. And a progressive training strategy is used to deal with large scaling factors (e.g. 4x and 8x) of super-resolution. We name our method as Progressive Attentional Learning (PAL). Experiments on a recently published underwater image super-resolution dataset, USR-248 [11], show the superiority of our method over other state-of-the-art methods.

Keywords: Super-resolution · Underwater image · Progressive learning · Attention mechanism

1 Introduction

Underwater robots are mechatronic systems designed to move around and perform specific tasks in the fluid environment. With the research community paying more and more attention to intelligence and miniaturization, unmanned underwater vehicle becomes a hot topic. Unmanned underwater vehicles can be classified into two major types: Remotely Operated Vehicle (ROV) and Autonomous Underwater Vehicle (AUV). Their difference lie in whether they are tethered to a surface support vessel.

Visual sensors are of great significance to underwater robots. These sensors can be used for both internal and external applications. Internal applications include ego-motion estimation [21], visual servo control [22], underwater object tracking [2], etc. External applications include underwater terrain reconstruction [28], ecology study [23], security applications [1], etc. However, images captured by visual sensors on the underwater robot are often affected by environment conditions like poor visibility, absorption, and scattering [12]. The miniaturization of robots limits the use of high-definition camera. Meanwhile, underwater robots can only survey objects at a large distance in some cases. These all degrade the

© Springer Nature Switzerland AG 2020
C. S. Chan et al. (Eds.): ICIRA 2020, LNAI 12595, pp. 233–243, 2020.
https://doi.org/10.1007/978-3-030-66645-3_20

quality of underwater images and finally impede the vision-based tasks. Single Image Super-Resolution (SISR) can be used to mitigate this problem.

The aim of single image super-resolution is to reconstruct the high-resolution (HR) image from its low-resolution (LR) counterpart. Image super-resolution (SR) is an ill-posed problem because there exist multiple possible high-resolution images for one input low-resolution image. This challenging task has been studied for over two decades. Early methods include example-based method [6], statistical methods [5,24], sparse representation method [29], patch-based method [7], etc. Deep learning has shown its capability in extracting representative features from images and achieving accurate classification or segmentation. As a result, more and more deep learning based image SR methods are being proposed. These methods have remarkable performance improvement compared with conventional methods. SRCNN [4] firstly used deep convolutional neural networks for image SR. VDSR [14] and DRCN [15] achieved great improvement over SRCNN by increasing the network depth. EDSR [18] and DRRN [25] further improved SR results by introducing residual block. RCAN [31] proposed a residual in residual structure and channel-wise attention mechanism to achieve accurate image SR.

However, image SR on underwater imagery is much less studied. Underwater images are different from normal images in many aspects such as dominating green or blue hue, non-linear distortions and degraded colors[12]. Image SR models trained on normal images fail to generate realistic HR underwater images from LR versions. Islam et al. [11] firstly provided an underwater image SR benchmark by publishing a large-scale dataset USR-248 composed of paired LR-HR images. They also proposed a fully-convolutional deep residual network-based generative model SRDRM and its adversarial version SRDRM-GAN. Several state-of-the-art methods and their proposed methods are tested on the USR-248 benchmark. The results of 2x SR and 4x SR is good. When the scaling factor is large (e.g. 8x), images generated by those methods all miss the finer details and lack the sharpness.

We introduce progressive attentional learning (PAL) to solve this problem. Progressive learning is firstly proposed in [13], which makes GAN generate realistic and high-resolution human face images. After that, several studies [19,27] applied progressive GAN to image SR. Attention mechanism can re-weight features to make the model focus on the most informative part of the input. Different attention modules [9,26] have been proposed for image classification task. Channel attention mechanism for image SR is firstly used in [31]. We combine progressive learning and attention mechanism, and propose a new method for underwater image super-resolution.

Overall, the contributions of this paper are as follows: (1) We propose a new pipeline for underwater image SR and introduce attention mechanism to it. (2) We present a new progressive training strategy to address information gap problem when the scaling factor of SR is large. (3) We conduct extensive experiments which confirms that our method outperforms state-of-the-arts without increasing computation complexity.

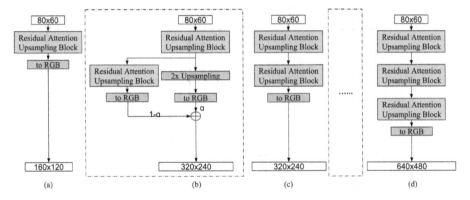

Fig. 1. Illustration of progressive learning for 8x SR, which includes 3 learning phases. Dash line boxes denote the transition process. (a) Network architecture of learning phase 1. (b) Network architecture of the transition process. (c) Network architecture of learning phase 2. (d) Network architecture of learning phase 3.

2   Approach

2.1   Progressive Learning

The proposed progressive learning strategy divides the 8x SR learning process into 3 phases as illustrated in Fig. 1. In learning phase 1, a simple network that can only achieve 2x SR is built. This network consists of one *Residual Attention Upsampling Block* (RAUB) and a *to RGB* convolutional layer. It is trained to generate accurate and realistic 2x HR images. In learning phase 2, an extra RAUB is inserted before the final *to RGB* layer. It is worth noting that the new RAUB is smoothly faded in, which is illustrated in Fig. 1(b). After learning phase 2 is finished, the network can achieve 4x SR. In learning phase 3, a new RAUB is added in the same way as in learning phase 2. The final network can achieve 8x SR.

As mentioned above, smooth fade-in is used when a RAUB is added in a new learning phase. In the beginning of the smooth fade-in, only a *2x Upsampling* layer is added to the network of the previous learning phase. This means that α in Fig. 1(b) is set as 1 and that the side stream does not affect the output. Then α decreases from 1 to 0 linearly with respect to the learning steps. Smooth fade-in avoids sudden shock of adding a bunch of layers into the previous well-trained super-resolution network.

When the scaling factor of SR is 4x, the progressive learning pipeline includes 2 learning phases. When the scaling factor of SR is 2x, progressive learning is not used and the method becomes just attentional learning.

Progressive learning allows the network to firstly focus on coarse structures and then progressively learn finer details. And it uses several learning phases to bridge the information gap between low resolution and very high resolution. Consequently, SR with large scaling factors using the proposed method will contain more details and looks better.

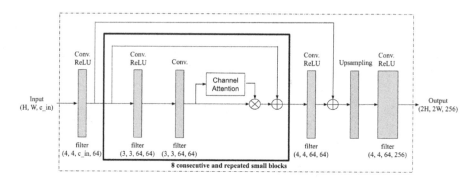

Fig. 2. Network architecture of the Residual Attention Upsampling Block (RAUB).

2.2 Residual Attention Upsampling Block

Residual Attention Upsampling Block (RAUB) is an essential component of the proposed neural network. After each RAUB, the height and the width of the input will be doubled. One RAUB consists of one pre-processing convolutional layer, eight consecutive and repeated small blocks, one post-processing convolutional layer, one upsampling layer and one final convolutional layer. The network architecture of a RAUB is presented in Fig. 2.

Residual structure is first proposed in [8]. This structure makes the network deeper and the training process easier. Residual structure in RAUB are realized in two ways: short skip connection and long skip connection. There are eight short skip connections in eight intermediate small blocks. One long skip connection links the output of the pre-processing convolutional layer and the post-processing convolutional layer. Such combined long and short skip connections further ease the learning process with the network going deeper.

Batch normalization [10] is a popular and effective trick in training convolutional neural networks. It solves the internal covariance shift problem by normalizing the mean and variance of intermediate features. It can stabilize and accelerate the training. However, some recent studies [18,30] find that batch normalization is not suitable for image SR task even though it can improve image classification. Therefore, we remove all batch normalization layers in the proposed network compared to SRDRM [11].

Channel attention mechanism is introduced in our RAUB. As illustrated in Fig. 3, it can re-weight channel-wise features. It includes two streams: an identity stream and a channel attention stream. The channel attention stream uses global average pooling to squeeze the feature map. Then two convolutional layers are used to calculate weights for channels. The outputs of the channel attention stream and the identity stream are multiplied to generate re-weighted feature map. This mechanism enhances the representation ability of the feature map, thus improving image super-resolution performance.

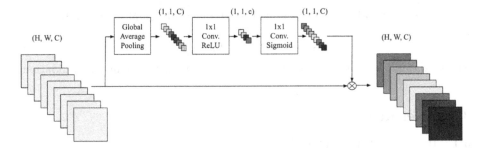

Fig. 3. Illustration of channel attention mechanism.

2.3   Objective Function

The objective of image SR is to obtain high-quality HR images from their LR versions. The proposed neural network for image SR is trained on a dataset using paired LR-HR images. The objective function measures how close and similar the generated HR images are to the ground-truth HR images. There exist many different objective functions for image SR, such as L1 loss, L2 loss, perceptual loss, content loss and adversarial loss. To show the effectiveness of the proposed progressive attentional learning method, we use the same objective function as previous work on underwater image super-resolution [11]. By considering pixel-wise similarity and global perceptual similarity, two loss functions are built.

MSE Loss: Mean squared error (MSE) is a widely-used loss function for learning-based image SR. It measures the pixel-wise similarity. MSE loss is formulated as

$$L_{MSE} = \frac{1}{N} \sum_{i=1}^{N} \frac{1}{HWC} \sum_{x,y,z=1}^{H,W,C} (G(I^{i-LR})_{x,y,z} - I^{i-HR}_{x,y,z})^2 \qquad (1)$$

where $G(\cdot)$ denotes the trained network. I^{i-LR} and I^{i-HR} are the i-th training image pair. N is the number of image pairs in a training batch. x, y and z are the coordinates along three image dimensions. H, W and C are the height, the width and the channel of the image.

Color Perceptual Loss: MSE loss only measures local pixel-wise similarity. It is also important to generate HR images with similar global color perception to ground-truth HR images. As suggested by [11], color perceptual loss is used to describe per-channel disparity, which is formulated as

$$L_{Perceptual} = \frac{1}{N} \sum_{i=1}^{N} \frac{1}{HWC} sum((512 \cdot \mathbf{1} + \mathbf{r}_+) \cdot \mathbf{r}_-^2 + 4 \cdot \mathbf{g}_-^2 + (767 \cdot \mathbf{1} - \mathbf{r}_+) \cdot \mathbf{b}_-^2) \quad (2)$$

where $\mathbf{r}_-$, $\mathbf{g}_-$ and $\mathbf{b}_-$ denotes normalized numeric channel difference matrix of a generated HR image and its ground-truth HR image. $\mathbf{r}_-$ is the mean of red channels. $sum(\cdot)$ denotes the sum operation of all elements of a matrix. $\mathbf{1}$ is an identity matrix of the same size as one channel of the image.

Combining two loss functions mentioned above, the final objective function is formulated as

$$L = \lambda_1 L_{MSE} + \lambda_2 L_{Perceptual} \tag{3}$$

where λ_1 and λ_2 are hyper-parameters that can be tuned in the validation process.

3   Experiments

3.1   Implementation Details

The experiments are conducted on a recently published large-scale underwater image super-resolution dataset, USR-248 [11]. The reason we choose this dataset is that it provides an underwater image super-resolution benchmark on which many state-of-the-art deep learning methods are tested. There are three sets of LR images of size 80×60, 160×120 and 320×240. HR images are of size 640×480. Training set has 1060 LR-HR image pairs. Testing set has 248 LR-HR image pairs.

We use Keras [3] to implement our network. Our codes are trained and tested on Google Colab using Tesla P100 GPU. Adam Optimizer [16] is used with a learning rate of 0.0002. Hyper-parameters λ_1 and λ_2 in the objective function are 0.8 and 0.2. In each phase of progressive learning, images with corresponding resolution are chosen from the dataset to train the network. And during the transition from previous phase to the following phase, weights trained in previous phase are set non-trainable, which means only the added block is trainable in each phase.

For the evaluation of SR results, we use both visual quality comparison and numeric quantity metrics. PSNR and SSIM are two common metrics for quantitative evaluation in SR task. PSNR stands for peak-signal-to-noise ratio. It is defined via MSE between generated images and ground-truth images. SSIM stands for structural similarity index. It is calculated from luminance, contrast and structural comparison functions. Higher PSNR and SSIM indicate better SR performance.

3.2   Ablation Study

In the ablation study, we design three experiments to see how the proposed method improve the performance. The first experiment is the proposed progressive attentional learning method for underwater image SR. The second experiment uses traditional one-phase learning instead of progressive learning to train the network, while keeping channel attention module. The third experiment removes channel attention module in the network, while using progressive learning. Three experiments are trained and tested for 8x image SR. The results are shown in Table 1. We can see that progressive learning and channel attention module both improve the results.

We also visualize the images generated from different phases in progressive learning process in Fig. 4. We can see that the proposed method can learn more and more details from learning phase 1 to learning phase 3.

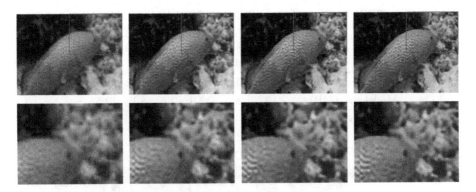

Fig. 4. Images generated from different phases in progressive learning. From left to right: LR image input (80 × 60), learning phase 1 (160 × 120), learning phase 2 (320 × 240), learning phase 3 (640 × 480). They are scaled for better comparison.

Table 1. SR performance in the ablation study

	PSNR	SSIM
PAL w/o Progressive Learning	22.39	0.6224
PAL w/o Channel Attention	22.50	0.6267
PAL	**22.51**	**0.6296**

3.3   Quantitative and Qualitative Evaluation

The results of our method are compared with other state-of-the-art methods, which are SRResNet [17], SRCNN [4], DSRCNN [20] and SRDRM [11]. The quantitative evaluation of SR using different methods is shown in Table 2. It is worth noting that PAL for 2x SR does not use progressive learning because the scaling factor 2x is small. We can see that the proposed PAL method outperforms other methods in both PSNR and SSIM metrics.

Table 2. Comparison of PSNR and SSIM for 2x/4x/8x SR results using different methods

	PSNR	SSIM
SRResNet [17]	25.98/24.15/19.26	0.72/0.66/0.55
SRCNN [4]	26.81/23.38/19.97	0.76/0.67/0.57
DSRCNN [20]	27.14/23.61/20.14	0.77/0.67/0.56
SRDRM [11]	28.36/24.64/21.20	0.80/0.68/0.60
PAL	**28.41/24.89/22.51**	**0.80/0.69/0.63**

Table 3. Computation efficiency of SRDRM and PAL

	SRDRM [11]			PAL		
	2x	4x	8x	2x	4x	8x
#Parameters (million)	0.83	1.90	2.97	0.83	1.92	2.99
Model size (MB)	3.5	8.0	12.0	3.5	8.1	12.5
Efficiency (fps)	7.11	6.86	4.07	–	–	–

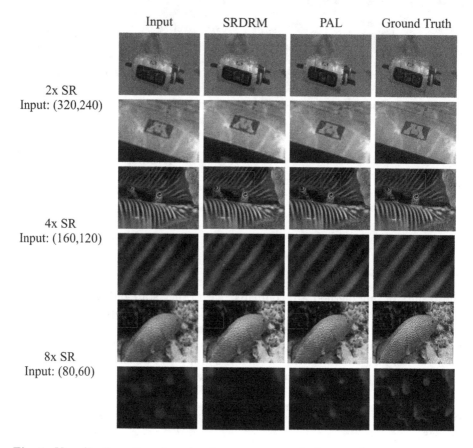

Fig. 5. Visualization of underwater image super-resolution results. Low-resolution images are scaled for better comparison.

We visualize some SR results in Fig. 5. We can observe that SRDRM [11] method and the proposed PAL method both can generate visually better images from low-resolution input. The generated images have more details than the input image, even though the generated images are not comparable with the ground-truth image. We can also see that image generated by the proposed PAL method have more details than those generated by SRDRM [11] method.

Overall, the proposed PAL method surpasses the state-of-the-art methods in both visual quality comparison and numeric quantity metrics.

3.4 Computation Efficiency

When we deploy underwater image super-resolution methods on real underwater robots, computation efficiency should be taken into consideration. NVIDIA Jetson TX2 is a fast and power-efficient embedded AI computing device that can be integrated with underwater robots. SRDRM [11] method achieves 4–7 fps on NVIDIA Jetson TX2. However, we do not have this computing device currently. We compare the proposed PAL with SRDRM [11] using the model size and the number of parameters. From Table 3, we can see that PAL barely increases computation complexity. The model size and the number of parameters using listed two methods are almost the same. These prove the feasibility of using the proposed PAL method for real-time onboard underwater image super-resolution.

4 Conclusion

In this paper, we present a progressive attentional learning method for underwater image super-resolution. We introduce channel attention mechanism to this task and propose a residual attention upsampling block (RAUB) to build the network. Progressive learning strategy is proposed to address information gap problem when the scaling factor of SR is large (e.g. 8x). Thorough experiments on USR-248 [11] dataset show the effectiveness of several components of the proposed method. And the results of the proposed method have great improvement over state-of-the-art methods.

Future research can be focused on adding adversarial loss to the proposed method, because GAN-based image SR method can make the generated images more perceptually realistic. Another research direction is to further compress the model without degrading performance, because light-weight model will fit in more robotic applications.

References

1. Anwar, B.M.M., Ajim, M.A., Alam, S.: Remotely operated underwater vehicle with surveillance system. In: 2015 International Conference on Advances in Electrical Engineering (ICAEE), pp. 255–258. IEEE (2015)
2. Balasuriya, B., Takai, M., Lam, W., Ura, T., Kuroda, Y.: Vision based autonomous underwater vehicle navigation: underwater cable tracking. In: Oceans' 1997. MTS/IEEE Conference Proceedings, vol. 2, pp. 1418–1424. IEEE (1997)
3. Chollet, F., et al.: Keras (2015). https://keras.io
4. Dong, C., Loy, C.C., He, K., Tang, X.: Learning a deep convolutional network for image super-resolution. In: Fleet, D., Pajdla, T., Schiele, B., Tuytelaars, T. (eds.) ECCV 2014. LNCS, vol. 8692, pp. 184–199. Springer, Cham (2014). https://doi.org/10.1007/978-3-319-10593-2_13

5. Fattal, R.: Image upsampling via imposed edge statistics. In: ACM SIGGRAPH 2007 Papers, pp. 95-07, Association for Computing Machinery, New York, NY, USA (2007). https://doi.org/10.1145/1275808.1276496
6. Freeman, W.T., Jones, T.R., Pasztor, E.C.: Example-based super-resolution. IEEE Comput. Graph. Appl. **22**(2), 56–65 (2002)
7. Glasner, D., Bagon, S., Irani, M.: Super-resolution from a single image. In: 2009 IEEE 12th International Conference on Computer Vision, pp. 349–356. IEEE (2009)
8. He, K., Zhang, X., Ren, S., Sun, J.: Identity mappings in deep residual networks. In: Leibe, B., Matas, J., Sebe, N., Welling, M. (eds.) ECCV 2016. LNCS, vol. 9908, pp. 630–645. Springer, Cham (2016). https://doi.org/10.1007/978-3-319-46493-0_38
9. Hu, J., Shen, L., Sun, G.: Squeeze-and-excitation networks. In: Proceedings of the IEEE Conference on Computer Vision and Pattern Recognition, pp. 7132–7141 (2018)
10. Ioffe, S., Szegedy, C.: Batch normalization: accelerating deep network training by reducing internal covariate shift. arXiv preprint arXiv:1502.03167 (2015)
11. Islam, M.J., Enan, S.S., Luo, P., Sattar, J.: Underwater image super-resolution using deep residual multipliers. In: IEEE International Conference on Robotics and Automation (ICRA). IEEE (2020)
12. Islam, M.J., Xia, Y., Sattar, J.: Fast underwater image enhancement for improved visual perception. IEEE Robot. Autom. Lett. **5**(2), 3227–3234 (2020)
13. Karras, T., Aila, T., Laine, S., Lehtinen, J.: Progressive growing of GANs for improved quality, stability, and variation. In: International Conference on Learning Representations (2018)
14. Kim, J., Kwon Lee, J., Mu Lee, K.: Accurate image super-resolution using very deep convolutional networks. In: Proceedings of the IEEE Conference on Computer Vision and Pattern Recognition, pp. 1646–1654 (2016)
15. Kim, J., Kwon Lee, J., Mu Lee, K.: Deeply-recursive convolutional network for image super-resolution. In: Proceedings of the IEEE Conference on Computer Vision and Pattern Recognition, pp. 1637–1645 (2016)
16. Kingma, D.P., Ba, J.: Adam: method for stochastic optimization. arXiv preprint arXiv:1412.6980 (2014)
17. Ledig, C., et al.: Photo-realistic single image super-resolution using a generative adversarial network. In: Proceedings of the IEEE Conference on Computer Vision and Pattern Recognition, pp. 4681–4690 (2017)
18. Lim, B., Son, S., Kim, H., Nah, S., Mu Lee, K.: Enhanced deep residual networks for single image super-resolution. In: Proceedings of the IEEE Conference on Computer Vision and Pattern Recognition Workshops, pp. 136–144 (2017)
19. Mahapatra, D., Bozorgtabar, B., Garnavi, R.: Image super-resolution using progressive generative adversarial networks for medical image analysis. Comput. Med. Imaging Graph. **71**, 30–39 (2019)
20. Mao, X.J., Shen, C., Yang, Y.B.: Image restoration using convolutional auto-encoders with symmetric skip connections. arXiv preprint arXiv:1606.08921 (2016)
21. Negahdaripour, S., Xu, X., Khamene, A., Awan, Z.: 3-D motion and depth estimation from sea-floor images for mosaic-based station-keeping and navigation of ROVS/AUVS and high-resolution sea-floor mapping. In: Proceedings of the 1998 Workshop on Autonomous Underwater Vehicles (Cat. No. 98CH36290), pp. 191–200. IEEE (1998)
22. Park, J.Y., Jun, B.H., Lee, P.M., Oh, J.: Experiments on vision guided docking of an autonomous underwater vehicle using one camera. Ocean Eng. **36**(1), 48–61 (2009)

23. Sheehan, E.V., Bridger, D., Nancollas, S.J., Pittman, S.J.: Pelagicam: a novel underwater imaging system with computer vision for semi-automated monitoring of mobile marine fauna at offshore structures. Environ. Monit. Assess. **192**(1), 11 (2020)

24. Sun, J., Xu, Z., Shum, H.Y.: Image super-resolution using gradient profile prior. In: 2008 IEEE Conference on Computer Vision and Pattern Recognition, pp. 1–8. IEEE (2008)

25. Tai, Y., Yang, J., Liu, X.: Image super-resolution via deep recursive residual network. In: Proceedings of the IEEE Conference on Computer Vision and Pattern Recognition, pp. 3147–3155 (2017)

26. Wang, F., et al.: Residual attention network for image classification. In: Proceedings of the IEEE Conference on Computer Vision and Pattern Recognition, pp. 3156–3164 (2017)

27. Wang, Y., Perazzi, F., McWilliams, B., Sorkine-Hornung, A., Sorkine-Hornung, O., Schroers, C.: A fully progressive approach to single-image super-resolution. In: Proceedings of the IEEE Conference on Computer Vision and Pattern Recognition Workshops, pp. 864–873 (2018)

28. Weidner, N., Rahman, S., Li, A.Q., Rekleitis, I.: Underwater cave mapping using stereo vision. In: 2017 IEEE International Conference on Robotics and Automation (ICRA), pp. 5709–5715. IEEE (2017)

29. Yang, J., Wright, J., Huang, T., Ma, Y.: Image super-resolution as sparse representation of raw image patches. In: 2008 IEEE Conference on Computer Vision and Pattern Recognition, pp. 1–8. IEEE (2008)

30. Yu, J., et al.: Wide activation for efficient and accurate image super-resolution. arXiv preprint arXiv:1808.08718 (2018)

31. Zhang, Y., Li, K., Li, K., Wang, L., Zhong, B., Fu, Y.: Image super-resolution using very deep residual channel attention networks. In: Proceedings of the European Conference on Computer Vision (ECCV), pp. 286–301 (2018)

Movie Genre Filtering for Automated Parental Control

Zuo Jun Yong[(✉)] and Wai Lam Hoo

Centre of Image and Signal Processing, Department of Information Systems, Faculty of Computer Science and Information Technology, Universiti Malaya, Jalan Universiti, 50603 Kuala Lumpur, Wilayah Persekutuan Kuala Lumpur, Malaysia
wva190009@siswa.um.edu.my, wlhoo@um.edu.my

Abstract. With cloud robotics, particularly robotic vision available within a household, human are able to live a convenient and safer life in an ambient assisted living environment. Recent advances in computational intelligence including neural network improves the computational capability of the robotic vision to better understand the environment. Recently, internet hoaxes that affected the social community greatly have raised strong awareness among public in parental control and the content that the youngster can view. Therefore, this paper focuses on filtering movies or videos that is not suitable for youngster by attempting to identify movie genre. Movie genre classification has been investigated in recent years, but there exist noise in normal videos referred as generic frames, as mentioned in [1], that makes differentiation movies with similar frame difficult. A filtering approach is proposed in this paper in order to identify generic frames within the video and discard them from genre classification process, in order to improve genre classification performance. Experiment shows that the filtering approach are able to improve action genre class, but have difficulties and improving other genre classes.

Keywords: Robotic vision · Movie genre filtering · Parental control · Generic frames

1 Introduction

Ambient assisted living has been crucial in making elderly live better life, and robotics support have played a major role in it. Cloud robotics making the internet of homes more intelligent with the help of connected devices with sensors, as well as computational intelligence algorithms such as neural network that classifies and detects the anomalies and response to it. Despite the recent advances in these area, parental control proves to be an important aspect due to the recent outbreak of internet hoaxes that harms the social community, especially youngsters. Internet hoaxes have affected the social media greatly, and it is difficult to stop these video to be uploaded from the internet. This is due to the huge amount of content within a video and the number of videos daily uploaded,

© Springer Nature Switzerland AG 2020
C. S. Chan et al. (Eds.): ICIRA 2020, LNAI 12595, pp. 244–253, 2020.
https://doi.org/10.1007/978-3-030-66645-3_21

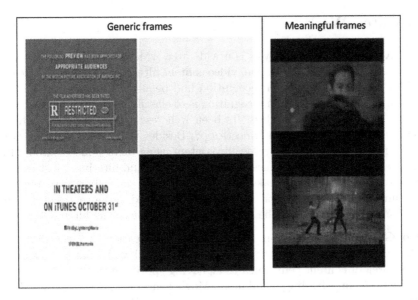

Fig. 1. Identifying generic frames from meaningful frames to reduce confusion between movie genre.

its difficult to have efficient technique that can filter all the irrelevant content or prevent the videos from uploading. Therefore, an efficient technique to filter these malicious content after uploaded to the social media and quickly disabled it becomes important.

Video learning area had been very active for the past several years with having the top 5 papers accepted in most of the conference. The most popular section of video learning is action recognition. Clips of human action that usually range between 10 to 20 s is used to perform action recognition and learning [2–4]. To tackle the problem of users uploading irrelevant or inappropriate content, video learning should be able to achieve learning general representation of video and its content. But videos is like arts, different artist will have different drawing style and video from different platforms will have different style of making their content. Besides, filtering will also need a list to refer to. The movie genre is one of the reasonable way to be use as a guidance, because movies come in different shape just like an art while also having age restriction applied to it. By leveraging movie genre classification, categorizing videos can be achieved. However, there exists clips of image that is generic and causing noise to genre classification [1].

Generic frames is described as human chatting and landscaping occurring in action genre [1], but after investigating the dataset, there exist other form of generic frames such as completely dark clips, trailer intro sections and too bright clips as shown in Fig. 1. To detect and filter these clips, motion difference between frames within a clip is take into account. Thus, this project will be doing low motion generic frames detection and filtering to enhance movie genre classification.

2 Related Work

Video Analysis. Video analysis is a wide area which can lead down to video classification, video summarization, video content filtering, and more. Video classification in the area of action recognition had been the fundamental for video classification, the aim of action recognition is to classify short clips of input video into targeted human action class. Its been well researched on action datasets, HMDB50 [4], UCF101 [3] and Kinectics [2]. Besides that, video summarization aim to generate summary of a video clip content using texts. Video content filtering consist of two parts, video segment classification and filtering. Most research on video content filtering breaks down a video clips into shorter segments or simply images, each segment or image will be classified into a target class. After that, filtering techniques will be use to remove or replace the filtered segments.

Video Genre Classification. Video genre classification is a specified topic which deals with video consist of more general and vast content. There are different area of content genre such as tv having sports, media and more, movie having action, comedy, drama and more. Movie genres is intangible, it is difficult to define a movie genre from a single frame. Classifying a movie genre need to take in a whole clip and multiple type of features. In this paper, several CNNs is used to extract and learn different aspect of a movie clip and in the end classify movie genre.

Video-Based CNN. Video based CNN are the state-of-the-art methods to use to classify a video. State-of-the-art CNN methods such as Res3D [5], Res2+1D [6], C3D [7] and I3D [8] performed well at action recognition problems. Although these 3D CNNs had achieve state-of-the-art at action recognition, but action datasets usually consist of video not longer than 10 s and maximum up to 1 min. Movie trailer dataset consist of video with duration up to 5 min. Thus to classify these trailers, a trailer will be divided into multiple short clips to allow these 3D CNNs performs genre classification.

Data Preprocessing. Dataset is collected through movielens 20 m dataset [9]. The dataset is going though preprocessing of collection, selection and cleaning. We narrow the dataset genre to only 4 genres action, comedy, drama and horror. To avoid problems faced in (movie genre classification through scene), we selectively pick trailer with only single genre. Each trailer consist of multiple cuts or scenes, using scene divide method, histogram of color and local minimum detection, each movie trailer is divided into multiple clips of images. Then by filter through each clips, we take a set of images with the length of 16 (the input size of 3D CNN), if a scene has 32 images, it will be divide futher again into 2 sets of 16 images clip, if a scene has less than 16 images then we ignore it and check the next scene.

Movie Genre Classification. To learn and classify movie trailer clips, multimodality is proposed by [10].Different modality feature is obtained from a single movie clips, such as motion,objects,scene and movie feature. This paper used C3D pretrained modal on sport 1m [11], googleNet [12] pretrained on imageNet

[13], googleNet pretrained on places 365 [14] and googleNet train on movie trailer from scratch. Then all classification results is combined to gether and feed into Support Vector Machine(SVM) for final result. Their paper shows that applying multi-modality can improve the classification results, besides they also apply different combination of modality combination experiment to improve classification result [1]. But these papers does not apply any noise removal to deal with generic frames [1]. In this paper, generic frames with low motion information will be taken into consideration and tackled to improve movie genre classification.

Content Filtering. Filtering content had been surrounding on issues of spams [15] [16], inappropriate content [17] and adult content [18,19]. There exist video spams and bag of visual feature(BoVF) is proposed to tackle it [15]. This paper explore low level image features by using handcraft image descriptor SIFT and STIP, then BoVF is built from features extracted, to determine whether the video is spam or not the BoVF feature is then projected to latent space and classify by Support Vector Machine(SVM) [16]. As for inappropriate content, this issues is highly concerns in YouTube as their content creator is worldwide and unpredictable, a lot of information exist in YouTube that needs to be taken in consider together to apply to their own guidelines, [17] takes in all the provided information of a YouTube video and construct a learning model from it. but the result is heavily text based classification. As for adult content, bag of visual words(BoVW) [19] and skin exposure [18] detection is used to tackle. This paper focus on issue of generic frames which does not fall into related works category of filtering task.

3   Methodology

The methodology flow is summarized as in Fig. 2.

Data Preprocessing. Dataset is collected through movielens 20 m dataset [9]. The dataset is going though preprocessing of collection, selection and cleaning. We narrow the dataset genre to only 4 genres action, comedy, drama and horror. To avoid problems faced in (movie genre classification through scene), we selectively pick trailer with only single genre. Each trailer consist of multiple cuts or scenes, using shot detection method [20]. A video shot is defined as a sequence of frames taken by a single camera without any major change in the color content of consecutive images. using color histogram space hue(H), saturation(S) and value(V) to obtain the representation of image in color histogram space. The representation vector (H) in HSV is 16 bin, 8 for hue, 4 for saturation and 4 for value. To obtain a value to represent an image on a time frame, all 16 bin values is sum up into H. The objective of scene splitting is to find the splitting point of a video to generate clips. Shot boundary is calculated by applying min function for each bins between two consecutive H vectors. Then the minimum function output of 16 bins is sum up in to shot boundary value S. To determine which section of the video to be separated into shots, a gaussian local minimum algorithm with distance of 1 is used. A list of shot separation is obtained and the video is divided into scenes respect to the shot boundary location. Then by

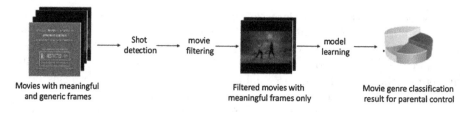

Fig. 2. Methodology flow.

filter through each clips, we take a set of images with the length of 16 (the input size of 3D CNN), if a scene has 32 images, it will be divide further again into 2 sets of 16 images clip, if a scene has less than 16 images then we ignore it and check the next scene.

$$S(i) = \sum_{j \in allbins} \min H_i(j), H_{i-1}(j)$$

Generic Frame Filtering. This clip exist in almost every video as a brief warning for the video might contain violence or sexual content. These clips does not provide information to help genre classification. Besides, these clips often exist in movies because of the splitting preprocessing aims to obtain low changes in HSV color space of video, there are nothing inside this clips except completely black. To tackle these two problems, motion difference [21] is used. Each clip of 16 images is converted into grey scale, pixel difference is the absolute difference between current frame and next frame, then by loop through all pixels on the frame difference image, if pixel value >1 then total pixel difference counter increase. There are two problems, thus two approaches is proposed to tackle each problem. By averaging the total pixel count between 15 motion images, trailer intro clips and black clips can be detected. Another approach is using minimum number of pixel difference exist within a clip. For each video V, there is multiple clips C, for each clip there is 16 frames F. Frame difference D is calculated by absolute difference between two frames F_n and F_{n+1}. A counting function $g(x)$ is used to count all pixel values in D above threshold of 0. First approaches is averaging within clip, each result P_y is the output of $g(x)$ is sum up and divided by 15, total pixel counts is obtain by adding all P. Second approach is similar to first, but instead of using average through clips, the lowest number of pixel difference is obtain as final total pixel counts. The total pixel counts is then use as a guidance value to find the best threshold points for filtering to occurs.

$$V = C_0...C_n$$
$$C_n = F_1, F_2, .., F_x$$
$$D_y = abs(F_x - F_{x+1})$$
$$P_y = \sum_{y=0}^{1} 5\, g(D_{yij}),$$

where n = total clip length, $x = 16$, $y = 15$, $i, j = [0, 255]$, and

$$g(D_{yij}) = \begin{cases} 1, & \text{if } D_{yij} > 0 \\ 0, & \text{otherwise.} \end{cases}$$

To find the most optimum filtering amount from dataset, there are two things to consider, percentage of clips to filter out and filtering threshold for total pixel count. By stating a specific percentage of clips to filter out of full dataset, the threshold of total pixel amount is adjusted with increment of 50. The threshold for nearest percentage will be taken as final and all the clips fall below the threshold will be remove before going through googleNet.

After filtering, The remaining of the dataset will be pass into imagenet pre-trained googleNet [12] to retrain. The setting of googleNet is batch size 128, learning rate of 1×10^{-4} and learning rate decay of $5e^{-4}$. The output of googleNet is 4 genre category, action, comedy, drama and horror.

4   Experiment Result

Dataset. The original dataset of baseline [1] is unavailable. Therefore, movie-Lens 20 m [9] is used. The preprocessing of dataset follows shot detection [20].

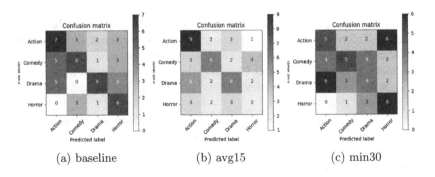

(a) baseline   (b) avg15   (c) min30

Fig. 3. Result analysis on video classification.

Table 1. Best results

Method	Action	Comedy	Drama	Horror	Overall accuracy
From Paper Baseline [10]	0.47	0.33	0.27	0.1	0.31
Reimplement Baseline	0.47	**0.40**	0.40	**0.60**	**0.46**
Average-15	**0.60**	0.40	0.40	0.20	0.39
Minimum-30	0.40	0.27	**0.47**	0.50	0.40

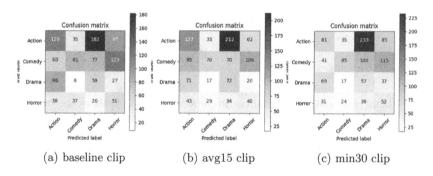

(a) baseline clip   (b) avg15 clip   (c) min30 clip

Fig. 4. Result analysis on extracted video clips.

The filtering process is depends on pixel threshold of a clip, so to find the most optimum threshold, a percentage of total clips to filter is considered, this is to prevent sample imbalanced. By going through experiments, the most optimum filtering threshold is pixel threshold 100, 150 and 200, the resulting percentage of clip filtered for average approach is 5%, 15% and 30%, for minimum approach is 5% 30% and 40%. Baseline did not go through the filtering process.

5 Discussions

This paper focus on Imagenet pretrained googleNet performance with and without filtering. Referring to Table 1, The baseline method without filtering achieve best in overall result.The first approach motion difference average with 15% filtering achieve best at Action and Comedy, the second approach is min motion difference with 30% filtering achieving best at drama classification but lacking in others.

Refer to Table 2, averaging approaches is achieving different outcome through different percentage of filtering. While Minimum approaches is achieving consistent result with the best being 30% removal.

To see how much major voting affecting the final classification, Table 3 shows clip voting result versus video classification result. Video base classification perform better than clips classification, but in clips classification, proposed average filtering approach is very close to baseline method. In clip based classification, average approaches achieve best result in action and drama, minimum approach achieve best in comedy and horror.

By referring to Fig. 3 video voted classification, baseline is getting overall best performance while have high probability to classify video as Action or Horror. Average approach have higher probability to classify video as Action but have less video classify as horror. Minimum approaches have high probability to classify video as horror compare to other genre.

Looking at Fig. 4 clip voted classification, the confusion matrix is not as clear as video ones, but baseline and average approach is very similar, while minimum approach has high probability to classify video as Drama and Horror.

Table 2. Different removal percentage setting

Approaches	Action	Comedy	Drama	Horror	Overall accuracy
Average-5	0.27	0.33	0.33	0.40	0.33
Average-15	**0.60**	**0.40**	0.40	0.20	0.39
Average-30	0.33	0.33	0.27	**0.60**	0.38
Minimum-5	0.25	0.36	0.36	0.33	0.32
Minimum-30	0.40	0.27	**0.47**	0.50	**0.40**
Minimum-40	0.27	0.33	**0.47**	0.40	0.36

Table 3. Clip classification vs video classification

Approaches	Action	Comedy	Drama	Horror	Overall accuracy
Video Baseline	0.47	**0.40**	0.40	**0.60**	**0.46**
Average-15	**0.60**	**0.40**	0.40	0.20	0.39
Minimum-30	0.40	0.27	**0.47**	0.50	**0.40**
Clip Baseline	0.28	0.24	0.33	0.35	**0.29**
Average-15	**0.29**	0.21	**0.40**	0.27	**0.29**
Minimum-30	0.19	**0.25**	0.32	**0.36**	0.27

6   Conclusion

In conclusion, two filtering approach namely average approach and minimum approach are investigated to remove generic frames within movie clips to improve genre classification performance. The proposed average approach is having high performance on Action and Comedy while minimum approach is having high performance on Horror and Drama. Average approach is able to detect clips with low average motion changes thus removing most of the clips with words foreground clips and clips with completely black and white. This reduced the probability of classification towards Horror genre as shown in experiment result. The minimum approaches in the other hand takes the lowest motion changes frame of each clip as consideration and remove the clip from learning, verifying with Fig. 4(c), it seems that clip base result will most probably classify as Horror with very high error rate.

In future work, more genres class can be investigated to observe and improve the effect of generic frames removal in overall genre classsification outcome. In addition, considering the significant result contrast between video-based and clip-based classification, improvement in formation of clips from existing work [20], can be a potential future direction.

Acknowledgements. Funding was provided by Fundamental Research Grant Scheme (FRGS) MoHE (Grant No. FP054-2019A) and Small Research Assistance (BKP) (Grant No. BK030-2018).

References

1. Simões, G.S., Wehrmann, J., Barros, R.C., Ruiz, D.D.: Movie genre classification with convolutional neural networks. In: 2016 International Joint Conference on Neural Networks (IJCNN), pp. 259–266. IEEE (2016)
2. Kay, W., et al.: The kinetics human action video dataset. arXiv preprint arXiv:1705.06950 (2017)
3. Soomro, K., Zamir, A.R., Shah, M.: Ucf101: a dataset of 101 human actions classes from videos in the wild. arXiv preprint arXiv:1212.0402 (2012)
4. Kuehne, H., Jhuang, H., Garrote, E., Poggio, T., Serre, T.: Hmdb: a large video database for human motion recognition. In: 2011 International Conference on Computer Vision, pp. 2556–2563. IEEE (2011)
5. Tran, D., Ray, J., Shou, Z., Chang, S., Paluri, M.: Convnet architecture search for spatiotemporal feature learning. arXiv preprint arXiv:1708.05038 (2017)
6. Tran, D., Wang, H., Torresani, L., Ray, J., LeCun, Y., Paluri, M.: A closer look at spatiotemporal convolutions for action recognition. In: Proceedings of the IEEE Conference on Computer Vision and Pattern Recognition, pp. 6450–6459 (2018)
7. Tran, D., Bourdev, L., Fergus, R., Torresani, L., Paluri, M.: Learning spatiotemporal features with 3D convolutional networks. In: Proceedings of the IEEE International Conference on Computer Vision, pp. 4489–4497 (2015)
8. Carreira, J., Zisserman, A.: Quo vadis, action recognition? a new model and the kinetics dataset. In: proceedings of the IEEE Conference on Computer Vision and Pattern Recognition, pp. 6299–6308 (2017)
9. Harper, F.M., Konstan, J.A.: The movielens datasets: history and context. ACM Trans. Interact. Intell. Syst. (TIIS) **5**(4), 1–19 (2015)
10. Wehrmann, J., Barros, R.C., Simões, G.S., Paula, T.S., Ruiz, D.D.: (Deep) learning from frames. In: 2016 5th Brazilian Conference on Intelligent Systems (BRACIS), pp. 1–6. IEEE (2016)
11. Karpathy, A., Toderici, G., Shetty, S., Leung, T., Sukthankar, R., Fei-Fei, L.: Large-scale video classification with convolutional neural networks. In: Proceedings of the IEEE Conference on Computer Vision and Pattern Recognition, pp. 1725–1732 (2014)
12. Szegedy, C., et al.: Going deeper with convolutions. In: Proceedings of the IEEE Conference on Computer Vision and Pattern Recognition, pp. 1–9 (2015)
13. Russakovsky, O., Deng, J., Hao, S., Krause, J., Satheesh, S., Ma, S., Huang, Z., Karpathy, A., Khosla, A., Bernstein, M., Berg, A.C., Fei-Fei, L.: ImageNet large scale visual recognition challenge. Int. J. Comput. Vis. (IJCV) **115**(3), 211–252 (2015)
14. Bolei, Z., Agata, L., Aditya, K., Aude, O., Antonio, T.: Places: a 10 million image database for scene recognition. IEEE Trans. Pattern Anal. Mach. Intell. **40**(6), 1452–1464 (2017)
15. Benevenuto, F., Rodrigues, T., Almeida, V., Almeida, J., Gonçalves, M.: Detecting spammers and content promoters in online video social networks. In: Proceedings of the 32nd International ACM SIGIR Conference on Research and Development in Information Retrieval, pp. 620–627 (2009)
16. da Luz, A., Valle, E., Araujo, A.: Content-based spam filtering on video sharing social networks. arXiv preprint arXiv:1104.5284 (2011)
17. Alcântara, C., Moreira, V., Feijo, D.: Offensive video detection: dataset and baseline results. In: Proceedings of The 12th Language Resources and Evaluation Conference, pp. 4309–4319 (2020)

18. Wayne, K., Andrew, D., Derek, M.: Screening for objectionable images: a review of skin detection techniques. In: International Machine Vision and Image Processing Conference, vol. 2008, pp. 151–158 (2008)

19. Moreira, D., Avila, S., Perez, M., Moraes, D., Testoni, V., Valle, E., Goldenstein, S., Rocha, A.: Pornography classification: the hidden clues in video space-time. Forensic Sci. Int. **268**, 46–61 (2016)

20. Rasheed, Z., Sheikh, Y., Shah, M.: On the use of computable features for film classification. IEEE Trans. Circ. Syst. Video Technol. **15**(1), 52–64 (2005)

21. Singla, N.: Motion detection based on frame difference method. Int. J. Inf. Comput. Technol. **4**(15), 1559–1565 (2014)

Bridging Explainable Machine Vision in CAD Systems for Lung Cancer Detection

Nusaiba Alwarasneh, Yuen Shan Serene Chow, Sarah Teh Mei Yan, and Chern Hong Lim[⊠]

Monash University, Selangor, Malaysia
{nalw0001,ycho0014,steh0004,lim.chernhong}@student.monash.edu

Abstract. Computer-aided diagnosis (CAD) systems have grown increasingly popular with aiding physicians in diagnosing lung cancer using medical images in recent years. However, the reasoning behind the state-of-the-art black-box learning and prediction models has become obscured and this resultant lack of transparency has presented a problem whereby physicians are unable to trust the results of these systems. This motivated us to improve the conventional CAD with a more robust and interpretable algorithms to produce a system that achieves high accuracy and explainable diagnoses of lung cancer. The proposed approach uses a novel image processing pipeline to segment nodules from lung CT scan images, and then classifies the nodule using both 2D and 3D Alexnet models that have been trained on lung nodule data from the LIDC-IDRI dataset. The explainability aspect is approached from two angles: 1) LIME that produces a visual explanation of the diagnosis, and 2) a rule-based system that produces a text-based explanation of the diagnosis. Overall, the proposed algorithm has achieved better performance and advance the practicality of CAD systems.

Keywords: Cancer detection · Machine vision · Deep learning · Image processing · Explainable AI

1 Introduction

Do you trust a computer system telling you that you have an aggressive cancerous tumor without any referencing facts? Artificial Intelligence (AI) based computer-aided diagnosis (CAD) system have been continuously reported with eye-catching performances in medical diagnosis [16]. However, its clinical value has not yet been realised, hindered partly by the lack of a clear understanding on how the reasoning is done in background due to non-transparent black-box AI model. This has motivated us to investigate further and improve the CAD system with proposing a more robust framework with interpretable features, but narrowed to only lung cancer detection.

In brief, lung cancer is a malignancy of the lungs that can arise anywhere in the respiratory tract from the cells lining the trachea to the cells of the alveoli.

© Springer Nature Switzerland AG 2020
C. S. Chan et al. (Eds.): ICIRA 2020, LNAI 12595, pp. 254–269, 2020.
https://doi.org/10.1007/978-3-030-66645-3_22

Globally, lung cancer is the most common cancer in men and the third most common cancer in women, and it remains the leading cause of cancer-related deaths in both men and women [1]. Therefore, diagnosing lung cancer at an early stage is of crucial importance, and various attempts at facilitating this early diagnosis have been conducted. One such attempt involves the use of CAD systems.

Studies conducted have shown that using CAD systems significantly improved the diagnostic accuracy of radiologists when compared to manual diagnosis [2]. Yet, recent advances in CAD systems caused by the transition to Convolutional Neural Network (CNN) and other black-box models has introduced a problem whereby physicians are less willing to trust the results due to a lack of transparency. Several authors cite the increasing need for transparency of CAD systems [17,18], and in fact, clinicians themselves have ranked explainability as their most desired feature in such systems [22].

This research aims to tackle this problem by bridging the gap between performance and explainability, through implementing a fully automated, end-to-end and explainable lung cancer detection system. This is achieved by integrating the novel image processing algorithms, classification model, and the interpretable features in the diagnosis pipeline.

Overall, we present three main contributions. First, our nodule detection algorithm is fully automated where all parameters that are used in the proposed image processing algorithm are automatically estimated which requires no external assistance. We then investigate the effectiveness of using vanilla and improved 2D and 3D AlexNet models to classify lung nodules as benign or malignant. In this step, further comparisons between different normalization techniques and different hyperparameter values are performed. Finally, the approaches that generate the explainable features of the overall system are presented.

The paper begins with a brief overview on related works in Sect. 2. Next, the proposed methodology is discussed, which sets out with image processing algorithms in Sect. 3.1 and then moves to incorporating the results of these algorithms into tuned CNN models in Sect. 3.2 as well as the different approaches for generating the explainable features in Sect. 3.3. We tested the proposed algorithms and models with the LIDC dataset and the results are discussed in Sect. 4. Finally, Sect. 5 concludes the research with suggestions on future works.

2   Related Work

To properly encompass the image processing and machine learning aspects used in the CAD with specifically in Lung cancer, we provide a review on Computer Assisted Detection of Lung Nodules (CADe) and Computer Assisted Diagnosis of Lung Nodules (CADx). Both systems implicitly involves the Image Processing and Machine Learning for the purposes. Apart from that, the review of explainable AI in this context takes place in the last part of this section.

2.1   Computer Assisted Detection of Lung Nodules (CADe)

Automated detection of pulmonary nodules in digital chest images using CAD schemes has been a popular field of research since the early 1990 s. In its early stages, image processing techniques are more often used as a standalone in automated lung detection schemes, assisted by a simple rule-based tests [6]. Currently, most research conducted within this area of study has since shifted to utilising hybrid lung detection models which combine image processing techniques with the usage of neural networks [24].

At its core, automated lung nodule detection systems are largely dependent on its image processing methodology. As such, most research within this field aims at proposing improved image processing algorithms, specifically, image segmentation techniques in order to segment the pulmonary nodules within the lung. One of the earliest works on the development of CAD schemes by Giger et al., first proposed to investigate the characteristic features of pulmonary nodules and its enclosing lung cavity background to distinguish them apart, after which feature extraction techniques is used to isolate suspected nodules. They reported that this detection scheme yields high true-positive (TP) rates and low false-positive (FP) rates, but only within the peripheral lung regions [6].

A novel concept combining the use of image processing methods with the aid of an artificial neural network (ANN) and adaptive rule-based tests was introduced by Xu et al. The paper reported that the ANN contributed to a significant reduction in false positives [24]. Not long after that, Li et al. pushed the research forward by including a lung cancer classification system within their CAD lung nodule detection scheme. Their proposed method includes nodule detection, selective nodule enhancement, feature extraction and classification. The selective nodule enhancement aided in the suppression of normal anatomic structures which resulted in a low false-positive rate. The proposed automated rule-based classifier minimised the overfitting effect with an improved classification performance [13].

Effectiveness of CAD lung nodule detection increased exponentially in 2019 when most proposed research reported a minimum of 90% sensitivity with low false-positive rates. For example, Ayyagari et al. proposed their CAD scheme which includes thresholding and watershed segmentation which reported a 94.92% sensitivity with zero false positive rate [4]. In the same year, Khan et al. presented their hybrid framework which combines segmentation and optimal feature extraction with a Support Vector Machine (SVM) which resulted in an impressive 97.45% sensitivity [10]. Their paper showed that their approach had outperformed many other state-of-the-art approaches such as hybrid models using CNN.

2.2   Computer Assisted Diagnosis of Lung Nodules (CADx)

On the other hand, numerous studies have been reported on the development and performance of CADx for lung nodule classification. The use of neural networks, specifically, has shown a shift from ANNs in the 1990 s and 2000 s, to CNNs following the breakthrough of AlexNet at the ImageNet challenge of 2012 [11], and finally, to explainable CNNs within the last few years.

Some of the earliest works on CADx in lung cancer involves the use of neural networks. For example, Henschke et al. tested a S-MODALS neural network on 28 lung nodules using 4 CT scan slices per nodule with manual feature extraction. They reported that the model misclassified three benign nodules, but no malignant ones [7]. Whereas their study involved fixed NN weights, Matsuki et al. reported using a back-propagation algorithm on a 3-layer, feed-forward ANN on 155 CT scan nodules and showed that the radiologist's diagnoses were significantly improved when aided by the ANN [15].

Yet overall, the use of CNN in early medical literature is extremely limited, and it was not until the recent years that they have gained popularity again, possibly due to the advances in datasets, GPUs and other computational resources. One of the earliest studies that investigated the use of a CNN specifically with lung nodule classification was done by Shen et al., where they presented a novel approach using the MC-CNN to automate nodule extraction followed by malignancy suspiciousness classification and achieved an accuracy of 86.24% [19]. Li et al. investigated using lung CT images in order to classify a nodule based on anatomy – solid, semi-solid and ground glass opacity. They proposed a 2-conv layer and 1-downsampling layer CNN and used images from the LIDC dataset, achieving an accuracy of 85.7% [14]. Similarly, Anthimopoulos et al. also proposed a model to classify lung nodules by anatomy, using a 8-layer CNN (5 conv layers and 3 FC layers). This differed from the previous study in that it used a deep CNN, and this is the first application of a deep CNN to classify lung nodule anatomy. They reported a classification accuracy of 85.5% [3].

A number of studies also examined the use of established CNN architectures, such as AlexNet and GoogleNet. For example, Chon et al. compared using a "vanilla" 3D CNN with a 3D GoogleNet architecture. They initially attempted to use the entire lung scans to train the CNNs to determine whether a 3D-whole lung scan had cancer or not, but the results were poor. This was mitigated by using U-Net to segment the lung nodules and only feed in these nodules into the CNNs. They reported an improved final accuracy of 70.5% for the vanilla 3D CNN vs. 75.1% for the 3D GoogleNet [5]. Hossain et al. used the conventional 3D-AlexNet architecture and the LIDC dataset to train their model to differentiate between cancerous and non-cancerous nodules, and reported an accuracy of 77.78% [8].

2.3 Explainability in CAD

Although much progress has been made in the field of automated classification of lung nodules as presented above, a fundamental problem persists in the black-box nature of such CNNs. It became clearer that as the accuracy of these models improved, their limitation shifted to their lack of human-understandable explanations. The need for transparency is increasingly being cited [17,18], but the availability of studies on explainable AI in medicine remain limited in literature. Lee et al. presented deep CNNs that were trained to classify different types of intracranial haemorrhage, and they used CAM to generate attention maps that highlights the most significant regions relevant to the model predictions. The results demonstrated a 95% agreement between their explained outputs and radiologist assessment, indicating that the model learned relevant classification features [12]. Using a similar approach, Chuang et al. trained CNNs to detect neuropathologies from slides of brain tissue, and explained their CNN results using Grad-CAM. Comparison with radiologist assessment showed that Grad-CAM was able to successfully demonstrate the features of importance towards diagnosis-making [21].

Palatnik et al. opted for a second approach to explainability by using LIME. They ran two CNN models, VCG16 and another publicly available Keras CNN model with a high AUC, and trained the models to detect malignant metastases in lymph node histopathology slides. They explained their output using heatmaps generated from LIME [20].

All in all, while CADe and CADx systems have evolved impressively over the years as discussed above, common limitations still persist in the form of the lack of 1) fully automated, end-to-end diagnostic systems, and 2) explainability. Our proposed system thus aims to combat these limitations by introducing a novel integrated framework for CAD system.

3   Methodology

To tackle the limitations in the current CAD systems, this research proposed the integration of three important modules into one underlying framework in obtaining a robust diagnosis outcome with reasoning. Firstly, a novel image processing algorithm is implemented to detect and crop the lung nodule from lung CT scans. The cropped images are then used to train a CNN model to diagnose the presence of lung cancer (either malignant or benign). In the processing pipeline, extensive experiments have been conducted on fine-tuning the CNN models with the fundamental architecture is based on the 2D and 3D Alexnet. Lastly, LIME algorithm as well as a rule-based reasoning are applied for the reasoning purposes. The details of these modules are in the following sections.

3.1   Nodule Detection

The stack of lung CT scans must be pre-processed to fit the input requirements of the CNN model before a prediction can be made. Pre-processing is required to return results such that: 1. Only slices that has an existing nodule will be returned, 2. Slices that are returned must be cropped to show only the nodule with some slight padding (without the rest of the lung).

Figure 1 depicts the entire pipeline for the proposed image processing algorithm. It begins with white noise removal by using edge preserving Median filtering technique. Then, thresholding is performed with a region growing algorithm which involve a series of erosion and dilation functions to eliminate the tiny features like pulmonary vessels. Next, segmentation is done to separate the distinct

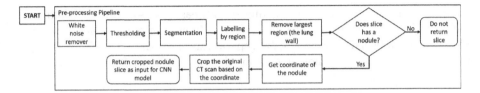

Fig. 1. Overview of image processing pipeline.

lung segments. Several segmentations techniques was tested and ACWE (Active Contours Without Edges) morphological segmentation algorithm was selected as it outperformed the others in this context. Lastly, is to label and identify the separate regions and then determining if the found region is a nodule. This is achieved by using color region labelling. An example image of the segmented lung segments and the cropped lung nodule (after performing the image processing algorithms) are shown in Fig. 2.

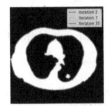

Result of ACWE segmentation.   Result of cropped lung nodule.

Fig. 2. Results from image processing pipeline.

A region is then considered as a nodule if it has: i. total area of less than 20000 pixels, and ii. solidity of 90% or more. This is because pulmonary nodules typically have high solidity. The solidity parameter of 90% is chosen after testing the image processing algorithm on various patients and realising that the solidity >90% is the most optimum parameter that produces the best result. Setting the solidity to be >90% reduces the possibility of tiny air pockets to be mis-classified as a nodule since air pockets usually have lower solidity. If a region has an area of more than 20000 pixels, it is likely not to be a nodule as it is almost the area size of the lung wall. No nodule can have an area that large. The values for the parameters were chosen after testing a range of values before observing that this pair of values returns the most optimum result across all the tests done.

3.2 Classification Using Deep Learning

The binary classification problem tackled is the differentiation between malignant and benign nodules. To this end, a 2D-AlexNet and a 3D-AlexNet were trained using data from the Lung Image Database Consortium (LIDC-IDRI) dataset. This data is in the form of individual images of cropped slices of nodules, which can be fed into the CNN as separate slices for the 2D-CNN, or stacked together and fed in as a whole nodule for the 3D-CNN (Fig. 3).

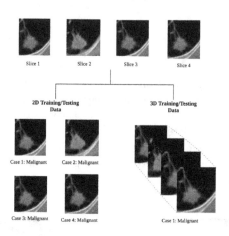

Fig. 3. The same nodule in 2D and 3D format. This nodule presents 4 training/testing cases for a 2D-CNN, and a single case for a 3D-CNN.

In the processing pipeline, first, the data must be normalized. Unlike normal RGB image data with pixels in the range [0, 255], DICOM pixel data spans a much wider range. For the sampled dataset, the pixel values range from −1454 to 3071. The data is thus normalized to the range [0, 1] using a min-max normalization (1).

$$x_{normalized} = \frac{x - x_{min}}{x_{max} - x_{min}} \tag{1}$$

This was initially done using $x_{min} = -1454$ and $x_{max} = 3071$. However, due to the sensitivity of this technique to outliers and the wide range of data, the distribution of the pixel values was examined in order to remove possible outliers that might be distorting the normalized distribution. Since the pixel distribution is concentrated between approximately -1100 and 600, it is expected that normalizing using the new values of $x_{min} = -1100$ and $x_{max} = 600$ will lead to a better CNN performance. Both methods of normalization were tested and discussed in Sect. 4.

After normalization, the 2D data was resized to 227 × 227, and due to memory constraints, the 3D data was resized to 111 × 111 with a depth of 20 slices per nodule. Segmentation of the nodules from the surrounding lung tissue was not done in order to preserve regional information, as surrounding structures can provide information as to the malignancy likelihood. Once the pre-processing steps are done, the samples are fed into the 2D and 3D CNN models for training and testing.

Table 1. Vanilla 2D CNN Model

Layer Type	Number of Kernels	Kernel Size	Stride	Output Size
Convolution 2D	96	11 × 11	4 × 4	(None, 55, 55, 96)
Max Pool 2D		2 × 2	2 × 2	(None, 27, 27, 96)
Convolution 2D	256	5 × 5	1 × 1	(None, 23, 23, 256)
Max Pool 2D		2 × 2	2 × 2	(None, 11, 11, 256)
Convolution 2D	384	3 × 3	1 × 1	(None, 9, 9, 384)
Convolution 2D	384	3 × 3	1 × 1	(None, 7, 7, 384)
Convolution 2D	256	3 × 3	1 × 1	(None, 5, 5, 256)
Max Pool 2D		2 × 2	2 × 2	(None, 2, 2, 256)
Dense	-	-	(None, 4096)
Dropout	-	-	(None, 4096)
Dense	-	-	(None, 4096)
Dropout	-	-	(None, 4096)
Dense (Softmax)	-	-	(None, 2)

Table 2. Changes of tuned 2D AlexNet

Parameter	Vanilla 2D-AlexNet	Tuned 2D-AlexNet
Optimization Algorithm	Stochastic Gradient Descent (SGD)	Adam
Learning Rate	Initialized at 0.01, automatically reduced by a factor of 0.1 with a 3-epoch plateau in validation accuracy	Initialized at 0.001, automatically reduced by a factor of 0.1 with a 3-epoch plateau in validation accuracy
Batch Size	128	32
Epochs	50	Initialized to 50, but early stopping was implemented upon a 5-epoch plateau in validation loss
Regularization	Weight decay (L2 0.0005) in the FC layer	No weight decay

2D Model. The architecture of the CNN model is depicted in Table 1. The model was initially implemented following all the original parameters as specified in Krizhevsky et al.'s paper [11]. This version of the 2D-CNN is referred to as the "vanilla" 2D-AlexNet. In an attempt to tune the model for this classification problem, some of the original parameters were changed to offer more robust results. This tuned version is referred to as the "tuned" 2D-AlexNet and these changes are summarized in Table 2.

3D Model. The 3D "vanilla" architecture was created by expanding the original 2D-AlexNet across one further dimension, and similar to the vanilla 2D-AlexNet, all the parameters were obtained from the original paper [11]. The architecture of this vanilla 3D-AlexNet is summarized in Table 3. In contrast to a 2D-CNN that captures spatial x, y differences, a 3D-CNN is able to extract three dimensional features, capturing additional information along the z-axis such as correlation between slices and volumetric differences. It is thus expected that the 3D-CNN will outperform the 2D-CNN in the classification task. Furthermore, tuning the vanilla 3D-AlexNet as described in Table 4 is expected to further improve classification performance. In order to ensure a fair comparison, the vanilla AlexNets

Table 3. Vanilla 3D CNN Model

Layer Type	Number of Kernels	Kernel Size	Stride	Output Size
Convolution 3D	96	11 × 11 × 11	4 × 4 × 4	(None, 26, 26, 3, 96)
Max Pool 3D		3 × 3 × 3	2 × 2 × 2	(None, 13, 13, 2, 96)
Convolution 3D	256	5× 5 × 5	1 × 1 × 1	(None, 13, 13, 2, 256)
Max Pool 3D		3 × 3 × 3	2 × 2 × 2	(None, 7, 7, 1, 256)
Convolution 3D	384	3 × 3 × 3	1 × 1 × 1	(None, 7, 7, 1, 384)
Convolution 3D	384	3 × 3 × 3	1 × 1 × 1	(None, 7, 7, 1, 384)
Convolution 3D	256	3 × 3 × 3	1 × 1 × 1	(None, 7, 7, 1, 256)
Max Pool 3D		3 × 3 × 3	2 × 2 × 2	(None, 4, 4, 1, 256)
Dense		-	-	(None, 4096)
Dropout		-	-	(None, 4096)
Dense		-	-	(None, 4096)
Dropout		-	-	(None, 4096)
Dense (Softmax)		-	-	(None, 2)

Table 4. Changes of tuned 3D AlexNet

Parameter	Vanilla 3D-AlexNet	Tuned 3D-AlexNet
Optimization Algorithm	Stochastic Gradient Descent (SGD)	Adam
Learning Rate	Initialized at 0.01, automatically reduced by a factor of 0.1 with a 3-epoch plateau in validation accuracy	Initialized at 0.0001
Batch Size	128	16
Epochs	50	Initialized to 50, but early stopping was implemented upon a 5-epoch plateau in validation loss
Regularization	Weight decay (L2 0.0005) in the FC layer	No weight decay
Dropout	Dropout after FC1 (0.5) and FC2 (0.5)	Dropout after FC1 (0.6) and FC2 (0.6)

(2D and 3D) and tuned AlexNets (2D and 3D) were trained, validated and tested on the same datasets.

3.3 Explainability of CAD System

Local Interpretable Model-Agnostic Explanation. LIME works by approximating the complex model's behaviour using a locally-faithful and simple interpretable model. This is done by generating perturbations of data around the original instance, weighting these data with respect to their proximity to the original sample, and learning how the complex model behaves with each perturbation [18]. This process is illustrated in Fig. 4.

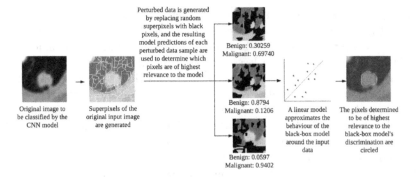

Fig. 4. Example output of LIME reasoning

Rule-Based System and Feature Extraction. The rule-based system is an extension of the proposed image processing algorithm. After a nodule is detected, feature extraction methods are used to extract determinant nodule details such as its spiculation and diameter Fig. 5. To extract the features of the nodules, firstly the nodule region's raw pixel properties are obtained, such as its area, diameter and equivalent diameter. These raw pixel values are obtained using scikit-image's region properties library.

The diameter of the region obtained is the diameter of the nodule at its widest line. While the diameter value can be directly used (after converting from pixel values to millimetre), the spiculation of the nodule must be manually calculated. Spiculation of a nodule is defined as the extent of distortion in the curvature of the nodule's edge. Thus, the more curved and smooth the edge of a nodule is, the less spiculated it is. The more distorted the curved edge, the more spiculated.

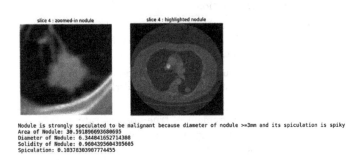

Fig. 5. Example output of the rule-based reasoning

To obtain the spiculation of the nodule, its equivalent diameter is first obtained. The equivalent diameter of a region is defined as the diameter of a circle with the same area as the region. Thus, the equivalent diameter of a nodule is the diameter of a circle that has the same area size as the nodule. Getting the difference between the equivalent diameter and the actual diameter of the nodule returns the extent of spiculation of the nodule. This works because if a nodule is less spiculated, its edge should be curved and round, much like a circle. Thus, the diameter of a less spiculated nodule should be similar to the diameter of a circle that has the same area (shows that there is less distortion in the nodule since the nodule resembles a circle). Contrarily, the diameter of a highly spiculated nodule will vary greatly from the diameter of an equivalent area circle (due to the edge distortion). Spiculation and diameter of nodule has shown in research to be determinant features when differentiating malignant and benign lung nodules [23].

The simple rule-based system outputs an assumption of the nodule's malignancy based on these extracted features. That is, if a nodule is detected to have diameter of <=3mm and is spiculated, it is strongly speculated that it is malignant [23]. The system then outputs the worded explanation based on the extracted features and their details.

4   Results and Discussions

4.1   Nodule Detection Results

To evaluate the outcomes of the proposed nodule detection algorithm, we quantify the performance metrics of the proposed algorithm by calculating various statistical measurement parameters (performance metrics) as shown in Table 5 and the quantitative values are recorded in Table 6.

Table 5. Summary of Performance Metrics.

Metric	Description
Accuracy	Measurement of how accurate (how well) the algorithm correctly identifies a nodule
Specificity	Measurement of the true negative rate (non-nodule slice is correctly classified as a non-nodule slice)
Sensitivity	Measurement of true positive rate (nodule slice is correctly classified as a nodule slice)
Precision	Measurement of how precise the algorithm is in detection of true positive results

Table 6. Summary of Performance Evaluation of Proposed Nodule Detection Algorithm.

Metric	Result
Accuracy	0.9791
Specificity	0.7650
Sensitivity	0.9920
Precision	0.8515
F score	0.8327

Based on the performance metrics of the nodule detection algorithm above, we can observe that the algorithm has high specificity and accuracy but low precision and sensitivity. Taking the performance metrics at face value, one might presume that the algorithm performs extremely well (since the accuracy is 97% and sensitivity is 99%). However, this is not the actual case because the performance metrics shows that the nodule detection algorithm is an overfit binary classifier. This is shown since the specificity, or recall, is significantly lower than any other metric.

In the case of our nodule detection algorithm, it can be deemed as a binary classification model. Since the average length of CT scan stack for one patient is typically around 150, and the average number of slices with a nodule is around less than 10, the negative cases (non-nodule slice) for this binary classification will forever heavily outweigh the positive cases (nodule slices). This means that the positive and negative classes are highly unbalanced which resulted in the performance metrics recorded above (especially accuracy and precision) can be considered delusive, where calculating the F measure (2) is considered to be a more reliable performance metric in such situations.

$$F_B = (1 + B^2) \cdot \frac{precision \cdot recall}{B^2 \cdot precision + recall} \quad where \quad B = 0.5 \tag{2}$$

From Eq. 2, a F score of 0.8327 was obtained. Thus, looking at the F score result, it can be said that a more reliable accuracy (considering false positives and false negatives) of the nodule detection algorithm is 83.27%. From the results, it can be concluded that the proposed algorithm is accurate and effective in lung nodule detection. Also, in comparison to most methods used in researches within this field which are semi-automatic, our algorithm is fully automatic. All parameters that are used in the proposed algorithm are automatically estimated which requires no external tweaking or assistance at any step of the algorithm. Thus, the proposed algorithm is concluded to be ideal for lung nodule detection.

4.2   Deep Learning Results

Normalization Techniques. Data was normalized using the two min-max ranges explained in Sect. 3.2, and a vanilla 2D-AlexNet was trained on both sets of data. The results are presented in Table 7.

Table 7. Summary of results achieved using two normalization ranges. Model: 2D-Vanilla AlexNet.

Normalization Range (min, max)	Training Accuracy	Testing Accuracy	Sensitivity	Specificity
−1454, 3071	89.3%	89.59%	93.47%	84.78%
−1000, 600	96.53%	91.93%	94.67%,	87.09%

In agreement with the expected outcome, using a narrower range of values to normalize the data slightly improved the model performance across all metrics. This is because the presence of a few extreme outliers, for example pixels with values over 3000, skewed the distribution of the data, and thus the high-frequency pixels within the range [−1000, 600] were "compressed" into a narrower range within [0, 1] for the sake of preserving the original scale. Since the outliers were very few in number, sacrificing these pixels and setting a custom normalization range handpicked for this dataset achieved an improve in performance.

Table 8. Performance comparison of the 4 models

Model	Training accuracy	Testing accuracy	Sensitivity	Specificity	Epochs to convergence
Vanilla 2D-AlexNet	96.53%	91.93%	94.67%	87.09%	50
Tuned 2D-AlexNet	99.49%	94.21%	96.66%	89.61%	17
Vanilla 3D-AlexNet	65.90%	61.54%	35.03%	86.90%	50
Tuned 3D-AlexNet	94.82%	85.86%	77.66%	93.69%	20

2D: Vanilla vs. Tuned AlexNet. Based on Table 8, tuning the model in accordance with the parameter changes described in Sect. 3.2 leads to an improve in performance across all metrics. Although there are concerns that Adam generalizes more poorly compared to SGD [9], this was not observed in the 2D models trained (generalization was measured by testing the accuracy of the model on the testing dataset, which was unseen to the trained models). This could be because learning rate annealing was implemented in an attempt to limit the adaptive learning rate tuning by Adam in the later stages of training. Both models have fully converged by the time the training ended. The tuned model had converged by 17 epochs and automatically quit training as part of the implemented callbacks, whereas the vanilla model converged by 50 epochs. Overall, it is observed that the tuned model not only delivers a higher classification accuracy, but also converges faster and requires less training time.

3D: Vanilla vs. Tuned AlexNet. Similar to the 2D models, tuning the 3D model in accordance with the parameter changes described in Sect. 3.2 led to an improvement in performance across all metrics. The difference between the vanilla and tuned 3D model performances, however, is much larger than that between the 2D models. This is possibly because the vanilla 3D-model has not converged yet within 50 epochs, while the tuned one had converged and automatically quit training by 20 epochs. For the tuned model, it is possible that it converged at a local minimum, limited in performance by the use of the Adam optimizer, due to its training speed and generalization trade off [9], and that for the 3D-data, SGD + momentum works better.

4.3   Explainability of the System

While the first two modules yield robust results in nodule detection and cancer classification, but the justification is hidden due to the nature of the black-box models. With the two approaches proposed in this research, we visualize and generate the reasoning parts that can be used to convince the users regarding the predicted outcomes. Figure 6 shows some reasoning results generated from the proposed approaches in Sect. 3.3.

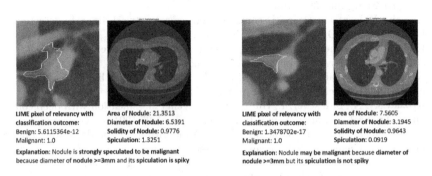

Fig. 6. Samples of reasoning result

5   Conclusion

In this paper, We discussed how our project aims to tackle the black-box issue by delivering a fully automated explainable lung cancer detection system. We presented our system through elaborating on its three main components integrated together: 1) the image processing component which presents an automated lung nodule detection system along with CNN preprocessing, 2) 2D and 3D Alexnet models that have been trained and tuned on the LIDC dataset, and 3) explainability algorithms, approached using LIME and a rule-based diagnosis system. In summary, our work presented a novel option for physicians looking for explainable CAD systems, and our results remain encouraging for possible future exploration.

References

1. Key statistics for lung cancer. Tech. rep., American Cancer Society (2020). https://www.cancer.org/cancer/lung-cancer/about/key-statistics.html

2. Amir, G.J., Lehmann, H.P.: After detection: the improved accuracy of lung cancer assessment using radiologic computer-aided diagnosis. Acad. Radiol. **23**(2), 186–191 (2016)

3. Anthimopoulos, M., Christodoulidis, S., Ebner, L., Christe, A., Mougiakakou, S.: Lung pattern classification for interstitial lung diseases using a deep convolutional neural network. IEEE Trans. Med. Imaging **35**(5), 1207–1216 (2016)

4. Ayyagari, M.R., Kumar Ahuja, D.G.: Lung cancer detection on computed tomography images using digital image processing techniques. J. Adv. Res. Dyn. Control Syst. **11**, 677–689 (2019)

5. Chon, A., Balachandar, N., Lu, P.: Deep convolutional neural networks for lung cancer detection. Standford University (2017)

6. Giger, M.L., Doi, K., MacMahon, H.: Image feature analysis and computer-aided diagnosis in digital radiography. 3. Automated detection of nodules in peripheral lung fields. Med. Phys. **15**(2), 158–166 (1988)

7. Henschke, C.I., Yankelevitz, D.F., Mateescu, I., Brettle, D.W., Rainey, T.G., Weingard, F.S.: Neural networks for the analysis of small pulmonary nodules. Clin. Imaging **21**(6), 390–399 (1997)

8. Hossain, K.J., et al.: Deep 3D convolutional neural network in early detection of Lung cancer. Ph.D. thesis, BRAC University (2018)

9. Keskar, N.S., Socher, R.: Improving generalization performance by switching from adam to sgd. arXiv preprint arXiv:1712.07628 (2017)

10. Khan, S.A., Hussain, S., Yang, S., Iqbal, K.: Effective and reliable framework for lung nodules detection from ct scan images. Sci. Rep. **9**(1), 1–14 (2019)

11. Krizhevsky, A., Sutskever, I., Hinton, G.: Imagenet classification with deep convolutional neural networks. Neural Inf. Process. Syst., 25 (2012)

12. Lee, H., et al.: An explainable deep-learning algorithm for the detection of acute intracranial haemorrhage from small datasets. Nat. Biomed. Eng. **3**(3), 173 (2019)

13. Li, Q., Li, F., Doi, K.: Computerized detection of lung nodules in thin-section ct images by use of selective enhancement filters and an automated rule-based classifier. Acad. Radiol. **15**(2), 165–175 (2008)

14. Li, W., Cao, P., Zhao, D., Wang, J.: Pulmonary nodule classification with deep convolutional neural networks on computed tomography images. Comput. Math. Methods Med. **2016**, 1–7 (2016)

15. Matsuki, Y., et al.: Usefulness of an artificial neural network for differentiating benign from malignant pulmonary nodules on high-resolution CT: evaluation with receiver operating characteristic analysis. Am. J. Roentgenol. **178**, 657–63 (2002)

16. Miller, D.D., Brown, E.W.: Artificial intelligence in medical practice: the question to the answer? Am. J. Med. **131**(2), 129–133 (2018)

17. Miotto, R., Wang, F., Wang, S., Jiang, X., Dudley, J.T.: Deep learning for healthcare: review, opportunities and challenges. Briefings in bioinform. **19**(6), 1236–1246 (2018)

18. Ribeiro, M.T., Singh, S., Guestrin, C.: "Why should i trust you?" explaining the predictions of any classifier. In: Proceedings of the 22nd ACM SIGKDD international conference on knowledge discovery and data mining, pp. 1135–1144 (2016)

19. Shen, W., et al.: Multi-crop convolutional neural networks for lung nodule malignancy suspiciousness classification. Pattern Recognit. **61**, 663–673 (2017)

20. Palatnik de Sousa, I., Maria Bernardes Rebuzzi Vellasco, M., Costa da Silva, E.: Local interpretable model-agnostic explanations for classification of lymph node metastases. Sensors, **19**(13), p. 2969 (2019)

21. Tang, Z., et al.: Interpretable classification of alzheimer's disease pathologies with a convolutional neural network pipeline. Nat. Commun. **10**(1), 1–14 (2019)

22. Teach, R.L., Shortliffe, E.H.: An analysis of physician attitudes regarding computer-based clinical consultation systems. Comput. Biomed. Res. **14**(6), 542–558 (1981)
23. Vlahos, I., Stefanidis, K., Sheard, S., Nair, A., Sayer, C., Moser, J.: Lung cancer screening: nodule identification and characterization. Transl. Lung Cancer Res. **7**(3), 288 (2018)
24. Xu, X.W., Doi, K., Kobayashi, T., MacMahon, H., Giger, M.L.: Development of an improved cad scheme for automated detection of lung nodules in digital chest images. Med. Phys. **24**(9), 1395–1403 (1997)

Extending Egocentric Vision into Vehicles: Malaysian Dash-Cam Dataset

Mahamat Moussa, Chern Hong Lim[(✉)], and KokSheik Wong

School of Information Technology, Monash University Malaysia, Selangor, Malaysia
{mahamat.moussaabbasali,lim.chernhong,wong.koksheik}@monash.edu

Abstract. Egocentric Vision (EV) has recently become one of the emerging areas in computer vision. It is a process of recording a person's activity by using a wearable camera installed on the head of that person to gain an exclusive viewpoint. This field has gained more attention recently due to the wide availability of wearable cameras commercially and the recent success of Deep Learning (DL). Such attention has resulted in publishing more EV datasets. In this paper, we argue that the concept of egocentric vision is similar to the recording of road activities by a portable camera installed on the dashboard of a vehicle (*commonly known as DashCam*) to gain an exclusive viewpoint. We attempt to extend the EV concept into vehicles and show how the two viewpoints are similar. To support this argument, we collect and annotate the first Malaysian DashCam dataset, which consists of 228 video clips about road activities classified into ten categories. We use this dataset to train a DL model on top of VGG16 model -*a DL model that is pre-trained on ImageNet*-, to perform an initial experiment (activity classification), which reveals some limitations of this work such as small dataset size and less distribution of examples among categories. From this work, we learn that such a dataset may lead to the discovery of valuable statistical knowledge about most frequently occurring behaviors that cause accidents and that we may leverage DL to understand such behaviors to save human life.

Keywords: Egocentric vision · Computer vision · Dashcam vision · Deep learning · Activity recognition · DashCam · Action recognition

1 Introduction

Egocentric Vision (EV) or first-person vision has recently become one of the emerging areas in Computer Vision (CV). It is portrayed by videos taken from the perspective of a person who is performing an activity, which produces an exclusive viewpoint of the scene [5]. The EV data is collected by installing a wearable camera on the head of a person referred to as first-person [2,11]. This field has gained more attention recently due to the wide availability of wearable cameras commercially and the recent success of activity and action recognition applications in deep learning [10,17]. Such attention has resulted in publishing

C. S. Chan et al. (Eds.): ICIRA 2020, LNAI 12595, pp. 270–281, 2020.
https://doi.org/10.1007/978-3-030-66645-3_23

more EV datasets, including [4,5,14,19,21]. These datasets allow researchers to observe, predict, and understand human activities and actions in a better way [18,23]. However, most EV studies are carried out from the perspective of a human wearing a portable camera on the head. Thus, we argue that such a concept is similar to mounting a camera (Dashcam) on a Dashboard of a vehicle, which produces a unique point of view in order to observe road activities from the perspective of this vehicle. In this paper, we attempt to extend the concept of EV to DashCams and show how the two concepts have a similar viewpoint, where Fig. 1 and 2 show examples of frames from EV and DashCams videos. It can be seen that Fig. 1 and 2 have a similar point of view, where the former represents images taken from a camera installed on a human head, while the latter represents images taken from a DashCam. Recently, the utilization of DashCam in Malaysia has increased. A survey we conducted on the total DashCam videos uploaded on YouTube by Malaysian suggested that there has been a steady rise in the number of videos recorded by DashCam. In the survey, we found that a total of 43 videos uploaded during the first four months of 2020, while only 28 videos uploaded in the entire year of 2019.

Fig. 1. Examples of Egocentric Vision (first-person) point of view.

As such, we anticipate the potential of DashCam data and its contribution to various domains such as behavior analysis of drivers, road safety enhancement, road action and activity recognition, as well as traffic risk assessment. Our main contributions of this work include: 1) To expose the CV community into a new direction of research on DashCam data from the perspective of Egocentric vision. This differs from the work of [24], which mainly focuses on DashCam videos as source of videos, not source of different point of view (egocentric to be specific), which requires different perspective of analysis, similar to the comparison of normal videos to egocentric videos. We believe that this would benefit the community as much as the egocentric vision is doing; 2) we collect, annotate, and publish the first Malaysian DashCam (MyDashCam) dataset to enable the community to push the boundary and invent more case studies such as road activity analysis and understanding, and; 3) we perform an initial video activity classification experiment on the dataset by developing a deep learning model to set a starting point for future performance comparison on such datasets. Despite DashCam datasets are used in different works such as [3,24], we believe that our localized dataset always has its own unique pattern of activity and behaviour.

The rest of this paper is organized as follows: In Sect. 2, we present related works to EV and their applications. Section 3 presents our datasets and explains the intuition behind the annotation process. Section 4 presents results of both datasets, and the experiment. Section 5 highlights future works to which this paper can be extended to, and finally, a conclusion is drawn in Sect. 6.

Fig. 2. Examples of DashCam's point of view.

2   Related Works

The literature on Egocentric Vision (EV) has shown the emergence of several exciting applications in different fields, including robotics [20], activity recognition [10], pedestrian path prediction by self-driving cars [18], action recognition [22], and personal locations recognition [8]. As mentioned in Sect. 1, more recent attention is channeled to producing EV dataset such as [4,5,14,19,21]. Although several applications of EV have been highlighted in the literature, activity as well as action recognition remain as the most popular subject of interest due to their various applications. For instance, Dey et al. [6] found that vehicle behavior has a pedestrian impact when the pedestrian is deciding to cross the street. We believe that such findings can be further investigated when an EV is brought into vehicles. Meanwhile, Li et al. [15] attempt to enhance the quality of life for people relying on Robots by adopting EV into these robots to eliminate the need to use their hands. In their study, they successfully integrate and synchronize robot wheelchair and EV of the user to send commands relying on EV point of view. In a similar study, Karuppiah [13] achieve such success by integrating a robotic arm into a wheelchair by utilizing computer vision.

Furthermore, Li et al. [16] address the challenge of applying action recognition techniques into EV data. They highlight that EV data carries rich data such as head movement, hand pose, and gaze information. Such cues increase the complexity of activity recognition. This point is further investigated by [12] when they explain the challenge of egocentric human action recognition. Specifically, the study focuses on estimating hand location and capturing their movement in egocentric vision. The study concludes that despite the difficulties in egocentric human action recognition, the valuable information in the scenes facilitates the process in a better way. Moreover, Huang et al. [9] propose a novel approach called temporal action proposals (TAPs), which advances the process of egocentric action recognition in a more generic way. Their experimental outcomes show that TAP performed better than other geocentric action recognition approaches.

Many of the techniques in the aforementioned studies of EV can be extended into or applied to DashCam videos because they share similar characteristics, as explained in Sect. 1. Some recent studies such as [3, 24] utilize DashCam videos in similar research. However, they focus on DashCam as another source of videos, while we look at it as a different perspective of recording videos (EV).

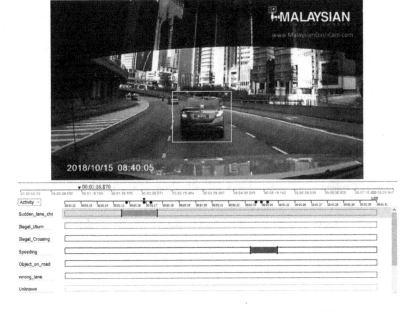

Fig. 3. A screenshot of the annotation process where the colored bars represent temporal segments (activity class) and the black dots on the top of each bar represent spatial annotation (object class). (Color figure online)

3　The Dataset

3.1　Data Collection

Malaysian DashCam (MyDashCam) dataset is an initiative proposed in this paper to bring researchers' attention to work on such a valuable source of data. The initial sample of our dataset consists of 228 video segments (clips) extracted from 21 long-videos that represent the total videos uploaded on the Malaysian Dash Cam Owners YouTube channel [1]. The 228 video clips represent road activities that have been recorded by using DashCam installed on different vehicles, commonly known as DashCam Video Compilation [1].

3.2　Class Definition

Before the annotation process, a set of classes have been defined to control this process. Specifically, we identified three types of classes: 1) activity class, 2) object class, and 3) time class. The detailed descriptions are presented below.

Activity Class: The Activity class is introduced to annotate road activities in each of the 228 video segments. We classified road activities into ten classes based on the activities commonly described by traffic rules such as illegal crossing and illegal U-turn, and also based on our observations and understanding, such as an object appear on the road and animal on the road. Each video clip is annotated and saved as temporal data. Table 1 explains the ten classes, and Fig. 3 illustrate the annotation process.

Table 1. The (10) Activity Classes.

ID	Class Name	Description
0	Sudden lane change	The driver suddenly changed lane
1	Illegal U turn	Making a U-turn while it is not allowed
2	Illegal Crossing	An object illegally crossed the road
3	Speeding	The speed of the vehicle causes unexpected event
4	Object on road	The vehicle confronts suddenly an object on the road
5	Wrong lane	The vehicle is on a lane that they are not supposed to be on
6	Unknown	None. Could be an accident, unexpected behaviour, etc.
7	Group of motorbikes	A group of motorbikes are riding together
8	Reverse drive	The vehicle is moving in the opposite direction of the road
9	Animal on road	The vehicle confronts an animal on the road

Table 2. The (5) Object Classes.

ID	Class Name	Description
0	Car	A generic class to annotate a vehicle, not necessarily a small car
1	Motorcycle	Refers to the standard motorbike on roads
2	Pedestrian	Refers to people walking or crossing road illegally
3	Bicycle	Refers to people who are cycling on the road
4	Another Object	Refers to any objects that falls out of the above classes

Object Class: We also defined five classes of object so that each activity class can be associated with an object according to the scene. The object class represents the main object involved in road activity. For instance, when a video segment labeled with the activity class [sudden lane change], we need to determine which object has suddenly changed its lane, e.g.., it is a car, or a motorbike? The object class will give an answer. The object is annotated and saved as spatial data within the temporal data. Table 2 explains the aforementioned five object classes, while Fig. 4 further illustrates how this task is done.

Time Type: The time here refers to either Day or Night. We annotated the time so that the activity class can be classified based on this. The annotation type of time class is considered as temporal data.

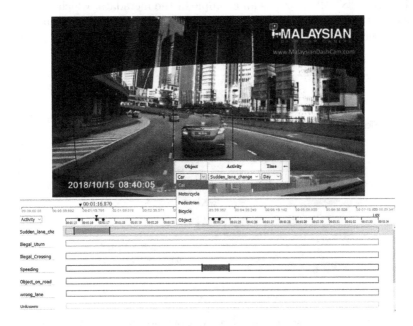

Fig. 4. A screenshot of the object annotation process shows how object class is being associated with its activity

3.3 Annotation Process

After collecting our sample data (i.e., 21 long videos) and defining the classes, we annotate them using a video annotation tool. In general, video annotation requires the extraction of a set of frames from the video and the frames can be collectively annotated together as temporal data, while a single frame can be labeled as spatial data. Therefore, our annotation process is divided into two processes: 1) temporal annotation, and 2) spatial annotation.

In the first process, we identify each activity segment in the input video (these segments eventually become the 228 video clips). Then, the appropriate class is assigned to that segment. On the other hand, in the spatial annotation process, we select three different frames within the temporal segment, annotate the main object in each frame (viz., *means the main object involved in that activity*) and assign the appropriate class to it. We repeat these two tasks for each of the 21 videos. To correctly perform the two tasks, we utilized the Visual Geometry Group (VGG) tool [7], which is an open-source manual annotation software for image, audio, and video. The output of this annotation process is a CSV file that carries the annotation information of the input video. Specifically, the output file consists of three important data fields: 1) temporal coordinates, 2) spatial coordinates, and 3) metadata. An example of the temporal coordinate data is [1.3492.741], which denotes a temporal segment from time 1.346 s to 2.741 s, and the temporal coordinate [4.633] denotes a single frame at 4.633 s. On the other hand, an example of spatial coordinates is [2, 10 20, 50 80], which denotes a rectangle whose shape ID is 2, and its size is 50×80 placed at the point $(10, 20)$. Finally, { "1" : "8", "2" : "0"} is an example of the metadata, which denotes the activity class with $ID = 8$ and the object class with ID = 0. Figure 5 summarizes the annotation process.

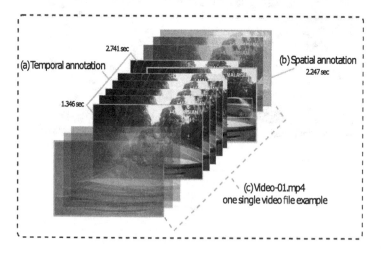

Fig. 5. Annotation of one single video file. (a) shows how temporal data is annotated, (b) shows how spatial data is annotated, and (c) shows the total frames of the sample.

3.4   Assumptions

For each video file, we annotate only the segments where the activity is explicit and appropriately match our defined classes. The rest of the clips in the given video, we treat them as typical scenes that are not directly related to our defined classes.

4   Results and Discussion

4.1   Data Annotation

The purpose of annotating and publishing this dataset is to draw researchers' attention to focus on the recent growth of DashCam videos. We believe the work performed on this data will help in addressing various aspects of our roads, including as driver behavior, traffic risk assessment, road action and activity recognition, road safety enhancement, and more. In this Section, we present the significant statistics related to the annotated dataset to show how such data can be utilized to understand road activities in depth. Performing an initial statistical analysis of the data reveals that the activity class [**sudden lane change**] is the most common road activity that has been recorded by DashCams, with a total of 61 video segments, given that the total activity segments we labeled in all of the 21 videos are 228 segments (activities). The average of video segments per class is 22.8; the geometric mean is 13.9, the median is 17, the largest is 61 (representing the first class), the smallest is 2 (representing Group of motorbikes class), and the average of activities per video is 11. It is observed that, whenever the activity [**sudden lane change**] occurs, it will result in either a road accident or near-miss, which means finding more solutions for such activity will potentially decrease road accidents, thus improving road safety. The second most common activity class is [**illegal crossing**], followed by [**speeding**] and [**wrong lane**]. The less common activities, on the other hand, are [**Group of motorbikes**] and [**Animal on Road**], which were reported for only 2 and 4 times, respectively. Figure 6 shows the distribution of activities per video, while Table 3 shows the summary of activity videos per class.

Table 3. Statistical Summary of Activity Videos Per Class

Activity Class	Sudden Lane Change	Illegal U-turn	Illegal Cross-ing	Speeding	Object on Road	Wrong Lane	Unknown	Group of Moto-bikes	Reverse drive	Animal On Road
# Clips	61	4	58	29	13	20	23	2	14	4

4.2   Road Activity Classification Using Deep Learning

We conducted an initial experiment of video activity classification on the dataset to see how it performs on a deep learning model. Firstly, we prepossessed the annotated videos by slicing each of them into clips based on the temporal data annotation. As a result, a total of 228 clips (activity class videos) are extracted. Out of these 228 videos, 43 clips (evenly distributed among classes) were separated for validation purposes, leaving only 185 to train our model. Next, we extracted the frames from each of the 185 training videos, and saved them into a training list. Then, we define our deep learning model architecture to start the training process. Since the dataset considered is very small for a video classification model, we used a pre-trained model (i.e., VGG16) to train our video classifier by fine-tuning the top layer of VGG16, and train the model on it. We use 20% of the 185 videos during the training as the validation data to assess the accuracy of the model. Various fine-tuned values are applied to the model. However, the best result is obtained when we set epochs=200 with a batch size of 256. After completing the training, we utilized the testing data (i.e., the 43 videos) to evaluate the mode's performance. The result of our model shows a weak performance, with only 32% of accuracy based on the f1-score measure. Two limitations can explain this output: 1) The size of dataset is tiny, and; 2) the number of examples per activity class is insufficient for training in some classes (namely, some activity class has only two examples). The limitations can further be interpreted from the full classification report in Table 4. However, this result can be utilized as a reference point for future experiments. When the dataset size is increased, various experiment techniques can be considered.

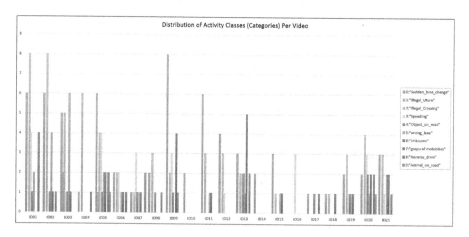

Fig. 6. Distribution of Activity Class per Video (the x-axis represents the ID of the 21 long videos)

Table 4. Accuracy of Road Activity Classification

Activity Class	Precision	Recall	f1-score	Support
Animal On Road	1.00	1.00	1.00	1
Group Of Motorbikes	0.00	0.00	0.00	1
Illegal Crossing	0.43	0.43	0.43	11
Illegal U-turn	0.00	0.00	0.00	1
Object On Road	0.00	0.00	0.00	2
Reverse Drive	0.00	0.00	0.00	2
Speeding	0.25	0.20	0.22	5
Sudden Lane Change	0.33	0.25	0.29	12
Unknown	0.50	0.25	0.33	4
Wrong Lane	0.33	0.50	0.40	4
micro avg	0.33	0.33	0.33	43
macro avg	0.28	0.27	0.27	43
Weighted avg	0.33	0.33	**0.32**	43

5 Future Works

This work is one step towards the future of DashCam. However, there is still much to be done on such a dataset. Some of the possible future works include annotating more data to increase the dataset so that we can get more examples of activity classes. Besides, the dataset can be used to train and test various deep learning models concerning activity recognition, action recognition, and safety measurement. We are also interested in understanding whether such a small dataset can be utilized to achieve better accuracy by applying state-of-the-art techniques on training deep learning models on a small dataset such as One-shot learning and Zero-shot learning. Furthermore, it is also possible to consider the inclusion of prediction software in future DashCams so that they can predict road activities based on historical data of DashCams, such as sudden lane change, thus improving road safety.

6 Conclusion

This paper has argued that although egocentric vision is observed from a human perspective, videos recorded by DashCam can also be viewed as egocentric from the perspective of a vehicle since they are sharing the same concept, and they produce exclusive viewpoints; therefore, the concept of Egocentric Vision can be extended into vehicles. As such, we believe that DashCam dataset is a rich source of data that researchers should capitalize from. We anticipate that it may contribute in understanding road activities in a better way. Therefore, we collect, annotate, and publish the first Malaysian DashCam (MyDashCam) dataset, which consists of 228 video clips representing ten different activity classes. The insights we gained from this study may be of assistance to push this direction of work to take advantage of the recent growth in DashCam videos, as well as the recent success of neural networks. This direction of research will be beneficial for both industry and government to improve road safety by investing in road activity and action recognition in deep learning.

References

1. Malaysian dash cam owners. https://www.youtube.com/channel/UCX_AZE75iRE SrlZ-sV-E19w. Accessed 10 May 2020
2. Buddubariki, V., Tulluri, S.G., Mukherjee, S.: Event recognition in egocentric videos using a novel trajectory based feature. In: Proceedings of the Tenth Indian Conference on Computer Vision, Graphics and Image Processing, pp. 1–8 (2016)
3. Chan, F.-H., Chen, Y.-T., Xiang, Y., Sun, M.: Anticipating accidents in dashcam videos. In: Lai, S.-H., Lepetit, V., Nishino, K., Sato, Y. (eds.) ACCV 2016. LNCS, vol. 10114, pp. 136–153. Springer, Cham (2017). https://doi.org/10.1007/978-3-319-54190-7_9
4. Damen, D., et al.: Rescaling egocentric vision. arXiv preprint arXiv:2006.13256 (2020)
5. Damen, D., et al.: Scaling egocentric vision: The epic-kitchens dataset. In: Proceedings of the European Conference on Computer Vision (ECCV), pp. 720–736 (2018)
6. Dey, D., Martens, M., Eggen, B., Terken, J.: The impact of vehicle appearance and vehicle behavior on pedestrian interaction with autonomous vehicles. In: Proceedings of the 9th International Conference on Automotive User Interfaces and Interactive Vehicular Applications Adjunct, pp. 158–162 (2017)
7. Dutta, A., Zisserman, A.: The VIA annotation software for images, audio and video. In: Proceedings of the 27th ACM International Conference on Multimedia. MM 2019, ACM, New York, NY, USA, pp. 2276-2279 (2019)
8. Furnari, A., Farinella, G.M., Battiato, S.: Recognizing personal locations from egocentric videos. IEEE Trans. Hum. Mach. Syst. $47(1)$, 6–18 (2016)
9. Huang, S., Wang, W., He, S., Lau, R.W.: Egocentric temporal action proposals. IEEE Trans. Image Process. $27(2)$, 764–777 (2017)
10. Imran, J., Raman, B.: Multimodal egocentric activity recognition using multi-stream CNN. In: Proceedings of the 11th Indian Conference on Computer Vision, Graphics and Image Processing, pp. 1–8 (2018)
11. Kapidis, G., Poppe, R., Van Dam, E., Noldus, L., Veltkamp, R.: Egocentric hand track and object-based human action recognition. In: IEEE SmartWorld, Ubiquitous Intelligence Computing, Advanced Trusted Computing, Scalable Computing Communications, Cloud Big Data Computing, Internet of People and Smart City Innovation (SmartWorld/SCALCOM/UIC/ATC/CBDCom/IOP/SCI), pp. 922–929 (2019)
12. Kapidis, G., Poppe, R., Van Dam, E., Noldus, L., Veltkamp, R.: Egocentric hand track and object-based human action recognition. In: IEEE SmartWorld, Ubiquitous Intelligence & Computing, Advanced & Trusted Computing, Scalable Computing & Communications, Cloud & Big Data Computing, Internet of People and Smart City Innovation (SmartWorld/SCALCOM/UIC/ATC/CBDCom/IOP/SCI), pp. 922–929. IEEE (2019)

13. Karuppiah, P., Metalia, H., George, K.: Automation of a wheelchair mounted robotic arm using computer vision interface. In: IEEE International Instrumentation and Measurement Technology Conference (I2MTC), pp. 1–5. IEEE (2018)
14. Lee, Y.J., Ghosh, J., Grauman, K.: Discovering important people and objects for egocentric video summarization. In: IEEE Conference on Computer Vision and Pattern Recognition, pp. 1346–1353. IEEE (2012)
15. Li, H., Kutbi, M., Li, X., Cai, C., Mordohai, P., Hua, G.: An egocentric computer vision based co-robot wheelchair. In: IEEE/RSJ International Conference on Intelligent Robots and Systems (IROS), pp. 1829–1836. IEEE (2016)
16. Li, Y., Ye, Z., Rehg, J.M.: Delving into egocentric actions. In: Proceedings of the IEEE Conference on Computer Vision and Pattern Recognition, pp. 287–295 (2015)
17. Ma, B., Reibman, A.R.: Dashcam video compression using historical data. In: Picture Coding Symposium (PCS), pp. 1–5. IEEE (2016)
18. Poibrenski, A., Klusch, M., Vozniak, I., Müller, C.: M2p3: multimodal multi-pedestrian path prediction by self-driving cars with egocentric vision. In: Proceedings of the 35th Annual ACM Symposium on Applied Computing, pp. 190–197 (2020)
19. Ragusa, F., Furnari, A., Battiato, S., Signorello, G., Farinella, G.M.: Ego-ch: dataset and fundamental tasks for visitors behavioral understanding using egocentric vision. Pattern Recogn. Lett. **131**, 150–157 (2020)
20. Reddy, R., Nagaraja, S.: Integration of robotic arm with vision system. In: IEEE International Conference on Computational Intelligence and Computing Research, pp. 1–5. IEEE (2014)
21. Spera, E., Furnari, A., Battiato, S., Farinella, G.M.: Egocentric shopping cart localization. In: 24th International Conference on Pattern Recognition (ICPR), pp. 2277–2282. IEEE (2018)
22. Tang, Y., Wang, Z., Lu, J., Feng, J., Zhou, J.: Multi-stream deep neural networks for rgb-d egocentric action recognition. IEEE Trans. Circuits Syst. Video Technol. **29**(10), 3001–3015 (2018)
23. Valsecchi, M., Akbarinia, A., Gil-Rodriguez, R., Gegenfurtner, K.R.: Pedestrians egocentric vision: individual and collective analysis. In: ACM Symposium on Eye Tracking Research and Applications, pp. 1–5 (2020)
24. Wu, M.C., Yeh, M.C.: Early detection of vacant parking spaces using dashcam videos. In: Proceedings of the AAAI Conference on Artificial Intelligence, vol. 33, pp. 9613–9618 (2019)

Robot Design, Development and Control

Kinematic Calibration for Industrial Robot Using a Telescoping Ballbar

Zeming Wu, Peng Guo, Yang Zhang[(✉)], and Limin Zhu

School of Mechanical Engineering, Shanghai Jiao Tong University, No. 800 Dongchuan Road, Minhang District, Shanghai 200240, China
meyzhang@sjtu.edu.cn

Abstract. Industrial robots are increasingly used in many applications where the positioning accuracy is of great importance. Kinematic calibration is an effective method to improve the positioning accuracy of industrial robots. In this paper, a new kinematic calibration method is proposed for six-axis serial industrial robots based on a single telescoping ballbar. The end of the robot to be calibrated is moved to a set of specific poses, and the actual distance between the tool center point (TCP) of the industrial robot and a fixed point in the world frame is measured by a telescoping ballbar. Through fitting the distance residual errors, the robot calibration problem is transformed into a nonlinear least-squares optimization problem. The optimization problem is solved using the Levenberg-Marquardt algorithm to derive the actual kinematic parameter errors of the robot. A simulation study demonstrates that the proposed method can effectively identify the kinematic parameter errors and the average position errors are reduced from 19.368 mm to 0.073 mm after calibration.

Keywords: Kinematic calibration · Industrial robot · Telescoping ballbar

1 Introduction

Industrial robots are increasingly used in many automation applications. It is well-known that the repositioning accuracy of industrial robots is relatively high, whereas the absolute positioning accuracy is relatively low due to the mismatch between the nominal and practical kinematic parameters of the robot. Although in some traditional applications such as pick and place, spray-painting and spot-welding, the absolute positioning accuracy is not strongly required, there are many other applications where the absolute positioning accuracy is very crucial such as drilling, milling, and robot offline programming.

The positioning accuracy of industrial robots can be affected by many factors such as geometric errors, ambient temperature and sensor errors. Among them, the geometric errors generating from the manufacturing and assembly of the robot account for about 90% of the total errors [1]. Therefore, compensating geometric errors is the dominant demand to improve robotic positioning accuracy. Kinematic calibration is a commonly used method to identify the robot geometric errors, which mainly includes the following modules: modeling, measurement, parameter identification, and error compensation.

© Springer Nature Switzerland AG 2020
C. S. Chan et al. (Eds.): ICIRA 2020, LNAI 12595, pp. 285–295, 2020.
https://doi.org/10.1007/978-3-030-66645-3_24

In the modeling module, a kinematic model is established to describe the mapping between the robot geometric parameters and the end pose. There are numerous available methods used for establishing the kinematic model of a robot, such as the classic Denavit-Hartenberg convention (DH) [2], Denavit-Hartenberg Modified convention (MDH) [3], the produce of exponentials (POE) [4, 5] and some other methods, e.g., the dual quaternions method [6].

During the calibration process, measurement needs to be performed for each pose. The measurement methods can be divided into two categories: direct measurement methods and indirect measurement methods. The laser tracker is widely used as the direct measurement method, due to its high measurement accuracy [7, 8]. However, it is expensive and complicated to operate. To reduce the measurement cost and improve the measurement operability, some indirect measurement methods are proposed based on geometric constraints, such as the distance and sphere constraints [9, 10], plane constraints [11–13], and the plane and distance constraints [14]. However, the measurement precision of these methods is relatively low.

Different from the devices used in the above researches [9–14], Albert Nubiola [15, 16] proposes a 6-dimensional pose measurement system using a high-precision telescoping ballbar and applies it to the robot calibration process. However, the measurement system can only measure a total of 72 poses, which is not enough for industrial robot calibration.

In this paper, an improved robot calibration method is proposed based on a telescoping ballbar. This method is achieved by measuring the distance between the end of the robot and a fixed point, which overcomes the deficiency that the 6-dimensional pose measurement system can only measure limited poses. And a telescoping ballbar has the advantage of being more accurate than laser trackers and yet cheaper than even the cheapest laser trackers. So that we can get more calibration data in different poses, and make the calibration process more accurate at a relatively low cost.

The remainder of this paper is organized as follows. Section 2 establishes the kinessmatic model of the industrial robot and introduces the parameter errors considered in the model. Section 3 describes the measurement system and the parameter identification process. In Sect. 4, the feasibility of the proposed calibration method is verified through simulation. Finally, conclusions are given in Sect. 5.

2 Robot Calibration Model

Establishing a kinematic model is the basis for robot calibration, and has a direct effect on the calibration results. The robot calibrated in this paper is a Yaskawa MOTOMAN-80 six-axis industrial robot. The coordinate system of each link is established according to the MDH modeling method as shown in Fig. 1. Frame $\{O_0\}$ is the base frame. Frame $\{O_6\}$ is the flange frame. Besides, a tool frame $\{O_{tool}\}$ with the same orientation as the flange frame $\{O_6\}$ is also considered, but not shown in the figure.

2.1 Kinematic Model

According to the MDH modeling method [3], the homogeneous transformation matrix $_i^{i-1}A$ between two adjacent link coordinate systems $\{0_{i-1}\}$ and $\{0_i\}$ is described by the

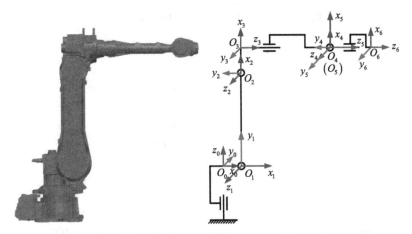

Fig. 1. MOTOMAN-80 and link coordinate system

four DH parameters θ_i, d_i, a_i, α_i shown in Eq. (1) where is the abbreviation of *cos* and s is the abbreviation of *sin*.

$$
{}^{i-1}_{i}A = Rot(z, \theta_i)Trans(o, o, d_i)Trans(a_i, 0, 0)Rot(x, \alpha_i)
$$
$$
= \begin{bmatrix} c\theta_i & -c\alpha_i\theta_i & s\alpha_is\theta_i & a_ic\theta_i \\ c\theta_i & c\alpha_i\theta_i & -s\alpha_ic\theta_i & a_is\theta_i \\ 0 & s\alpha_i & c\alpha_i & d_i \\ 0 & 0 & 0 & 1 \end{bmatrix} \tag{1}
$$

However, when the adjacent two joint axes are parallel or near parallel, a slight non-parallelism will cause a huge change in the kinematic parameters [17]. In this case, Eq. (1) is no longer appropriate. So a modified method used another four DH parameters θ_i, a_i, α_i, β_i is proposed as follows:

$$
{}^{i-1}_{i}A = Rot(z, \theta_i)Trans(\alpha_i, 0, 0)Rot(x, \alpha_i)Rot(y, \beta_i)
$$
$$
= \begin{bmatrix} c\beta_ic\theta_i - s\alpha_is\beta_is\theta_i & -c\alpha_i\theta_i & s\beta_is\theta_i + c\beta_is\alpha_is\theta_i & a_ic\theta_i \\ c\beta_is\theta_i + s\alpha_is\beta_is\theta_i & c\alpha_ic\theta_i & s\beta_is\theta_i - c\beta_is\alpha_ic\theta_i & a_is\theta_i \\ -c\alpha_is\beta_i & s\alpha_i & c\alpha_ic\beta_i & 0 \\ 0 & 0 & 0 & 1 \end{bmatrix} \tag{2}
$$

This transformation can ensure that small joint location and orientation may be modeled by small parameter variations when the adjacent two joint axes are parallel or near parallel.

The nominal values of the DH parameters of the MOTOMAN-80 are shown in Table 1. According to Eqs. (1) and (2), the homogeneous transformation matrix between adjacent link coordinate systems can be calculated. Then the transformation from the base frame $\{O_0\}$ to the tool frame $\{O_{tool}\}$ can be expressed as follows:

$$
{}^{0}_{tool}A = {}^{0}_{1}A{}^{1}_{2}A{}^{2}_{3}A{}^{3}_{4}A{}^{4}_{5}A{}^{5}_{6}A{}^{6}_{tool}A \tag{3}
$$

Table 1. Nominal values of the DH parameters

i	$\theta_i(rad)$	$d_i(mm)$	$a_i(mm)$	$a_i(rad)$	$\beta_i(rad)$
1	θ_1	0	145	$\pi/2$	
2	$\theta_2 + \pi/2$		870	0	0
3	θ_3	0	210	$\pi/2$	
4	θ_4	1025	0	$-\pi/2$	
5	θ_5	0	0	$\pi/2$	
6	θ_5	175	0	0	

where ${}_{tool}^{6}\mathbf{A}$ represents a homogeneous coordinate transformation from $\{O_6\}$ to $\{O_{tool}\}$, and it can be expressed as:

$$
{}_{tool}^{6}\mathbf{A} = \begin{bmatrix} 1 & 0 & 0 & x_{tool} \\ 0 & 1 & 0 & y_{tool} \\ 0 & 0 & 1 & z_{tool} \\ 0 & 0 & 0 & 1 \end{bmatrix} \tag{4}
$$

2.2 Kinematic Parameter Errors Considered

There are usually small deviations between the actual and the nominal DH parameters due to the manufacturing and assembly of the robot. The purpose of kinematic calibration is to identify these small parameter errors and to compensate them. The robotic DH parameters with kinematic parameter errors are shown in Table 2.

Table 2. DH parameters with kinematic errors

i	$\theta_i(rad)$	$d_i(mm)$	$a_i(mm)$	$a_i(rad)$	$\beta_i(rad)$
1	$\theta_1 + \delta\theta_1$	δd_1	$145 + \delta a_1$	$\pi/2 + \delta a_1$	
2	$\theta_2 + \pi/2 + \delta\theta_2$		$870 + \delta a_2$	δa_2	$\delta\beta_2$
3	$\theta_3 + \delta\theta_3$	δd_3	$210 + \delta a_3$	$\pi/2 + \delta a_3$	
4	$\theta_4 + \delta\theta_4$	$1045 + \delta d_4$	δa_4	$-\pi/2 + \delta a_4$	
5	$\theta_5 + \delta\theta_5$	δd_5	δa_5	$\pi/2 + \delta a_5$	
6	$\theta_6 + \delta\theta_6$	$175 + \delta\theta_5$	δa_6	δa_6	

In addition to the DH parameter errors, there are also small errors in the homogeneous coordinate transformation from $\{O_6\}$ to $\{O_{tool}\}$. As shown in Eq. (5).

$$
{}^{6}_{tool}\mathbf{A} = \begin{bmatrix} 1 & 0 & 0 & x_{tool} + \delta x \\ 0 & 1 & 0 & y_{tool} + \delta y \\ 0 & 0 & 1 & z_{tool} + \delta z \\ 0 & 0 & 0 & 1 \end{bmatrix}
\tag{5}
$$

Considering all the above small errors, the position of TCP can be expressed in the base frame as follows:

$$
\mathbf{P_t} = f(\theta, \Delta, e)
\tag{6}
$$

Where $\theta = [\theta_1, \theta_2, \ldots, \theta_6]$ is the collection of the joint angels.

$e = [\delta_x, \delta_y, \delta_z]$ is the collection of the tool coordinate system errors.
$\Delta = [\delta\theta_1, \delta d_1, \delta a_1, \delta\alpha_1, \ldots, \delta\theta_6, \delta d_6, \delta a_6, \delta\alpha_6]$ is the collection of the 24 DH parameter errors.

3   Measurement and Parameter Identification

The measurement configuration is shown in Fig. 2. Point P_0 is a fixed point at the origin of the world frame. Point P_t is a moving test point which is at the TCP of the robot. The actual distance between P_0 and P_t for each pose is measured by a telescoping ballbar. To identify more kinematic parameter errors, the distance is measured under plenty of different poses, and the joint angles are recorded at each pose.

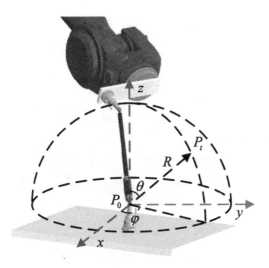

Fig. 2. The measurement configuration

3.1 Description of Measurement System

In this paper, a Renishaw QC20-W telescoping ballbar is used in the calibration process since it is compact and wireless. The nominal length of the QC20-W is 100 mm. The extension bars and calibrator allow highly accurate measurement of lengths near 100 ± 1 mm, 150 ± 1 mm, and 300 ± 1 mm. The accuracy of QC20-W can be up to 600 ± 1 mm. Therefore, high-precision distance measurement can be achieved.

The position of the test point P_t should be near to a specified spherical surface due to the limitation of the measuring range of the telescoping ballbar. Note that the nominal positions are first generated by a program, and then the robot is controlled to move to the given position through a programming interface.

To make the calibration result more accurate, the test points should be distributed at different positions on the spherical surface. A distribution model is developed in this paper as shown in Fig. 3, which can be expressed as Eq. (7).

$$\begin{cases} x = R \sin \theta \cos \varphi \\ y = R \sin \theta \cos \varphi \\ z = R \cos \theta \end{cases} \tag{7}$$

where $\theta \in [-\pi/2, \pi/2]$, $\varphi \in [-\pi/2, \pi/2]$ and they are chosen every $\pi/6$, R is the radius of the specified spherical surface which can be respectively set as 100 mm, 150 mm, or 300 mm. For each measurement position, random five different tool postures are selected to increase the number of measured poses.

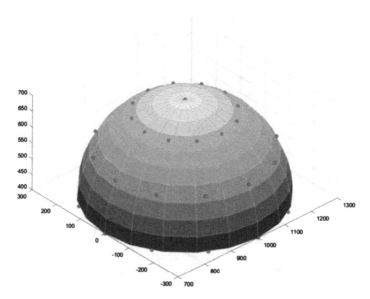

Fig. 3. Distribution of test points

3.2 Kinematic Parameters Identification

The kinematic parameter identification is carried out based on the data obtained at different poses. The identification can be regarded as a nonlinear least-square optimization problem, in which the deviation between the nominal and the actual distance from P_0 to P_t reaches a minimum. The nominal distance can be expressed as:

$$L_i = \left\| \mathbf{P}_{t,i} - \mathbf{P}_0 \right\|_2 \tag{8}$$

where $P_{t,i}$ is the coordinate of the i-th test point and it can be calculated from Eq. (6) with the i-th set of the nominal joint angles. $\| \cdot \|_2$ represents the Euclidean norm of a vector.

The actual distance measured by the telescoping ballbar is recorded as $L_{m,i}$ and the objective function is described as follows:

$$F(X) = \sum_{i=1}^{n} (L_i - L_{m,i})^2 \tag{9}$$

where $\mathbf{X} = [\Delta, \mathbf{e}, \mathbf{P}_0]$ is a 30-dimensional error vector. Δ is the 24 DH parameter error vector, $\mathbf{e}$ is the TCP error vector and $\mathbf{P}_0$ is the coordinate of the fixed point.

The robot calibration problem is finally transformed into an optimization problem. The objective function is optimized by the Levenberg-Marquardt algorithm. This algorithm is sensitive to the initial value. The initial value should be close to the true value of the error parameter. In this paper, the initial value of Δ, $\mathbf{e}$ is zero. As for $\mathbf{P}_0$, it can be read from the robot's teaching pendant when the TCP moves to the position of the fixed point.

In the identification process, some parameters may not be identified due to the limitation of the calibration model and the selected robot poses. Therefore, an identifiability analysis needs to be performed to find the unidentifiable parameters.

Following the existing method [9], the identifiability analysis process can be achieved base on the rank of $\mathbf{J}$, which is the Jacobian matrix of the distance error. Let $g_i(\mathbf{X})$ be the distance error at pose i and it can be expressed as:

$$g_i(\mathbf{X}) = L_i - L_{m,i} \tag{10}$$

The Jacobian matrix $\mathbf{J}$ of the distance error is obtained as follows by differentiating Eq. (10).

$$\mathbf{J} = \left[\frac{\partial g_1(\mathbf{X})}{\partial X} \frac{\partial g_2(\mathbf{X})}{\partial X} \cdots \frac{\partial g_n(\mathbf{X})}{\partial X} \right]^T \tag{11}$$

$\mathbf{J}$ is a matrix of $n \times m$ where n is the number of poses and m is the number of error parameters. The rank of $\mathbf{J}$ represents the number of identifiable parameters.

In this paper, the coordinate of TCP in $\{O_{tool}\}$ is $\mathbf{P}_{tool} = [100\ 0\ 50]$ and the location of the fixed point in the base frame is $\mathbf{P}_{tool} = [1000\ 0\ 400]$ The rank of $\mathbf{J}$ is 24, so only 24 parameters among 30 parameters can be identified. The linearly dependent parameters can be found by removing the column related to each parameter. Finally, the following 5 groups of linearly dependent parameters are found, i.e., $[\delta d_1, z_1]$, $[\delta\theta_1, y_0]$, $[\delta d_6, z]$, $[\delta a_6, \delta x]$, $[\delta\theta_6, \delta a_6, \delta y]$.

For each set of linearly dependent parameters, only one of the parameters can be chosen, and the others are assigned to a fixed value. In this paper, the values of these parameters $[\delta d_1, \delta\theta_1, \delta d_6, \delta a_6, \delta\theta_6, \delta\alpha_6]$ are fixed to zero. The 24 identifiable parameters are finally retained and divided into three groups.

$\boldsymbol{\Delta} = [\delta a_1, \delta\alpha_1, \delta\theta_2, \delta a_2, \delta\alpha_6, \delta\beta_2, \delta\theta_2, \delta d_2, \ldots, \delta a_5, \delta\alpha_5]$ is an 18-dimensional DH parameter error vector.

$\mathbf{e} = [\delta x, \delta y, \delta z]$ is the end tool coordinate system error.

$\mathbf{P}_0 = \begin{bmatrix} x_0 & y_0 & z_0 \end{bmatrix}$ is the position of the fixed point.

4 Simulation Study

The simulation study aims to investigate the effectiveness of the proposed method on identifying the actual parameter values, which mainly consists of three parts: data acquisition, parameter identification, and method validation.

The data required for the parameter identification process includes the joint angles of the robot and the corresponding actual distance of the telescoping ballbar at each calibration pose. To get the above data, the joint angles corresponding to the ideal position of the test points as described in Sect. 3.1 are calculated through the inverse kinematics of the robot without considering parameter errors. After that, the calculated joint angles and the kinematic model with predefined parameter errors showed in Table 3 are used to calculate the actual positions of the test points through the forward kinematics. Then the actual distance values of the telescoping ballbar can be simulated from the actual positions of the test points and the fixed point.

Once the simulation data is obtained, the Levenberg-Marquardt algorithm is used to solve the objective function, i.e., Equation (9), and the parameter errors can then be calculated. The identified kinematic parameter errors are shown in Table 3, which is nearly the same as the predefined parameter errors.

The positioning accuracy after calibration is also analyzed under the condition that a Gaussian noise with $\sigma = 0.002$ mm is added to the measurement data. The positioning error is defined as the deviation between the actual position and the theoretical position of the TCP. The actual position can be calculated through Eq. (6) with the predefined parameter errors, whereas the theoretical position is calculated with the identified parameter errors. The position errors of 20 poses are shown in Fig. 4, which shows that the end position errors are significantly reduced after calibration.

The accuracy of the calibration result depends on the accuracy of the measurement system, which is mainly affected by the noise of the measurement sensor. The telescoping ballbar we used in this paper is Renishaw QC20-W, whose measurement accuracy can reach± 0,006 mm under our experimental conditions. According to the principle of Gaussian distribution, the actual noise of the telescoping ballbar is around $\sigma = 0.002$ mm. Therefore, Gaussian noises with σ ranging from 0.002 mm to 0.012 mm are introduced to the simulation data. The average position errors under different noise conditions are shown in Fig. 5. As can be seen from the figure, the average position error is equal to 0.073 mm when $\sigma = 0.002$ mm and the method can still maintain high precision when the noise is further increased. This demonstrates that our method is robust to the noise.

Table 3. Results of parameter identification

Parameters	Normal values	Actual errors introduced	Errors identified without noise	Errors identified with $\sigma = 0.002$
$\theta_1 (rad)$	0	*	*	*
$d_1 (mm)$	0	*	*	*
$a_1 (mm)$	145	0.556	0.5560	0.5348
$\alpha_1 (rad)$	$\pi/2$	0.008	0.0080	0.0079
$\theta_2 (rad)$	0	-0.011	-0.0110	-0.0111
$a_2 (mm)$	870	0.547	0.5470	0.5541
$\alpha_2 (rad)$	0	0.007	0.0070	0.0070
$\beta_2 (rad)$	0	0	0	-0.00006
$\theta_3 (rad)$	0	0.005	0.0050	0.0050
$d_3 (mm)$	0	0.622	0.6220	0.6557
$a_3 (mm)$	210	0.213	0.2130	0.1903
$\alpha_3 (rad)$	$\pi/2$	0.010	0.0100	0.0099
$\theta_4 (rad)$	0	-0.008	-0.0080	-0.0080
$d_4 (mm)$	1025	0.418	0.4180	0.4162
$a_4 (mm)$	0	0.688	0.6880	0.6903
$\alpha_4 (rad)$	$-\pi/2$	0.012	0.0120	0.0120
$\theta_5 (rad)$	0	0.007	0.0070	0.0070
$d_5 (mm)$	0	0.233	0.2330	0.2309
$a_5 (mm)$	0	0.545	0.5450	0.5424
$\alpha_5 (rad)$	$\pi/2$	0.007	0.0070	0.0070
$\theta_6 (rad)$	0	*	*	*
$d_6 (mm)$	175	*	*	*
$a_6 (mm)$	0	*	*	*
$\alpha_6 (rad)$	0	*	*	*
$x (mm)$	100	0.5	0.5000	0.5000
$y (mm)$	0	0.5	0.5000	0.4993
$z (mm)$	50	0.5	0.5000	0.5006

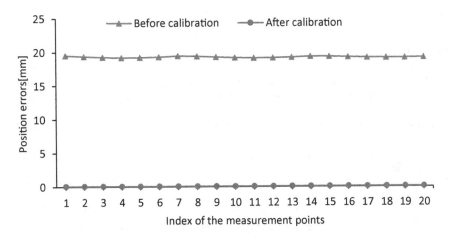

Fig. 4. Position errors before and after calibration

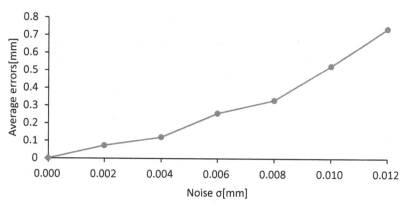

Fig. 5. Average position errors with noise increase

5   Conclusion

In this paper, a new calibration method for six-axis industrial robots is proposed based on a telescoping ballbar. A telescoping ballbar has the advantage of being more accurate than laser trackers and yet cheaper than laser trackers, whereas the laser tracker is the most popular robotic calibration system. The proposed method performs calibration with numerous poses so that high precision calibration can be achieved at a relatively low cost. The feasibility of this method is verified through simulation on a Yaskawa MOTOMAN-80 six-axis industrial robot. After calibration, the average position errors are reduced from 19.368 mm to 0.073 mm when the Gaussian noise with $\sigma = 0.002$ mm is added to the sensor data. Besides, a robustness analysis shows that the proposed method is robust to the sensor noise.

Acknowledgments. This research was supported by the National Nature Science Foundation of China (Grant No. 51905346, No. 91648202 and No. 91948301)

References

1. Shiakolas, /.P.S., Conrad, K.L., Yih, T.C.: On the accuracy, repeatability, and degree of influence of kinematics parameters for industrial robots. International journal of modeling and simulation **22**(4), 245–254 (2002)
2. David, J., Hartenberg, R.S.: A kinematic notation for lower-pair mechanisms based on matrices. J. Appl. Mech. **22**(1), 215–221 (1955)
3. Craig, J.J.: Introduction to Robotics: Mechanics and Control, 3rd edn. Pearson Prentice Hall, Upper Saddle River (2005)
4. Okamura, K., Park, F.C.: Kinematic calibration using the product of exponentials formula. Robotica **14**(4), 415–421 (1996)
5. Xiong, G., Ding, Y., Zhu, L.M.: A product-of-exponential-based robot calibration method with optimal measurement configurations. Int. J. Adv. Robot. Syst. **14**(6), 1729881417743555 (2017)

6. Özgür, E., Mezouar, Y.: Kinematic modeling and control of a robot arm using unit dual quaternions. Robot. Auto. Syst. **77**, 66–73 (2016)
7. Nubiola, A., Bonev, I.A.: Absolute calibration of an ABB IRB 1600 robot using a laser tracker. Robot. Comput. Integr. Manuf. **29**(1), 236–245 (2013)
8. Sun, T., Zhai, Y., Song, Y., et al.: Kinematic calibration of a 3-DoF rotational parallel manipulator using laser tracker. Robot. Comput.-Integr. Manuf. **41**, 78–91 (2016)
9. Joubair, A., Bonev, I.A.: Kinematic calibration of a six-axis serial robot using distance and sphere constraints. Int. J. Adv. Manuf. Technol. **77**(1–4), 515–523 (2015)
10. Wang, R., Wu, A., Chen, X., et al.: A point and distance constraint based 6R robot calibration method through machine vision. Robot. Comput.-Integr. Manuf. **65**, 101959 (2020)
11. Ikits, M., Hollerbach, J.M.: Kinematic calibration using a plane constraint. In: International Conference on Robotics and Automation,Vol. 4, pp. 3191–3196. IEEE (1997)
12. Hanqi, Z., Motaghedi, S., Roth, Z.: Robot calibration with planar constraints. In: IEEE International Conference Robotics Automation, pp. 805–810. IEEE (1999)
13. Joubair, A., Bonev, I.A.: Non-kinematic calibration of a six-axis serial robot using planar constraints. Precis. Eng. **40**, 325–333 (2015)
14. Lembono, T.S., Suárez-Ruiz, F., Pham, Q.C.: SCALAR: simultaneous calibration of 2-D laser and robot kinematic parameters using planarity and distance constraints. IEEE Trans. Autom. Sci. Eng. **16**(4), 1971–1979 (2019)
15. Nubiola, A., Slamani, M., Bonev, I.A.: A new method for measuring a large set of poses with a single telescoping ballbar. Precis. Eng. **37**(2), 451–460 (2013)
16. Nubiola, A., Bonev, I.A.: Absolute robot calibration with a single telescoping ballbar. Precis. Eng. **38**(3), 472–480 (2014)
17. Hayati, S.A.: Robot arm geometric link parameter estimation. In: The 22nd IEEE Conference on Decision and Control, pp. 1477–1483. IEEE (1983)

Variable Impedance Control
of Manipulator Based on DQN

Yongjin Hou, Hao Xu, Jiawei Luo, Yanpu Lei, Jinyu Xu,
and Hai-Tao Zhang[✉]

School of Artificial Intelligence and Automation,
Huazhong University of Science and Technology, Wuhan 430074, China
zht@mail.hust.edu.cn
http://imds.aia.hust.edu.cn/

Abstract. For traditional constant impedance control, the robot suffers from constant stiffness, poor flexibility, large wear and high energy consumption in the process of movement. To address these problems, a variable impedance control method based on reinforcement learning (RL) algorithm Deep Q Network (DQN) is proposed in this paper. Our method can optimize the reference trajectory and gain schedule simultaneously according to the completion of task and the complexity of surroundings. Simulation experiments show that, compared with the constant impedance control, the proposed algorithm can adjust impedance in real time while manipulator is executing the task, which implies a better compliance, less wear and less control energy.

Keywords: Variable impedance control · Reinforcement learning · Compliance · Control energy · Manipulator

1 Introduction

To complete specific tasks like assembly, polishing, deburring etc., the end-effector of robot must move from the initial position to the target position as desired, such as in [1,3,4]. In general, when a robot completes a specified task, in order to better interact with the environment, manipulator needs to have a certain degree of flexibility. [4] introduce adaptive variable impedance control in unstructured environment to make the end of manipulator better interact with the environment. However, the introduction of impedance control is usually effective only if manipulator contacts with environment. When it moves in free space, its stiffness remains unchanged, so the flexibility cannot be introduced before contact. For high precision assembly task [3], when manipulator is getting closer to the goal, its work environment becomes more and more complex, for robot with constant stiffness in free space, its flexibility is poor. At this point, if the arm and environment collision will cause irreversible damage to these fine components. In this regard, we need to introduce variable impedance control in free space, and let the robot actively learn this strategy through reinforcement

© Springer Nature Switzerland AG 2020
C. S. Chan et al. (Eds.): ICIRA 2020, LNAI 12595, pp. 296–307, 2020.
https://doi.org/10.1007/978-3-030-66645-3_25

(a) Assemble memory strip   (b) Assemble CPU

Fig. 1. High precision assembly task.

learning. When the robot completes an assembly task in Fig. 1, it can change its compliance in real time according to the situation near target and the complexity of environment.

Recently, RL has gained tremendous success in solving all kinds of problems, for instance, manipulation [9–12], robot motion [13–17], and playing video games [18–20]. RL can be divided into two methods, model-based and model-free [21,22]. While model-based policy search is computationally more expensive than model-free methods. Recent progress in Deep Neural Networks suggests deploying them for functions in RL methods [23], that is, Deep Reinforcement Learning (DRL). Impedance control could realize intelligent behavior and improve interaction between robot and environment [24,25]. Recently, it has been used to solve for constrained robots [26], finger-arm robots [27], and quadruped robots [28], as well as the application in interaction [29]. Variable impedance control could regulate the dynamic relationship between contact force and movement, and give the flexibility to change these dynamics during interaction tasks.

Scholars all over the world have conducted in-depth studies on the optimal control, impedance characteristics and reinforcement learning [1–3], and [8] summarize the application of reinforcement learning in robotics detailedly. With Policy Improvement with Path Integrals (PI2) algorithm [1,6] propose a variable impedance control of the robot, which can simultaneously optimize the reference trajectory and gain schedule, during the completion of task, the gain should be minimized as soon as posible so as to achieve optimal control of energy. Of course, when a large force is needed at the end of the robot, a large stiffness can be generated quickly. However, the overall response speed is slow due to excessive emphasis on reducing gain while completing the task. [2] design a variable impedance actuator and modele the dynamics and random characteristics of the system. Based on the trade-off of task accuracy and control energy, optimal control command is generated. However, the variable impedance characteristic is realized by hardware, and the actuator is difficult to be generalized to general manipulators, which has certain limitations. [3] combine reinforcement learning and force sensor information, where, a new neural network structure is proposed, which integrates the sensor information into neural network and learns an adaptive and compliant behavior. However, its limitation lies in that force sensor information cannot play a role in free space, only in contact with environment, can the corresponding flexibility be shown.

To solve the above problems, we propose a variable impedance control for manipulator based on reinforcement learning. When robot away from goal, the gain K_p is larger, the system has faster response speed, which makes mechanical arm close to the goal as soon as possible, and in general, when the distance from target is far, working environment is relatively simple, it need not to consider the flexibility of robot too much. When the robot gets closer to the target, the gain K_p will gradually decrease, and the response speed of system will gradually slow down, which reduces the control energy effectively, moreover, when robot is close to target, the working environment becomes more and more complex, good compliance can reduce the damage to environment. The main contributions of this paper are as follows.

- During the movement of manipulator, we weigh the relationship between completion of task and gain, and optimize the reference trajectory and gain schedule concurrently.
- The mechanical arm can adjust the gain in real time according to the completion of tasks and changes in external environment, showing different flexibility. While optimizing the gain, it can effectively reduce wear and control energy consumption.

The rest of this paper is organized as follows. In Sect. 2, we introduce the related work, including principle of reinforcement learning algorithm DQN and variable impedance control. Section 3 describes how to apply DQN algorithm to variable impedance control of manipulator. To further analyze the behavior of the proposed algorithm, experimental results are presented in Sect. 4. Finally, in Sect. 5, we conclude the method proposed in this paper.

2   Preliminaries

The traditional method to robot control is negative feedback control with fixed proportional-derivative (PD) gains. However, fixed gain is not ideal for many tasks. In this paper, the variable impedance control of robot is realized based on DQN algorithm.

2.1   Deep Reinforcement Learning: DQN Algorithm

In this paper, DQN algorithm, a value-based method, is used to realize the variable impedance control of manipulator. The algorithm outputs values of all actions using network, and selects actions based on the highest value. It does not need to represent the environment with a model, that is, it does not try to understand the environment, but gets feedback from the environment and learns from it. The algorithm is one-step update and outputs discrete actions.

DQN is developed based on Q-learning, the following is added to the framework of Q-Learning, which is why DQN has better performance. First of all, neural network is used to estimate Q value, instead of looking up the table like Q-learning, which can be better applied to high-dimensional cases without worrying about the difficulty of updating the Q table due to high dimension. Secondly, use experience pool to store the latest states, actions, rewards and next states for updating the network, so as to cut off the correlation between experiences and use them for repeated learning. What is more, temporarily freeze the parameters of Q_target Network and assign the parameters of Q_eval Network to Q_target Network every fixed steps, which is also a method to cut the correlation between experiences.

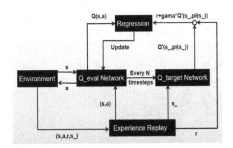

Fig. 2. The principle of DQN algorithm.

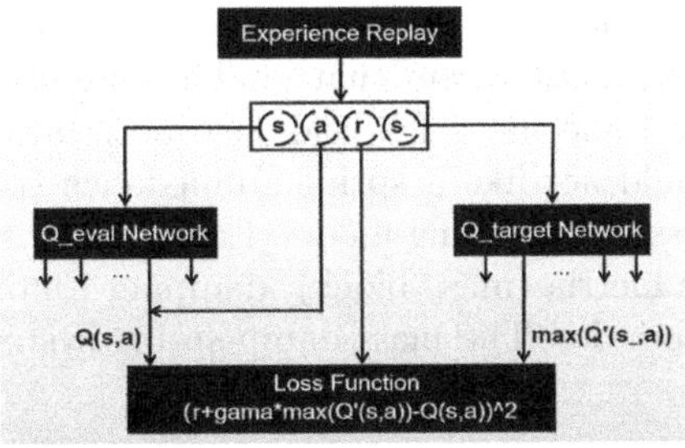

Fig. 3. DQN network update mode.

The principle of DQN algorithm is shown in Fig. 2.

(1) Import the environment state s into Q_eval Network, and the output of Network is Q value of every actions. Select an action a corresponding to the maximum Q_value, or randomly generate an action according to a certain probability, that is, introduce exploration. Then apply resulting action to environment to get reward r and next state of the environment s_, and store the current state s, action a, reward r, and next state s_ in experience pool. The s_ is then entered into Q_eval Network, and step (1) is iterated until experience pool is filled with data, during which Q_eval Network is not updated, but only used for acquisition actions.

(2) Sample a part (s, a, r, s_) from experience pool, and input the sampled state s into Q_eval Network, then get Q(s, a), the estimated value of corresponding action a.

(3) Input the sampled state s_ into Q_target Network, multiply the obtained Network output value Q '(s_, pi (s_)) by the discount coefficient γ, then add corresponding sampled reward r, and the obtained value is the actual Q value of corresponding action a.

(4) The estimated Q value is regressed with actual Q value to update the parameters of Q_eval Network.

(5) Loop step (1) to (4). After every certain number of steps, assign parameter of Q_eval Network to Q_target Network.

The specific update method of Q_eval Network is shown in Fig. 3.

2.2   Variable Impedance Control

Impedance control [5] analyzes the dynamic relationship between the end of manipulator and environment, comprehensively considers force and position control, and uses same strategy to realize it. The purpose of impedance control is to build a system that enables the manipulator to control force and position simultaneously. Traditional robotic arms are too "hard", a good robotic arm should act like a spring. Impedance control is established on the basis of the mass-damp-spring model. Each part of the mass-damp-spring system represents the inertia (mass block), damping (damping block) and stiffness (spring) characteristics. The mass-damp-spring system is shown in Fig. 4.

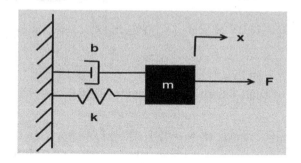

Fig. 4. Mass-damping-elastic system.

The relationship between force and position is adjusted by adjusting three parameters of impedance controller (inertia coefficient, damping coefficient and stiffness coefficient). In Fig. 4, m, b, and k respectively represent inertia coefficient, damping coefficient and stiffness coefficient, and F represents contact force between the end of manipulator and environment.

The impedance control mentioned in this paper is different from traditional impedance control, which is a kind of generalized impedance control, we only consider the characteristics of stiffness (spring) and damping (damping block).

The so-called variable impedance control is to regulate the stiffness coefficient and damping coefficient of system.

Consider a robot that controls the joint through torque. After joint trajectory is planned, feedforward control term is obtained through system inverse dynamics, and the joint position and velocity are modified by PD controler, so as to calculate the expected control torque, which is shown in Eq. (1).

$$T = -K_p(q - q_d) - K_d(\dot{q} - \dot{q}_d) + T_{ff} \tag{1}$$

$$T_{ff} = M(q)\ddot{q} + V(q, \dot{q}) + G(q) \tag{2}$$

Where K_p and K_d represent the gain matrix of joint position and velocity respectively, q, $\dot{q}$, and $\ddot{q}$ represent actual joint position, velocity and acceleration respectively, q_d and $\dot{q}\_d$ represent expected joint position and velocity respectively. Therefore, the stiffness and damping of the system can be changed by adjusting K_p and K_d, so as to realize variable impedance control. T_{ff} represents the feedforward control term, where M represents the mass matrix, V represents centrifugal force and Coriolis force vector, and G represents gravity vector.

3   Variable Impedance Control Based on DQN

Generally, reinforcement learning framework is shown in Fig. 5. State, Action and Reward are respectively observed states, actions performed and rewards obtained. In reinforcement learning, Agent interacts with environment, each action performed will affect the next state of environment. Meanwhile, Agent can obtain rewards from environment to measure satisfaction degree of completing task.

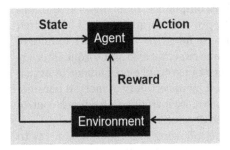

Fig. 5. The framework of Reinforcement learning.

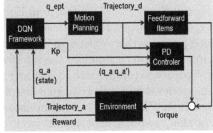

Fig. 6. Block diagram of variable impedance control based on DQN.

In reinforcement learning, we only learn position gain matrix in PD controller, but not velocity gain matrix directly. In this paper, we use experience

[1] to calculate velocity gain K_d, that is, the velocity gain is equal to the square root of position gain and then multiplied by a factor, as shown in Eq. (3).

$$K_d = \xi \sqrt{K_p} (0 \leq \xi \leq 1) \tag{3}$$

The control block diagram of variable impedance based on DQN mentioned in this paper is shown in Fig. 6. We take actual state of robot, namely its joint position, as input of the DQN framework. After analyzing the environment and other information in the reinforcement learning part, desired position and joint gain are outputed. Then, the joint space trajectory between actual position and desired position is planned, and the expected trajectory and joint gain are taken as input of controller. Where, the feedforward control part of controller [7] is calculated through inverse dynamics Eq. (2), and then combined with PD control term, it is taken as the control output of controller, namely the desired joint torque Eq. (1). Then, obtained joint torque is applied to the robot to obtain actual trajectory, and the joint position and velocity in the actual trajectory are fed back to PD controller. Finally, we continue to input final joint position into DQN framework as new state until the robot completes specified task or reaches specified time.

4   Experiment

The proposed method is applied to a two-manipulator robot, and use simulink and robot toolbox in MATLAB for simulation, so as to verify the effects of proposed method. The simulation experiment is set as a via-point experiment in joint space, that is: within specified number of steps, each joint reaches predetermined position. The experimental results are used to illustrate effectiveness of variable impedance learning strategy and compare it with constant impedance control to illustrate advantages of our method in terms of compliance, reducing wear and control energy. Feedforward control item Eq. (2) provides torque required for desired joint space trajectory. Because feedforward term linearizes nonlinear dynamics of working point, so there exist errors. At the same time, we also add torque interference in simulation to simulate the interference of external environment in actual work. In order to compensate the errors caused by uncertainty of inertial parameters, unmodeled forces and external disturbances, we introduce a feedback control term, namely PD controller, to eliminate control errors.

For this via-point experiment, we use the following loss function Eq. (4) to describe the task, and add terminal loss Eq. (8) (only relevant to task and independent of gain) to describe completion of task, so as to better accelerate convergence speed.

$$loss_{all}(t) = \omega_{gain} * loss_{gain}(t) + \omega_{task} * C(t) \tag{4}$$

$$loss_{gain}(t) = |K_{p1}(t) - K_{p1\_min}| + |K_{p2}(t) - K_{p2\_min}| \tag{5}$$

$$C(t) = loss_{task\_temp}(t) + loss_{task\_term}(t) \tag{6}$$

Where, gain loss term Eq. (5) is the sum of absolute value of the difference between proportional gain and the preset lower limit of gain. Since smaller gain yields some desired properties, such as better compliance, less wear, and lower control energy, we expect a lower gain by adding a gain penalty term to loss function. Task loss item Eq. (6) is used to describe completion of the task by manipulator within the specified number of steps. When robot is far from target point or does not complete the task within specified number of steps, it will be punished, including transient loss and terminal loss. ω_{gain} and ω_{task} are used to weigh gain loss against task loss. What is more, t stands for nominal time, which is actually the number of steps taken by the robot.

The maximum number of steps in whole movement is 20. The $loss_{task\_temp}$ and $loss_{task\_term}$ respectively represent transient loss and terminal loss of the task, as shown in Eq. (7) and Eq. (8):

$$loss_{task\_temp} = \sqrt{(q_1(t) - q_{1\_goal})^2} + \sqrt{(q_2(t) - q_{2\_goal})^2} \tag{7}$$

$$loss_{task\_term} = \varphi * \delta(t - 20) * (|q_1(t) - q_{1\_goal}| + |q_2(t) - q_{2\_goal}|) \tag{8}$$

Where, q_1 and q_2 respectively represent current actual positions of two joints, q_{1_goal} and q_{2_goal} respectively represente target positions, φ is used to regulate the penalty for terminal loss. The reward of environmental feedback can be obtained after the loss is treated, as shown in Eq. (9), $E(loss_{all})$ represents expected value of total loss, we can replace the expected value of the loss with the average value of previous loss.

$$r = E(loss_{all}) - loss_{all} \tag{9}$$

Figure 7 is learning curve. Since DQN can only select actions on discrete values, the losses in each round obtained are discrete values, so we use scatter plot to represent learning situation, including gain loss and overall loss.

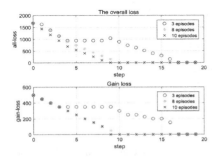

Fig. 7. Learning curve.

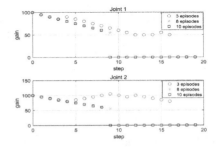

Fig. 8. Gain curve.

Figure 7 shows that overall loss and gain loss can converge to zero finally, that is, our algorithm is convergent, and with the increase of episode, the convergence rate is faster and faster. As we can see, when the episode is 10, gain loss and overall loss can be reduced to zero when step 9, showing that the robot finished the task within stipulated 20 steps, and the control gain gradually decreases from initial values, eventually to achieve minimum. What's more, we also plot gain curve Fig. 8, when the gain changes to zero, the task has been completed.

It can be seen from Fig. 9 and Fig. 10 that for constant impedance control and variable impedance control, the variation trend of joint torque and error is basically same. It should be noted that when the robot completes specified task, joint torque and error keep a constant value, corresponding to the loss value above reduce to zero. In addition, since joint 2 needs to overcome its own gravity when it stays at rest, its torque is not zero. In order to better compare control effect of two methods, we plot joint torque and error variation curves of 10 episodes of constant impedance control and variable impedance control into the same figure. As shown in the figure below.

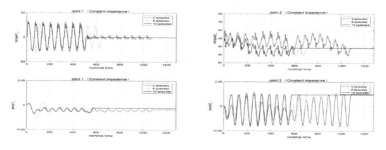

Fig. 9. Learning Constant impedance control joint torque, error change curve.

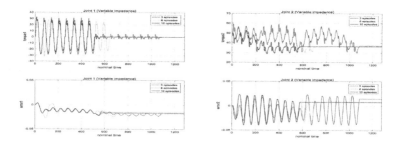

Fig. 10. Gain Variable impedance control joint torque, error change curve Constant impedance control joint torque, error change curve.

The predicted output of DQN is position and gain of next step. As manipulator approaches target position, its joint gain decreases until reaches the specified minimum value, which is manifested as overall control energy changes from large to small, and the flexibility of manipulator gradually gets better. As can be seen from Fig. 11, compared with constant impedance control, the error of joint 1 increases in the later stage of motion, while the error of joint 2 decreases in the whole process. As far as joint torque is concerned, in the whole process, control toque of joint 1 does not change significantly, while that of joint 2 decrease significantly. In short, variable impedance control can effectively reduce joint torque, that is, reduce control energy, increase manipulator's flexibility, and reduce system wear while ensuring joint error basically unchanged. Its comprehensive control effect is obviously better than that of constant impedance control.

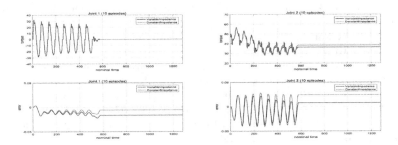

Fig. 11. Comparison of joint torque and error variation with constant impedance and variable impedance.

5 Conclusion

In this paper, we presente a reforcement learning approach that could learn variable impedance control for robot. The proposed algorithm makes it suitable to optimize both reference trajectory and gain schedule simultaneously, and it can display different impedance characteristics according to the completion of task, so as to improve compliance, reduce wear and reduce control energy.

For task-oriented robots, such as assembly, grinding, cutting, etc., the task can be described as follows. The robot move from initial position to target position and then complete assigned task. In this kind of task, the method proposed in this paper can be used for control, so that robot can reach target position as soon as possible. At the same time, its flexibility changes gradually as it approaches target position, avoiding damage to target devices due to excessive stiffness, especially for some delicate assembly tasks. Due to the assembly device is small and prone to damage, excessive robot rigidity may cause irreparable damage to target device, such as memory strip fracture, CPU contact surface wear, etc. By adopting control method proposed in this paper, the established tasks can be accomplished better.

Funding Information. This work is supported by National Natural Science Foundation (NNSF) of China under Grant U1713203.

References

1. Buchli, J., Stulp, F., Theodorou, E., Schaal, S.: Learning variable impedance control. Int. J. Rob. Res. **30**(7), 820–833 (2011)
2. Mitrovic, D., Klanke, S., Vijayakumar, S.: Learning impedance control of antagonistic systems based on stochastic optimization principles. Int. J. Robot. Res. **30**(5), 556–573 (2011)
3. Luo, J., Solowjow, E., Wen, C., Ojea, J. A., Agogino, A. M., Tamar, A., Abbeel, P.: Reinforcement learning on variable impedance controller for high-precision robotic assembly. In: International Conference on Robotics and Automation (ICRA), pp. 3080–3087. IEEE (2019)
4. Ya-hui, G., Jin-jun, D., Xian-zhong, D.: Adaptive variable impedance control for robot force tracking in unstructured environment. Control and Decision, p. 10 (2019)
5. Lynch, K.M., Park, F.C.: Modern Robotics: Mechanics, Planning, and Control. Cambridge University Press (2017)
6. Theodorou, E.A., Buchli, J., Schaal, S.: A generalized path integral control approach to reinforcement learning. J. Mach. Learn. Res. **11**(2010), 3137–3181 (2010)
7. O'Regan, Gerard: Robotics. The Innovation in Computing Companion, pp. 221–226. Springer, Cham (2018). https://doi.org/10.1007/978-3-030-02619-6_47
8. Kober, J., Bagnell, J.A., Peters, J.: Reinforcement learning in robotics: a survey. Int. J. Robot. Res. **32**(11), 1238–1274 (2013)
9. Sergey, L., Wagener, N., Abbeel, P.: Learning contactrich manipulation skills with guided policy search. In: Proceedings of the 2015 IEEE International Conference on Robotics and Automation (ICRA), Seattle, WA, USA, pp. 26–30 (2015)
10. Chebotar, Y., Kalakrishnan, M., Yahya, A., Li, A., Schaal, S., Levine, S.: Path integral guided policy search. In: IEEE International Conference on Robotics and Automation (ICRA), pp. 3381–3388. IEEE (2017)
11. Peters, J., Mulling, K., Altun, Y.: Relative entropy policy search. In: AAAI, Atlanta, pp. 1607–1612 (2010)
12. Fu, J., Levine, S., Abbeel, P.: One-shot learning of manipulation skills with online dynamics adaptation and neural network priors. In: IEEE/RSJ International Conference on Intelligent Robots and Systems (IROS) pp. 4019–4026. IEEE (2016)
13. Abbeel, P., Coates, A., Quigley, M., Ng, A.: An application of reinforcement learning to aerobatic helicopter flight. In: International Conference on Neural Information Processing Systems, pp. 1–8 (2006)
14. Luo, J., Edmunds, R., Rice, F., Agogino, M.: Tensegrity robot locomotion under limited sensory inputs via deep reinforcement learning. In: IEEE International Conference on Robotics and Automation (ICRA), pp. 6260–6267. IEEE. (2018)
15. Schulman, J., Levine, S., Abbeel, P., Jordan, M., Moritz, P.: Trust region policy optimization. In: International Conference on Machine Learning, pp. 1889–1897 (2015)
16. Levine, S., Abbeel, P.: Learning neural network policies with guided policy search under unknown dynamics. In: Advances in Neural Information Processing Systems (NIPS), pp. 1071-1079 (2014)
17. Zhang, T., Kahn, G., Levine, S., Abbeel, P.: Learning deep control policies for autonomous aerial vehicles with mpc-guided policy search. In: IEEE International Conference on Robotics and Automation(ICRA), pp. 528–535 (2016)
18. Mnih, V., et al.: Playing atari with deep reinforcement learning. arXiv preprint arXiv: 1312.5602 (2013)

19. Mnih, V., et al.: Asynchronous methods for deep reinforcement learning. In: International Conference on Machine Learning, pp. 1928–1937 (2016)
20. Lillicrap, T.P., et al.: Continuous control with deep reinforcement learning. arXiv preprint arXiv: 1509.02971 (2015)
21. Sutton, R.S., Barto, A.G.: Reinforcement Learning: An Introduction. Vol. 1. 1. MIT press Cambridge (1998)
22. Deisenroth, M.P., et al.: A Survey on Policy Search for Robotics. Foundations and Trends in Robotics, pp. 1–142 (2013)
23. Levine, S., Finn, C., Darrell, T., Abbeel, P.: End-to-end training of deep visuomotor policies. J. Mach. Learn. Res. **17**(1), 1334–1373 (2016)
24. Hogan, N.: Impedance control: an approach to manipulation. In: American Control Conference, pp. 1–24. IEEE (1985)
25. Jung, S., Hsia, T.C., Bonitz, R.G.: Force tracking impedance control of robot manipulators under unknown environment. IEEE Trans. Control Syst. Technol. **12**(3), 474–483 (2004)
26. Yi, S.: Stable walking of qauadruped robot by impedance control for body motion. Int. J. Control Autom. **6**(2), 99–110 (2013)
27. Sano, Y., Hori, R., Yabuta, T.: Comparison between admittance and impedance control method of a finger-arm robot during grasping object with internal and external impedance control. Nihon Kikai Gakkai Ronbunshu C Hen/Trans. Japan Soc. Mech. Eng. C, **79**(807), 4330–4334 (2013)
28. He, W., Dong, Y.: Adaptive fuzzy neural network control for a constrained robot using impedance learning. IEEE Trans. Neural Netw. Learn. Syst. **29**(4), 1174–1186 (2017)
29. Huang, L., Ge, S.S., Lee, T.H.: Fuzzy unidirectional force control of constrained robotic manipulators. Fuzzy Sets Syst. **134**(1), 135–146 (2003)

Simulation of Human Upright Standing Push-Recovery Based on OpenSim

Ting Xiao[✉], Biwei Tang, Muye Pang, and Kui Xiang

School of Automation, Wuhan University of Technology, Wuhan, Hubei, China
15271155966@163.com

Abstract. Investigating the human standing balance mechanisms under push-recovery task is of great importance to the study of biped robot balance control. Under human push-recovery mission, the passive stiffness, stretch reflex and short-range stiffness control mechanisms of human ankle joint are the main components in the internal mechanism of human body. To this end, this paper dedicates to evaluating the roles of the three aforementioned mechanisms during human upright standing push-recovery mission. Firstly, based on the simulation platform OpenSim4.0, this paper chooses a simplified lower-limb musculoskeletal model as the research object. Subsequently, this paper completes the design of the passive stiffness, stretch reflex and passive stiffness controller, and completes the static standing test and upright push-recovery simulation of the selected musculoskeletal model. Finally, in order to verify the effectiveness of the simulation, this paper uses electromyography, force plate and dynamic capture system to collect the relevant data of the human upright push-recovery. The experimental and simulation results reveal that the selected musculoskeletal model can basically simulate the process of human upright push-recovery under the joint actions of the three mechanisms noted above, which, to some degree, can reflect the effectiveness of the established method. Thus, the established method may provide some insights on the balance control of the bipedal robot.

Keywords: Musculoskeletal model · Human upright push-recovery · OpenSim · Stretch reflex · Short-range stiffness · Passive stiffness

1 Introduction

In an uncertain and complicated environment with a small external disturbance, human body has strong adaption which can apply muscle as the source of compliance to drive joint rotation, change joint stiffness and maintain body balance [1]. On one hand, this compliance can increase the elasticity of human joint, so that the joint can bend flexibly in a certain range and slow down the damage to the body structure caused by the disturbance. On the other hand, the energy generated by the collision can be stored in the elastic element of muscle to improve the movement efficiency of the joint [2]. The mentioned compliance of human joint can provide valuable insights both on the mechanical design and control system design for different biped robots, such that the balance control

© Springer Nature Switzerland AG 2020
C. S. Chan et al. (Eds.): ICIRA 2020, LNAI 12595, pp. 308–319, 2020.
https://doi.org/10.1007/978-3-030-66645-3_26

ability of the biped robot can be enhanced as far as possible. In terms of applying the aforementioned compliance characteristics of human ankle joint to design the control system for the biped robot, discovering the internal balance mechanism of human ankle joint remains a significant issue that must be first addressed.

As one of the most important load-bearing organs of the human body, the ankle joint plays an important role in maintaining human upright balance in the case where human body subjects to small external disturbance. Among all the human balance regulation mechanisms, muscle reflex is known as a bottom-level controller which can help human body to maintain upright balance via the motor nerve of spinal cord. Also, it has been discovered that human body can quickly maintain upright balance via stretch refection when human body suffers from small disturbances [3]. Moreover, the authors in [4] have found that too high level of muscle activation (30%–40% MVC) can lead to a relatively large swing of the ankle, which is unbeneficial to the upright balance of human body. In fact, the small-angle human ankle swing is always preferred under real-world human push-recover mission. Thanks to the regulation and control mechanism of the ankle joint stiffness, the human ankle joint can absorb the external impact on the human body and slow down the impact of disturbance on upright balance [5].

Based on the study of ankle dynamics, Kearney et al. have divided the ankle impedance model into intrinsic and reflex components [6]. The author in [7] have pointed out that the joint intrinsic impedance model mainly consisted by passive stiffness provided by the physiological structure, like the tendons, ligaments, cartilage, etc., of the joint. Through the simulation study, De et al. found that there is a delay in nerve transmission, and the muscle activity does not change in a short period of time after being disturbed [8]. By providing a short-range stiffness, the ankle joint makes the muscle force increase rapidly when the muscle fiber is stretched, and increases the stability of the system. Short-term stiffness can not only provide an effective output torque for the ankle at the start of being disturbed, but also affect the muscle reflex in the later stage.

At present, the research on various mechanisms of human ankle joint is gradually improved. However, the majority of the currently-existing studies only investigate the effect of a single mechanism on human balance. The research on joint stiffness is usually based on the measurement of disturbance device. Although this approach can explain the partial characteristics of the ankle joint, it has limitations when the human body is not disturbed and it is difficult to ensure the real-time of the data. Generally, based on the best knowledge of the authors, evaluating the effects of muscle reflexes and joint stiffness on disturbance responses under the combined action of multiple mechanisms has not been extensively studied or addressed. Moreover, the simulation software on the market is only for commercial use and usually closes the access to the open source.

OpenSim is an open source biomechanical simulation system, which can be applied to the development, simulation, and motion analysis of human musculoskeletal models [9]. With the help of this platform, this paper mainly studies the mechanism of stretch reflection, passive stiffness and short-range stiffness in ankle balance strategy. Then, we build a musculoskeletal model simulation platform and adjust the parameters of the model. Using this model, we carry out simulation experiments under static standing and dynamic interference. Finally, we combine the data of simulation and human upright push-recovery experiment to compare and analyze the effects of different control models on joint kinematics. This paper reveals the biomechanical properties of human ankle joint and lay a theoretical foundation for the development of biped robot and exoskeleton equipment.

2 Single-Degree-of-Freedom Inverted Pendulum Model of Human Body

Because of the complexity of neural control system and muscle structure, the joint motion of human body is unable to be completely simulated by a single model. In order to simulate the real process of human upright push-recovery, each control strategy is added to the model in different periods. As shown in Fig. 1, the control mechanism based on COM and passive stiffness is added at the initial stage of the model standing. The short-range stiffness control is added after being disturbed, and the stretch reflex control mechanism is added when the short-range stiffness control ends at the first 50 ms. The 'Disturbance' represents the external force on the back of the experimenter. The 'Motor command' comes from the high level of the nervous systems. 'Time lag' contains the reflex pathway and muscle dynamic time lag [10].

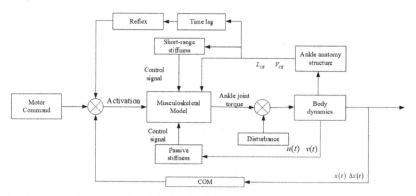

Fig. 1. Neuromuscular control model of ankle joint. Muscle fiber parameters L_{VE} and V_{CE} are converted from 'body dynamics' via 'ankle anatomy structure'. Ankle angle parameters $u(t)$ and $v(t)$, COM parameters $x(t)$ and $\Delta x(t)$ are both from 'body dynamics'.

On the mechanical level, the balance of the human body can be judged by the angle of the center of mass (COM) relative to the base of support (BOS). As a result, the key to the balance control of the human body lies in the control of the COM. We can simplify the human body into a single-chain inverted pendulum model around the ankle [3], as displayed in Fig. 2. In the balance position, the COM of the human body falls within the BOS, and the gravity of the human body and the supporting force produced by the ankle cancel each other out. When human body is disturbed by the force in the horizontal direction, the COM shifts in the opposite direction and makes angle with the BOS, resulting in the imbalance of the human body.

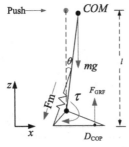

Fig. 2. Schematic of upright standing push-recovery. COM is the center of mass. F_{GRF} is the z-axis ground reaction force. D_{COP} is the equivalent pressure center point. τ for ankle output torque. θ is the ankle joint angle. l is the vertical distance between the center of mass and the ground. F_m is the calf muscle force.

At the initial stage of standing (that is, before being disturbed), the main instability of the model comes from the contact between the soles of the feet and the ground. Model can maintain static standing balance just by controlling the value of the COM in the x direction. The control consists of two parts. One part is the control of the current location value, and the other is the control of the change in position. According to these two parts, the corresponding muscles are activated when the model leans forward (soleus, SOL) or backward (tibialis anterior, TA).

The control model based on COM is expressed as follows:

$$a_{COM}(t) = K_{COM}x(t) + D_{COM}\Delta x(t) \tag{1}$$

where $\Delta x(t)$ is determined by the following equation:

$$\Delta x(t) = x(t_n) - x(t_{n-1}) \tag{2}$$

where K_{COM} is the gain for the COM in the x direction, and D_{COM} is the gain for COM change in the x direction. $a_{COM}(t)$ is the degree of muscle activation provided by the COM control at time t. $\Delta x(t)$ is the change of the COM in the x direction at time t.

When human is at the static standing and subjects to external interference, the muscle reflex caused by spinal nerve transmission has a time delay (about 50 ms), which is called the initial mechanical response period. During this period, muscle activity remains at the same level as the model is standing statically. The disturbance causes the muscle fibers to be stretched rapidly, leading to the deformation of the attached muscle cross – bridges and resulting in short-range stiffness. Due to the actin-myosin overlap, the short-term stiffness is the highest in the optimal fiber length. Referring to [8], this paper establishes a control model of short-term stiffness: when the muscle fiber of the model is detected to be stretched, the control signal of the torque actuators increases rapidly. This provides a reverse control torque for the ankle joint to prevent the muscle fiber length from being overstretched and the ankle angle changing greatly. When the muscle is stretched to the optimal length, the control signal reaches to the peak, based on the model established as follows:

$$T_s(t) = - \begin{cases} 0 & \text{if } \Delta \tilde{l}_M < 0 \\ k_\gamma F_M^0 a_b f_{act}\left(\tilde{l}_M\right)\Delta \tilde{l}_M & \text{if } 0 < \Delta \tilde{l}_M < \delta \\ k_\gamma F_M^0 a_b f_{act}\left(\tilde{l}_M\right)\delta & \text{if } \Delta \tilde{l}_M > \delta \end{cases} \tag{3}$$

where k_γ is the coefficient of short-range stiffness. F_M^0 is maximal isometric muscle force. a_b is the degree of muscle activation. $f_{act}(\tilde{l}_M)$ is muscle force-length relation. $\tilde{l}_M$ is fiber length normalized by optimal fiber length. $\Delta \tilde{l}_M$ is normalized fiber stretch, and δ is the normalized critical stretch. $T_s(t)$ is the torque provided by short-range stiffness at time t.

Biologists briefly describe the stretch reflex as the following process: under the joint action of musculoskeletal, neural and sensory systems, interference information is first transmitted to the cerebellum through the sensory system and then transmitted to the muscle system in the form of electrical signals by the nervous system to stimulate muscle contraction. Muscles rotate the bones around the joints by pulling the tendons, resulting in instep extension and metatarsal flexion to control the balance of body in the sagittal plane. As the receptor of stretch reflex, muscle spindle is sensitive to the changes of muscle fiber length and contraction velocity, so the stretch reflex can be expressed in a form similar to the PD control model.

In this paper, the current muscle length is collected and compared with the muscle fiber length when the model is standing still, and the difference is taken as part of the control value of muscles' activation. The control of muscle's length is similar to the proportional control, which can quickly deal with the angle deviation caused by disturbance. The control of the rate change in muscle length is similar to differential control, which can predict the change of muscle length, respond in advance, and shorten the time for the model to reach equilibrium [10]. In this paper, the following stretch reflex model is mathematically established as follows:

$$a_{stretch}(t) = K_a(L_{CE}(t - t_0) - L_{CEO}) + K_b V_{CE}(t - t_0) \tag{4}$$

where K_a and K_b are, respectively, the gains for muscle spindle length change and the rate of the change in length. L_{CEO} indicates the optimal fiber length. $L_{CE}(t)$ is the fiber length at time t. t_0 represents the combined time delay caused by neural transmission and muscle electromechanical delays. $a_{stretch}(t)$ is the degree of muscle activation provided by the stretch reflex control at time t. V_{CE} is the current rate of change of the fiber length and obtained as follows:

$$V_{CE} = \frac{V_{CE}(t_n) - V_{CE}(t_{n-1})}{t_n - t_{n-1}} \tag{5}$$

The passive stiffness is caused by the viscoelastic properties of joints, passive tissue and active muscle fibers. In article [6], joint stiffness can be defined as the dynamic relationship between the angular position of a joint and the torque acting about it, and the intrinsic stiffness is described by a linear second-order system under stationary conditions. The passive stiffness belongs to the intrinsic stiffness, and has similar properties to intrinsic stiffness: it is only affected by the organizational structure of the ankle joint, and whether it exists or not has nothing to do with external interference; it is not controlled by the muscles, and even if the muscles are not activated, the joints have a degree of elasticity. Therefore, in the whole simulation process of this paper, passive stiffness has been playing a role.

Referring to the intrinsic stiffness model established by Kearney et al. [6], this paper simulates the torque control mechanism of passive stiffness through the feedback of ankle angle and angular velocity. The reason why angular acceleration is not used as input is that the acquisition of acceleration requires quadratic differentiation and produces noise. For disturbances acting in a small range, the passive stiffness mechanism can be described by the following model as:

$$T_p(t) = K_p(u(t) - u_0) + B_p v(t) \tag{6}$$

where $u(t)$ is the joint angular position at time t, $v(t)$ velocity. u_0 is the angle of ankle at static standing. B_p and K_p are the viscous and elastic parameters respectively. $T_p(t)$ is the torque provided by passive stiffness at time t.

3 Experimental Setup and Data Analysis

3.1 Forward Dynamics Experiment

Based on the fact that the focus of this paper is on the contribution of the musculoskeletal structure around the ankle joint in the process of upright standing push-recovery with disturbance, it does not involve the study of the movement of the upper limb, so this paper adopts a simplified three-dimensional musculoskeletal model from OpenSim-gait10dof18 model, which is centered on the structure of the lower limb. As shown in Fig. 3, the upper limb of the model omits muscles and joint components and is a rigid structure composed only of the torso. Moreover, the lower limb contains 7 bone segments, which are driven by 18 muscles to rotate around the joint, with a total of 10 degrees of freedom. In order to reflect the effect of the ankle joint more purely, only one degree of freedom of the ankle joint and part of the motion characteristics of the hip joint

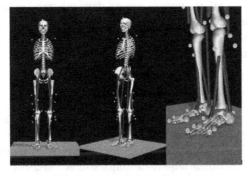

Fig. 3. Musculoskeletal model

Table 1. Parameters of muscle

Parameters	Soleus	Tibialis anterior
Maximum tension/N	5137.0	3000.0
Muscle fiber length/m	0.1	0.1
Muscle-tendon length/m	0.2514	0.2228

Table 2. Parameters of contact properties

Parameters	Rigidity	Dispersion	Static friction	Dynamic friction	Viscous friction
Value	10^8	0.5	0.9	0.9	0.6

are retained. The parameters of the tibialis anterior and soleus are shown in Table 1, and the contact properties between the foot and the ground are shown in Table 2.

We use the above model to analyze the forward dynamics of the standing balance mechanism during the disturbance. In the process of solving the movement of the model, the excitation signals of muscle and actuator are firstly calculated respectively according to formulas from chapter 2, to replace the neural control signal. Then, the controls of muscle force and actuator are added to the system-wide model controller, and are used as the input of the controller to determine the force and torque applied on the model joint. Finally, the position information of joint movement can be calculated by multi-joint dynamic analysis, and the resulting motion can be determined.

The simulation time is set to 30 s. The control based on COM and passive stiffness is added to the controller in t = 0 s. Under the above two mechanisms, if the body can maintain long-term stability, and muscle activity is basically constant. We apply a forward force to the pelvis of the model at 9.9 s, which linearly increases to a peak of 30N in the first 100 ms and decreases to 0N in the second 100 ms. A short-range stiffness control is added to the model during 9.9 s–9.95 s, and stretch reflex control is during 9.95 s–30 s. After the end of the simulation, the data of ankle angle, force and torque are derived. Combining with the muscle control data, the stability and anti-disturbance mechanisms of human body are compared and analyzed, and the roles of passive stiffness and short-range stiffness in disturbance response are verified. After the simulation, the data of ankle angle, force and torque are exported.

3.2 "Sandbag Disturbance" Experiment

The ankle torque of the subjects is calculated from the data by six-axis force platform, the ankle angle and angular acceleration information are measured by the motion capture system, and muscle sEMG data are recorded using an electromyograph (ELONXI EMG 100-Ch-Y-RA). This experiment requires testing the three muscles of the calf: the tibialis anterior, the gastrocnemius and the soleus. In order to avoid the muscle fatigue effecting the experimental results and ensure the objectivity of the experimental data, all the subjects don't do any strenuous exercise in the 24 h before the experiment.

First of all, according to the requirements of motion capture, a custom marker set of 39 points is defined, and reflective markers are installed at the key anatomical points of the body. Then, we remove the leg hair with fine sandpaper, and place electrodes on the selected muscles. Finally, the subject is asked to stand at a stance width equal to their hips on the force platform, an interference device with 750 g sandbags is placed behind the subject which to be sufficient to stimulate the ankle joint strategy without involving the hip joint strategy. The overall experimental layout is shown in Fig. 4.

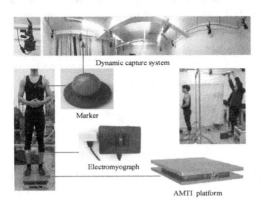

Fig. 4. Human upright push-recovery experiment

After starting the experiment process noted above, we collect the relevant data for 10 s under the static posture. Next, the experimental helper releases the sandbag which is pulled by a rope to move in a circle, hitting the subject's back at the lowest point. While keeping the body as straight as possible, the subject restores himself to an upright state by contracting the calf muscles. Each subject carries out three groups of experiments and the interval of each experiment is about 10 s.

The motion of the whole body is recorded in a NOKOV optical three-dimensional dynamic capture system with eight high-definition cameras at a frequency of 50 Hz. The sampling frequency of force plate and EMG is 1000 Hz. Since the time of human muscle reflex is very short, it is necessary to collect the relevant data of each system synchronously to ensure the timeliness of the experimental data. The preliminary scheme of this project is: through on-site guidance of different instructions, the use of real-time simulator to collect EMG data, force plate data and the TTL of dynamic capture system to achieve synchronization.

3.3 Experimental Results and Analysis

This part is mainly divided into two stages to analyze the effects of three control mechanisms on human balance, namely, the static standing stage (shown in Fig. 5) and after interference (displayed in Figs. 6, 7, 8 and 9). In the initial stage of the static standing, the main purpose is to compare the influence of adding passive stiffness and without passive stiffness on the stability of the model. The comparison of the angle of the ankle joint is shown in Fig. 5(A). We can be seen that the two curves are finally stable at about 2.5°. Because the rotation center of the model ankle joint is closer to the heel than the toe, the model has enough space to adjust the ankle torque when tilting forward. When the model is tilted back, the adjustment space is relatively small, so that the soles of the feet are flexion, the toes are raised, and the model falls back. Therefore, the model is more stable in the forward tilting state, which is consistent with the measured results of the actual human body (0–5°). Under passive stiffness control, the final stable angle of the model is relatively smaller, the time is reduced by about 25 ms when the ankle start to change, so the response speed is faster. And the fluctuation after the decrease of angle may be due to the small damping term of the control parameters. the rapid change of actuator force leads to system oscillation.

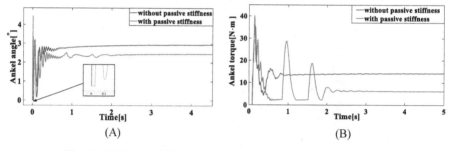

Fig. 5. Ankle angle(A) and torque(B) of the model in static standing

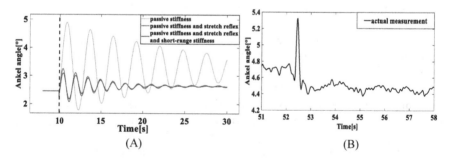

Fig. 6. Ankle angle of the model(A) and human(B) after interference

Figure 5(B) is a comparative diagram of the changes of soleus torque at the initial stage of the model standing. When there is no passive stiffness and the model is tilted forward, the activation of tibialis anterior is low, the model mainly depends on the pulling force of soleus muscle to resist gravity, and the output torque of soleus muscle is approximately equal to the torque of ankle joint. Under the condition that the model gravity is constant, the ankle torque required to keep the body stable is almost constant. When the actuator of passive stiffness detects and resists the change of angle, to produce

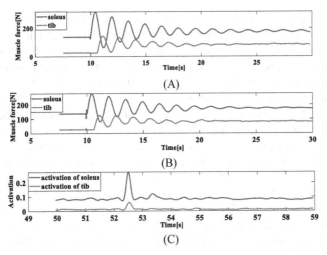

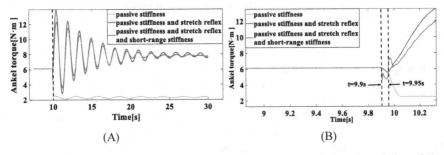

Fig. 7. Muscle force and activation of the soleus and tibial anterior muscles after interference. Fig. A is the muscle force without short-range stiffness in the model; Fig. B shows the muscle force after short-range stiffness is added to the model; Fig. C shows the degree of muscle activation during the experiment.

Fig. 8. Ankle torque of the model after interference, (B) is the partially enlarged view of (A)

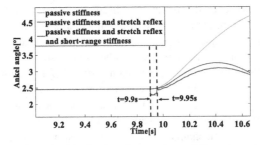

Fig. 9. Ankle angle of the model after interference (a partially enlarged view of Fig. 6(A))

a reverse control torque which shares part of the output torque of the soleus, resulting the decrease of torque for soleus.

In order to explore the effect of short-range stiffness and stretch reflex on the model after being disturbed, this paper draws the ankle angle diagram after adding passive stiffness and stretch reflection and short-range stiffness in sequence, as shown in Fig. 6(A). The ankle angle increases rapidly under the disturbing, and finally shows a stable trend after a certain period of vibration with the action of each controller. With the exception that the adjustment time of the model with only the passive stiffness added is too long, the model with the other two controls added is basically consistent with the change of ankle angle measured in the human upright push-recovery experiment in Fig. 6(B).

The stretch reflex controller takes the change that soleus muscle is stretched as input, controlling the activation degree of the soleus muscle to increase rapidly. It can be seen from Fig. 7(A) that the force of soleus increases rapidly to 278N, which drives the ankle joint to rotate and resist forward interference. On the other hand, the muscle force produced by the tibialis anterior is relatively small. And this overall trend is highly similar to the activation of human muscle in Fig. 7(C). After adding this mechanism, the amplitude of the angle shock is obviously reduced, and the time to return to the stable state is also shortened a lot. This shows that the stretch reflex is the main mechanism of human upright push-recovery.

Even if there is no outrageous angle change in the model without short-range stiffness, and the stationarity and rapidity of the angle curve are only improved by a small part after adding short-range stiffness to the model, the role of short-range stiffness can't be ignored. It is similar to passive stiffness in that it can produce a reverse resistance torque and share part of the output torque of the soleus, as shown in Fig. 8. But there are some differences that short-range stiffness is extremely sensitive to changes in muscle length, and the magnitude of the force provided by it is proportional to the change in muscle length. As seen from Figs. 8(B) and 9, the action time of short-range stiffness is only 50 ms in the whole process, but it can make an obvious feedback response to the disturbance at the first time.

Comparing the force of soleus under stretch reflex (Fig. A) before and after adding short-range stiffness (Fig. B) in the Fig. 7, it is clear that the force decreases after adding short-range stiffness, which is slightly different from the fact that short-range stiffness can help the force of muscle increase rapidly in human body. This would probably because the short-range stiffness of the human body is a historically dependent property of muscles. However, this paper only discusses its role as a torque actuator that contributes to the overall torque and angle of the model ankle joint.

Note that the above comparative analysis of simulation data and experimental data only makes a reference on the changing trend rather than absolute values. This could be interpreted by the following two facts. The first one is that different people show different motion characteristics even in the same external environment because of individual differences. The second one is that human body has a nervous system to participate in control, and the control mechanism is diverse. People often use an optimal distribution scheme to regulate various mechanisms. At present, what machine control can do is to slowly improve these mechanisms, so that the motion characteristics of the machine are closer to those of the human body.

4   Conclusion

In the process of static standing of human body, passive stiffness has a good control effect on the change of angle in a small range, but it cannot be used alone for control after disturbance. This paper applies the passive stiffness alone on the disturbed model, just to compare the effect of stretch reflex. As the most important component of muscle reflex, stretch reflex can quickly respond to disturbance by feeling the change of muscle fiber length, and make the model return to the equilibrium state. Short-range stiffness has little effect on the standing of the model in the experiment, partly because the stretch reflex in the model has restored it to stable state, and the model is in an almost saturated state. However, after adding short-range stiffness, the workload of muscle is reduced and the stability of ankle joint is increased. Under the overall control of these mechanisms, the model basically simulates the process of upright push-recovery and can achieve the upright balance effect of the actual human body.

In the follow-up research, on the one hand, we will consider to combine with the characteristic that the actual parameters of the human body are non-linear, to optimize the controller;on the other hand, we must improve the measurement methods to ensure the accuracy and representativeness of the data.

Acknowledgments. This work was supported in part by the National Natural Science Foundation of China under Grant 61603284 and 61903286.

References

1. Basic Biomechanics of the Musculoskeletal System. Wolters Kluwer/Lippincott Williams & Wilkins Health (2012)
2. Lee, H., Rouse, E.J., Krebs, H.I.: Summary of human ankle mechanical impedance during walking. IEEE J. Transl. Eng. Health Med. **4**, 1–7 (2016)
3. Yin, K., et al.: Artificial human balance control by calf muscle activation modelling. IEEE Access **PP**(99), 1 (2020)
4. Warnica, M.J., Weaver, T.B., Prentice, S.D., et al.: The influence of ankle muscle activation on postural sway during quiet stance. Gait and Posture **39**(4), 1115–1121 (2014)
5. Misgeld, B.J.E., Zhang, T., Lüken, M.J., et al.: Model-based estimation of ankle joint stiffness. Sensors **17**(4), 713 (2017)
6. Guarin, D.L., Jalaleddini, K., Kearney, R.E.: Identification of a parametric, discrete-time model of ankle stiffness. In: Proceedings of the 35th Annual International Conference of the IEEE Engineering in Medicine and Biology Society (EMBC 2013), pp. 5065–5070. IEEE (2013)
7. Weiss, P.L., Kearney, R.E., Hunter, I.W.: Position dependence of ankle joint dynamics—I. Passive mechanics. J. Biomech. **19**(9), 727–735 (1986)
8. De Groote, F., Allen, J.L., Ting, L.H.: Contribution of muscle short-range stiffness to initial changes in joint kinetics and kinematics during perturbations to standing balance: a simulation study. J. Biomech. **55**, 71–77 (2017)
9. Delp, S.L., Anderson, F.C., Arnold, A.S., et al.: OpenSim: open-source software to create and analyze dynamic simulations of movement. IEEE Trans. Biomed. Eng. **54**(11), 1940–1950 (2007)
10. Pang Muye, X., Xiangui, T.B., Kui, X., Zhaojie, J.: Evaluation of calf muscle reflex control in the 'Ankle Strategy' during upright standing push-recovery. Appl. Sci. **9**(10), 2085 (2019)

Design and Development of sEMG-Controlled Prosthetic Hand with Temperature and Pressure Sensory Feedback

Chenxi Li[1] and Nianfeng Wang[2(✉)]

[1] Guangzhou Zhixin School, Guangzhou 510080, People's Republic of China
[2] School of Mechanical and Automotive Engineering, South China
University of Technology, Guangzhou 510640, People's Republic of China
menfwang@scut.edu.cn

Abstract. In this paper we deal with the design and development of a prosthetic hand using sEMG control, temperature and pressure sensors. Through the acquisition of the sEMG produced by the movement of the flexor digitorum superficialis, the prosthetic hand controlled the contraction and relaxation of the prosthetic hand. At the same time, the prosthetic hand can automatically detect the pressure and temperature of objects being held. It can pass the real-time pressure and temperature of the object held by the prosthetic hand to users, who can independently control the force, perceive the temperature of the object, and heat the surface of the prosthetic hand to the same temperature as the body by using the heating films. The prosthetic hand can provide better satisfaction to users and promote the humanistic care of disabled people. At present, the prosthetic hand has achieved the functions below: the acquisition of the sEMG; movement control of finger opening and closing; the acquisition of pressure and temperature of the object being held; pressure and temperature perception in prosthetic hand users; thermostatic control of prosthetic hand surface.

Keywords: Surface electromyography signals · Anthropomorphic prosthetic hand · Pressure acquisition · Temperature acquisition

1 Introduction

There are many commercial companies or research institutions that have produced many anthropomorphic prosthetic hands [1–6]. The surface electromyography (sEMG) signal is bioelectrical signal collected on the skin surface when the human body is in motion [7], which is widely used in motion recognition [8–13]. Threshold control is a simple and effective way in myoelectric control [14] and pattern recognition is also widely used in myoelectric control[15]. In some researches, temperature signals are also collected simultaneously for human-machine interaction [19,20]. Based on the pressure and sEMG signals, grasping can be achieved by the prosthetic hand [21].

© Springer Nature Switzerland AG 2020
C. S. Chan et al. (Eds.): ICIRA 2020, LNAI 12595, pp. 320–331, 2020.
https://doi.org/10.1007/978-3-030-66645-3_27

This paper designs an anthropomorphic prosthetic hand that can shake hands and grasp things based on sEMG signals. In order to achieve grasp, a sEMG acquisition module is used to collect sEMG signals and the sEMG signal of stretching or tightening is analyzed. The handshake has become the most acceptable gesture of greeting in many cultures. In order to achieve the strength and temperature control during the handshake process, multi-level control is used. The strength of the grasp is controlled according to the sEMG, and at the same time the strength of the prosthetic hand is collected by the pressure sensor. The user can use the pressure feedback to adjust the force. When the pressure is too large, the prosthetic hand will stretch automatically to prevent injury to the gripped object. When shaking hands with others, the temperature acquisition module is used to collect the temperature of the gripped object and the temperature will be transmitted to the user through the temperature adjustment system. In the handshake process, the surface of the prosthetic hand is always kept at a constant temperature close to the body temperature, which avoids others feeling like that they are holding a cold machine during the handshake process.

2   Overall Structure and Principle Analysis of Prosthetic Hand

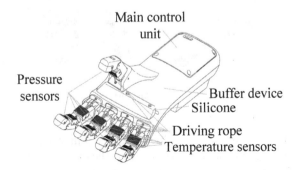

Fig. 1. Mechanical structure of prosthetic hand.

The overall structure of the prosthetic hand is divided into two parts. One is the mechanical structure, including the anthropomorphic prosthetic hand and the feedback device, as shown in Fig. 1. The second part is the electronic control logic, which is mainly processed by the microcontroller based on the signals collected by the sEMG sensor, pressure sensor, and temperature sensor and fed back to the anthropomorphic prosthetic hand and feedback device. The overall control system logic diagram is shown in Fig. 2.

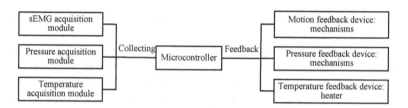

Fig. 2. Schematic diagram of prosthetic hand control.

2.1   Mechanical Structure

The modeling of the prosthetic hand is shown in Fig. 3a. The structural design is carried out through the analysis of the difficulty of implementation, adjustable moderation, manufacturability, and reliability. The driving motor is selected according to the motor torque, control method, voltage, and current. The flexibility and expandability of the prosthetic hand and the grasping reliabilityare achieved by choosing the transmission wire and adjusting the torque of the micromotor. Multiple micromotors are controlled to perform grasp through a single channel of the sEMG signal.

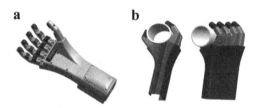

Fig. 3. Prosthetic modeling.

As shown in Fig. 3b, in order to achieve the reliability of grasp, a plurality of micro motors are used to control the thumb and the other four fingers separately. Each moving finger uses a buffer structure to compensate and control the force of the finger. The structure used to drive the fingers to grasp the objects is not the same so that the grasping objects are mainly stressed on the thumb, index finger, and middle finger. The buffer structure can protect the mechanism, and can also compensate the motion accuracy of the micro motor to make the grasp more flexible and stable.

As shown in Fig. 4, silica gel is installed on all finger surfaces that are in contact with objects to play a non-slip effect. The fingers all adopt a multi-joint structure to ensure the flexibility of movement and the reliability of grasp.

The finger surface is equipped with pressure and temperature sensors. Multiple pressure sensors can ensure the strength of grasping objects and shaking hands. Through multiple temperature sensors, the temperature of the contacting

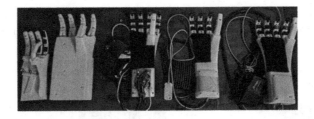

Fig. 4. Prosthetic hand made by 3D printing technology.

object can be sensed. It is verified that the prosthetic hand can achieve grasp and shaking hands.

2.2 Acquisition and Processing of sEMG Signal

The main muscles that control the movement of the human fingers are the flexor pollicis longus and the flexor digitorum superficialis located on the forearm. When the motion commands issued by the brain are transmitted to the flexor pollicis longus and the flexor digitorum superficialis through the central nervous system, continuous action potentials are generated at the muscles and muscle contraction occurs. The sEMG signal can reflect the contraction state of the muscle. The greater the contraction amplitude, the greater the signal amplitude. The potential signal is transmitted through the subcutaneous tissue and is detected by the sEMG acquisition module at the skin surface. The contraction force of the prosthetic hand is controlled according to the strength of sEMG signals.

MyoWare muscle sensor is used as sEMG acquisition module and it includes a sEMG sensor, an electrical signal acquisition rectification circuit, and a power supply to the circuit. The module is shown in Fig. 5.

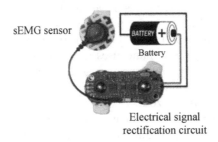

Fig. 5. Electromyographic signal acquisition module.

The sEMG sensor collects the sEMG signal and perform amplification and rectification processing. After integrating the signal, an integrated signal will be obtained. By comparing the integrated sEMG signal and the state of the muscle, two different movement states of grasp and relaxation are determined. Muscle states and integrated signals are shown in Fig. 6.

Fig. 6. The signal waveform of the relaxation state (a) and grasping state (b).

The signals collected by the sEMG acquisition module will vary according to the personal constitution, environment, and sensor attachment position. It is necessary to perform calibration on the signal waveform to obtain a reliable signal to Perform movement control of the prosthetic hand.

2.3   Collection and Feedback for Pressure Signal

The pressure sensor detects the pressure of grasp by the prosthetic hand and transmits the pressure signal to the microcontroller. After the arithmetic processing of the microcontroller, the action of the steering engine on the feedback device is controlled to transmit the detected pressure of the gripped object to users in real-time. The user can adjusts the strength of the muscles according to the feedback to control the grasping force of the prosthetic hand.

Pressure sensors are divided into two different types, namely the piezoelectric pressure acquisition module and the piezoresistive pressure acquisition module. The response speed of the piezoelectric pressure acquisition module is fastest in the working state. However, its circuit is relatively complicated, and it is difficult to ensure the reliability of the assembly structure during processing. Correspondingly, to the piezoresistive pressure acquisition module, the impedance of the pressure sensor will change under pressure. As shown in Fig. 7a, the module can be made into a flexible and bendable form. It can be stably installed in the prosthetic hand. The prosthetic hand uses a piezoresistive pressure acquisition module, as shown in Fig. 7b. The sensing range is 20g~2kg, and the pressure can correspond to the resistance value.

As shown in Fig. 8, The horizontal axis is the pressure value, and the vertical axis is the impedance of the pressure acquisition module. The microcontroller can calculate the pressure value according to the corresponding relationship between the voltage value and the resistance value.

The pressure feedback device is shown in Fig. 9a, which uses a motor to drive mechanism in order to provide pressure. The force feedback steering engine is shown in Fig. 9b, which applies pressure to the arm wearing the pressure feedback device. Due to the direct drive of the micro motor, the pressure feedback is rapid and direct.

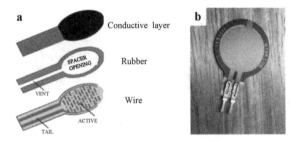

Fig. 7. Mechanical structure of pressure sensors.

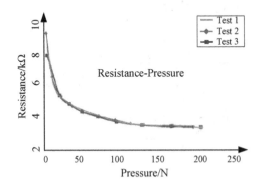

Fig. 8. The relationship between pressure and resistance.

2.4 Collection and Feedback for Temperature Signal

The temperatures received by the object held by the prosthetic hand and on the feedback device are detected by temperature sensors. Temperature signals are transmitted to the microcontroller in the form of a resistance value, and those two temperature values are compared by the microcontroller to control the feedback device. The heating film heats up so that the temperature of the feedback device reaches the same temperature as the object being held by the prosthetic hand. The entire collection and feedback process for the temperature signal is shown in Fig. 10.

When selecting the temperature sensor, there is a high requirement for temperature acquisition speed. Thus, the chip thin-film thermistor with a small size is suitable for applying, as shown in Fig. 11a. In actual use, it is not conspicuous due to its small size, and can also respond to the actual temperature while absorbing less temperature.

The accuracy of the temperature sensor is $\pm 1 \,^{\circ}C$, which means that every time the temperature changes by one degree Celsius, the corresponding resistance value of the temperature sensor will change accordingly. And the higher the temperature, the smaller the resistance change. This module uses a voltage divider circuit to respond to changes in resistance, as shown in Fig. 11b. The AD conversion function of the microcontroller reads the resistance value of the

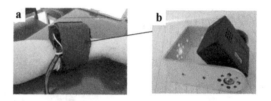

Fig. 9. Force feedback device.

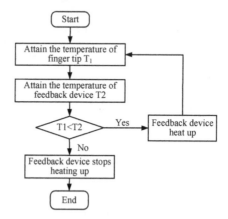

Fig. 10. The flow chart of temperature feedback control.

thermistor corresponding to the current temperature and then the temperature of the thermistor in the current state is obtained according to the resistance temperature correspondence table.

The flexible heating module includes carbon fiber, carbon film, heating wire, graphene heating film, Flexwarm flexible heating film, etc. Because of the fast heating speed, low power consumption, and the constant temperature characteristics, Flexwarm flexible heating film is finally selected as the heating material.

3   Implementation Process

3.1   Workflow of the System

As shown in Fig. 12, the system first enters the self-test program to check whether sensors and control devices work normally after powered on. Then, the system subsequently obtains the feedbacks of sEMG, pressure and temperature. Finally, the system combines these three signals to determine the ultimate output action in a completed action.

3.2   Control of the System

The system uses an independent sEMG acquisition module to collect and analyze the signal. The processed data is transmitted to the microcontroller for

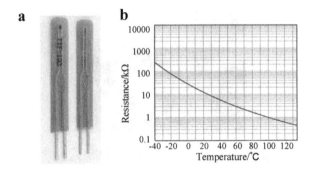

Fig. 11. The relationship between temperature and resistance.

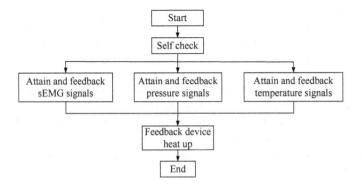

Fig. 12. The flowchart of the system.

processing. In order to ensure the stability of the data, the muscle location must be correct and the skin should stay clean.

The collecting point of the sEMG signal is in the flexor digitorum superficialis, as shown in Fig. 13. The flexor digitorum superficialis is used to collect the sEMG signals, and the elbow bone is used as the reference point. The tests are carried out to observe the signal of sEMG acquisition module under different muscle contraction force. And the process is shown below:

1. Relax the arm until the sEMG signal is stable.
2. Contract the flexor digitorum superficialis slightly to obtain a signal about 1.5V.
3. Contract the flexor digitorum superficialis greatly to obtain a signal about 3.5V and an obvious waveform.
4. Contract and relax the flexor digitorum superficialis 6 times to obtain 6 waves for use.

The experimental results show that the flexor digitorum superficialis will produce sEMG signals during muscle contraction and the magnitude depends on the muscle contraction force. The larger contraction force, the larger sEMG signal.

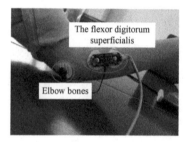

Fig. 13. The placement of sEMG sensor.

The pressure sensors are attached on the surface of the prosthetic hand. When the prosthetic hand grasps the object, the sensors will detect the pressure of fingers and transmit the signal to the microcontroller. Then, the microcontroller will control the steering engine on the feedback device and transmits the pressure to the human skin.

When the prosthetic hand holds the object, the temperature sensors will detect its temperature and feed it back to the microcontroller. Then, the microcontroller will control the heating system on the feedback device to transmit the real-time temperature of the object to the human body, as shown in Fig. 14.

A multi-point thermometer is used to record the temperature of the fingertip and the sensing zone of the prosthetic hand in the grasp of the heat source, and the result is shown in Fig. 15.

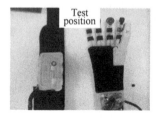

Fig. 14. The placement of temperature sensor.

The temperature sensor parameters of prosthetic hand and feedback device are 10K/3435b, the accuracy is $\pm 1^\circ C$, the temperature of the object is $39.1^\circ C$, the temperature of feedback device is from $38^\circ C$ to $42^\circ C$, the total temperature error of prosthetic hand and feedback device is $\pm 4^\circ C$ after tolerance superposition.

4 Experimental Performance and Analysis

The experimental results show that there is a relationship between sEMG signals and the actions of the hand. Through the analysis of the waveform of the electrical signal, it is found that the grasp force and waveform strength are positively

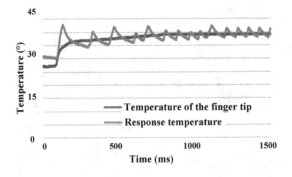

Fig. 15. The results of the temperature test.

correlated. The waveform can be extracted and used to control the prosthetic hand.

Figure 16(a) shows the experiments in which the users drink the water of a stainless-steel water cup, a 500ml bottle, and a ceramic cup through the sEMG signal of the prosthetic hand.

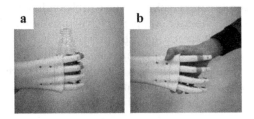

Fig. 16. The test of grasping a cup and handshake.

The tests of handshaking with the prosthetic hand are shown in Fig. 16b. As the temperature of the prosthetic hand is set to the body temperature in default, it will not make the hand feel cold in the touch even if the weather is a little cold. Also, the user can feel the temperature of the hand through the temperature feedback. At the same time, the user can feel the force of the hand through the pressure feedback. As a result, the designed prosthetic hand is more user-friendly.

Besides, two individuals with arm disability participle the test of the prosthetic hand. As shown in Fig. 17, they are equipped with the prosthetic hand and finish a series of basic tasks, and the tests of inductive feedback and wearing comfortness.

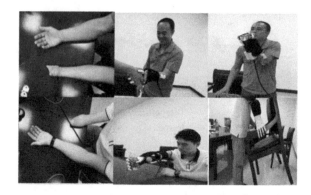

Fig. 17. Trial of the anthropomorphic prosthetic hand for the disabled.

5   Conclusion

In this paper, a anthropomorphic prosthetic hand based on sEMG signals is designed and fabricated. The sEMG acquisition module is used to collect sEMG signals. Synchronous movement of prosthetic hand can be achieved by using sEMG signals. A pressure acquisition module is added to the prosthetic hand to obtain the pressure on the fingertip, and the corresponding pressure is generated by the feedback device. The user can perceive the pressure and make the corresponding response. At the same time, the prosthetic hand adopts a multi-level strength control strategy, which can realize the handshake and grasp using different strengths. A temperature acquisition module is added to the prosthetic hand. Meanwhile, a constant temperature heating module is added on the surface of the prosthetic hand to synchronize the body temperature.

Acknowledgements. The authors gratefully acknowledge the reviewers' comments.

References

1. Dalley, S.A., Wiste, T.E., Withrow, T.J., Goldfarb, M.: Design of a multifunctional anthropomorphic prosthetic hand with extrinsic actuation. IEEE-ASME Trans. Mechatron. **14**(6), 699–706 (2009)
2. Connolly, C.: Prosthetic hands from touch bionics. Ind. Robot Int. J. Robot. Res. Appl. **35**(4), 290–293 (2008)
3. Cipriani, C., Controzzi, M., Carrozza, M.C.: The smarthand transradial prosthesis. J. Neuroeng. Rehabil. **8**(1), 1–4 (2011)
4. Wang, N., Lao, K., Zhang, X.: Design and myoelectric control of an anthropomorphic prosthetic hand. J. Bionic Eng. **14**(1), 47–59 (2017)
5. Yang, B., Jiang, L., Hu, J., Li, C., Liu, H.: A compact control system and a myoelectric control method for multi-dofs prosthetic hand. In: IEEE International Conference on Real-time Computing and Robotics (RCAR), pp. 592–597 (2019)
6. Xu, K., Liu, Z., Zhao, B., Liu, H., Zhu, X.: Composed continuum mechanism for compliant mechanical postural synergy: an anthropomorphic hand design example. Mech. Mach. Theory **132**, 108–122 (2019)

7. Oskoei, M.A., Hu, H.: Myoelectric control systems-a survey. Biomed. Signal Process. Control **2**(4), 275–294 (2007)

8. Ding, Q., Xiong, A., Zhao, X., Han, J.: A review on researches and applications of semg-based motion intent recognition methods (in chinese). Acta Automatica Sinica **42**(01), 13–25 (2016)

9. Man, Z., Qiao, Y., Li, l., Rong, H.: Classification of surface emg signals based on lda (in chinese). Comput. Eng. Sci. **38**(11), 2321–2327 (2016)

10. Zhang, F., Yang, D., Liu, T.: Gesture recognition based on emg signal and attitude signal (in chinese). Transducer Microsyst. Technol. **38**(07), 46–49+52 (2019)

11. Wang, N., Chen, Y., Zhang, X.: The recognition of multi-finger prehensile postures using lda. Biomed. Signal Process. Control **8**(6), 706–712 (2013)

12. Wang, N., Lao, K., Zhang, X., Lin, J., Zhang, X.: The recognition of grasping force using lda. Biomed. Signal Process. Control **47**, 393–400 (2019)

13. Wang, N., Chen, Y., Zhang, X.: Realtime recognition of multi-finger prehensile gestures. Biomed. Signal Process. Control **13**, 262–269 (2014)

14. Bottomley, A.H.: Myo-electric control of powered prostheses. J. Bone Joint Surg. Br. **47**, 411–415 (1965)

15. Gu, Y., Yang, D., Huang, Q., Yang, W., Liu, H.: Robust emg pattern recognition in the presence of confounding factors: features, classifiers and adaptive learning (in chinese). Expert Syst. Appl. **96**, 208–217 (2018)

16. Yang, J.: Research on the multi-sensor fusion sensing system for the lower exoskeleton robot (in chinese). Master (2017)

17. Xiao, S.: Control system design and research of an intelligent rehabilitation robot (in chinese). Master (2015)

18. Wu, C.: Research on emg based control and force tactile perception feedback of the prosthetic hand (in chinese). Doctor (2016)

19. Guo, X., Huang, Y., Teng, K., Liu, P., Liu, C., Tian, H.: Modular design and implementation of flexible artificial skin with temperature and pressure sensors (in chinese). Robot **37**(04), 493–498 (2015)

20. Tian, Y., Zhang, X., Zhang, l.: Research on temperature sensing technology for bionic prosthetic hand (in chinese). Mach. Electron. (10), 12–16 (2015)

21. Luo, Z., Yang, G.: Prosthetic hand fuzzy control based on touch and myoelectric signal (in chinese). Robot **02**, 224–228 (2006)

Two-Wheel Balancing Robot Foot Plate Control Using Series Elastic Actuator

Yeong-keun Kwon[1], Jin-uk Bang[1], and Jang-myung Lee[2(✉)]

[1] Department of Electrical Engineering, Pusan National University, Busan, Republic of Korea
yeongkeun1696@pusan.ac.kr
[2] Department of Electronics Engineering, Pusan National University, Busan, Republic of Korea
jmlee@pusan.ac.kr

Abstract. Controlling foot plate on two-wheel balancing robot, system can compensate outer force and it makes system safer. Compared to a wide usage of personal mobility, it has large possibility of getting accident. So, system proposed in this paper attached Series Elastic Actuator (SEA) between robot body and foot plate to control foot plate and it can deal with several situations. The experiment conducts with 20 kg weight on the right side of footplate. The result of experiment shows that system can compensate the outer force. We expect that controlling foot plate can deal well with hard driving situations (uneven road drive, high speed curve drive, etc.).

Keywords: Series Elastic Actuator (SEA) · Personal mobility · Two-wheel balancing robot · Control

1 Introduction

New kind of transportation is developing with growth of fourth industrial revolution. Personal Mobility is one of them and widely used in short distance transfer. Two-wheel balancing robot, electric kick board, etc. are classified as PM. But compare to four-wheel car, PM does not have enough safety device on it. For this reason, PM driver is more vulnerable to accident [1] So this paper suggests two-wheel balancing robot with attached Series Elastic Actuator for safety.

When driving two-wheel balancing robot, driver face two problems: uneven ground, centrifugal force from high speed turning driving. Series Elastic Actuator (SEA) is active suspension which define as "a computerized system in automobiles that actively adjusts the suspension in response to driving conditions" [2]. SEA can act as cushion on uneven ground. Also, attaching two set of SEA and one passive spring to two-wheel balancing robot, it detects roll, pitch angle changes in footplate and compensate it to deal with centrifugal force.

The rest contents of this paper are as follow. Section 2 discuss design and control of system. Section 3 describe foot plate angle control by two SEA attaching to PM. Section 4 show and analyze the result of experiment.

© Springer Nature Switzerland AG 2020
C. S. Chan et al. (Eds.): ICIRA 2020, LNAI 12595, pp. 332–340, 2020.
https://doi.org/10.1007/978-3-030-66645-3_28

2 Design and Control of Two-Wheel Balancing Robot with Attached SEA

2.1 System Configuration

Two-wheel balancing robot mainly consist of two wheel and IMU sensor for self-leveling whether it stop or driving. [3] Two-wheel balancing robot has a foot plate which is linking with SEA, it is shown in Fig. 1. Driving part, SEA, Power part are connecting around control board as shown in Fig. 2. Specification and part list of system are listed in Table 1.

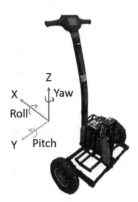

Fig. 1. Overall structure of two-wheel balancing robot with attached SEA

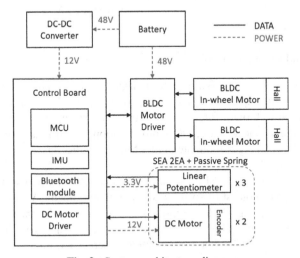

Fig. 2. System architecture diagram

Table 1. Specification and part list of two-wheel balancing robot with SEA.

HW	Size	L390 × W820 × H1200 mm
	Weight	30 kg
Driving	Driving method	Two-wheel differential drive
	BLDC In-wheel motor	13 inch, 48 V, 500 W
	BLDC motor driver	DC12~48 V, 20 A
	Speed	Max 40 km/h
SEA	Slide potentiometer	10 KOhm, 60 mm (2EA), 30 mm (1EA)
	SEA motor	DC geared motor + Encoder 12 V, 41.3 W
	SEA spring	Spring constant: 3 kgf/mm
Power	Battery	48 V, 13S5P, 17.5 Ah
	DC-DC converter	60 V to 12 V, 10 A, 120 W
Control board	MCU	STM32F407
	IMU	EBIMU-9DOF
		3-axis Accelerometer
		3-axis Gyroscope
		3-axis Magnetometer
	Bluetooth module	100 m, UART
	DC motor driver	BTN8982TA
		8~18 V, Max 55 A

2.2 Operational Principle of SEA

SEA is proposed by Pratt and Williamson in 1995. [4] It is using for walking robot, suspension, etc. In this paper, SEA's roles are suspension and linear actuator. In case of suspension, SEA absorb outer force and vibration with spring which is located inside of SEA. In case of linear actuator, DC motor rotation would make ball-screw linear motion and finally it changes footplate angle. More specific algorithms and equations are described in Sect. 2.3.

There are two main part in SEA, sensing part and actuating part which are shown in Fig. 3. Sensing part consist of spring and slide potentiometer. When outer force is given in SEA, SEA's spring would compress and compressing distance is measured by slide potentiometer. Applying hook's law, outer force can estimate with spring distance. Actuating part consist of DC motor, timing belt, ball-screw. DC motor rotation translate ball screw linear motion with connecting of timing belt between them. So, ball screw's moving make SEA's plate up and down. [5–7].

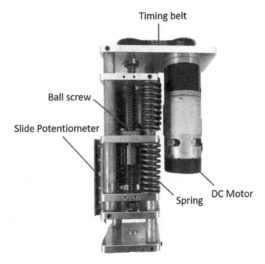

Fig. 3. Structure of SEA

2.3 Foot Plate Angle Control Algorithm

There are two-way of angle change, roll and pitch. These two angles are formed from outer force or SEA's force. In Fig. 4, passive spring is locating center of footplate and two SEA are locating around passive spring. Passive spring hold standard center point of roll and pitch angle. SEA1 is sensing and control roll angle and SEA2 is sensing and control pitch angle. Since forming roll and pitch angle mechanism is same, this paper mainly dealing with roll angle.

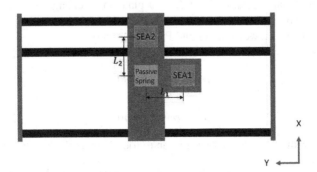

Fig. 4. SEA arrangement on foot plate

To see the relation between outer force (F_0) and SEA force (F_{SEA}), we use torque equilibrium equation and conservation of energy which is drawn from Fig. 5. L_1 represent distance between *SEA1 and Passive Spring*. L_3 represent distance between F_0 and Passive Spring. v is ball-screw linear velocity. ω is SEA motor rotational speed. T_m represent SEA Motor torque. K_s is spring coefficient. d_1 is SEA1 ball-screw moving distance. Ph is Ball screw lead.

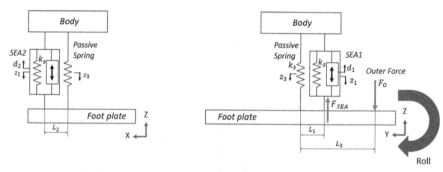

Fig. 5. Schematic of assembled SEA with body and foot plate

The equation of roll angle torque equilibrium in foot plate is as follow:

$$F_{SEA} * L_1 = F_0 * L_3 \ [kgf * m] \tag{1}$$

The equation of law of conservation of energy with rotating and linear motion is as follow:

$$F_{SEA} * v = T_m * \omega [W] \tag{2}$$

$$F_{SEA} = T_m * \frac{\omega}{v} = K_s * d_1 \left[kgf \right] \tag{3}$$

$$v = \omega * Ph [m/s] \tag{4}$$

From relation between motor and SEA ball-screw motion, we can find several equations about distance change of SEA, foot plate angle change. rev_1, rev_2 represent SEA1, SEA2 Motor's number of revolutions. d_2 is SEA2 ball-screw moving distance. L_2 represent distance between SEA2 and Passive Spring.

The equation of translating rotational motion to linear motion is as follow:

$$d_1 = rev_1 / Ph [m] \tag{5}$$

$$d_2 = rev_2 / Ph [m] \tag{6}$$

The equation of translating linear motion to foot plate angle is as follow:

$$Compensated \ Roll \ angle = \arctan(d_1/L_1) \left[° \right] \tag{7}$$

$$Compensated \ Pitch \ angle = \arctan(d_2/L_2) \left[° \right] \tag{8}$$

The equation of roll and pitch angle change from outer force is as follow:

$$Measured \ Roll \ angle = \arctan((z_1 - z_3)/L_1) \left[° \right] \tag{9}$$

$$Measured \ Pitch \ angle = \arctan((z_2 - z_3)/L_2) \left[° \right] \tag{10}$$

Figure 6 present the outer force compensating algorithms in this system. When outer force exerts in footplate, slide potentiometer measure spring distance to get angle change. Then in an opposite way SEA make linear motion to compensate the measured angle change. In this compensating algorithm, PD controller is used for SEA control. Because in real-time target angle changing situation, it has less necessity of accuracy and want to avoid accumulate errors.

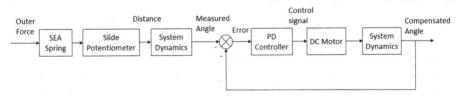

Fig. 6. Outer force compensating control block diagram

3   Experiment

To verify foot plate angle control algorithm, two kinds of experiment was carried out. First one is about controlling footplate with respect to arbitrary step input which is set by person. Second one is verifying outer force compensating algorithm shown in Fig. 6.

3.1   Foot Plate Angle Control with SEA

First experiment conduct to find optimizing PD controller gain for advanced experiment. There is no load on foot plate. Control signal input is 5° step input. Every 5 ms, system sample the outcome from SEA slide potentiometer. From Eq. (9), roll angle measured,

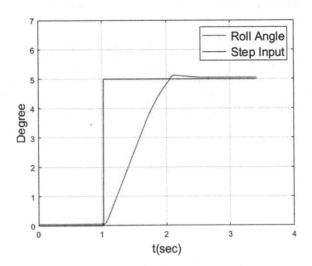

Fig. 7. Step input (5°) response result

and it would compare to step input. The control system target is set by whole system environment. Settling time is less than 1 s, %OS is less than 10%, steady-state error is less than 0.1°. When PD gain is $K_p = 3500$, $K_d = 100$, the result satisfies target performance. So, it means that system has enough ability to compensate outer force (Fig. 7).

3.2   Compensation of Outer Force Change

Second experiment main purpose is to compensate outer force. When we are doing experiment, 20 kg load is used. And it is placed in right side of foot plate. Controlling Algorithm is represented on Fig. 6. From outer force, roll angle changes and it is measured by SEA in every 5 ms. Compare to first experiment target performance, only steady-state error is 0.5 since it need to deal with real-time changing situation (Figs. 8 and 9).

Fig. 8. Compensating 20 kgf force experiment enviroment.

There are some disturbances in measured roll angle, but result meet the target control performance. And steady-state error is more generous than first experiment. So, it is not too much sensitive to disturbance. Thinking of real situation, outer force can exert from driver, ground uneven condition, etc. As presented system well dealing with outer situation, we can expect that SEA attached two-wheel mobility is safer than original two-wheel mobility.

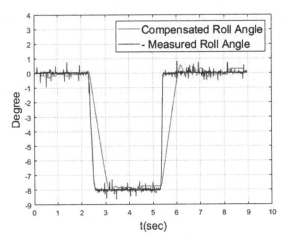

Fig. 9. Compensating 20 kgf force experiment result.

4 Conclusions

This paper mainly deals two-wheel balancing robot foot plate angle changing for rider's safety. System has two SEA to control roll and pitch angle. Firstly, when outer force gets into foot plate, spring is going to compress, and it is measured by slide potentiometer. Then angle would measure and directly compensating angle from SEA's linear motion. From experiment, target performance of SEA was verified. Even though the experiment conducts with 20 kg load on just one side of foot plate, real world driving situation does not make that much one-sided load. So, it can be used in common situation. Future research would integrate foot plate angle changing with various driving condition. For example, uneven road driving, slope way driving, high speed turning driving, etc.

Acknowledgment. This paper was supported by Korea Institute for Advancement of Technology (KIAT) grant funded by the Korea Government (MOTIE) (P0008473, HRD Program for Industrial Innovation).

References

1. Zagorskas, J., Burinskienė, M.: Challenges Caused by increased use of E-powered personal mobility vehicles in European cities. Sustainability **12**(1), 273 (2020)
2. Merriam-Webster.com Dictionary, Merriam-Webster: Active Suspension. https://www.merriam-webster.com/dictionary/active%20suspension. Accessed 18 Apr 2020
3. Ruan, X., Li, X., Xing, X.: Dynamic modeling and simulation of a flexible two-wheeled balancing robot. In: Xiong, C., Huang, Y., Xiong, Y., Liu, H. (eds.) ICIRA 2008. LNCS (LNAI), vol. 5314, pp. 1011–1020. Springer, Heidelberg (2008). https://doi.org/10.1007/978-3-540-88513-9_108
4. Pratt, G.A., Williamson, M.M.: Series elastic actuators. In: Proceedings 1995 IEEE/RSJ International Conference on Intelligent Robots and Systems. Human Robot Interaction and Cooperative Robots, vol. 1. IEEE (1995)

5. Tseng, H.E., Hrovat, D.: State of the art survey: active and semi-active suspension control. Veh. Syst. Dyn. **53**(7), 1–29 (2015)
6. Pratt, J., Krupp, B., Morse, C.: Series elastic actuators for high fidelity force control. Ind. Robot Int. J. **29**, 234–241 (2002)
7. Yun, H., Bang, J., Kim, J., Lee, J.: High speed segway control with series elastic actuator for driving stability improvement. J. Mech. Sci. Technol. **33**(11), 5449–5459 (2019). https://doi.org/10.1007/s12206-019-1039-x

Kinematics Analysis of a New Spatial 3-DOF Parallel Mechanism

Yang Chao, Li Duanling$^{(\boxtimes)}$, and Jia Pu

Beijing University of Posts and Telecommunications, Beijing 100876, China
buptyangchao@bupt.edu.cn, liduanling@126.com

Abstract. In response to the increasing demand for low-DOF parallel mechanisms in industry, this paper proposes a new 2TPT-2RPR parallel mechanism, which has three degrees of freedom (one rotation, two translation). In this paper, helix theory is used to verify the degree of freedom of the parallel mechanism, and the kinematics of the parallel mechanism is studied. The positive and inverse solutions of the position of the 2TPT-2RPR parallel mechanism are analyzed, and the theoretical results are verified by numerical examples. At the same time, the velocity and acceleration characteristics of the parallel mechanism are analyzed and deduced, and the velocity and acceleration are simulated by core5.0.

Keywords: Parallel mechanism · Position analysis · Velocity analysis · Acceleration analysis

1 Introduction

In recent years, with the continuous improvement of the related theoretical research of parallel mechanisms and the continuous expansion of application fields, the low-DOF parallel mechanism has become the main object of research in the field of parallel mechanisms [1]. Compared with the 6-DOF parallel mechanism, the low-DOF parallel mechanism has the advantages of low manufacturing cost, simple mechanical structure, large working space, easy driving, and small cumulative position error. It occupies an increasingly important position in various industries and academic research.

At present, there are few low-DOF parallel mechanisms that have been developed at home and abroad with better comprehensive performance. Including Clavel's famous DELTA 3-DOF mobile mechanism, which was designed by using 4S parallelogram mechanism in the branch of parallel robot in 1988 [2]; In 1996, Tsai built a branch through a 4R parallelogram mechanism, and proposed a 3-DOF mobile parallel robot mechanism [3]; Tsai and Kim studied the kinematic synthesis of spatial 3-RPS parallel robots in 2003[4]. In 1995, Huang Zhen proposed several new 3-DOF cubic parallel robots [5], and Liu Xinjun proposed a new 3-DOF parallel robot with one rotation and two translations [6]. Zlatanov and Gosselin proposed the 3R1T 4-DOF parallel mechanism [7] in 2001 and the 2R2T 4-DOF parallel mechanism in 2003 IEEE conference [8]. In the study of parallel mechanisms, spiral theory is already a mainstream mathematical analysis tool. Literature [9] based on spiral theory has studied the type synthesis of a

© Springer Nature Switzerland AG 2020
C. S. Chan et al. (Eds.): ICIRA 2020, LNAI 12595, pp. 341–350, 2020.
https://doi.org/10.1007/978-3-030-66645-3_29

non-overconstrained 4-DOF parallel mechanism. Reference [10] made corresponding analysis and research on the kinematics, working space and transmission performance of a class of 4DOF 3T1R parallel mechanism. Literature [11] studied the kinematics performance of a 4-UPS&UP parallel mechanism with redundant drive branch chain, and proposed local kinematics evaluation indicators based on the Jacobian matrix. This paper presents a new type of 2TPT-2RPR parallel mechanism (T for Hook hinge, R for revolute joint, and P for prismatic pair), and uses the helix theory to analyze its degree of freedom characteristics, as well as its kinematics characteristics.

2 Analysis of DOF

The structure of the 2TPT-2RPR parallel mechanism is shown in Fig. 1. L_1 and L_3 are TPT branch chains, and L_2 and L_4 represent RPR branch chains. The coordinate system shown in Fig. 1 is established.

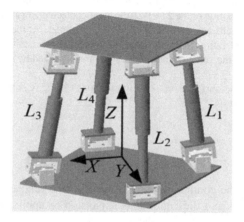

Fig. 1. Structure diagram of 2TPT-2RPR parallel mechanism.

The spiral system of the L_2 branch is Formula (1)

$$\begin{cases} \$_1 = \left(1\ 0\ 0;\ 0\ b_1\ c_1 \right) \\ \$_2 = \left(0\ 0\ 0;\ 0\ b_2\ c_2 \right) \\ \$_3 = \left(1\ 0\ 0;\ 0\ b_3\ c_3 \right) \end{cases} \quad (1)$$

The spiral system of the L_3 branch is Formula (2)

$$\begin{cases} \$_4 = \left(0\ 1\ 0;\ a_4\ 0\ c_4 \right) \\ \$_5 = \left(0\ 0\ 0;\ 0\ 0\ 0 \right) \\ \$_6 = \left(0\ 0\ 0;\ 0\ b_6\ c_6 \right) \\ \$_7 = \left(1\ 0\ 0;\ 0\ b_7\ c_7 \right) \\ \$_8 = \left(0\ 1\ 0;\ a_8\ 0\ c_8 \right) \end{cases} \quad (2)$$

In formula (1), $\$_1$, $\$_2$, $\$_3$ are the kinematic pair spinor of the L_2 branch from bottom to top. In formula (2), $\$_4$, $\$_5$, $\$_6$, $\$_7$, $\$_8$ are the kinematic pair spinor of L_3 branch chain from bottom to top.

By finding the union of the anti-helix of branches L_2 and L_3, a constrained helix system of TPT branch and RPR branch is obtained, which is Eq. (3).

$$
\begin{cases}
\$_{m1} = \left(1\,0\,0;\ 0\,0\,0 \right) \\
\$_{m1} = \left(0\,0\,0;\ 0\,1\,0 \right) \\
\$_{m1} = \left(0\,0\,0;\ 0\,0\,1 \right)
\end{cases}
\tag{3}
$$

Similarly, list the anti-helix of L_1 and L_4, and find their union. The result obtained is the same as the reverse helix of L_2 and L_3, as shown in Eq. (3). Therefore, the constraints of the entire platform are the three constraints of translation about the X-axis, rotation about the Y-axis, and rotation about the Z-axis. It can be seen that the 2TPT-2RPR parallel mechanism has three degrees of freedom which rotate around the X-axis and translational around the Y-axis and Z-axis.

3 Position Analysis

3.1 Position Inverse Solution

For the 2TPT-2RPR parallel mechanism shown in Fig. 1, the position analysis model of the mechanism shown in Fig. 2 is established. $B_iA_i (i = 1 \sim 4)$ means 4 branches, when $i = 1$ or $i = 3$, it means TPT branch; when $i = 2$ or $i = 4$, it means RPR branch.

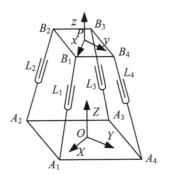

Fig. 2. Position analysis model of 2TPT-2RPR parallel mechanism.

In Fig. 2, the platform $A_1A_2A_3A_4$ is a square, and the distance from the coordinate origin O to the four vertices is set to R_1; the moving platform $B_1B_2B_3B_4$ is also a square, and the distance from the coordinate origin P to the four vertices is set to R_2. (α, β, γ) is the angle of rotation of the $Pxyz$ coordinate system around the X-axis, Y-axis, and Z-axis. Since the platform can only rotate around the X-axis, (β, γ) is always zero. After projecting point B_i on the dynamic coordinate system to the fixed coordinate system, there is Eq. (4).

$$L_i = B_i - A_i \tag{4}$$

From this, the lengths of the four branched chains can be obtained. Taking the single-open chain B_1A_1 as an example, the coordinate of point B_1 in the $Pxyz$ coordinate system is $(R_2, 0, 0)$. When mapping B_i in $Pxyz$ coordinates to $OXYZ$, the conversion matrix T is Eq. (5).

$$T = \begin{vmatrix} 1 & 0 & 0 & x_p \\ 0 & \cos\alpha & -\sin\alpha & y_p \\ 0 & \sin\alpha & -\cos\alpha & z_p \\ 0 & 0 & 0 & 1 \end{vmatrix} \tag{5}$$

According to the coordinate conversion, the coordinate of point B_1 in the fixed coordinate system $OXYZ$ is $(R_2 + x_p, y_p, z_p)$. The coordinate of point A_1 $(R_1, 0, 0)$ is known, so Eq. (6) is obtained by Eq. (4).

$$L_1 = \sqrt{(R_2 + x_p - R_1)^2 + y_p^2 + z_p^2} \tag{6}$$

Similarly, formula (7), formula (8), and formula (9) can be obtained.

$$L_2 = \sqrt{(-R_2\sin\alpha + z_p)^2 + (-R_2\cos\alpha + y_p + R_1)^2 + x_p^2} \tag{7}$$

$$L_3 = \sqrt{(R_1 - R_2 + x_p)^2 + y_p^2 + z_p^2} \tag{8}$$

$$L_4 = \sqrt{(R_2\cos\alpha + y_p - R_1)^2 + (R_2\sin\alpha + z_p)^2 + x_p^2} \tag{9}$$

3.2   Position Positive Solution

Equations (6)–(9) can be used to obtain expressions for the four unknowns of x_p $y_p z_p$, and α, which are Eq. (10), Eq. (11), Eq. (12), and Eq. (13), respectively.

$$x_p = \frac{(L_1^2 - L_3^2)}{4(R_2 - R_1)} \tag{10}$$

$$Z_p = m\frac{-2nf + \left[4n^2f^2 - 4(m+n)(f^2 + t^2 - L_1^2)\right]^{\frac{1}{2}}}{2(m+n)} \tag{11}$$

$$y_p = n\left\{\frac{-2nf + \left[4n^2f^2 - 4(m+n)(f^2 + t^2 - L_1^2)\right]^{\frac{1}{2}}}{2(m+n)} + \frac{L_2^2 - L_4^2}{4nm}\right\} \tag{12}$$

$$\alpha = \arccos\frac{L_2^2 + L_4^2 - 2(R_1^2 + R_2^2)}{4R_1R_2}. \tag{13}$$

In the above formula, $m = -R_2\cos\alpha + R_1$, $n = R_2\sin\alpha$, $t = \left(R_2 + \frac{L_1^2 - L_3^2}{4(R_2 - R_1)} - R_1\right)^2$, $f = \frac{L_2^2 - L_4^2}{4mn}$

3.3 Calculation Examples of Positive and Inverse Solutions

In Set $R_1 = 400$ mm for the lower platform and $R_2 = 300$ mm for the upper platform, and then use Matlab to calculate the positive and inverse solutions. The position inverse solution corresponding to the reference point P of the upper platform is shown in Table 1, and the position positive solution is shown in Table 2.

Table 1. Examples of position inverse solutions of 2TPT-2RPR parallel mechanism.

x_p (mm)	y_p (mm)	z_p (mm)	α (°)	L_1 (mm)	L_2 (mm)	L_3 (mm)	L_4 (mm)
0	0	400	0	412.311	412.311	412.311	412.311
0	100	400	0	424.264	412.311	424.264	400
0	100	400	30	424.264	346.687	424.264	551.467
0	−100	400	30	424.264	551.467	424.264	364.687

Table 2. Examples of position positive solutions of 2TPT-2RPR parallel mechanism.

L_1 (mm)	L_2 (mm)	L_3 (mm)	L_4 (mm)	x_p (mm)	y_p (mm)	z_p (mm)	α (°)
412.311	412.311	412.311	412.311	0	0	400.000	0
424.264	412.311	424.264	400	0	100	399.999	0
424.264	346.687	424.264	551.467	0	100.001	400.000	30
424.264	551.467	424.264	364.687	0	−100.001	400.000	30

4 Analysis of Velocity and Acceleration

In order to quickly solve the velocity and acceleration of the 2TPT-2RPR parallel mechanism, the more commonly used method of motion influence coefficient is used in this paper. This method does not use the time derivative method. The derivation process is easy to understand and the calculation speed is fast.

4.1 Velocity Analysis

As shown in Fig. 2, L_i is used to denote the vector between points B_i and A_i on the i-th branch chain, and the formula for calculating L_i is shown in Eq. (14).

$$L_i = B_i - A_i \tag{14}$$

Let l_i be the length of the i-th rod, then the unit vector of L_i is shown in Eq. (15).

$$q_i = L_i / l_i \tag{15}$$

Known:

$$l_i^2 = L_i \cdot L_i \tag{16}$$

V_{bi} and B_{bi} represent the velocity and acceleration of point B_i on the moving platform, and V and ω represent the velocity and angular velocity of the moving platform. Equation (17) can be obtained from the relevant theory of rigid body kinematics.

$$V_{bi} = V + \omega \times r_{bi} \tag{17}$$

In Eq. (17), r_{bi} is the position vector of the point B_i on the moving platform in the moving coordinate system.

Derivation of the first and second order of time by formula (16) can get the following relationship.

$$l_i \dot{l}_i = L_i \cdot V_{bi} \tag{18}$$

$$l_i \ddot{l}_i + \dot{l}_i^2 = L_i \cdot B_{bi} + V_{bi} \cdot V_{bi} \tag{19}$$

Therefore, the relationship between the input velocity and acceleration of the i-th rod can be expressed as Eq. (20) and Eq. (21).

$$\dot{l}_i = q_i \cdot V_{bi} \tag{20}$$

$$\ddot{l}_i = q_i \cdot B_{bi} + \left(V_{bi} \cdot V_{bi} - \dot{l}_i^2\right)/l_i \tag{21}$$

Equation (17) can be rewritten into the matrix form of Eq. (22).

$$\{V_{bi}\} = \left[G_P^{bi}\right]\{\dot{P}\} \tag{22}$$

In the above formula, $\{\dot{P}\} = \{\omega_x \ \omega_y \ \omega_z \ V_x \ V_y \ V_z\}^T$, $[G_P^{bi}] = [i \times r_{bi} \ j \times r_{bi} \ k \times r_{bi}]$. $i \ j \ k]$. i, j, and k are unit vectors in the X, Y, and Z directions, respectively.

Bring Eq. (22) into Eq. (20) and write it into the matrix form of Eq. (23).

$$\dot{l}_i = \{q_i\}^T\{V_i\} = \{q_i\}^T\left[G_P^{bi}\right]\{\dot{P}\} \ (i = 1 \sim 4) \tag{23}$$

Let $\{\dot{l}\} = \{\dot{l}_1 \ \dot{l}_2 \cdots \dot{l}_6\}^T$, $[G_P^L] = \left[\{q_1\}^T[G_P^{b1}] \ \{q_2\}^T[G_P^{b2}] \cdots \{q_4\}^T[G_P^{b4}]\right]^T$. Then the four equations of Eq. (23) can be written as Eq. (24).

$$\{\dot{l}\} = \left[G_P^L\right]\{\dot{P}\} \tag{24}$$

From Eq. (24), the velocity positive solution equation of the parallel mechanism can be obtained.

$$\{\dot{P}\} = \left[G_L^P\right]\{\dot{l}\} \tag{25}$$

In Eq. (25), $\left[G_L^P\right]$ is the first-order influence coefficient of the output speed of the moving platform on the input speed of the branch chain.

As shown in Fig. 1, assuming that the rod length $l_1 = 370$ mm, $l_3 = 370$ mm, $l_4 = 370$ mm, the rod length l_2 is extended from 360 mm to 370 mm. At this time, the position change curve of the L_2 rod simulated by core5.0 is shown in Fig. 3, and the velocity change curve is shown in Fig. 4.

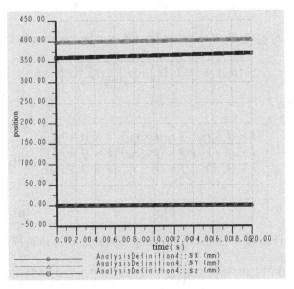

Fig. 3. Position change curve.

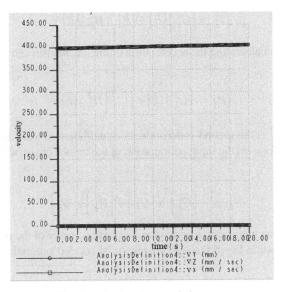

Fig. 4. Velocity characteristic curve.

4.2 Acceleration Analysis

Let $B = \begin{pmatrix} B_x & B_y & B_z \end{pmatrix}^T$ and $\boldsymbol{\varepsilon} = \begin{pmatrix} \varepsilon_x & \varepsilon_y & \varepsilon_z \end{pmatrix}^T$ be the acceleration and angular acceleration of the moving platform respectively, then there is Eq. (26).

$$B_{bi} = B + \boldsymbol{\varepsilon} \times r_{bi} + \boldsymbol{\omega} \times (\boldsymbol{\omega} \times r_{bi}) \tag{26}$$

Let $\{\ddot{P}\} = \{ \varepsilon_x \ \varepsilon_y \ \varepsilon_z \ B_x \ B_y \ B_z \}^T$, then Eq. (26) can be rewritten as Eq. (27).

$$\{B_{bi}\} = \left[G_P^{bi} \right]\{\ddot{P}\} + \{\dot{P}\}^T \left[H_P^{bi} \right]\{\dot{P}\} \tag{27}$$

In the formula:

$$\left[H_P^{bi} \right] = \begin{bmatrix} \boldsymbol{i} \times (\boldsymbol{i} \times r_{bi}) & \boldsymbol{j} \times (\boldsymbol{i} \times r_{bi}) & \boldsymbol{k} \times (\boldsymbol{i} \times r_{bi}) & \vdots & \\ \boldsymbol{i} \times (\boldsymbol{j} \times r_{bi}) & \boldsymbol{j} \times (\boldsymbol{j} \times r_{bi}) & \boldsymbol{k} \times (\boldsymbol{j} \times r_{bi}) & \vdots & [0]_{3\times3} \\ \boldsymbol{i} \times (\boldsymbol{k} \times r_{bi}) & \boldsymbol{j} \times (\boldsymbol{k} \times r_{bi}) & \boldsymbol{k} \times (\boldsymbol{k} \times r_{bi}) & \vdots & \\ \cdots & \cdots & \cdots & \vdots & \\ & [0]_{3\times3} & & \vdots & [0]_{3\times3} \end{bmatrix} \tag{28}$$

Equations (29) can be obtained by bringing Eqs. (22), (23) and (17) into Eq. (21).

$$\ddot{l}_i = \{q_i\}^T \left[G_P^{bi} \right]\{\ddot{P}\} + \{\dot{P}\}^T \left[\{q_i\}^T * \left[H_P^{bi} \right] + \frac{1}{l_i} \left(\left[G_P^{bi} \right]^T \left[G_P^{bi} \right] - \left[G_P^{bi} \right]^T \{q_i\}\{q_i\}^T \left[G_P^{bi} \right] \right) \right]\{\dot{P}\} \quad (i = 1 \sim 4) \tag{29}$$

Formula (29) is sorted into formula (30).

$$\{\ddot{l}\} = \left[G_P^L \right]\{\ddot{P}\} + \{\dot{P}\}^T \left[H_P^L \right]\{\dot{P}\} \tag{30}$$

The formula for calculating the acceleration of the mechanism can be obtained from Eq. (24).

$$\{\ddot{P}\} = \left[G_L^P \right]\{\ddot{l}_i\} + \{\dot{l}_i\}^T \left[U_L^P \right]\{\dot{l}_i\} \tag{31}$$

In Eq. (31), U_L^P is the second-order influence coefficient of the output acceleration of the moving platform on the input acceleration of the branch chain, which is expressed as Eq. (32).

$$U_L^P = -\left[G_P^L \right]^T \left(\left[G_P^L \right] * \left[H_P^L \right] \right) \left[G_P^L \right] \tag{32}$$

The acceleration change curve corresponding to the change of the position of the L_2 bar in Fig. 3 is shown in Fig. 5 through the results of core5.0 simulation.

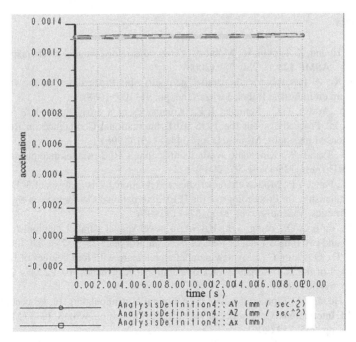

Fig. 5. Acceleration characteristic curve.

5 Summary

This paper proposes a new spatial 3-DOF parallel mechanism. The 2TPT-2RPR parallel mechanism has one rotation degree of freedom and two movement degrees of freedom. The spiral theory is used to analyze and verify the number of degrees of freedom and the characteristics of the degrees of freedom of the parallel mechanism. At the same time, the positive and inverse position solutions of the parallel mechanism are analyzed, and its velocity and acceleration characteristics are also derived. The velocity and acceleration are simulated by core5.0. The advantages of the 2TPT-2RPR parallel mechanism are as follows: (1) The structure is relatively simple, consisting only of several easily processed parts such as the prismatic pair, the revolute joint, and the Hook hinge; (2) Combining the freedom of movement and the freedom of rotation in a 3-DOF space parallel mechanism. The article analyzes the positive and inverse kinematics solutions of the mechanism. The analysis results show that the parallel mechanism has a unique kinematics positive solution for a certain set of drive inputs and has a large working space. The parallel mechanism can be widely used in industries such as industrial robots and parallel machine tools.

Acknowledgement. This work was supported partly by the National Natural Science Foundation of China (51775052), and the Natural Science Basic Research Program of Shaanxi (2019JM-181).

References

1. Dai, J.S., Huang, Z., Lipkin, H.: Mobility of over constrained parallel mechanisms. J. Mech. Des. Trans. ASME **128**(1), 220–229 (2006)
2. Delta, C.R.: A fast robot with parallel geometry. In: Proceedings of the International Symposium on Industrial Robot, Switzerland, pp. 91–100 (1988)
3. Tsai, L.W., Walsh, G.C., Stamper, R.E.: Kinematics of a novel three DOF translational platform. In: Proceedings opf the 1996 IEEE International Conference on Robotics and Automation, Minneapolis, Minnesota, pp. 3446–3451 (1996)
4. Kim, H.S., Tsai, L.W.: Kinematic synthesis of spatial 3–RPS parallel manipulator. J. Mech. Des. ASME Trans. **125**(1), 92–97 (2003)
5. Huang, Z., Fang, Y.F.: Motion characteristics and rotational axis analysis of 3–DOF parallel robot mechanisms. In: Proceedings of the IEEE International Conference on Systems, Man and Cybernetics, Vancouver, vol. 1, pp. 67–71 (1995)
6. Liu, X.J., Tang, X.Q., Wang, J.S.: HANA: a novel spatial parallel manipulator with one rotational and two translational degrees of freedom. Robotica **23**(2), 257–270 (2005)
7. Zlatanov, D., Gosselin, C.M.: A new parallel architecture with four degrees of freedom. In: Proceedings of the 2nd Workshop on Computational Kinematics, 19–22 May 2001, Seoul Korea, pp. 57–66 (2001)
8. Li, Q.C., Huang, Z.: Type synthesis of 4-DOF parallel manipulators. In: Proceedings of the 2003 IEEE International Conference on Robotics & Automation, WM13, 12–17 May, Taiwan, P.R China, pp. 755–759 (2003)
9. Sheng, G., Yue-fa, F., Hai-bo, Q.: Type synthesis of 4–DOF nonoverconstrained parallel mechanisms based on screw theory. Robotica **30**, 31–37 (2012)
10. Ling-min, X., Qiao-hong, C., Lei-ying, H.: Kinematic analysis and design of a novel 3T1R 2-(PRR)(2) RH hybrid manipulator. Mech. Mach. Theory **112**, 105–122 (2017)
11. Chenglin, D., Haitao, L., Tian, H.: Kinematics performance analysis of 4-UPS&UP parallel mechanism with redundant drive branch chain. J. Mech. Eng. **52**(5), 124–129 (2016)

Design of USV for Search and Rescue in Shallow Water

Chew Min Kang, Loh Chow Yeh, Sam Yap Ren Jie, Tan Jing Pei, and Hermawan Nugroho$^{(\boxtimes)}$

Electrical and Electronic Engineering Department, University of Nottingham, Nottingham, Malaysia

Minkangchew@gmail.com, lohchowyeh3898@gmail.com, samyap1997@gmail.com, peitjp98@gmail.com, hermawan.nugroho@nottingham.edu.my

Abstract. Growing interests in Unmanned Surface Vehicle (USV) for Search and Rescue missions associated with both oceans and shallow waters have increased for last five years. Researchers found that these disaster robotic boats can contribute to the successful rate of search mission. Currently there are no standards structure of the SAR boat in regards to its speed, stability and load capacity. In this project, we propose a modified single chine twin hull catamaran design for the USV. In the proof-of-concept prototype, an adaptation of PID control system integrated in the robotics algorithms is applied to provide light speed maneuverability and stability. The USV is equipped with an artificial intelligence module to seek and pinpoint the location of drowning victims and a localization system that allows the USV to reach targeted location. Synthesis of these systems and technology proves a successful deployment of the USV. Experiment shows that the prototype has a good potential to be further developed.

Keywords: Unmanned surface vehicles (USVs) · Shallow water · Search and rescue · Drowning detection

1 Introduction

Catastrophic accidents on water surfaces e.g. river, lakes, open water are usually happened silently [1]. The accidents occur out of prediction, even though we have the information (i.e. weather information, waves prediction) and prepare some prevention actions. Securing the golden window for the victim is crucial for the success of the search & rescue (SAR) mission. However, involvement of human in SAR missions when there is heavy storm & arbitrary tidal or current for example can be dangerous. In this paper, we propose and develop the prototype of an Unmanned Surface Vessel (USV) for SAR mission on the clear and shallow water surface. Sending an untrained person for SAR mission may cause more casualties, especially during strong current, wave & stormy weather. Moreover, response time is critical for the success of the SAR mission. The SAR mission should be executed before the targeted objects are being drawn away from the incident site.

© Springer Nature Switzerland AG 2020
C. S. Chan et al. (Eds.): ICIRA 2020, LNAI 12595, pp. 351–363, 2020.
https://doi.org/10.1007/978-3-030-66645-3_30

The adaptation of USV technologies will reduce response time and increase productivity of search operations on water bodies. And most importantly, it will ensure the safety of all parties involved. Thus, it is important for the USV to have a proper response capability while reducing the risk for first responders. To provide aid for the victims, in the project, we also develop and equipt the USV with drowning detection module.

2   Platform Design

2.1   Structure of the USV

The USV drawing shown in Fig. 1. In Fig. 1, a free body diagram is drawn in which parameter were taken into consideration for the vessel optimum initial stability when immersed in water body.

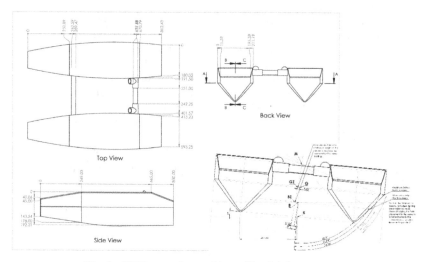

Fig. 1.   USV general assembly and its righting moment

As initial stability is important, especially when the USV performs the mission (i.e. drowning detection), we verify the design of catamaran by calculating its metacentric and righting arm. Table 1 shows the formula used in the calculation. Results of the calculations are illustrated in Fig. 2.

Figure 2 shows that the simulation result. It shows that prototpye vessel has an allowable rolling angle of 47.7°, from that angle onward, reaching 62.79° or 15.1° from the keel of the ship. It will be the minimum safety, and 74.96° will be the maximum rolling angle. Considering these allowable safety angle, the placement of the components is distributed so that internal rolling is minimal. External rolling however is unavoidable as high tidal waves can cause the vessel to capsize. This is the current limitation of the USV. Self-righting technology can be implemented to prevent the vessel from capsizing.

Table 1. Formula for metacentric and righting arm

Metacentric		Righting arm
The wake line is set as T to the keel of the boat		*Righting arm before critical* [2–5] $Righting\ Arm = GM\ \mathrm{Sin}\,\theta$ (2) This formula applied when the hull do not exceed the buoyant point at 47.7° based on Figs. 1b and 2b. When the rolling angle exceeds that point the formula is calculated as followed:
T	Transverse line or wake line	
K	Keel to centre of buoyant	
BM	Center of buoyant to metacenter	
GM	Center of gravity of metacenter	
Metacentric Height Equation [2–4] $$T = 0.105\,m$$ $$KB = \frac{T}{2}$$ $$= 0.0525\,m$$ $$B = 0.41323\,m \quad (1)$$ $$BM = \frac{B^2}{12T}$$ $$KM = KB + BM$$ $$GM = KM - KG$$		*Righting arm after critical* [2–5]
		Righting Arm = $$\left[GM + \frac{BM}{2}\tan^2\theta\right]\mathrm{Sin}\,\theta \quad (3)$$ The formula changed because beyond that point, more force needs to be applied on the other side creating moment great enough to counter the weight disturbances

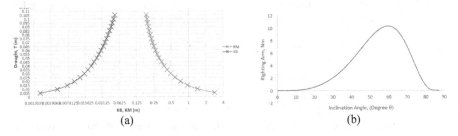

(a)   (b)

Fig. 2. a) Metacentric, b) Righting moment

2.2 Electronics Components

The main electronics system of the USV is made up of a Nvidia Jetson Nano which connect with two Arduino boards. One of the Arduino board is used for navigation (connected with a Pixhawk GPS module). And the other is used for maneuvering, positioning, executing commands coming from object detection module (run on Jetson Nano) and monitoring the heat exchanger system.

The Jetson Nano is mainly used to run the drowning detection using input from a Pi Camera. The wiring configuration between each components of the electronics system and the corresponding PCB can be seen in Fig. 3.

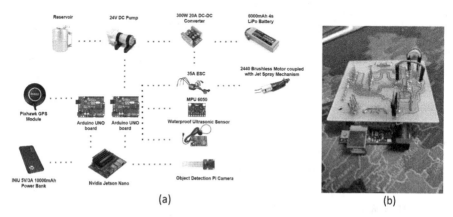

(a)   (b)

Fig. 3. a) Wiring configuration, b) PCB board

3 Robotics Algorithms and Control System

3.1 Robotics Algorithm

Robotics algorithm is an integral part of designing an autonomous system. For the USV to have automation capabilities, the system needs to have the abilities to sense, perceive, plan and act to interact. To simplify the automation process, we categorize the modules into three main areas, namely; mapping, localization and path planning.

Mapping Module. Autonomous navigation is the ability for a robotic system to determine its location and plan a path without assistance. In order to accomplished autonomous navigation, the system needs to have a map of the environment to figure out an optimal path to reach the goal. In our USV, the algorithm for mapping application uses the Gaussian grid map which utilizes the earth sciences as a gridded horizontal coordinate system for constructing the map. The algorithm then converts the map into Cartesian coordinates from a 2D matrix of Boolean values and represents the map for the user interface. For experiment, the outdoor swimming pool of University Nottingham Malaysia is used as the target mapping location.

Localization Module. Robot localization is the process of determining its own location with respect to the environment. For the system to obtain accurate position information (a less than few meters), GPS signals are not sufficient and must be paired with additional sensors. Here, the USV uses data from IMU. We apply Kalman Filter to fuse the sensor data. Kalman Filter is an estimation algorithm based on uncertain measurement.

For our USV, the factors causing the uncertainties are satellite's position, atmospheric effects and sensor inaccuracy. The Kalman Filter works by performing estimation of the current state and prediction for the next state using the state update equation and state extrapolation equation respectively. Table 2 shows the equations related to Kalman Filter used in the USV.

Table 2. Kalman filter formula

State extrapolation Eq. (4)	State update Eq. (5)
$$\hat{x}_{n+1,n} = \hat{x}_{n,n} + \hat{\dot{x}}_{n,n}\Delta t + \hat{\ddot{x}}_{n,n}\frac{\Delta t^2}{2}, \ s$$ $$= s_0 + ut + a\frac{t^2}{2}$$ $$\hat{\dot{x}}_{n+1,n} = \hat{\dot{x}}_{n,n} + \hat{\ddot{x}}_{n,n}\Delta t, \ v = u + at$$ $$\hat{\ddot{x}}_{n+1,n} = \hat{\ddot{x}}_{n,n}$$	$$\hat{x}_{n,n} = \hat{x}_{n,n-1} + \alpha(z_n - \hat{x}_{n,n-1})$$ $$\hat{\dot{x}}_{n,n} = \hat{\dot{x}}_{n,n-1} + \beta\left(\frac{z_n - \hat{x}_{n,n-1}}{\Delta t}\right)$$ $$\hat{\ddot{x}}_{n,n} = \hat{\ddot{x}}_{n,n-1} + \gamma\left(\frac{z_n - \hat{x}_{n,n-1}}{0.5\Delta t^2}\right)$$

Path Planning. The A* algorithm is applied for path planning. It reads the map, which is in the form of a Python dictionary having keys as tuples (Cartesian coordinates of current position) and values as a vector of tuples (Cartesian coordinates of neighbors), from a 2D matrix of Boolean values obtained from the mapping process [6]. The heuristic approach is based on the Euclidean distance. It is assumed that the distance between two neighbors is considered as 1 unit [6]. The algorithm provides the shortest path by comparing different paths taken from the starting point to end point.

3.2 Control System

Boat Dynamics. The dynamics of the boat are generically expressed in the Fossen's nonlinear differential equations and matrix-vector form [7] where M, C and D denote the system inertia, Coriolis and damping matrices (Fossen's Differential Equation).

$$\tau = M\dot{v} + C(v)v + D(v)v \tag{6}$$

where $v = [u \ v \ r]^\tau$ denotes the boat velocity which consists of the boat surge velocity (u), sway velocity (v) and yaw rate (r). Matrix M is the system inertia matrix which is the summation of a rigid body mass matrix and an added mass matrix as shown in Eq. (7).

$$M = M_{rigidbody} + M_a$$

$$= \begin{bmatrix} m - X_{\dot{u}} & -X_{\dot{v}} & -my_G - X_{\dot{r}} \\ -X_{\dot{u}} & m - Y_{\dot{v}} & mx_G - Y_{\dot{r}} \\ -my_G - N_{\dot{u}} & mx_G - N_{\dot{v}} & I_z - N_{\dot{r}} \end{bmatrix} \tag{7}$$

$C(v)$ is a Coriolis and entripetal matrix (Eq. 8) which includes the rigid body Coriolis matrix and the added mass Coriolis matrix [7].

$$C(v) = C_{rigidbody} + C_a$$

$$= \begin{bmatrix} 0 & 0 & -m(x_Gr + v) \\ 0 & 0 & -m(y_Gr + u) \\ m(x_Gr + v) & m(y_Gr - u) & 0 \end{bmatrix} + \begin{bmatrix} 0 & 0 & -\left(\left(Y_{\dot{p}}v + \frac{Y_{\dot{r}}+N_{\dot{v}}}{2}r\right)\big/200\right) \\ 0 & 0 & X_{\dot{u}}u \\ Y_{\dot{p}}v + (Y_{\dot{r}} + N_{\dot{v}})r/200 & -X_{\dot{u}}u & 0 \end{bmatrix} \tag{8}$$

$D(v)$ is damping matrix which sums up the hydrodynamic damping matrix and the non-linear damping matrix as shown in Eq. 9 [7].

$$D(v) = D + D_N$$

$$= - \begin{bmatrix} X_u & 0 & 0 \\ 0 & Y_v & Y_r \\ 0 & N_v & N_r \end{bmatrix} - \begin{bmatrix} X_{u|u|}|u| & 0 & 0 \\ 0 & Y_{v|v|}|v| + Y_{|r|v}|r| & Y_{|v|r}|v| + Y_{r|r|}|r| \\ 0 & N_{v|v|}|v| + N_{|r|v}|r| & N_r|v| + N_{r|r|}|r| \end{bmatrix} \tag{9}$$

$J(\eta)$ denotes the transformation matrix to relate the body-fixed frame with the earth-fixed frame in Eq. (10) [7].

$$J(\eta) = \begin{bmatrix} \cos\varphi & -\sin\varphi & 0 \\ \sin\varphi & \cos\varphi & 0 \\ 0 & 0 & 1 \end{bmatrix} \tag{10}$$

Trajectory Tracking Control. Development of control methods is oriented by different aims while the main objective of the boat is kept. The boat design configuration consists of two independent jet thrusters controlled by two motors. The two distinct inputs are propulsion force and yaw moment. The control strategies provide adequate maneuvering input to control the direction and speed of the boat for navigation along a pre-determined path. A simple boat trajectory tracking cascaded control system is designed which comprises of two feedback loops, inner loop and outer loop, with PID control strategies implemented as shown in Fig. 4. The inner loop is employed to regulate the drive voltage applied to the motor and thus maintaining constant speed. The outer loop feedback works by comparing the current yaw rate with the desired yaw rate to produce a cross-track error.

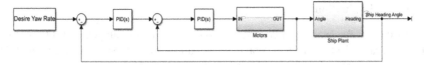

Fig. 4. Cascaded PID control structure

4 Drowning Detection

The drowning detection module consists of a camera connected and integrated with a Deep Learning Neural Network (DLNN) which is run on Jetson Nano. Using Tensor-Flow, the model was trained until the loss is dropped consistently at an average of 3.20, which requires approximately 200,000 steps and 30 h training time on GPU 1070Ti.

4.1 Data Collection

One of the main challenges for drowning detection is that there are no datasets available online and it is hard to build such datasets with actual person drowning on various weather and wavy conditions.

In this project, the dataset is obtained from two methods as discussed below:

1. MS COCO Dataset

Using COCO API, a python script was written to download the dataset of human. There were altogether 64115 images contributing to the dataset. Test images for model evaluation were prepared in the similar way. There were 2593 images in total.

2. Unity

Despite a myriad of images available from COCO, there are still limited images of a drowning person in the sea, which may impact the accuracy of model. To obtain more relevant images, Unity, a gaming design software, is used to simulate the scene of sea. Scenes of 'drowned human' with various postures under the simulated environment were exported as images. There were altogether 500 and 100 images created for training and test dataset respectively.

(a)   (b)

Fig. 5. Images created from unity

4.2 Model Evaluation

The test dataset for evaluation consists of 2693 images with altogether 11004 ground truth (the real object boundary). For this evaluation, the Intersection of Union (IoU) threshold is predefined as 0.5 to classify if the prediction is a true positive or a false positive. Using TensorFlow Object Detection API, the model is evaluated by finding Average Precision (AP), which is area under Precision-Recall graph (Fig. 5).

From the model evaluation shown in Fig. 6, Average Precision with IoU = 0.5 (AP@0.5IoU) is about 0.509 (50.9). Comparing to AP@0.5IoU of SSD300 which is 41.2 [8], it can be concluded that AP of our trained model is fairly high and acceptable.

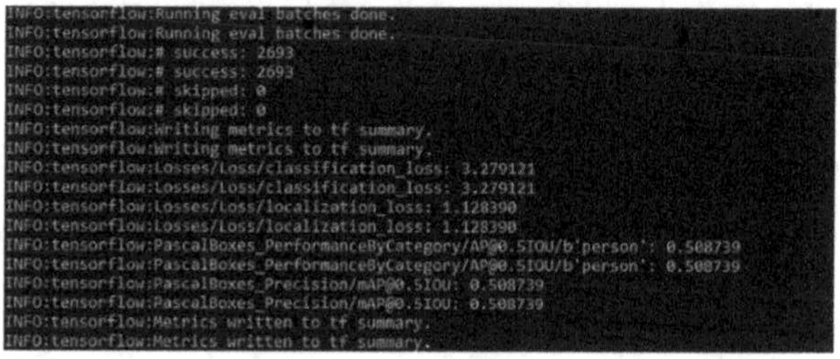

Fig. 6. Model evaluation

4.3 Model Optimization

Although TensorFlow is instructed to run all the inference operations on GPU, there are still operations which are run on CPU. This is due to unregistered GPU kernels for these operations, such as NonMaxSuppressionV3. As these operations cannot be processed on GPU, TensorFlow must transfer the intermediate output from GPU memory to CPU memory, process it on CPU and transfer the result back to GPU. As a result, the program becomes slower due to data transfer [9].

To optimize the model, all operations in 'NonMaxSuppresion' block are placed in CPU because most of flow control operations are in this block. As a result, the total inference time are decreased from approximately 50 to 30 ms. Time cost for GatherV2 decreases significantly by about 65% from 7.516 to 2.646 ms. Besides, time cost for ConcatV2 is also reduced from 3.588 to 1.422 ms. Furthermore, there was less data transfer between GPU and CPU after modification of model.

5 Simulation and Experimental Analysis

5.1 Structural Analysis

To evaluate the catamaran hull design stability, SOLIDWORK 2019 Flow Simulation is used to simulate practical cases of the hull subjected to liquid and airflow. Figure 7 show examples of the simulation steps. Results are tabulated in Table 3. The simulations are conducted for two types of water velocity, 2 and 8 m/s.

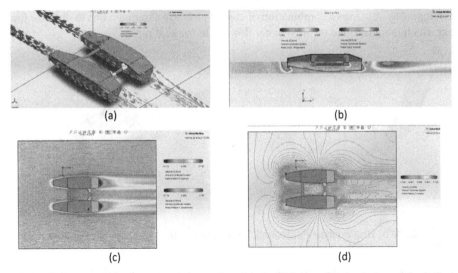

(a)   (b)

(c)   (d)

Fig. 7. a) Flow trajectory simulation in (X) direction; b) Wake cut (Z) simulation of the hull; c) Wake pattern (X) simulation of the hull; d) Wake pattern (Z) simulation of the hull

Table 3. Simulation results

Velocity	Simulation											
	Flow trajectory		Wake pattern						Wake cut			
			X		Y		Z		X		Z	
	Min	Max	Min	Max	Min	Max	Min	Max	Min	Max	Min	Max
2 m/s	−0.24	1.999	−0.18	2.150	−0.32	0.25	−1.13	1.120	−0.03	2.06	−0.08	0.068
8 m/s	−0.14	8.241	−0.31	8.618	−1.39	1.056	−4.16	4.153	−0.14	8.24	−0.32	0.273

From the simulation, we can conclude that the USV can maneuver on water surfaces easily (having velocity less than 8 m/2). It is observed that when water bodies are in contact with the USV, positive gain in water flow is perceived, indicating the hydrodynamic of the design. Aerodynamic are also observed when the USV moves at high speed. The higher the velocity, the larger the air particles is pushed down. The simulation of flow trajectory, wake pattern and wake cut on the catamaran hull design show that the theorical designs of the USV are correct.

5.2 Thermal Analysis

During the USV prototyping, we find that heat is generated due to the confined space which causes the computer to shut down. Since air cooling is impractical, we use heat exchanging system-based on liquid cooling as shown in Fig. 8a. From simulation (Fig. 8b), we know the optimised parameters and adopt it to the system. The heat exchanging system uses a combination of copper pipe and aluminum fins to disperse the heat which is transmitted through water flowing in the heat pipes.

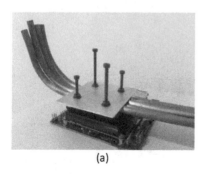
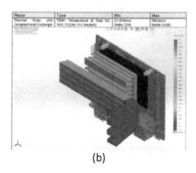

(a)   (b)

Fig. 8. a) Heat exchanging system, b) Simulation of the heat exchanging system

Based on the thermal reading, the cooling system is able reduce the heat from 70 to 56 °C with reservoir water intake on room temperature of an average of 24.4 °C. After installation of the heat exchanging system, the drowning detection system runs smoother and the computer do not have shutdown issues.

5.3   Robotics Algorithm and Control System Simulation Results

Robotics Algorithm. In the experiments conducted on the pool of the University of Nottingham Malaysia, the USV position is tracked in two dimensions using the α-β-γ filter. The simulation illustrates a two-dimensional world and the USV moves from the starting point to the end point of the swimming pool, as shown on Fig. 9. During the journey, there is a sudden disturbance windblown causing the boat coordinate to move from (8,18) to (6,18). With the application of Kalman Filter, the algorithm can automatically self-correct its location with the GPS sensor and accelerometer measurement.

Fig. 9. Localization results (on the pool of Univ. of Nottingham Malaysia)

From the result, it shows that the GPS measurement value fluctuates consistently from 0 to 48 s with respect to the true value as shown in Fig. 10. The y-axis range estimated value is identical as the true value. For the x-axis, due to the external disturbance (i.e. wind-blown), the estimated value has differences from the true value. With Kalman Filter, the estimated value can be corrected.

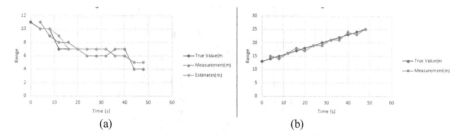

(a)   (b)

Fig. 10. a) x-axis vs time and b) y-axis vs time

Control System. The control system is simulated in which the control loop compares the boat's heading angle with the reference heading angle. A reference trajectory is defined as a step change from 0 to 180 degrees, with a step time of 200 s, rise time of 75 s and percentage of overshoot of 5%. By tuning of the controller's gains, the USV is able to match the reference trajectory. From the experiments, the variables to be tuned by the optimization tool are the PID controller gains and the gains. The variables are K_p = 0.1, K_i = 1.742, K_d = 0 and a noise filter coefficient of N = 100.

Experimental validation plays a significant role in bridging the gap between the theory and practice. Yaw rate data from zig-zag test obtained experimentally is used to estimate and fine-tuned the hydrodynamic coefficients that fill the system inertia, Coriolis and damping matrices presented in the mathematical model. A parameter estimation tool in MATLAB is used to estimate and validate the derived mathematical model with the experimental model obtained. Tuning of the hydrodynamic coefficients are completed as the validation data and simulated data matches as shown in Fig. 11.

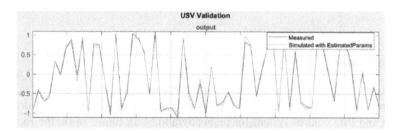

Fig. 11. Validation output of USV

5.4 Model Deployment and Pool Site Testing

Once reaching the target area, the USV will perform drowning detection by running machine learning model using camera stream. Once human detected, the USV will move until the target is located at the center of camera. Then, it will slowly approach and stop at a distance (Fig. 12).

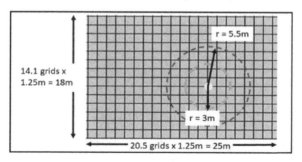

Fig. 12. Testing at swimming tool in grid

Based on the testing at swimming pool, as illustrated in Fig. 13, the USV has a capability to detect human within 5.5 m. In other words, the target will have to be located within a maximum of 4.4 grids. The confidence hower is relatively low (0.4). The performance can be significantly improved to 0.7 within a range of 3 m. Figure 13 shows the actual USV on testing ground.

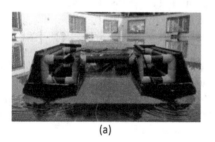
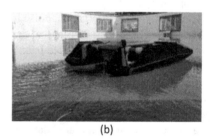

(a)   (b)

Fig. 13. USV on testing ground

6 Conclusion

This paper proposed an autonomous approach for SAR mission. The USV successfully demonstrated some basic capabilities for navigation, deep-learning-based-drowning detection and PID control system for manoeuvrability. In future work, the focus will be on enhancement on the performance of the USV by improving the inferencing time of deep learning model and implementing an adaptive PID control algorithm which is capable in adapting to variations such as a sudden incoming wave.

References

1. Unmanned Vehicles Could Aid Search and Rescue. https://www.maritime-executive.com/edi
   torials/unmanned-vehicles-could-aid-search-and-rescue. last accessed 2016/12/16

2. Benedek, Z.: Stability of Catamaran. PhD Thesis. Department of Aero and Thermotechnics, Technical University Dudapest, Budapest (1974)
3. Larssons, L., Eliasson, R.E.: Principles of Yatch Design, 2nd edn. Adlard Coles Nautical, London (2000)
4. Lewis, E.V.: Principles of Naval Architecture, 2nd edn., p. 601. The Society of Naval Architects and Marine Engineers, Pavonia Avenue Jersey City, NJ (1988)
5. Dubrovsky, V.A.: Decreasing of wet deck slamming. Int. J. Recent Develop. Eng. Technol. **6**(4), 1–7 (2017)
6. A_star-algorithm-in-python. https://github.com/VaibhavSaini19/A_Star-algorithm-in-Python. last accessed 2020/04/30
7. Klinger, W.B.: Adaptive controller design for an autonomous twin-hulled surface vessel with uncertain displacement and drag. Msc Thesis. Florida Atlantic University, Boca Raton, Florida (2014)
8. Liu, W., Anguelov, D., Erhan, D., Szegedy, C., Reed, S., Fu, C.-Y., Berg, A.C.: SSD: Single shot MultiBox detector. In: Leibe, B., Matas, J., Sebe, N., Welling, M. (eds.) Computer Vision—ECCV 2016, LNCS, vol. 9905. Springer, Cham (2016)
9. Optimize NVIDIA GPU performance for efficient model inference. https://towardsdatascience.com/optimize-nvidia-gpu-performance-for-efficient-model-inference-f3e9874e9fdc. last accessed 2020/02/14

4-Leg Landing Platform with Sensing for Reliable Rough Landing of Multi-copters

Dong-hun Cheon, Ji-wook Choi, and Jang-myung Lee[✉]

Department of Electrical Engineering, Pusan National University, Busan, Republic of Korea
{Donghun7379,jmlee}@pusan.ac.kr

Abstract. In this paper we propose a landing system in which multi-copters can make stable landings even in rough terrain. In order to utilize multi-copters in various environments, it is necessary to develop a landing platform that can be landed in rough terrain. The 4-leg landing platform of 2-link structure was analyzed kinematically and produced through 3D modeling. The landing platform detects contact with the ground through the Force Sensor upon landing and estimates the slope angle with the IMU (Inertial Measurement Unit) sensor. Using the formula proposed in this paper, maintain horizontality by controlling the angle value of each leg joint. We presented a rough environment and tested the proposed landing platform to verify its effectiveness.

Keywords: Multi-copter · Landing platform · 2-Link kinematics · Control

1 Introduction

With the recent growth of the Multi-copter market, the scope of use is also expanding. It is used for the purpose of solving the shortage of labor in the agricultural field or performs aerial surveying and field monitoring at the construction site. Furthermore, it is being used in various industries, such as shipping goods from logistics services or using Multi-copters in places where humans are hard to reach in military operations.

The UAV has the shape of a rotary wing and a fixed wing according to the design method, and the rotary wing has a feature of vertical takeoff/landing, but it is greatly affected by the condition of the landing point. To solve this problem, many studies have been conducted in the past [1, 2], and studies on mechanical design are also being conducted. [3] We propose 4-Leg landing Platform of type 2-Link structure. Section 2 deals with 3D modeling and kinematics and inverse kinematics of the landing platform, and formulates equations for maintaining level. Section 3 assumes the rough environment and proves the proposed formula and level maintenance through the landing platform experiment. Section 4 Finally, we conclude with a summary of the results.

2 Design and Control of 4-Leg Landing Platform

2.1 Landing Platform Modeling

In order for Multi-copter to land in rough environments, it must be able to cope with the situation of the landing point. Multi-copter landing gear used in the past has a limitation

© Springer Nature Switzerland AG 2020
C. S. Chan et al. (Eds.): ICIRA 2020, LNAI 12595, pp. 364–373, 2020.
https://doi.org/10.1007/978-3-030-66645-3_31

in a fixed form. It works according to the ground condition through 4 landing legs of 2-Link structure. Model as shown in Fig. 1 below using CATIA 3D.

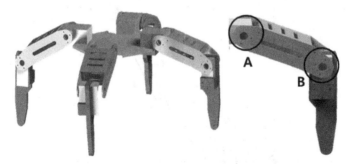

Fig. 1. Landing leg 3D modeling

In Fig. 1, the revolute joints A and B corresponding to θ_1 and θ_2 use high-torque servomotors that can withstand the load of the multi-copter. And θ_1, θ_2 angle control according to the inclination of the ground keeps it level even on rough terrain. Based on 3D Modeling, 3D Print was used to produce the landing platform as shown in Fig. 2.

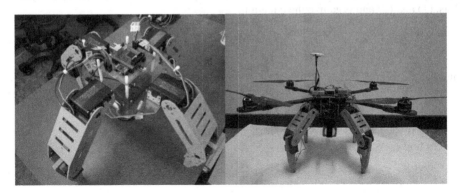

Fig. 2. Landing platform assembly

2.2 Kinematics Model

To maintain the horizontal movement, you need to know the angle and coordinates of each joint. Figure 3 shows the kinematics analysis of 2-DOF (Degrees of Freedom) Landing Leg with 2-Link structure. [4]

The coordinates of the landing leg End-effector are as shown below through Forward Kinematics.

$$x = l_1 cos\theta_1 + l_2 cos(\theta_1 + \theta_2) \tag{1}$$

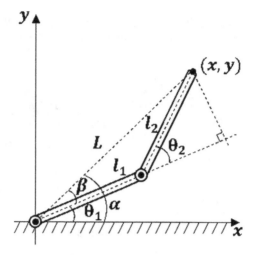

Fig. 3. 2-DOF structure

$$y = l_1 sin\theta_1 + l_2 sin(\theta_1 + \theta_2) \tag{2}$$

Inverse kinematics are required to induce θ_1 and θ_2 of the landing leg. If you square Eqs. (1) and (2) and organize the equation,

$$cos\theta_2 = \frac{x^2 + y^2 - l_1^2 - l_2^2}{2l_1 l_2} \tag{3}$$

and the theorem of θ_2 is as follows.

$$\theta_2 = \pm atan2(sin\theta_2, cos\theta_2) \tag{4}$$

$-\theta_2$ is used in this study considering the motion radius of the leg.
To find θ_1, α, β is

$$\alpha = atan2(y, x) \tag{5}$$

$$\beta = atan\left(\frac{l_2 sin\theta_2}{l_1 + l_2 cos\theta_2}\right) \tag{6}$$

and the theorem of θ_1 is as follows.

$$\theta_1 = atan2(y, x) - atan\left(\frac{l_2 sin\theta_2}{l_1 + l_2 cos\theta_2}\right) \tag{7}$$

2.3 Slope Correction Formula

When a multi-copter lands on rough terrain, the multi-copter tilts depending on the ground conditions. For a stable landing, it is necessary to maintain the level of the multi-copter connected to the landing platform through the control of the landing leg by the inclined angle. To do this, adjust the angle of the Lading Leg based on the current tilt through the IMU sensor to make it horizontal. [5]

To derive the equation, we assume a two-dimensional situation in which the landing platform is tilted by Ø as shown in Fig. 4. In this study, the landing platform was devised to be horizontal if the A point of the body was moved up to the B point to make it horizontal, and the leg end-effector was moved in the opposite direction by the distance the body was moved.

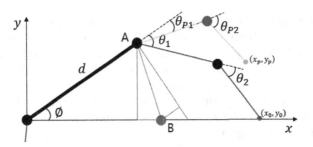

Fig. 4. θ_{P1}, θ_{P2} coordinate system for formula induction

When the length of the landing platform body is d and the length of the link is l_1, l_2, the movement distance of the x, y axis of the body A point is as follows.

$$x_\emptyset = 2dsin^2\left(\frac{\emptyset}{2}\right) \tag{8}$$

$$y_\emptyset = dsin\emptyset \tag{9}$$

The initial position of the leg end-effector in the tilted state can be obtained using Eqs. (1) and (2). In the initial landing position, θ_1 and θ_2 of each leg are known.

$$x_0 = l_1cos\theta_1 + l_2cos(\theta_1 + \theta_2) \tag{10}$$

$$y_0 = l_1sin\theta_1 + l_2sin(\theta_1 + \theta_2) \tag{11}$$

In order to move the leg end-effector by the moving distance of the body, the coordinates moved by the Eqs. (8) and (9) are as follows.

$$x_p = x_0 - x_\emptyset \tag{12}$$

$$y_p = y_0 + y_\emptyset \tag{13}$$

θ_{P1} and θ_{P2} for placing the leg end-effector at x_p, y_p can be calculated using the inverse kinematic Eqs. (4) and (7).

$$\theta_{P2} = \pm atan2(sin\theta_{P2}, cos\theta_{P2}) \tag{14}$$

$$\theta_{P1} = atan2(y_p, x_p) - atan\left(\frac{l_2 sin\theta_{P2}}{l_1 + l_2 cos\theta_{P2}}\right) \tag{15}$$

If the value of each joint of the leg is controlled by θ_{P1} and θ_{P2}, the A point of the body moves to B and is leveled.

3 Experiment

The formula derived in Sect. 2 was applied to the landing platform and the experiment was conducted. STM32F4 MCU(Micro Controller Unit) was used for control, and each joint was equipped with a servo motor of 60kgf.cm. In order to detect the inclination, roll and pitch values are received using one 9DOF EBIMU sensor without using an IMU for each leg. [6] A flexible force sensor is attached to the leg end-effector, and the moment of landing leg contacting the ground is identified through ADC data. [7]

The operation process of the landing platform when the multi-copter lands on the rough terrain is shown in Fig. 5.

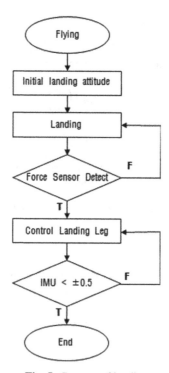

Fig. 5. Process of landing

3.1 Matlab Simulation

Matlab Simulation was conducted to verify the proposed equation. The equation derived in Sectionmulti-copters 2 was applied to the 2-leg landing platform to test it. A situation inclined by 20° in the rough terrain was presented, and the coordinates of the end-effector were moved by the inclined angle to compensate for the tilt. The simulation results are shown in Fig. 6 below. It was confirmed that the coordinates moved by the end-effector were applied to the equation to control the joint angle of the right landing leg to make it horizontal.

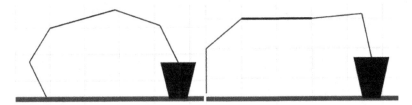

Fig. 6. Matlab simulation

3.2 Landing Platform Test 1

In this paper, we presented the inclined situation as shown in Fig. 7(a) without flying. The 7.5 cm height box was placed on the 3rd and 4th legs, and it was assumed that the landing platform landed in the initial attitude state, and the landing platform tilt was measured.

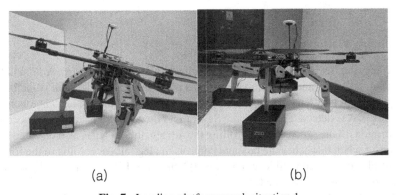

(a)   (b)

Fig. 7. Landing platform rough situation 1

In Fig. 7(a), the roll was measured at about −10.5° and the pitch at about 9.8°. When the signal is transmitted to the MCU, the IMU data and the proposed equation are calculated with a control period of 100 ms and operate until level. Two Legs that operate until the roll reaches zero and two legs that operate until the pitch becomes zero are specified. Priority was given to processing Force Sensor data according to the inclined situation even when in contact with the ground. Among the legs where the force sensor is detected, the leg with the larger slope is first controlled by the formula. As a result, the figure when leveled is as shown in Fig. 7(b).

In the proposed situation, the end-effector coordinates of the landing leg in contact with the rough land were calculated, and the angles of θ_1 and θ_2 of each leg were controlled through the proposed formula. As a result, in Fig. 8, it was confirmed that the roll, pitch converged horizontally to 0.12° and 0.14°, respectively.

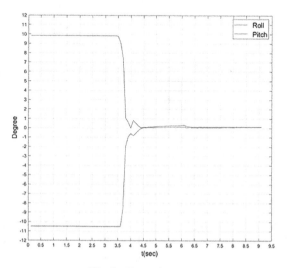

Fig. 8. Test 1 data plot

3.3 Landing Platform Test 2

In Test2, the rougher terrain than Test 1 is assumed as shown in Fig. 9(a). A 1 cm tall box was placed on Leg 1, and a 15 cm tall box was placed on Leg 4. The roll is inclined by 21.2° and the pitch by 8.5°. In 2 boxes with different heights, the 4th leg that touches the higher box first is detected by the Force Sensor and the leg joint is controlled according to the formula. In the process of controlling the 4th leg, the 1st leg is also detected by the force sensor when it touches the box and controls the leg joint according to the formula. Leg control is continued until leveling with 100 ms control cycle. Figure 9(b) shows the picture when the level is aligned.

In Fig. 10, the Roll and Pitch data of the process where the landing platform is leveled are received and plotted. The moment the signal is given to the landing platform, it can be seen that the joint of the landing leg changes according to the formula, and the roll and pitch are 0.05 and 0.12°, respectively.

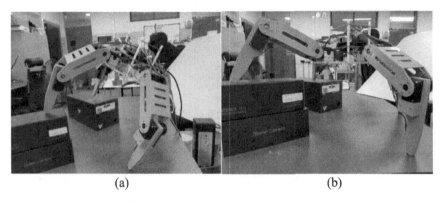

(a)   (b)

Fig. 9. Landing platform rough situation 2

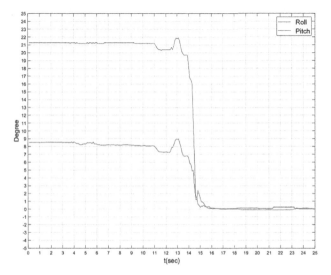

Fig. 10. Test 1 data plot

Figure 9 (a) In the initial attitude of the landing platform, the joints θ_1 and θ_2 of each leg are equal to 47°. In order to level it, joint control of Leg 1 and Leg 2 is required. θ_1 of Leg 1 decreases to $-17°$, and θ_2 of Leg 1 increases to 91° (Fig. 11 a, b). θ_1 of Leg 4 decreases to $-16°$, and θ_2 of Leg 4 increases to 86°. (Figure 11 c, d) According to the proposed formula, the validity of the formula could be confirmed by bending and leveling the leg in contact with the rough terrain.

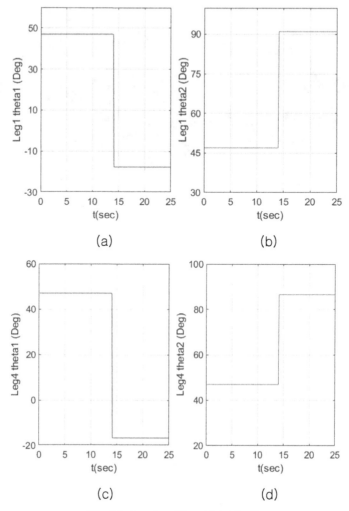

Fig. 11. Leg 1 and Leg 4 joint θ_1, θ_2

4   Conclusions

In this paper, we designed and proposed a landing platform that can maintain a level so that the Multi-copter can stably land when landing on rough terrain. Through the IMU sensor and the Force Sensor, the Multi-copter was detected to be inclined when landing on a rough land, and θ_1 and θ_2 for moving the End-effector of the landing leg were calculated accordingly. The calculated θ_1 and θ_2 were applied to each leg so that the Multi-copter was horizontal. The simulation and experiment were conducted by applying the proposed formula to the produced landing platform, and it was confirmed that it reached the level in a tilted situation. In the future, tests will be conducted to maintain level even in various rough environments, and research will be conducted to improve the accuracy of motion through feedback control.

Acknowledgements. This research was supported by the National Research Foundation of Korea(NRF) grant funded by the Korea government(MSIT) (No. 2019R1A2C2088859)

References

1. Mason, S.: Helicopter self-leveling landing gear, ed: Google Patents (1974)
2. Felder, D.W.: Slope landing compensator system, ed: Google Patents (1977)
3. Baker, S., Soccol, D., Postula, A., Srinivasan, M.V.: Passive landing gear using coupled mechanical design. In: Proceedings of Australasian Conference on Robotics and Automation, pp. 1–8 (2013)
4. Muslimin, S., Istardi, D.: Inverse kinematics analysis for motion prediction of a hexapod robot. In: 2018 International Conference on Applied Engineering (ICAE), pp. 1–5. IEEE (2018)
5. Choi, Y.H., Jung, H.S., Park, J.B.: Posture control of quadruped robot using gyroscope. In: The Korean Institute of Electrical Engineers Conference (2010)
6. Sarkisov, Y.S., Yashin, G.A., Tsykunov, E.V., Tsetserukou, D.: Drone gear: A novel robotic landing gear with embedded optical torque sensors for safe multicopter landing on an uneven surface. IEEE Robot. Autom. Lett. **3**(3), 1912–1917 (2018)
7. Jia, R., Jizhen, W., Xiaochuan, L., Yazhou, G.: Terrain-adaptive bionic landing gear system design for multi-rotor UAVs. In: 2019 Chinese Control And Decision Conference (CCDC), pp. 5757–5762. IEEE (2019)

Determination of Singularity Occurrence and Characteristic Analysis of Dual 6-DOF Manipulator Using Manipulability and Manipulability Ellipsoid

Jong-Hak Lee[1], Jin-Uk Bang[1], and Jang-Myung Lee[2(✉)]

[1] Department of Electrical and Computer Engineering, Pusan National University,
2, Busandaehak-ro 63beon-gil, Geumjeong-gu, Busan, Republic of Korea
{jonghak1696,jinuk1696}@pusan.ac.kr
[2] Department of Electronic Engineering, Pusan National University,
2, Busandaehak-ro 63beon-gil, Geumjeong-gu, Busan, Republic of Korea
jmlee@pusan.ac.kr

Abstract. In this paper, we judge whether the singularity occurs among the path of dual 6-DOF manipulator using manipulability and analyze the characteristics in the velocity space of the straight path and alternative path what avoids the singularity using manipulability ellipsoid. To do this, we solved the inverse kinematics of the manipulator and made simulator in MATLAB. In this simulator, the path can be set in advance and change of manipulability and manipulability ellipsoid can be observed. This allows better path to be determined in advance before following it. Experiments has confirmed that singularity occurs in straight path and eigenvalue in the direction of yaw is zero. In addition, the alternative path is identified as an improved path due to the large eigenvalue in the direction of pitch and yaw. Since this result can be used as an indicator for evaluating the agent's state, we intend to apply to reinforcement learning and research to create a singularity avoidance path by the robot itself.

Keywords: Manipulator · Manipulability · Manipulability ellipsoid · Path analysis

1 Introduction

Various studies have been conducted for robots to behave like humans. Among them, research on the movement path of the manipulator is an important part [1]. Posture variation what occurs in following path is a method in which skillfully acting robot motion and strength capabilities along specific cartesian space in order to perform complex tasks [10]. In addition, because the change of the end effector is a result of a function expressing the change of manipulator joint, so researchers should pay attention to the end effector and its path because it plays an important role in robot manipulation. However, singularity is a problem when researchers study end effector.

© Springer Nature Switzerland AG 2020
C. S. Chan et al. (Eds.): ICIRA 2020, LNAI 12595, pp. 374–386, 2020.
https://doi.org/10.1007/978-3-030-66645-3_32

In singularity where there is no Jacobian inverse matrix, the degree of freedom decreases, causing problems in control [5]. Manipulability can be used to solve the problem of singularity in the path. The singularity of the path can be found by using the property that this value is 0 [6, 7]. It is also used as a useful parameter for analyzing and controlling the characteristics of the end effector in posture deformation, which plays an important role in robot operation [10]. And manipulability ellipsoid is used as an effective tool to analyze the speed of the manipulator end effector [9].

In this paper, manipulability and manipulability ellipsoid calculated from Jacobian are used to determine whether singularity occurs among paths, and to analyze the characteristics of existing and alternative paths in the velocity coordinate system.

2   Kinematics of Manipulator

2.1   Forward Kinematics

First, kinematic analysis of dual 6-DOF manipulators is performed. Figure 1 shows coordinate for each joint, red line is Z-axis, the rotation axis of the motor, and blue line is X-axis.

Fig. 1. Joint coordinates of dual manipulator robot (Color figure online)

When Z-axis and X-axis are specified, Y-axis is automatically set. Complete DH table through the set joint coordinate as shown in Table 1 and Table 2. As shown in Fig. 1, since the two manipulators have a symmetrical structure, the parameters of Table 1 and Table 2 also appear symmetrically.

When one row in the Table 1 and Table 2 represents shift the coordinate from i-th joint to (i + 1)-th joint, and this is expressed as T_{i+1}^i, homogenous matrix for one manipulator is as Eq. (1). E.E. means End-Effector.

$$T_2^1 T_3^2 T_4^3 T_5^4 T_6^5 T_{E.E.}^6 = T_{E.E.}^1 \tag{1}$$

In other words, the manipulator calculates kinematics from the 1st joint. On the other hand, reference coordinate of robot is the coordinate what represents XYZ axis

Table 1. DH table of right manipulator

	θ	d	a	α
1	θ_1	25.2	0	90°
2	θ_2	0	0	90°
3	θ_3	−25.1	0	−90°
4	θ_4	0	0	90°
5	θ_5	−22.7	0	−90°
6	θ_6	14	0	−90°

Table 2. DH table of left manipulator

	θ	d	a	α
1	θ_1	25.2	0	−90°
2	θ_2	0	0	−90°
3	θ_3	−25.1	0	90°
4	θ_4	0	0	−90°
5	θ_5	−22.7	0	90°
6	θ_6	14	0	90°

in the lower part of Fig. 1. Therefore, to calculate the manipulator kinematics based on the reference coordinate of robot, Eq. (2) and Eq. (3) should be multiplied before the kinematics of right and left manipulator respectively.

$$T_1^{ref.} = Tr_z(96) * R_z(-90°) * R_x(90°) * R_z(-90°) \tag{2}$$

$$T_1^{ref.} = Tr_z(96) * R_z(-90°) * R_x(-90°) * R_z(90°) \tag{3}$$

Meanings of Tr and R in expressions Eq. (2) and (3) are shown in Table 3.

Finally, total homogenous matrix of each manipulator based on reference frame of robot is shown as Eq. (4).

$$T_1^{ref.} T_2^1 T_3^2 T_4^3 T_5^4 T_6^5 T_{E.E}^6 = T_{E.E.}^{ref.} \tag{4}$$

2.2 Inverse Kinematics

Inverse kinematics is essential to command posture of robot from human perspective. Inverse kinematics is opposite of forward kinematics by using cartesian space information to find joint space information.

Table 3. Meanings of expression in Eq. (2) and Eq. (3)

Expression	Meaning	Matrix
$Tr_z(t)$	Transport by t along z-axis	$\begin{bmatrix} 1 & 0 & 0 & 0 \\ 0 & 1 & 0 & 0 \\ 0 & 0 & 1 & t \\ 0 & 0 & 0 & 1 \end{bmatrix}$
$R_z(\theta)$	Rotate by θ on z-axis	$\begin{bmatrix} cos\theta & -sin\theta & 0 & 0 \\ sin\theta & cos\theta & 0 & 0 \\ 0 & 0 & 1 & 0 \\ 0 & 0 & 0 & 1 \end{bmatrix}$
$R_x(\theta)$	Rotate by θ on x-axis	$\begin{bmatrix} 1 & 0 & 0 & 0 \\ 0 & cos\theta & -sin\theta & 0 \\ 0 & sin\theta & cos\theta & 0 \\ 0 & 0 & 0 & 1 \end{bmatrix}$

To solve Inverse kinematics, change Eq. (4) to Eq. (5).

$$T_4^3 T_5^4 = \left(T_1^{ref.} \cdot T_2^1 T_3^2\right)^{-1} T_{E.E.}^{ref.} \cdot (T_6^5 T_{E.E}^6)^{-1} \tag{5}$$

Equation (5) is the simplest form of solving inverse kinematics among the possible variations of equation.

Organizing (2, 4) components of both sides of Eq. (5) with respect to θ_1, Eq. (7) can be obtained. θ_6 is left unknown. c_6 represents $cos\theta_6$, and s_6 represents $sin\theta_6$ and n, o, a, p is components of $T_{E.E.}^{ref.}$, it is shown in Eq. (6).

$$T_{E.E.}^{ref.} = \begin{bmatrix} n_x & o_x & a_x & p_x \\ n_y & o_y & a_y & p_y \\ n_z & o_z & a_z & p_z \\ 0 & 0 & 0 & 1 \end{bmatrix} \tag{6}$$

$$\theta_1 = tan^{-1}\frac{-p_y + 14a_y + 22.7(a_yc_6 - n_ys_6)}{96 - p_z + 14a_z + 22.7(a_zc_6 - n_zs_6)} \tag{7}$$

By organizing the components (1, 4) of Eq. (5) in same way, equation for θ_2 can be obtained as Eq. (8)

$$\theta_2 = tan^{-1}\frac{25.2 - 14a_x + p_x - 22.7a_xc_6 + 22.7n_xs_6}{96c_1 + 22.7c_1(a_zc_6 - n_zs_6) + 22.7s_1(a_yc_6 - n_ys_6) + 14(a_zc_1 + a_ys_1) - p_zc_1 - p_ys_1} \tag{8}$$

By organizing the components (3, 2) of Eq. (5) in same way, equation for θ_5 can be obtained as Eq. (9)

$$\theta_5 = tan^{-1}\frac{o_x s_2 - o_z c_1 c_2 - o_y s_1 c_2}{-c_2\left(n_z c_1 c_6 + a_y s_1 s_6 + a_z c_1 s_6 - n_y s_1 c_6\right) + s_2(a_x s_6 + n_x c_6)} \tag{9}$$

Organizing the component (1, 2) of Eq. (5) for $sin\theta_3$, and component (2, 2) for $cos\theta_3$. Then by dividing these two, Eq. (10) for θ_3 can be obtained.

$$\theta_3 = tan^{-1}\frac{o_x c_2 c_5 - c_2 s_5(n_x c_6 + a_x s_6) + s_2 c_5(o_y s_1 + o_z c_1) - s_2 s_5(a_z c_1 s_6 + n_y s_1 c_6 + n_z c_1 c_6 + a_y s_1 s_6)}{s_5(n_y c_1 c_6 + a_y c_1 s_6 - a_z s_1 s_6 - n_z s_1 c_6) + c_5(o_z s_1 - o_y c_1)} \tag{10}$$

Organizing the component (3, 1) of Eq. (5) for $sin\theta_4$, and component (3, 3) for $cos\theta_4$. Then by dividing these two, Eq. (11) for θ_4 can be obtained.

$$\theta_4 = tan^{-1}\frac{o_x s_2 s_5 - s_2 s_5(n_x c_6 + a_x s_6) + c_2 s_5(o_y s_1 + o_z c_1) - c_2 c_5(a_z c_1 s_6 + n_y s_1 c_6 + n_z c_1 c_6 + o_y s_1 s_6)}{c_2(a_y s_1 s_6 + n_y s_1 s_6 + a_z c_1 c_6 - n_z c_1 s_6) + s_2(n_x s_6 - a_x c_6)} \tag{11}$$

To determine θ_6, calculate homogenous matrix from x, y, z, roll, pitch, yaw information firstly what user input. Let call this H1. Then substitute the angles in order from 0 to $\frac{\pi}{2}$ in θ_6 and calculate inverse kinematics using Eq. (7) to (11). After that make homogeneous matrix using this inverse kinematics result. Let call this H2. θ_6 which makes H2 equal to or closest to H1, is define as correct θ_6.

3 Parameters and Singularity

3.1 Determination of Singularity

Velocity of joint space of manipulator is $\dot{\theta}$ what contains $\dot{\theta}_1 \sim \dot{\theta}_6$ as elements, and velocity of cartesian space is $\dot{X}$ what contains $\dot{x}, \dot{y}, \dot{z}, \dot{\phi}, \dot{\vartheta}, \dot{\psi}$.

$\dot{x}, \dot{y}, \dot{z}$ means x, y, z linear velocity and $\dot{\phi}, \dot{\vartheta}, \dot{\psi}$ means roll, pitch, yaw rotation velocity. $\dot{\theta}$ and $\dot{X}$ have relationship as Eq. (12), $J(\theta)$ is called Jacobian.

$$\dot{X} = J(\theta)\dot{\theta}$$

To find joint velocity vector corresponding to given end effector velocity vector, differential kinematic inverse solution of Eq. (12) is expressed as a below

$$\dot{\theta} = J(\theta)^{\dagger}\dot{X} \tag{12}$$

where $J(\theta)^{\dagger}$ is $(m \times n)$ Moore-Penrose pseudoinverse Jacobian matrix. If $n = m$, $J(\theta)^{\dagger}$ is replaced by inverse Jacobian matrix representing $J(\theta)^{-1}$.

To understand the concept of kinematic singularity, consider the singular value decomposition of $J(\theta)$ at certain point of path that can be written as a below

$$J(\theta) = U\Sigma V^T = \sum_{i=1}^{r}\lambda_i u_i v_i^T \tag{13}$$

where U is $(m \times m)$ orthonormal matrix of output singular vectors u_i, V is the $(n \times n)$ orthonormal matrix of input singular vector v_i, and $\Sigma = (S\ O)$ is the $(m \times n)$ matrix whose $(m \times m)$ diagonal submatrix S contains the singular values λ_i of the matrix $J(\theta)$ [11].

At kinematic singularity, r which is rank of $J(\theta)$, becomes $r < m$ and $(m - r)$ singular values become zero [11]. This means that end effector loses its feasible velocity capabilities which is related to Sect. 3.2. As a result, if one or more singular values are 0 at a certain point of path, it can be determined that singularity has occurred, and Jacobian is of non-full rank.

The kinematic singularity configuration occurs when axes of rotation are aligned. There are typically wrist, elbow and shoulder singularity. Wrist singularity occurs when the rotation axes of the two wrist joints of the manipulator are aligned, elbow singularity occurs when the rotation axis of the wrist joint is located in the plane consisting of the rotation axes of joint2 and joint3, and the shoulder singularity occurs when the rotation axes of joint1 and rotation axes of other joint are aligned [12, 13]. Each case is shown in Fig. 2 and joints that satisfy the singularity condition are marked in red.

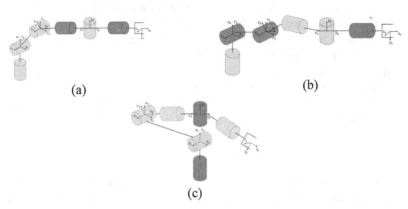

(a)   (b)

(c)

Fig. 2. (a) Wrist singularity configuration at joint 4 and 6 (b) Elbow singularity configuration at joint 2, 3 and 6 (c) Shoulder singularity configuration at joint 1 and 6 (Color figure online)

3.2 Manipulability

Yoshikawa defined extent which end effector can change position and orientation in the current configuration as manipulability measure and is also called manipulability. The expression Manipulability is as shown in Eq. (15) [6].

$$M(\theta) = \sqrt{\det\left(J(\theta)J^T(\theta)\right)} \tag{15}$$

However, in the case of non-redundant manipulator that has no marginal degree of freedom, it is as shown in Eq. (16).

$$M(\theta) = |\det J(\theta)| \tag{16}$$

Sine 6-DOF manipulator used in this thesis, Eq. (16) is used as manipulability. Because Yoshikawa said Eq. (15) is same as $\prod_{i=1}^{k} \lambda_i^2$ [6], it is trivial Eq. (16) same as $\prod_{i=1}^{k} \lambda_i$. So we calculate manipulability as Eq. (17).

$$M(\theta) = \prod_{i=1}^{k} \lambda_i \tag{17}$$

Since $\lambda_1 \sim \lambda_6$ match $\dot{x}, \dot{y}, \dot{z}, \dot{\phi}, \dot{\vartheta}, \dot{\psi}$ in order, for example, when λ_3 becomes 0, it means that velocity vector in z direction is 0, so it cannot move in z direction. This is related to manipulability that follows.

3.3 Singularity and Manipulability Ellipsoid

There is a point in the trajectory that cannot be reached due to the limitation of joint angle or shape of the manipulator. This point is called singularity, and if the singularity is included in the trajectory, the manipulator may be malfunctioning and cause trajectory defects at an inappropriate velocity [5].

In the formula, $J(\theta)^{-1}$ does not exist, which means $|\det J(\theta)| = 0$. Therefore, connect with Eqs. (16) and (17), singularity is the point where one or more of the eigenvalues of $J(\theta)$ become zero. This means that the velocity of the joint space does not propagate completely into the cartesian space.

Manipulability ellipsoid is a set of velocity vectors that end effector can have. It has the eigenvalue of $J(\theta)$ as ellipsoid's axis radius and ellipsoid closer to sphere, it has better velocity propagation $\dot{\theta}$ to $\dot{X}$. Eigenvalues can be obtained by solving singular value decomposition of $J(\theta)$, and eigenvalues are diagonal component of S in Eq. (18).

$$[U \ S \ V^T] = J(\theta) \tag{18}$$

Since $\dot{X}$ has components as $[\dot{x}, \dot{y}, \dot{z}, \dot{\phi}, \dot{\vartheta}, \dot{\psi}]^T$, the manipulability ellipsoid exists in the 6 dimensional coordinate, but for convenience, think separately as linear velocity coordinate $[x, y, z]^T$, angular velocity coordinate $[\dot{\phi}, \dot{\vartheta}, \dot{\psi}]^T$. Figure 3 is example of the manipulability ellipsoid in the $[x, y, z]^T$ coordinate.

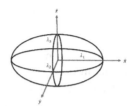

Fig. 3. Manipulability ellipsoid

If λ_3 becomes 0, the radius in $\dot{z}$ direction becomes 0, and manipulability ellipsoid becomes 2-dimensional ellipse.

4   Simulation

4.1   Simulator

Simulator is constructed using MATLAB to check the characteristics of the manipulator while following path in advance. The simulator outputs configuration of manipulator by directly input the joint angle or adjusting the joint angle with the slider and also outputs information such as the position and roll, pitch, yaw of the end effector, manipulability, and eigenvalue and so on (Fig. 4).

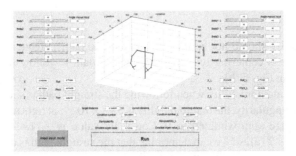

Fig. 4. MATLAB simulator

4.2   Analysis of Path

A straight path what start point $(-33.00, -48.99, 99.52)$ and end point $(-33.00, -38.97, 50.81)$ from top to bottom is set as shown in Fig. 5 to carry rod down. As a result of calculating the manipulability of all points on the path, the value became 0 at $(-33.00, -39.21, 52.15)$. It means that point is singularity and it shown as red point in Fig. 6 (a). Because $\theta_4 = 0$ in singularity, joint3 and 5 are aligned. Therefore, it is wrist singularity among the cases described in Sect. 3.1. Referring to motor arrangement direction shown in Fig. 1, we can see that elbow and shoulder singularity can't be come out.

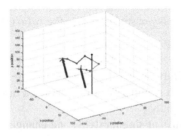

Fig. 5. Straight path

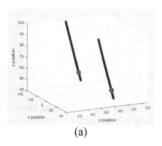 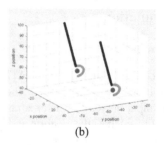

(a)   (b)

Fig. 6. (a) Occurrence of singularity in original path (b) Singularity and changing path (Color figure online)

Since singularity should not be included among the paths, create semicircle in the direction of progress centering on singularity, and a path to follow the semicircle is created as shown in Fig. 6 (b).

The change in manipulability when following the straight path is shown in Fig. 7 (a) on a log scale. At singularity, the manipulability is 3.48×10^{-12} at right arm, 6.09×10^{-12} at left arm what are approximated to 0. Log scale value is -11.458 and -11.215, respectively, in Fig. 7 (a), so we can determine that the straight path is a bad path. Since both arms follow a symmetrical path, the manipulability is almost similar, so the graphs overlap.

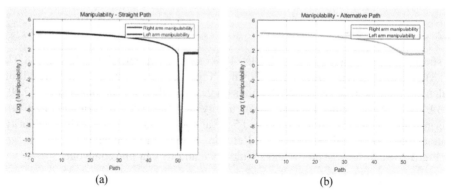

(a)   (b)

Fig. 7. (a) Log scale manipulability of straight path in Fig. 6. (a) (b) Log scale manipulability of straight path in Fig. 6. (b)

The result of manipulability when following path of Fig. 6 (b) is shown in Fig. 7 (b). Because it moved away from singularity, it maintained a large manipulability without manipulability decreased rapidly.

Table 4 shows eigenvalues at singularity(S) and changing point(C) what has same z position with singularity in changing path. It indicates the extent to which angular velocity is propagated to cartesian velocity.

In singularity, we can see that λ_6 becomes 0 and λ_5 also has a value close to 0. This means that even if the joint moves, it cannot rotate in the yaw direction. In addition,

Table 4. Eigenvalues at singularity (S) and changing point (C)

Point		λ_1	λ_2	λ_3	λ_4	λ_5	λ_6
S	R	73.2	61.8	1.42	0.81	0.01	0.00
	L	73.2	61.8	1.42	0.81	0.01	0.00
C	R	73.2	61.8	1.43	0.91	0.18	0.14
	L	73.2	61.8	1.43	0.91	0.18	0.14

if the joint is moved a lot, end effector can be rotated very little in pitch direction. Next, looking at the eigenvalue of the change point, λ_6 value increased to 0.14 and λ_5 also increased to 0.18 to deviate from singularity. It means that pitch and roll rotation of the end effector can be made easier than in the singularity.

Figure 8 shows the output of the velocity space manipulability ellipsoid at singularity, and it looks like a line segment because the λ_6 is zero and λ_5 is very small. Because the position and orientation of two end effectors are symmetric so manipulability ellipsoid also came out symmetrically. Therefore, only ellipsoid of right end effector is represented as a representative.

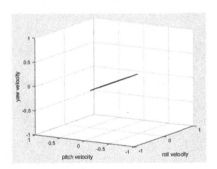

Fig. 8. Manipulability ellipsoid at singularity

For a closer look, projection of ellipsoid on the roll-pitch plane are shown in Fig. 9 (a), and projection ellipsoid on the pitch-yaw plane are shown in Fig. 9 (b).

In Fig. 9 (a) and (b), the end effector can have a velocity in the roll direction and the pitch direction, but not yaw direction, so the ellipsoid becomes a two-dimensional ellipse with a small radius in the pitch direction.

The manipulability ellipsoid at the changing point is shown in Fig. 10. As it deviated from the singularity, it has three-dimensional ellipsoid shape, and the ellipsoid projection on the planes shown in Fig. 11 (a) and (b) also show that pitch and yaw velocity are improved compared to Fig. 9 (a) and (b) (Fig. 10).

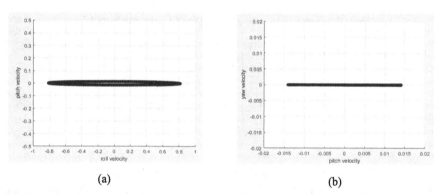

(a)   (b)

Fig. 9. (a) Projection of Fig. 9 on roll-pitch velocity plane (b) Projection of Fig. 9 on pitch-yaw velocity plane

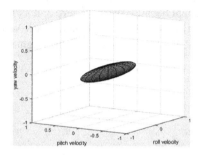

Fig. 10. Manipulability ellipsoid at changing point

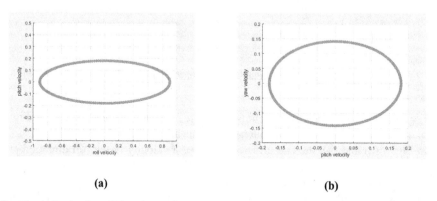

(a)   (b)

Fig. 11. (a) Projection of Fig. 10 on roll-pitch velocity plane (b) Projection of Fig. 10 on pitch-yaw velocity plane

5 Conclusion

In this paper, we used the characteristics of manipulability becoming zero when the manipulator falls into singularity to determine whether there is Singularity in the path and conduct analysis of the path what has singularity and avoids it.

The presence or absence of singularity in the path can be confirmed by finding the point where the value becomes 0 by calculating manipulability for each path point.

The singularity occurrence point is found in the end effector's straight path in the experiment, and changing path is designed to avoid singularity by drawing a semicircle in the direction of progress near the singularity. In addition, through the manipulability ellipsoid, the characteristic of singularity is examined, and by examining the characteristics of the changing path. As a result, it is confirmed that the changing path is better path than straight path by path analysis.

We plan to research the path planning algorithm that avoids singularity by using the analysis in this paper as an index to evaluate the state of the agent by applying it to reinforcement learning.

Acknowledgement. This research is based upon work supported by the Ministry of Trade, Industry & Energy (MOTIE, Korea) under Industrial Technology Innovation Program. No. 10073147

References

1. Kim, D.-E., Park, D.-J., Park, J.-H., Lee, J.-M.: Collision and singularity avoidance path planning of 6-DOF dual-arm manipulator. In: Chen, Z., Mendes, A., Yan, Y., Chen, S. (eds.) ICIRA 2018. LNCS (LNAI), vol. 10985, pp. 195–207. Springer, Cham (2018). https://doi.org/10.1007/978-3-319-97589-4_17
2. Liu, X., Zhao, F., Liu, B., Mei, X.: Multi-point interaction force estimation for robot manipulators with flexible joints using joint torque sensors. In: Yu, H., Liu, J., Liu, L., Ju, Z., Liu, Y., Zhou, D. (eds.) ICIRA 2019. LNCS (LNAI), vol. 11742, pp. 499–508. Springer, Cham (2019). https://doi.org/10.1007/978-3-030-27535-8_45
3. Xu, Y., Hu, J., Zhang, D., Yao, J., Zhao, Y.: A five-degree-of-freedom hybrid manipulator for machining of complex curved surface. In: Huang, Y.A., Wu, H., Liu, H., Yin, Z. (eds.) ICIRA 2017. LNCS (LNAI), vol. 10463, pp. 48–58. Springer, Cham (2017). https://doi.org/10.1007/978-3-319-65292-4_5
4. Zheng, W., Chen, M.: Tracking control of manipulator based on high-order disturbance observer. IEEE Access **6**, 26753–26764 (2018)
5. Reboucas Filho, P.P., Da Silva, S.P.P., Praxedes, V.N., Hemanth Jude, De Albuquerque V.H.C.: Control of singularity trajectory tracking for robotic manipulator by genetic algorithms. J. Comput. Sci. **30**, 55–64 (2019)
6. Yoshikawa, T.: Analysis and control of robot manipulators with redundancy. In: Robotic Research. The First International Syposium, pp. 735–747 (1984)
7. Merlet, J.P.: Jacobian, manipulability, condition number, and accuracy of parallel robots. ASME J. Mech. Des. **128**(1), 199–206 (2005)
8. Vahrenkamp, N., Asfour, T., Metta, G., Sandini, G., Dillmann, R.: Manipulability analysis. In: 2012 12th IEEE-RAS International Conference on Humanoid Robots, Osaka, pp. 568–573 (2012)

9. Chiacchio, P.: A new dynamic manipulability ellipsoid for redundant manipulators. Robotica **18**(4), 381–387 (2000)
10. Rozo, L., Jaquier, N., Calinon, S., Caldwell, D.G.: Learning manipulability ellipsoids for task compatibility in robot manipulation. In: 2017 IEEE/RSJ International Conference on Intelligent Robots and Systems, Vancouver, BC, pp. 3183–3189 (2017)
11. Chiaverini, S.: Singularity-robust task-priority redundancy resolution for real-time kinematic control of robot manipulators. IEEE Trans. Robot. Autom. **13**(3), 398–410 (1997)
12. Nakai, K., et al.: Control of robot in singular configurations for human-robot coordination. In: Proceedings. 11th IEEE International Workshop on Robot and Human Interactive Communication. IEEE (2002)
13. Robohub Homepage, https://robohub.org/3-types-of-robot-singularities-and-how-to-avoid-them. Accessed 27 Sept 2020

Rapid Actuation for Soft Pneumatic Actuators Using Dynamic Instability Mechanism

Xin-Yu Guo, Wen-Bo Li, and Wen-Ming Zhang[✉]

State Key Laboratory of Mechanical System and Vibration, School of Mechanical Engineering,
Shanghai Jiao Tong University, 800 Dongchuan Road, Shanghai 200240, China
wenmingz@sjtu.edu.cn

Abstract. The response speed of soft pneumatic actuators is usually limited by the intrinsic material and unstressed stable form. This paper describes a design strategy for improving the response speed of pneumatic actuators by employing the dynamic instability mechanism of the actuator. An adjustable elastic cord is incorporated into the pneumatic actuator to form the snapping pneumatic actuator (SPA). Furthermore, the bottom layer of the pneumatic actuator is accompanied by a higher elasticity modulus to improve the recovery performance of SPA. In the inflation and deflation processes, the torque reversal triggering occurs due to the instability of SPA, releasing the elastic energy stored in elastic cord quickly. The dynamic instability mechanism of SPA improves the response speed and amplifies the performance of SPA. Theoretical modeling and experiments are implemented to reveal the fast response characteristics of SPA. Experimental results show that the maximum inflation bending angular velocity of SPA reaches 41.88 rad/s, the maximum deflation bending angular velocity is 36.65 rad/s. The saltation that occurs multiple times due to the application of pre-loading force increases the maximum response speed of SPA by more than 9 times.

Keywords: Dynamic instability · Rapid actuation · Pneumatic actuator

1 Introduction

The large deformation ability of elastomeric materials enables the design of the soft actuators that achieve complex motion and interact well with the environment [1]. However, drawbacks still exist under certain circumstances wherein rapid response is required to deform and move fast [2, 3]. A number of approaches to regulate the response speed of soft actuators have been proposed recently. Inspired by flea, a catapult mechanism is proposed to realize the fast jumping robots [4]. This mechanism stores a large amount of elastic energy and release it quickly by torque reversal triggering. Furthermore, the instability caused by the round shell structure improves the performance of the actuator [5]. Then, a bistable jumping mechanism that use elastic components to store elastic energy to jump is proposed [6]. Energy generated by the actuator is gradually stored as elastic energy in a spring, which when triggered instantly releases the stored energy and enables jumping. A surface tension-dominated jumping robotic insect also uses

© Springer Nature Switzerland AG 2020
C. S. Chan et al. (Eds.): ICIRA 2020, LNAI 12595, pp. 387–397, 2020.
https://doi.org/10.1007/978-3-030-66645-3_33

the stored elastic energy of the spring to achieve high-performance resilience motion [7]. Moreover, utilizing the characteristics of bistable structure to store elastic energy, a simple design principle for an untethered, soft swimming robot with preprogrammed, directional propulsion without battery or onboard electronics is proposed [8].

The elastic energy storage structure is also suitable for pneumatic actuators which response relatively slow due to the long inflation and deflation time. Johannes et al. employ the harnessing snap-through instabilities to amplify the response of soft actuators [9]. Then, the prestressed elastomeric layer and spring are employed to achieve a high-speed recovery due to the release of the elastic energy stored by soft actuators [10–13]. What's more, the bistable structure can also improve the performance of the pneumatic actuators. Tang et al. demonstrate a generic design principle to realize high-speed locomotion robots by employing leveraging elastic instabilities of the spine mechanism [14]. However, actuators that use elastic energy storage either use bistable state which the requires external driving or only improve one-way response speed.

In this paper, a new design strategy for improving the inflation and deflation response speed of pneumatic actuators by employing the dynamic instability mechanism of the actuator is proposed. The elastic cord with pre-loading force and the pneumatic actuator with a larger elasticity modulus at the bottom layer form a dynamic unstable actuator. During the deformation of SPA, the elastic cord and the actuator release elastic energy quickly multiple times due to the torque reversal triggering, improving the response speed of the actuator. This design strategy can increase the deformation speed and amplify the performance of the all-flexible actuator during the inflation and deflation process.

2   Design and Working Principle

Figure 1(a) shows the conceptual structure of SPA. It consists of two components: a Pneu-Nets bending actuator and an elastic element (rubber elastic cord with a diameter of 2 mm). The elastic cord is pre-tensioned to store the elastic potential energy to improve the response speed and amplify the performance of SPA. The components that make up the SPA are all flexible parts, maintaining the flexibility of the pneumatic actuator. The bottom layer of pneumatic actuator is made of silicone with a larger elasticity modulus (E_2) to store elastic potential energy, while the top layer uses silicone with a smaller elasticity modulus (E_1) to reduce the air pressure required for deformation.

As shown in Fig. 1(b), in the initial position, the air chambers of the actuator are pressed together under the tension of the elastic cord, which causes the initial bending state of the actuator. When the compressed air is input to the SPA, the air chambers expand and squeeze each other, the air pressure resists the pretension of the elastic cord, resulting in a stretch and wide range of the bending angle change of SPA. State I indicates that the SPA is stretched to a straight state. When the SPA continues to be inflated and bend beyond state I, the torque reversal occurs, and the elastic potential energy stored in the elastic rope drives the SPA to bend quickly until it reaches the extreme position. Since the bottom layer of the pneumatic actuator has a large elasticity modulus, the pneumatic actuator especially the bottom layer stores a large amount of elastic potential energy when the SPA is at the extreme position. When the SPA is deflated, the extreme position of SPA is unstable because the elastic potential energy stored in the pneumatic

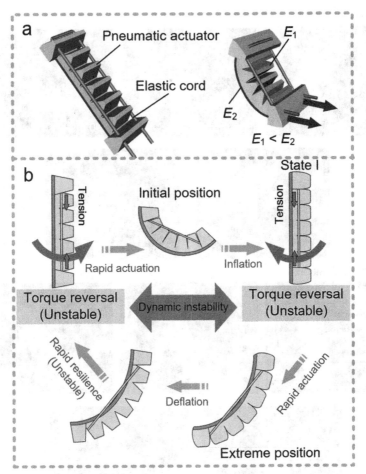

Fig. 1. Design and working principle of SPA. (a) Conceptual structure design of SPA. (b) Working principle and deformation process of SPA.

actuator forces the SPA to rebound to achieve a rapid recovery. After SPA rebounds beyond the straight state, the torque reversal occurs again, elastic potential energy stored in the elastic rope results in the dynamic instability of SPA, improving the performance of SPA. Dynamic instability in both of the inflated bending and deflated resilience processes improves the response speed of SPA.

The comparison of response speeds of SPA under the pre-loading force of 0 N and 2 N is shown in Fig. 2. As shown in Fig. 2(a), the SPA under 2 N pre-loading force needs 490 ms to reach the extreme position from initial position, as shown in Fig. 2(b), the SPA without pre-loading force needs 470 ms. However, the SPA under 2 N pre-loading force achieves a larger bending angle of approximately 330°, the average bending angular velocity is about 11.75 rad/s, the SPA without pre-loading force only reaches about 4.65 rad/s.

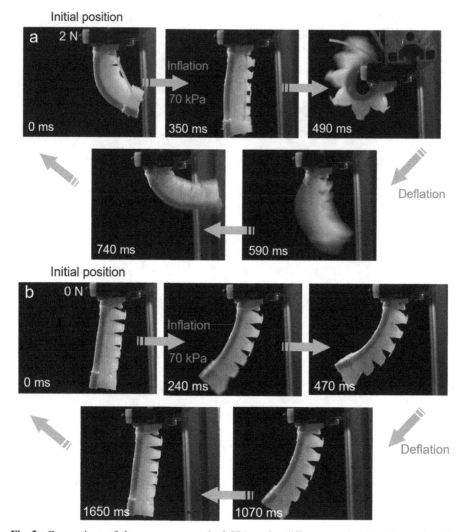

Fig. 2. Comparison of the response speed of SPA under different pre-loading forces. (a) The response process of SPA under 2 N pre-loading force. (b) The response process of SPA without pre-loading force.

During the response process, SPA under 2 N pre-loading force takes 250 ms to reach the initial position, achieving a large bending at an average angular velocity of 20.94 rad/s. But the SPA without pre-loading force takes 1180 ms, the average angular velocity is about only 2.54 rad/s. Therefore, the dynamic instability of SPA not only increases the bending angle range, but also greatly improves the response speed of SPA.

3 Modeling

When the actuator is inflated, the air chambers squeeze each other to overcome the elastic force of the elastic cord. In order to realize the complete deformation process of SPA, sufficient air pressure is needed to realize the bending action of the actuator. Therefore, a model is established to predict the relationship between the internal pressure of SPA and the bending angle when the actuator is between initial position and state I.

Some assumptions are made to simplify the model and neglect the minor factors: (1) the squeezing force between the air chambers overcomes the elasticity of the elastic rope; (2) all air chambers are deformed uniformly and produce equal pressure. As shown in Fig. 3(a), it can be concluded from the geometric relationship that the angle θ_F between the extrusion force F_P acting on the terminal air chamber and the pre-loading force F exerted by the elastic cord is

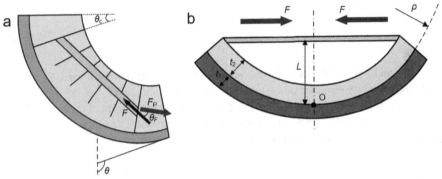

Fig. 3. Modeling and analysis of SPA. (a) The static mechanical analysis of SPA. (b) The resilience performance analysis of SPA.

$$\theta_F = \frac{\theta}{2} - \theta_c \tag{1}$$

where θ is the bending angle of SPA, θ_c denotes the angle between the side walls of an air chamber. Therefore, the relationship between the extrusion force F and the pre-loading force F_P can be written as

$$F_p = F \cdot \cos\left(\frac{\theta}{2} - \theta_c\right) \tag{2}$$

Since the extrusion force F_P between the air chambers is generated by the internal air pressure of the actuator, F_P can be written as

$$F_p = A \cdot P \tag{3}$$

where A represents the contact area between air chambers of the actuator, P is the internal pressure of the actuator. According to Eqs. (1)–(3), the relationship between the internal pressure P of SPA and the bending angle θ is

$$P = \frac{F}{A} \cos\left(\frac{\theta}{2} - \theta_c\right) \tag{4}$$

In addition, the SPA in the extreme position requires the elastic force of the actuator to overcome the pretension force exerted by the elastic rope to achieve resilience. The theoretical model is established to analyze the resilience performance of SPA. To simplify the model, the bottom and top layers of the actuator are respectively simplified into beams with different thickness and elasticity modulus, as shown in Fig. 3(b). What's more, the contact surface of the double-layer beam is regarded as the neutral layer of the composite beam [15]. The spring-back moment of the layers of SPA due to bending can be written as

$$M_1 = \frac{E_1 \cdot I_1}{\rho + \frac{t_1}{2}} \tag{5}$$

$$M_2 = \frac{E_2 \cdot I_2}{\rho + \frac{t_2}{2}} \tag{6}$$

where M_1 and M_2 are the bending moments of the top and bottom layers, E_1 and E_2 denote the elasticity modulus of the top and bottom layers, I_1 and I_2 are the moments of inertia of the top and bottom layers, ρ denotes the radius of curvature of the neutral layer, t_1 and t_2 present the thickness of the top and bottom layers. The torque M_F produced by the pre-loading force F on SPA can be written as

$$M_F = F \cdot L \tag{7}$$

where L is the force arm of the pre-loading force about the center point O of the neutral plane. According to the Eqs. (5)–(7), the condition for SPA to achieve resilience is

$$\frac{E_1 \cdot I_1}{\rho + \frac{t_1}{2}} + \frac{E_2 \cdot I_2}{\rho + \frac{t_2}{2}} > F \cdot L \tag{8}$$

4 Experiments and Discussion

4.1 Bending Angle Test

As shown in Fig. 4, the bending angle of the inflation and deflation processes of SPA under different pre-loading forces and air pressures is compared. Figure 4(a, b) show that the inflation response speed of SPA without pre-loading force increases with the air pressure. However, in the deflation process of the SPA, the time-varying slopes of the bending angle under different pressures are the same, which means that the deflation response speed of SPA without pre-loading force is not greatly affected by the air pressure. The deflation speeds of SPA without pre-loading force are basically the same. However, as the pre-loading force increases to 1 N, SPA under 40 kPa pressure cannot reach state I, as shown in Fig. 4(c, d). The inflation response speed of SPA with higher pressure

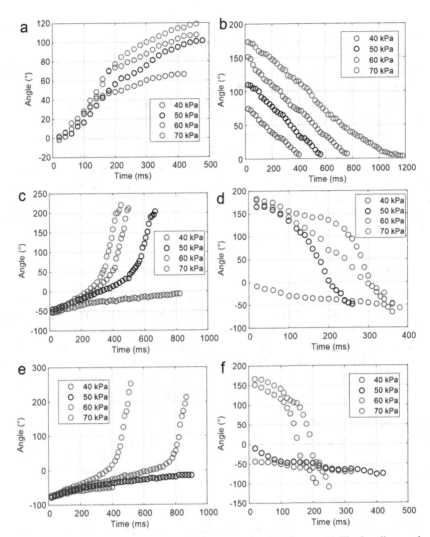

Fig. 4. The bending angle of SPA under different pre-loading forces. (a) The bending angle of inflated SPA under 0 N pre-loading force. (b) The bending angle of deflated SPA under 0 N pre-loading force. (c) The bending angle of inflated SPA under 1 N pre-loading force. (d) The bending angle of deflated SPA under 1 N pre-loading force. (e) The bending angle of inflated SPA under 2 N pre-loading force. (f) The bending angle of deflated SPA under 2 N pre-loading force.

also increases with the air pressure, which is faster than that of SPA without pre-loading force. As the pre-loading force increases to 2 N, the SPA under the pressure of 40 kPa and 50 kPa cannot reach state I. As shown in Fig. 4(e, f), the inflation and deflation response speed of SPA increases with the air pressure, which is faster than that of SPA with smaller pre-loading force. The results show that the inflation and deflation response speed of SPA increase with the pre-loading force. Therefore, in addition to the positive influence of air pressure on the response speed, the unstable state caused by the pre-loading force also improves the response speed of the SPA. At the same time, the pre-tightening force greatly increases the bending angle range of SPA. Although the pressure has an effect on the response speed of the SPA, the effect is small compared to the pre-loading force.

What's more, as shown in Fig. 4(c–f), the response speed of SPA after torque reversal triggering is greatly improved due to the rapid release of elastic energy. As shown in Fig. 4(e, f), under the action of 70 kPa pressure and 2 N pre-loading force, the inflation response angular velocity of SPA between 400 ms and 500 ms can reach 41.88 rad/s, the deflation response angular velocity of SPA between 150 ms and 250 ms can reach 36.65 rad/s. The dynamic instability mechanism not only shortens the response time of SPA, but also increases its bending angle range, which further improves the response speed of SPA. The dynamic instability mechanism due to the application of pre-loading force increases the maximum response speed of SPA by more than 9 times. In general, the dynamic instability mechanism greatly improves the response speed and performance of the SPA.

4.2 Response Speed

To compare the effect of pre-loading force on the response speed of SPA more intuitively, the total response time and response speed of SPA under different pre-loading forces are illustrated in the histogram. As shown in Fig. 5(a), the total response time of SPA with 70 kPa decreases with the pre-loading force. The response time of SPA with pre-loading force is much shorter than that without pre-loading force, but the response time of SPA with 1 N pre-loading force is relatively close to that of 2 N. Moreover, as shown in Fig. 5(b), the inflation speed and deflation response speed of SPA both increase with the pre-loading force. The results indicate that the application of pre-loading force greatly increases the inflation and deflation response speed and bending angular velocity of SPA, thereby greatly reducing the total response time. Due to the snapping actuation caused by the dynamic instability mechanism, the response speed of the SPA with pre-loading force is significantly higher than that without pre-loading force. However, achieving

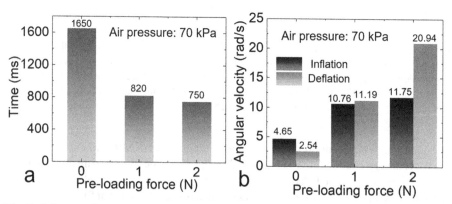

Fig. 5. Influence of pre-loading force on response performance of SPA. (a) Comparison of the total response time of SPA under different pre-loading forces. (b)Comparison of the angular velocity of SPA under different pre-loading forces.

torque reversal does not require excessive pre-loading force, and continuing to increase the pre-tightening force has relatively little effect on shortening the response time.

4.3 Motion Trajectory

To research the morphological characteristics of SPA, the end trajectories during the inflation and deflation processes of SPA with different pressures and pre-loading forces are compared, as shown in Fig. 6. With the pressure of 70 kPa, the movement distance of the end of SPA without pre-loading force is shorter but farther from the initial position, as shown in Fig. 6(a, b). The trajectory length of the end of the pre-tightened SPA is longer, but it is limited to a range closer to the fixed point. As the pre-loading force increases, the trajectory changes relatively small. As shown in Fig. 6(c, d), the trajectories of the end of SPA with 1 N pre-loading force almost completely coincide under different pressures. The experimental results indicate that the pre-loading force has an obvious limiting effect on the trajectory of the end of SPA. Besides, the air pressure has relatively weak influence on the trajectory of SPA.

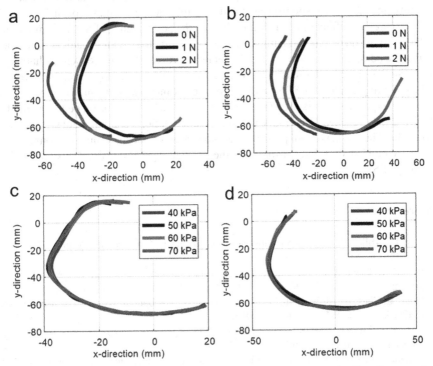

Fig. 6. The motion trajectories of the end of SPA under different pre-loading forces and air pressures. (a) The trajectories of the end of inflated SPA under the air pressure of 70 kPa. (b) The trajectories of the end of deflated SPA under the air pressure of 70 kPa. (c) The trajectories of the end of inflated SPA under 1 N pre-loading force. (d) The trajectories of the end of deflated SPA under 1 N pre-loading force.

In addition, the trajectories of the inflated and deflated SPA are different. The kinetic energy of SPA increases rapidly during the saltation, and the change of angle greatly increases under the influence of inertia, making the trajectory extended. At the same time, the saltation positions of inflated and deflated SPA are different, therefore, the shapes of the trajectories of inflated and deflated SPA are different under the influence of saltation.

5 Conclusion

In this paper, we present a quick response pneumatic actuator by employing the dynamic instability mechanism. To further improve the storage of elastic potential energy, the bottom layer of the pneumatic actuator is accompanied by a higher elasticity modulus. More importantly, the elastic cord is incorporated into the pneumatic actuator to form the SPA. In the process of inflating and deflating the SPA, due to the change of the geometric position of the pneumatic actuator and the elastic rope, the SPA has a torque reversal. The change of position quickly releases the elastic potential energy stored by the pneumatic actuator and the elastic rope, resulting in the saltation multiple times. The instability improves the response speed and amplifies the performance of SPA.

To research the effect of pre-loading force on the performance of SPA, the modeling of SPA is established. The modeling analyzes the geometric and mechanical properties of SPA. Besides, experiments are implemented to reveal the fast response characteristic of SPA. Experimental results show that the pre-tightening force greatly increases the bending angle range of SPA, and the response time is significantly reduced. As a result, the response speed is greatly improved. The maximum inflation bending angular velocity of SPA reaches 41.88 rad/s, the maximum deflation bending angular velocity is 36.65 rad/s. The application of pre-loading force increases the maximum response speed of SPA by more than 9 times. Moreover, the pre-loading force also affects the motion trajectory of SPA.

References

1. Rus, D., Tolley, M.T.: Design, fabrication and control of soft robots. Nature **521**(7553), 467–475 (2015)
2. Jiang, Y., Chen, D., Liu, C., et al.: Chain-like granular jamming: a novel stiffness-programmable mechanism for soft robotics. Soft Rob. **6**(1), 118–132 (2019)
3. Yang, Y., Chen, Y., Li, Y., et al.: Novel variable-stiffness robotic fingers with built-in position feedback. Soft Rob. **4**(4), 338–352 (2017)
4. Noh, M., Kim, S.W., An, S., et al.: Flea-inspired catapult mechanism for miniature jumping robots. IEEE Trans. Rob. **28**(5), 1007–1018 (2012)
5. Pezzulla, M., Stoop, N., Steranka, M.P., et al.: Curvature-induced instabilities of shells. Phys. Rev. Lett. **120**(4), 048002 (2018)
6. Jung, S.P., Jung, G.P., Koh, J.S., et al.: Fabrication of composite and sheet metal laminated bistable jumping mechanism. J. Mech. Rob. **7**(2), 021010 (2015)

7. Koh, J.S., Yang, E., Jung, G.P., et al.: Jumping on water: surface tension–dominated jumping of water striders and robotic insects. Science **349**(6247), 517–521 (2015)
8. Chen, T., Bilal, O.R., Shea, K., et al.: Harnessing bistability for directional propulsion of soft, untethered robots. Proc. Natl. Acad. Sci. **115**(22), 5698–5702 (2018)
9. Overvelde, J.T., Kloek, T., D'haen, J.J., et al.: Amplifying the response of soft actuators by harnessing snap-through instabilities. Proc. Natl. Acad. Sci. **112**(35), 10863–10868 (2015)
10. Pal, A., Goswami, D., Martinez, R.V.: Elastic energy storage enables rapid and programmable actuation in soft machines. Adv. Funct. Mater. **30**(1), 1906603 (2020)
11. Chi, Y., Tang, Y., Liu, H., et al.: Leveraging monostable and bistable pre-curved bilayer actuators for high-performance multitask soft robots. Adv. Mater. Technol. **5**, 2000370 (2020)
12. Cafferty, B.J., Campbell, V.E., Rothemund, P., et al.: Fabricating 3D structures by combining 2D printing and relaxation of strain. Adv. Mater. Technol. **4**(1), 1800299 (2019)
13. Qin, L., Liang, X., Huang, H., et al.: A versatile soft crawling robot with rapid locomotion. Soft Rob. **6**(4), 455–467 (2019)
14. Tang, Y., Chi, Y., Sun, J., et al.: Leveraging elastic instabilities for amplified performance: spine-inspired high-speed and high-force soft robots. Sci. Adv. **6**(19), eaaz6912 (2020)
15. Xiao, R.: Modeling mismatch strain induced self-folding of bilayer gel structures. Int. J. Appl. Mech. **8**(07), 1640004 (2016)

Robotic Vision, Recognition, and Reconstruction

The Point Position and Normal Detection System and Its Application

Jin Yun[1], Li Chen[2], Zhang Xu[1], and Tu Dawei[1(✉)]

[1] Shanghai University, School of Mechatronic Engineering and Automation,
Shanghai, China
tdw@shu.edu.cn
[2] Huazhong University of Science and Technology,
School of Mechanical Science and Engineering, Wu Han, China

Abstract. Mixed reflection objects have the characteristics of specular reflection and diffuse reflection, so it is difficult to reconstruct and the accuracy is low. In order to solve such problems, a point and normal detection system is proposed. In this paper, an industrial camera, a point laser and a holographic film are used to form a point position and normal sensor, and based on this, the incident light and reflected light of the sampling point on the surface of the mixture are calculated and the normal information of the sampling point is obtained. The reconstruction of objects through Zernike polynomials have highly accuracy. Compared with the traditional method of reconstructing mixed reflection objects, the experimental results show that the point position and normal detection system has high reconstruction accuracy and robust. It can reconstruct both specular reflection and mixed reflection objects with easy operation, and diffuse reflection can achieve scanning modeling. The accuracy of the model can reach micron level.

Keywords: Specular reflection · Mixed reflection · Diffuse reflection · Normal detection · 3D reconstruction · Gradient integration

1 Introduction

In many cases, 3D information is needed to assess the quality of a product [1,2], such as solar concentrator, solar cell silicon wafer, car paint shell and optical mirror. At present, the main measuring methods are the tangent three-coordinates measuring machine and laser scanning, which have the advantages of accurate measurement and no requirement for the smoothness of the measuring surface [3,4].

For the 3D reconstruction of the surface of mixed reflection objects, the point structured light is proposed by Xu et al. [5] Firstly, the Plücker matrices of the laser rays are determined by the laser points, then the center of the laser point is contributed by the singular value decomposition. The laser photolysis are obtained by the algebraic solution of the joint point in the 2D-reference coordinate frame and the reconstruction is achieved by the calibrated laser rays of the projector. It uses 3D points cloud to achieve reconstruction, but the research

© Springer Nature Switzerland AG 2020
C. S. Chan et al. (Eds.): ICIRA 2020, LNAI 12595, pp. 401–414, 2020.
https://doi.org/10.1007/978-3-030-66645-3_34

results show that only apply to the non-specular objects and the measurement procedure is complicated. The specular surface topography measurement and reconstruction have researched by Xiao et al. [6] which based on specular fringe reflection. They proposed that the fringe images reflected by different phases are obtained by four step phase-shifting method. And based on the phase difference map obtained by time unwrapping operationthe reconstruction of the mirror surface is realized. It uses the height difference to achieve reconstruction, however the camera fails to obtain a clear image of the display screen through the reflection of the mixed reflective surface which causes low phase accuracy. The reconstruction of ceramic-bowl-surface and proposed an algorithm of local point cloud reconstruction based on multiple images have researched by Guo et al. [7] Binocular camera with Kirsch and Canny operator are used to detect and extract feature points, to get high precision matching point pairs. The motion recovery structure algorithm and reconstruction algorithm are used to achieve the local 3D reconstruction of 2D surfaces defects. In this research, feature points of images are matched with low cost and the reconstruction accuracy of mixed reflection surface is limited, so the reconstruction accuracy is not as high as that of laser detection. A low-cost 3D reconstruction method have proposed by Zabalza et al. [8] which uses line laser and mesh camera to collect pictures. Pictures contain the surface of the object. Then the 3D shape are obtained through image processing and software comparison processing. It uses phase and height differences to achieve reconstruction. Although this method is simple and feasible, it can only be applied to measure and reconstruction with low accuracy objects. A method to reconstruct the surface shape of the object to be measured based on the gradient integral of spherical harmonic function have proposed by Vázquez-OteroAlejandro et al. [9] which improves the speed and accuracy of reconstruction at the same cost.

The most of traditional structured light 3D sensors only obtains 3D information and lack in reconstructing diffuse reflection objects because the mixed reflective surfaces have both diffuse and specular reflections, which interfere with each other. This paper proposes a point position and normal detection method, which can get not only 3D points but also normal of objects and reconstruct objects with mixed reflections by gradient integration. The main works in this paper include: First, the incident and reflection light on the surface of the object to be measured are calculate by using point laser and holographic film, and then the normal information and the three-dimensional coordinates of the point are obtained. Second, the object surface is reconstructed with high fitting precision through the gradient integral of sampling points by using Zernike polynomial [10–12]. The advantage of this simple method is that the surface with mixed reflection object can be reconstructed accurately with low cost. The experimental results show that the proposed method can reconstruct of specular and diffuse surfaces effectively.

2   The Point Position and Normal Detection System

2.1   3D Point Calculation

As shown in Fig. 1, this system consists of a camera, a point laser and a holographic film. The incident light AE on the surface of the object is emitted by a point laser firstly, and then reflected on the surface of the object to form the reflected light BC. Lefting the incident light spot B on the surface of the object. At the same time, the reflected light BC is received by the holographic film, leaving the reflected light spot C on the holographic film. Spot B' and spot C are captured by the area-array cameras and imaged on a CCD sensor.

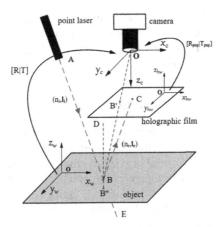

Fig. 1. Schematic diagram of point position and normal detection system.

The spot image is filtered first. According to the calculation formula of pixel value [13] and the center of gravity of gray algorithm [14] extracting the light spots in gray level images. The coordinates of the light spot (u_B, v_B) and spot C as (u_C, v_C), in the pixel coordinate system, are obtained.

Firstly, the camera is calibrated with the calibration board, and the rotation matrix R and the translation matrix T are obtained. According to the camera model:

$$R^{-1}M^{-1}s \begin{bmatrix} u_B \\ v_B \\ 1 \end{bmatrix} = \begin{bmatrix} X_{Bw} \\ Y_{Bw} \\ Z_{Bw} \end{bmatrix} + R^{-1}T \tag{1}$$

Where M is intrinsic parameters of cameras, (X_{Bw}, Y_{Bw}, Z_{Bw}) is the coordinates of spot B in the world coordinate system and Z_{Bw} is 0, s is proportional coefficient. The coordinates of the spot in the camera coordinate system is:

$$\begin{bmatrix} X_{Bc} \\ Y_{Bc} \\ Z_{Bc} \end{bmatrix} = R \begin{bmatrix} X_{Bw} \\ Y_{Bw} \\ Z_{Bw} \end{bmatrix} + T \tag{2}$$

Secondly, In order to obtain the pose of camera coordinate in the world coordinate with the plane of the holographic film as XOY plane, the function in OpenCV library and calibration board are used to directly solve the rotation matrix R_{pnp} and the translation matrix T_{pnp}. According to the formula (1) and (2), the coordinates of the spot C in the camera coordinate system is:

$$
\begin{bmatrix} X_{Cc} \\ Y_{Cc} \\ Z_{Cc} \end{bmatrix} = R_{pnp} \begin{bmatrix} X_{Cw} \\ Y_{Cw} \\ Z_{Cw} \end{bmatrix} + T_{pnp}
\tag{3}
$$

Where (X_{Cw}, Y_{Cw}, Z_{Cw}) is the coordinates of the spot C in the world coordinate system and Z_{Cw} is 0. About calculating normal, suppose the direction vector of the incident and reflected light equation is (n_{iw}, m_i, l_i) and (n_r, m_r, l_r) respectively, so normal can be obtained.

2.2   Calculating the Incident Light

The calibration board is used to calculate the incident light. The object to be measured was replaced with a calibration board, and pictures of light spots are taken under different poses. The center of gravity of gray level method is used to extract the coordinates the spot of each image in the pixel coordinate system.

From the formula (2), the coordinates of spots of all images are obtained as $(X_{c1}, Y_{c1}, Z_{c1}), ...(X_{ci}, Y_{ci}, Z_{ci})$. According to the least square method, the equation of incident light in camera coordinate system is fitted by using the obtained coordinates.

$$
\begin{bmatrix} m\ X_0 \\ n\ Y_0 \end{bmatrix} = \begin{bmatrix} \sum X_i Z_i & \sum X_i \\ \sum Y_i Z_i & \sum Y_i \end{bmatrix} \begin{bmatrix} \sum Z_i^2 & \sum Z_i \\ \sum Z_i & n \end{bmatrix}^{-1}
\tag{4}
$$

Where (X_i, Y_i, Z_i) is the coordinate value, $n_i : (X_0, Y_0, 0)$ is the point on incident light, and $\overrightarrow{l_i} : (m, n, 1)$ is the direction vector of the line.

2.3   Calculating the Reflection Light

As shown in Fig. 1, the camera will photograph light spot B' and C on the holographic film. By calculating the three dimensional space coordinates of the two points, the linear equation of reflection light can be obtained by using the two-point form equation. Light spot B on the surface of the object can be considered as the intersection point of the coordinates of the pixel in camera coordinates and line BO and incident light AE between the centers of the camera.

However, spot B is only the intersection point of the incident light and the reflected light under ideal conditions. Since the incident light is fitted by the least square method, there is a distance between spot B and the incident light AE. In order to improve the accuracy and precision of calculation, spot B is treated as follows:

Step 1: According to the pixel coordinates of the center of the spot on the calibration plate, the two-dimensional linear equation of the incident light in the pixel coordinates is fitted by the least square method.

Step 2: Calculating the coordinate b of B'' in the pixel coordinate system.

Step 3: Point b is projected vertically onto 2D line, and the projection point is and convert the point b' to the 3D coordinate B'' in the camera coordinate system.

Setting the light center of the camera as the origin of the camera coordinate system, the vector of the incident light AE is $\overrightarrow{n}(a,b,c)$, the vector of $B''O$ is $\overrightarrow{m}(X_{B''}, Y_{B''}, Z_{B''})$, the point on the line AE is (X_{ir}, Y_{ir}, Z_{ir}). The truly intersection spot $B(X_{Bc}, Y_{Bc}, Z_{Bc})$ of line $B''O$ and line AE is obtained by using the following equation:

$$\begin{cases} \begin{bmatrix} a\ X_{B''} \\ b\ Y_{B''} \\ c\ Z_{B''} \end{bmatrix} \begin{bmatrix} K1 \\ -K2 \end{bmatrix} = \begin{bmatrix} X_{ir} \\ Y_{ir} \\ Z_{ir} \end{bmatrix} \\ \begin{bmatrix} X_{Bc} \\ Y_{Bc} \\ Z_{Bc} \end{bmatrix} = K_1 \overrightarrow{m} = K_2 \overrightarrow{n} \end{cases} \tag{5}$$

So n_r : (X_{Bc}, Y_{Bc}, Z_{Bc}) is the point on reflected light, and $\overrightarrow{l_r}$:$(X_{Cc} - X_{Bc}, Y_{Cc} - Y_{Bc}, Z_{Cc} - Z_{Bc})$ is the direction vector of the line.

3   Three-Dimensional Reconstruction

3.1   Zernike Integration

The reconstruction method of gradient data by Zernike orthogonal polynomial is selected. It can be considered that the surface of the object to be tested is smooth and continuous, and its surface shape can be expressed as:

$$w(\text{x,y}) = a_1 Z_1(x,y) + a_2 Z_2(x,y) + \ldots + a_n Z_n(x,y) = \sum_{k=1}^{n} a_k Z_k(x,y) \tag{6}$$

Where $w(x,y)$ is a Surface function, $a_k Z_k(x,y)$ is Zernike polynomials under cartesian coordinate system, a_k is a coefficients of Zernike polynomials, n is the number of terms of Zernike polynomials. Take the derivation of X and Y respectively according to the above formula:

$$\begin{aligned} w^x(x,y) &= a_1 Z_1^x(x,y) + a_2 Z_2^x(x,y) + \ldots + a_n Z_n^x(x,y) \\ &= \sum_{k=1}^{n} a_k Z_k^x = \sum_{k=1}^{n} a_k Z_k^x \frac{\partial Z_k(x,y)}{\partial x} \\ w^y(x,y) &= a_1 Z_1^y(x,y) + a_2 Z_2^y(x,y) + \ldots + a_n Z_n^y(x,y) \\ &= \sum_{k=1}^{n} a_k Z_k^y = \sum_{k=1}^{n} a_k Z_k^y \frac{\partial Z_k(x,y)}{\partial y} \end{aligned} \tag{7}$$

Where $Z_k^x(x,y)$ and $Z_k^y(x,y)$ are derivatives of Zernike polynomials in X and Y directions, respectively, the slope in the X and Y directions, I. E. $w^x(x,y)$ and $w^y(x,y)$, are obtained by the point position and normal detection system.

Covert the upper form in matrix form:

$$
S = \begin{bmatrix} w^x(1) \\ \vdots \\ w^x(m) \\ w^y(1) \\ \vdots \\ w^y(m) \end{bmatrix} = \begin{bmatrix} Z_1^x(1) & Z_2^x(1) & \cdots & Z_{n-1}^x(1) & Z_n^x(1) \\ \vdots & \vdots & \vdots & \vdots & \vdots \\ Z_1^x(m) & Z_2^x(m) & \cdots & Z_{n-1}^x(m) & Z_n^x(m) \\ Z_1^y(1) & Z_2^y(1) & \cdots & Z_{n-1}^y(1) & Z_n^y(1) \\ \vdots & \vdots & \vdots & \vdots & \vdots \\ Z_1^y(m) & Z_2^y(m) & \cdots & Z_{n-1}^y(m) & Z_n^y(m) \end{bmatrix} \begin{bmatrix} a_1 \\ a_2 \\ a_3 \\ a_4 \\ \vdots \\ a_n \end{bmatrix} = Aa \quad (8)
$$

Among which S is the surface shape gradient of the object to be measured, that is, the measured gradient data matrix, the size of the Matrix is 2m*1, m is the number of sampling points. $w^x(m)$ and $w^y(m)$ are the X and Y slope of the M sampling point of Zernike, respectively. A is the gradient value Matrix of Zernike polynomials. $Z_n^x(m)$ and $Z_n^y(m)$ are the slope in the X and Y directions of Zernike nth term at the m sampling point. a is the coefficients of Zernike polynomials that need to be solved, then the least square solution of a according to formula $a = [a_1, a_2, \cdots, a_n]^T$ is $a = (A^T A)^{-1}(A^T S)$.

After the coefficients of Zernike polynomials, the shape of the measured surface can be reconstructed by substituting formula (6).

3.2   Zernike Simulation

The computer simulation method is used to verify the precision of the Zernike algorithm used in the actual measurement experiments. The continuous smooth curves and surfaces are selected as the simulation measurement surfaces. By using Matlab, the original simulation surfaces are directly generated by setting the known parameters, and then the gradient data along the horizontal and vertical directions are obtained for the surface shape. Adding different degrees of Gaussian noise, analog error of normal detection system detection is calculated. And then the surface is rebuild.

The following surface equation is used for 3D reconstruction

$$
x^2 + y^2 + z^2 = 200, (-5 \leq x \leq 5, -5 \leq y \leq 5) \quad (9)
$$

Calculating the gradient of Z direction, as shown in Fig. 2 gradient distribution:

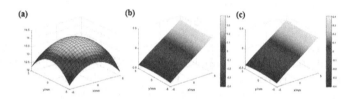

Fig. 2. (a) The original surface (b) X direction gradient (c) Y direction gradient

For this gradient, the expectation is 0 and the standard deviation is 0.0005. The value of Gaussian noise is further enlarged to make the gradient change larger. The 36 terms of Zernike polynomial are selected for reconstruction. The error between the reconstruction result and the true value is shown in Fig. 3:

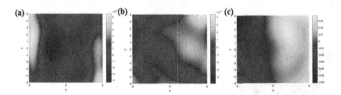

Fig. 3. The error between the reconstruction result and the original surface (a) noise with a standard deviation of 0.0005 (b) noise with a standard deviation of 0.005 (c) noise with a standard deviation of 0.05

The noise with a standard deviation of 0.0005 is added, the standard deviation of the error is 2.364×10^{-4} and the average error is 2.073×10^{-5} (Fig. 4).

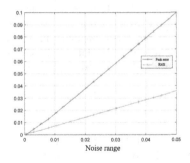

Fig. 4. Two-dimensional reconstruction of peak error and standard deviation

When the error of gradient data is less than the Gaussian noise with a standard deviation of 0.005, the standard deviation of one - and two-dimensional reconstruction is stable on the order of 10^{-3}, which is considered stable. The accuracy of reconstruction is higher than the error of gradient data of this order of magnitude. The average error is on the micron scale. Therefore, as long as the normal detection accuracy is high enough, the reconstructed surface shape accuracy will be correspondingly improved.

4   Experiment

In this experiment, the entire measurement system is designed based on the point position and normal detection theory. The system composition diagram is shown in the following Fig. 5. A point laser with a diameter of 0.02 mm is used as the light source, and the point laser is projected onto the surface of the object and reflected onto the holographic film. The industrial CCD camera CM3-U3-13Y3M is used together with an 8 mm ordinary lens to shoot and collect the image. In order to facilitate the surface sampling of the object to be measured, the whole system is installed on a two-dimensional moving guide rail.

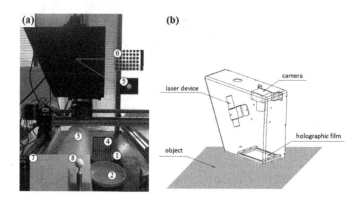

Fig. 5. Point position and normal detection system equipment (a) physical system, object no. 1 is the a precision mirror with a surface precision of 1 μ, object no. 2 is a ordinary ceramic bowl, object no. 3 is a ordinary plastic cylinder, and no. 4 is the calibration board, object no. 7 is a ceramic plate with dent, object no. 8 is a precision ceramic ball with surface accuracy of 0.5 μ (b) model system

4.1   Calibraiton

The calibration board is used to replace the object to be measured. Ten images are collected. As shown in no. 5 of Fig. 5(a), the center of gravity of grey level is extracted. According formula (2), the 3D coordinates of camera coordinate system are obtained. Then obtain the equation of the incident light by formula (4). n_i : $(-113.88385, -0.49733, 0)$ is the point on incident light, $\overrightarrow{l_i}$:$(0.42799, -0.00422, 1)$ is the direction vector of the line.

The calibration board is placed on the plane where the holographic film is located, and the image of calibration board is obtained with the camera, as shown in no. 6 of Fig. 5(a).

Due to the defocus of the image, it is necessary to preprocess the image. As shown in Fig. 6, the images is filtered first, the circle boundary is identified by using hough transform second, the coordinates of the center of each circle in the image in the pixel coordinate system is obtained by using OpenCV function third.

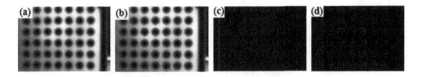

Fig. 6. Center extraction of circular calibration board (a) Function filter (b) Hough detection (c) Boundary extraction (d) Centroid extraction

The $[R_{pnp}|T_{pnp}]$ is obtained according PnP theory [15]:

$$\left[\begin{array}{ccc|c} -0.01209 & 0.99982 & -0.01447 & -24.40727 \\ 0.99988 & 0.01223 & 0.00942 & -22.93705 \\ 0.00961 & -0.01435 & 0.99985 & 146.64147 \end{array}\right] \tag{10}$$

4.2   Single Point Experiment

As shown in no. 1 of Fig. 5(a), a plane mirror is used as the object (Fig. 7).

Fig. 7. (a) Original image (b) Center of gravity of gray level extraction (c) Contour centroid extraction

Because the blurring of the edges of the light spots on the holographic film, the brightness of the light spots is not uniform. The center of gravity of gray level method by area can be used to extract the coordinates of the spot more accurately. The spot B in the camera coordinate system is $(-23.7542, -1.76196, 212.34)$. $[R_{pnp}|T_{pnp}]$ is used to calculate the light spot C in the camera coordinate system is $(11.7542, -2.7619, 146.34)$. The reflected light in the camera coordinate system has been determined according to the two-point linear equation. Among which the direction vector of the line is $\overrightarrow{l_r}$:(-34.3202, 1.36425, 67.019). The spatial positions of the incident and reflected lights are shown in the following Fig. 8:

Sampling points arranged in 5540 on the hollow plastic cylinder of object no.3 as shown in Fig. 5 are selected, and the spacing of each sampling point is 0.2mm. After the point cloud is generated, the cylindrical surface is synthesized and the error between the point cloud and the cylindrical surface is calculated (Fig. 9).

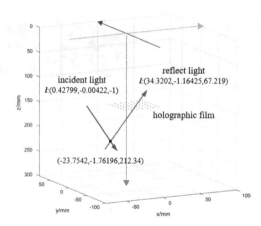

Fig. 8. The spatial positions of the incident and reflected lights

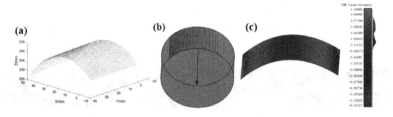

Fig. 9. (a) Point could of sampling spot (b) Fitting cylinder (c) Error between point cloud and cylinder

The maximum error is 0.13741 mm, the diameter of the fitting cylinder is 85.277 mm, and the actual measured cylinder diameter is 85 mm.

4.3   Experiment of Specular Reflection

According to the experiment above, the point position and normal detection system device is installed on the two-dimensional moving guide rail. Sampling points arranged as 60100 are selected from the plane mirror shown in no. 1 of Fig. 5(a), with the spacing of each sampling point being 0.2 mm.

As shown in Fig. 10, the sample points for each row and each column are not exactly on the same line and are slightly shifted, respectively. These are caused by errors caused by mechanical processing of device and assembly, but this does not affect the calculation accuracy of normal line. In the actual measurement, only the actual light spot position on the surface of the object to be measured is calculated, and the reconstructed surface is also the surface composed of the actual light spot.

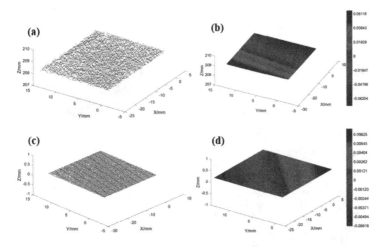

Fig. 10. Point cloud and the reconstruction error (a) Point could of sampling spot (b) Fitting error of sampling point plane (c) Point could of calculated point (d) Fitting error of Calculated point plane

About 3D point fitting plane obtained by actual measurement, the maximum distance from point to plane is 0.0806 mm, the average distance from point to surface is 0.0059 mm, the RMS is 0.0715. And about 3D point fitting plane obtained by calculation, the maximum distance from point to plane is 0.00631 mm, the average distance from point to surface is 0.0003 mm, the RMS is 0.0028. So the plane precision reconstructed by Zernike polynomial is higher than that obtained by direct measurement, indicating that the system can well meet the required accuracy requirements.

4.4   Experiment of Mixed Reflection

According to the single-point experiment, firstly, sampling points arranged in 2525 in the middle of the ceramic bowl bottom of object no. 8 as shown in Fig. 5 are selected, and the spacing of each sampling point is 0.2 mm.

Secondly, sampling points arranged in 6036 in the middle of the ceramic bowl bottom of object no. 2 as shown in Fig. 5 are selected, and the spacing of each sampling point is 0.5 mm.

Fourthly, sampling points arranged in 6570 in the middle of the ceramic bowl bottom of object no. 7 as shown in Fig. 5 are selected, and the spacing of each sampling point is 1 mm.

As shown in Figs. 11, 12 and 13, the point cloud plane composed of 3D points obtained by calculation has higher accuracy and reduction degree, so the system can also reconstruct low-reflection objects well.

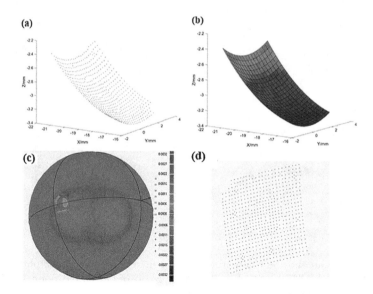

Fig. 11. Results of precision ceramic ball (a) Point could of calculated point (b) 3D surface composed of calculated point

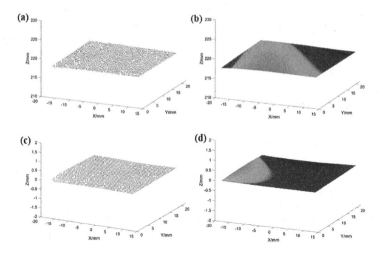

Fig. 12. Results of bowl bottom experiment (a) Point could of sampling spot (b) 3D surface composed of sampling points (c) Point could of calculated point (d) 3D surface composed of calculated point

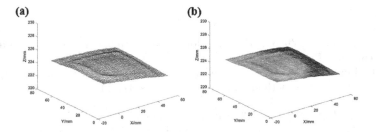

Fig. 13. Results of ceramic plate with dent (a) Point could of sampling spot (b) 3D surface composed of sampling points

5 Conclusion

In this paper, a point position and normal detection system is proposed and designed, which is used to reconstruct the surface shape of highly reflective and lowly reflective objects. The equations of incident light and reflection light at the sampling point in camera coordinate system are calculated respectively. Then the gradient is calculated and the 3D shape of the object to be measured is calculated using Zernike polynomials. The variation trend of standard deviation of 1D and 2D gradient reconstruction results in different Gaussian noise ranges is analyzed. The results show that the higher the noise is, the higher the gradient information error is, and then the reconstruction accuracy is lower and the standard deviation is larger. In the allowable range of Gaussian noise, the 3D reconstruction of Zernike polynomials is accurate. The 3D reconstruction experiment of plane mirror is carried out by using the designed system. The experimental results show that the precision error between the reconstructed plane mirror and fitting plane is within 320 nm. Then the system is used to measure and reconstruct the ceramic ball and the precision error of the reconstructed sphere is within 3.2μ. About bottom of the ceramic bowl, it has good and stable results. Compared with the traditional contact measurement, the system has a wider range of measurement, lower cost and higher precision.

Because the prototype system in this paper uses common laser and low resolution industrial camera, there are still some errors in the calculation of center of gravity and coordinate results, and holographic film size is limited, so unable to reconstruct larger size objects. However, the experiment still shows its advantages in accuracy and ability to measure both highly and lowly reflective objects. In the future, the equipment precision will be improved, and the point position and normal detection system and its application will be further studied.

Acknowledgement. This study is supported by the National Natural Science Foundation (NSFC) (Grant No. 61673252), China Postdoctoral Science Foundation (Grant No. 2019M662591), and (NSFC) (Grant No. 61673252, No. 51975344, No. 51535004).

References

1. Li, X.F., Yao, W., Zhao, X.H., et al.: Joint calibration of multi-sensor measurement system. Opt. Precis. Eng. **21**(11), 2877–2884 (2013). (in Chinese)
2. Li, F., Liu, J.T., Cai, J.J.: Surface shape measurement of mirror-like objects based on structured light method. J. Electron. Devices **37**(05), 882–886 (2014). (in Chinese)
3. Zhang, J.M.: Object three-dimensional contour measuring system based on laser triangulation. Wuhan University of Technology (2015). (in Chinese)
4. Bothe, T., Li, W., von Kopylow, C., et al.: High-resolution 3D shape measurement on specular surfaces by fringe reflection. Proc. Spie Int. Soc. Opt. Eng. **5457**, 411–422 (2014)
5. Xu, G., Shen, H., Li, X., Chen, R.: A flexible 3D point reconstruction with homologous laser point array and monocular vision. Optik **205** (2020)
6. Lai, X., Mi, X.L., Chen, M.C., Wang, W.M.: Research and application of mirror surface shape detection algorithm based on machine vision. Automat. Instr. **34**(12), 68–74 (2019). (in Chinese)
7. Guo, M., Hu, L.L., Li, J.: Local point cloud reconstruction of ceramic bowl surface defects based on multiple images. Acta Optica Sinica **37**(12), 247–253 (2017). (in Chinese)
8. Zabalza, J., RenLow, J., et al.: Low cost structured-light based 3D surface reconstruction. **12**(1), 1–11 (2019)
9. Manuel, S.-A.J., Alejandro, V.-O., Danila, K., Raquel, D., Natividad, D.: Using spherical-harmonics expansions for optics surface reconstruction from gradients. Sensors (Basel, Switzerland), **17**(12) (2017)
10. Lyu, H., Huang, Y., Sheng, B., et al.: Absolute optical flatness testing by surface shape reconstruction using Zernike polynomials. **57**(9), 094103 (2018)
11. Han, L., Tian, A.L., Nei, F.M., et al.: Research on 3D reconstruction algorithm of stripe reflection based on Zernike polynomial. J. Xi'an Technol. Univ. **39**(02), 21–28 (2019). (in Chinese)
12. Li, G., Li, Y., Liu, K., et al.: Improving wave front reconstruction accuracy by using integration equations with higher order truncation errors in the southwell geometry. J. Optical Soc. Am. **30**(7), 1448 (2013)
13. Dang, W.J.: Research and application of image region segmentation algorithm. J. Anhui Univ. Sci. Technol. (2018). (in Chinese)
14. Huang, L.F., Liu, G.D., Zhang, C., Gan, H., Luo, W.T., Li, L.: Algorithm for extracting laser stripe center based on gray weight model. Laser Technol. **44**(02), 190–195 (2020)
15. Geiger, A., Moosmann, F., Car, O., et al.: Automatic camera and range sensor calibration using a single shot. In: Proceedings IEEE International Conference on Robotics & Automation, pp. 3936–3943 (2012)

6D Pose Estimation for Texture-Less Industrial Parts in the Crowd

Dexin Zhou[1], Ziqi Chai[1], Chao Liu[1,2], Peng Li[3], and Zhenhua Xiong[1(✉)]

[1] State Key Laboratory of Mechanical System and Vibration, School of Mechanical Engineering, Shanghai Jiao Tong University, Shanghai 200240, China
{aalon,mexiong}@sjtu.edu.cn
[2] Hai'an Institute of Intelligent Equipment, SJTU, Nantong 226000, China
[3] Technology Development Department, Faw JieFang Group Co., Ltd., Changchun 130000, China

Abstract. Recovering the 6D object pose of industrial part is a common but challenging problem in many robotic applications. In this paper, an accurate 6D pose estimation approach is proposed for texture-less industrial part in the crowd. The proposed method consists of three stages: object detection, pose hypotheses generation, and pose refinement. Firstly, the bounding boxes of object instances in an RGB image are detected by a convolution neural network. The training dataset is automatically synthesized using an efficient image rendering method. Then, highlight detection and removal are employed to eliminate noise edges. The coarse pose hypotheses are generated using an edge-based fast directional chamfer matching algorithm. After that, the accurate 6D poses are obtained by applying a non-linear optimization to these pose hypotheses. A re-weighted least-squares loss function is utilized to suppress outlier noise in optimization. Finally, an edge direction consistency score is used to evaluate these obtained poses and eliminate outliers. The proposed method only relies on single RGB image to recover the 6D object pose in the crowd. Experimental results of texture-less industrial parts show the accuracy and robustness of the proposed method.

Keywords: 6D pose estimation · Pose refinement · Highlight removal

1 Introduction

6D pose estimation is a crucial technique in many robotics applications such as robot-aided manufacturing, bin picking, and service robotics. Although many 6D pose estimation methods have been proposed, it still remains a challenging problem especially for texture-less industrial parts in the crowd. In this scenario, the pose estimation is more difficult to perform due to metal surface reflection, occlusion and background clutter.

As for sufficiently textured objects, many methods based on local feature descriptors demonstrate good results. For instance, SIFT [1], SURF [2], ORB [3] are widely used feature descriptors. These methods rely on surface texture to extract invariant local feature descriptors to establish correspondences between training descriptors and the extracted descriptors of the scene image. However, for texture-less objects, the local

© Springer Nature Switzerland AG 2020
C. S. Chan et al. (Eds.): ICIRA 2020, LNAI 12595, pp. 415–428, 2020.
https://doi.org/10.1007/978-3-030-66645-3_35

descriptor is not discriminative enough to establish reliable correspondences. The cluttered background also leads to mismatches. Template-based methods utilize object contour and surface shape information, which are applicable to the pose estimation problem of texture-less objects such as industrial parts. However, these methods underperform in heavy occlusion scenes.

With the development of consumer-level depth cameras, many methods utilize 3D point clouds for object recognition and 6D pose estimation. These methods need to extract feature points of the scene point cloud and generate local or global feature descriptors, then estimate the 6D object pose through point cloud registration. For non-Lambertian materials such as metal surfaces, the acquired point clouds usually suffer serious distortion, missing and fragmentation due to occlusion or specular reflection. Thus, these methods cannot estimate the 6D pose accurately in the case of poor point cloud quality.

In recent years, CNN-based methods have been proposed to estimate object 6D poses, some of which train the end-to-end model to regress the pose directly while others extract high-dimensional features and estimate the pose combining some machine learning methods, such as Random Forest, Hough forest, Random Fern and so on. Most of these methods need both RGB and depth images as input and rely on point cloud registration to obtain refined poses.

As introduced above, since these industrial parts have poor texture features and 3D point cloud quality, the RGB image edge information is more suitable for their detection and pose estimation. Motivated by the edge-based fast directional chamfer matching (FDCM) approach in [4] and [5], we propose an accurate and robust 6D pose estimation method of texture-less industrial parts. The YOLOv3 convolutional neural network [6] is applied to detect the bounding box of the object instance in the scene image, which separates the target object from the clutter background and reduces candidate poses. An efficient data rendering method is adopted to automatically synthesize the training dataset. Then, the FDCM method is further employed to generate initial pose hypotheses. A non-linear optimization procedure is performed to optimize these coarse poses and obtain the final accurate poses. The iterative reweighted least squares (IRLS) loss function is used in optimization to suppress outliers.

To summarize, the main contributions of this paper include: (1) We propose a pipeline that only relies on single RGB image to recover accurate 6D pose of texture-less industrial part in the crowd. (2) We design an efficient approach to automatically synthesize training dataset with ground-truth labels. This approach significantly reduces the cost for training deep convolutional neural network. (3) We adapt a robust loss function in pose optimization process to suppress the outliers, which significantly improves the accuracy of refined 6D pose.

The paper is organized as follows. Section 2 reviews recent 6D pose estimation methods. In Sect. 3, the proposed method is introduced. Section 4 presents the experimental results. Finally, the conclusion is made in Sect. 5.

2 Related Work

This section provides a review of recent 6D pose estimation methods. These approaches can be divided into three main categories including template-based methods, point cloud registration methods and learning-based methods.

2.1 Template-Based Methods

Hinterstoisser et al. [7] proposed the LINEMOD template matching algorithm. The template consists of binary-coded gradient feature and surface normal feature. The SSE parallel computing instructions are used in feature matching stage to improve the real-time performance. The matching results are further refined with the Iterative Closest Point (ICP) algorithm to get an accurate pose. Since only the color gradient feature of the boundary is utilized, this method can handle the pose estimation problem of texture-less objects. After that, [8, 9] extend the LINEMOD method to further improve matching efficiency and accuracy. Hodan et al. [10] created a template with multiple triples. The randomly selected triplet from the depth map is described by depth differences and normal relationships. Cascade-style evaluation and hash-coded voting framework are employed to improve the accuracy. Liu et al. [4] proposed the fast directional chamfer matching method which includes edge direction information to improve the registration accuracy. And in the matching stage, a directional chamfer distance (DCD) tensor is used to improve the efficiency. Imperoli et al. [5] extended the FDCM and proposed Direct Directional Chamfer Optimization method, which refines the pose by non-linear optimization procedure. However, noise edges generated by background clutter, occlusion and specular reflection will cause an incorrect match.

2.2 Point Cloud Registration Methods

Methods based on point cloud registration need to extract local or global feature descriptors from the object model and scene. The 6D pose of an object can be estimated by searching the correspondences between two feature point sets and aligning them. Many discriminative point cloud feature descriptors have been proposed, such as PFH (Point Feature Histogram) [11], FPFH (Fast Point Feature Histogram) [12], VFH (Viewpoint Feature Histogram) [13] and so on. Drost et al. [14] proposed the PPF (Point Pair Feature) that combines the Euclidean distance of two points and the neighborhood normal relationship. A hash table is established to speed up the search for similar point pairs. Candidate poses are obtained through the Hough Voting method. Choi et al. [15] proposed a point pair feature descriptor describing both color and shape and applied it to a voting-based 6D pose estimation algorithm. As is introduced above, when applying these methods to industrial parts, the quality of point cloud will strongly limit the effectiveness.

2.3 Learning-Based Methods

Brachmann et al. [16] extracted pixel-wise features and trained a regression forest. Each pixel predicts its category and 3D coordinates in the object frame. Then, the 6D pose is estimated using a PnP method based on these 2D-3D correspondences. Doumanoglou et al. [17] trained a Sparse Auto Encoder to extract the features of local image patches and used them to train a Hough Forest. The object pose is predicted by Hough voting. Kehl et al. [18] adopted a framework similar to Doumanoglou. A feature codebook is established in the training process and the pose is recovered by searching similar features in the codebook. Recently, Xiang et al. [19] proposed the Pose-CNN. This network uses RGB-D data as input and predicts the object's rotation quaternion directly. The translation is recovered by semantically segmenting the object and calculating the mask centroid. Kehl et al. [20] proposed SSD-6D to regress the 6D pose in an end-to-end way. The regressed pose needs to be refined by the ICP algorithm.

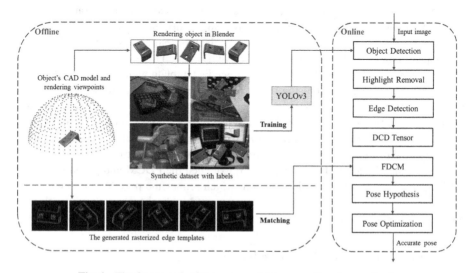

Fig. 1. The framework of the proposed 6D pose estimation method.

3   The Proposed Method

In this section, a detailed description of the proposed 6D pose estimation method for texture-less industrial parts will be introduced. Figure 1 shows the framework of the proposed method. Firstly, a YOLOv3 convolutional neural network is trained to detect the bounding box of the object. An efficient data rendering method is adopted to automatically synthesize the training dataset. Secondly, the highlight detection and removal are employed to eliminate noise edges. Then, a DCD tensor of the edge map is computed to speed up the fast directional chamfer matching process. Candidate poses are generated using the IOU filter and matching score filter. Finally, the coarse poses are refined by minimizing an IRLS loss function. The IRLS loss will suppress occluded or mismatched 3D contour points. A pose evaluation scheme is applied to select out final accurate poses.

3.1 Object Detection

The purpose of object detection is to obtain the bounding box of all instances in the scene image, so that the target object can be separated from background clutter and the searching area of FDCM is significantly reduced. In this paper, we use the YOLOv3 to detect the object. YOLOv3 replaces the backbone network structure of classic YOLO and makes some optimization design, which improves real-time performance while maintaining detection accuracy.

To train a YOLOv3 network with better detection performance, large amounts of labeled training data are required. We designed an efficient approach to automatically synthesize training data. The synthetic image can simulate different lighting conditions, background clutter, reflectance properties, and foreground occlusion. Furthermore, the ground truth bounding box label can be generated automatically and avoid manual labeling.

Firstly, we use the software Blender to render the CAD model of texture-less industrial part. We set the CAD model's material properties and appropriate virtual background environment in Blender. Then the model is rendered automatically with various position and orientation. For each template image generated by rendering, the mask of the object and its location in image can be calculated. Finally, we render these object masks to randomly selected background images at random locations. These background images are selected from the validation images of the COCO dataset 2017. We perform random rotation, scaling and color adjustment on object masks to enhance the dataset. We also randomly render multiple object masks to some background images to simulate the occlusion situation. Figure 2 shows the synthetic training images and detection results.

Fig. 2. (a) and (b): Examples of synthetic images. (c): Object detection result of synthetic image. (d): Object detection result of real image.

3.2 Highlight Detection and Removal

FDCM only uses the edge information of the scene image. Although the detected bounding box can help eliminate clutter background edges, the specular reflection of non-Lambertian surface will cause local highlight areas in RGB image and these highlight areas will generate disturbing edges. Too many disturbing edges seriously affect the accuracy of FDCM.

There are both diffuse reflection and specular reflection on the surface of the industrial part. Given the scene image $\mathbf{I}$, $\mathbf{I}(x) = [\mathbf{I}_r(x), \mathbf{I}_g(x), \mathbf{I}_b(x)]$ stores red, green and blue color channels at pixel x. Each pixel can be divided into an over-saturated pixel or unsaturated pixel according to the intensity of specular reflection. For the unsaturated pixel area, we use the intensity ratio in the chromaticity space to separates the specular reflection from the diffuse reflection. Then, we follow the method proposed in [21] to remove the specular reflection area. As for over-saturated pixels, only specular reflection components exist and make it difficult to repair directly in the chromaticity space. A Poisson reconstruction algorithm is used to predict the pixel value of the over-saturated area according to neighboring pixels.

According to the dichromatic reflection model, $\mathbf{I}(x)$ can be represented by the linear combination of its diffuse $\mathbf{D}(x)$ and specular $\mathbf{S}(x)$ reflection components.

$$\mathbf{I}(x) = \mathbf{D}(x) + \mathbf{S}(x) = w_d(x)\mathbf{\Lambda}(x) + w_s(x)\mathbf{\Gamma} \tag{1}$$

where w_d and w_s are weights related to diffuse and specular reflections over the surface geometry. $\mathbf{\Lambda}(x)$ is the diffuse chromaticity and $\mathbf{\Gamma}$ is the illumination chromaticity that is usually considered fixed. Let $\mathbf{I}^{\min}(x)$ and $\mathbf{I}^{\max}(x)$ be single-channel images that represent the maximum and minimum values of red, green and blue intensities.

$$\mathbf{I}^{\min}(x) = \min(\mathbf{I}_r(x), \mathbf{I}_g(x), \mathbf{I}_b(x)) = w_d(x)\mathbf{\Lambda}^{\min}(x) + w_s(x)\mathbf{\Gamma} \tag{2}$$

$$\mathbf{I}^{\max}(x) = \max(\mathbf{I}_r(x), \mathbf{I}_g(x), \mathbf{I}_b(x)) = w_d(x)\mathbf{\Lambda}^{\max}(x) + w_s(x)\mathbf{\Gamma} \tag{3}$$

Then, we can define the intensity ratio $\mathbf{I}^{ratio}(x)$ as:

$$\mathbf{I}^{ratio}(x) = \frac{\mathbf{I}^{\max}(x)}{\mathbf{I}^{\max}(x) - \mathbf{I}^{\min}(x)} \tag{4}$$

For a pixel with pure diffuse reflection, the intensity ratio is represented by:

$$\mathbf{I}^{ratio}(x) = \frac{w_d(x)\mathbf{\Lambda}^{\max}}{w_d(x)(\mathbf{\Lambda}^{\max} - \mathbf{\Lambda}^{\min})} = \frac{\mathbf{\Lambda}^{\max}}{\mathbf{\Lambda}^{\max} - \mathbf{\Lambda}^{\min}} \tag{5}$$

However, for a pixel with diffuse and specular reflections, the intensity ratio is:

$$\mathbf{I}^{ratio}(x) = \frac{w_d(x)\mathbf{\Lambda}^{\max} + w_s(x)\mathbf{\Gamma}}{w_d(x)(\mathbf{\Lambda}^{\max} - \mathbf{\Lambda}^{\min})} \tag{6}$$

From (4) and (5), for pixels contain specular reflection component, the intensity ratio is higher than that of purely diffuse pixels. Therefore, the intensity ratio can be used as an effective metric to separate diffuse and specular pixels. In order to separate diffuse and specular pixels, we need the diffuse chromaticity image $\mathbf{\Lambda}(x)$, a pseudo specular-free image $\mathbf{I}^{psf}(x)$ is used to estimate $\mathbf{\Lambda}(x)$. $\mathbf{I}^{psf}(x)$ is defined in (7), where the $\bar{\mathbf{I}}^{\min}$ is the mean of $\mathbf{I}^{\min}(x)$ that can suppress the noise.

$$\mathbf{I}^{psf}(x) = \mathbf{I}(x) - \mathbf{I}^{\min}(x) + \bar{\mathbf{I}}^{\min} \tag{7}$$

After that, we cluster the pixel with close diffuse chromaticity in image $\mathbf{I}^{psf}(x)$ and compute the intensity ratio of each pixel. We sort the intensity ratio per cluster and select the median value as the diffuse reflections intensity ratio $\mathbf{I}^{ratio}(x)$. The final diffuse and specular reflection are given by (8) and (9).

$$\mathbf{S}(x) = \mathbf{I}^{\max}(x) - \mathbf{I}^{ratio}(x)(\mathbf{I}^{\max}(x) - \mathbf{I}^{\min}(x)) \tag{8}$$

$$\mathbf{D}(x) = \mathbf{I}(x) - \mathbf{S}(x) \tag{9}$$

As for over-saturated pixels in $\mathbf{D}(x)$, a morphological operation is performed to obtain the mask of the over-saturated region. Then, a Poisson reconstruction algorithm is used to repaint the inner pixel values of this region. The highlight removal experimental results are shown in Fig. 3.

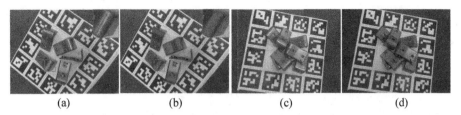

Fig. 3. Hightlight removal results. (a) and (c): The input images. (b) and (d): The output of the highlight removal method.

3.3 Pose Hypothesis Generation

The pose hypotheses are generated by employing FDCM. FDCM includes two stages: offline edge template generation and online template matching.

The edge templates are generated by OpenGL rendering, as shown in Fig. 4. Firstly, several virtual camera viewpoints in different poses are generated. These viewpoints are the vertices of a regular polyhedron which is generated by decomposing an icosahedron recursively. The object coordinate axes are aligned with the OpenGL world coordinate axes and the camera z-axis is aligned with the vector pointing from the current viewpoint to the origin. At each viewpoint, the camera will rotate around the z-axis at a fixed step. The projection matrix in OpenGL is set from the intrinsic parameters of the camera in the experiment. Rendering uses the 3D wireframe model of the object. The z-buffer is used to deal with occlusions and only edges that belong to high curvature parts or to the external object shape are preserved. Then we perform rasterization operation that sampling along the preserved 3D edges in a fixed step ds to get discrete edge points. The finally generated template is a set of discrete points $\mathcal{T} = \{(\mathbf{x}_1, \mathbf{d}_1), (\mathbf{x}_2, \mathbf{d}_2), \cdots, (\mathbf{x}_m, \mathbf{d}_m)\}$, $\mathbf{x}_i$ is the position of the raster edge point in object frame, $\mathbf{d}_i$ is unit orientation vector of the 3D edge which the raster point belongs to.

Given the transform $[\mathbf{r}, \mathbf{t}]$ from the object frame to the camera frame and the camera intrinsic parameters $\mathbf{k}$, we can project the raster points on the 2D image frame. We define

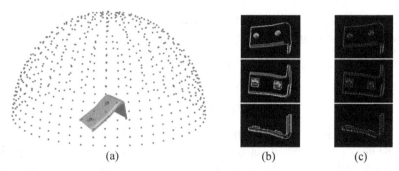

Fig. 4. Templates generation. (a) Target object and rendering viewpoints. (b) Examples of edge templates. (c) Examples of rasterized templates.

the projection function $\pi(\mathbf{x}_i, \mathbf{r}, \mathbf{t})$ and $\sigma(\mathbf{d}_i, \mathbf{r}, \mathbf{t})$ as follows. The projected pixel location is given by:

$$x_i = [\mathbf{k}, \mathbf{0}] \begin{bmatrix} \mathbf{r} & \mathbf{t} \\ \mathbf{0}^T & 1 \end{bmatrix} \begin{bmatrix} \mathbf{x}_i \\ 1 \end{bmatrix} = \pi(\mathbf{x}_i, \mathbf{r}, \mathbf{t}) \tag{10}$$

The projected edge point direction is given by $d_i = \sigma(\mathbf{d}_i, \mathbf{r}, \mathbf{t})$.

$$\sigma = \text{Normalize}(\pi(\mathbf{x}_i + \mathbf{d}_i \cdot ds, \mathbf{r}, \mathbf{t}) - \pi(\mathbf{x}_i, \mathbf{r}, \mathbf{t})) \tag{11}$$

In the online matching stage, the proposed RGB-D edge detection method is used to extract the scene edge map $\mathcal{S} = \{(x_1, d_1), (x_2, d_2), \cdots, (x_s, d_s)\}$, x_i is the edge pixel location and d_i is the edge direction. The DCM matching cost between template $\mathcal{T}$ and scene edge map $\mathcal{S}$ is given by:

$$d_{DCM}(\mathcal{T}, \mathcal{S}) = \frac{1}{n} \sum_{\mathbf{x}_i, \mathbf{d}_i \in \mathcal{T}} \min_{x_j, d_j \in \mathcal{S}} (\|\pi(\mathbf{x}_i) - x_j\| + \lambda \|\phi(\sigma(\mathbf{d}_i)) - \phi(d_j)\|) \tag{12}$$

where $\phi(\cdot)$ is an operator that provides the nearest quantized angle of an orientation vector. The parameter λ is a weighting factor between the location and orientation terms.

The matching cost given in (12) requires finding the minimum matching cost over location and orientation terms for each template edge point. Liu et al. [4] proposed the Directional Chamfer Distance tensor to speed up template matching. As shown in Fig. 5, The DCD tensor is 3D tensor represented by an ordered sequence of distance transform maps. The edge direction is evenly quantized into n channels. Each edge pixel (x, d) on the edge map corresponds to an element $(x, \phi(d))$ in DCD tensor. The element encodes the minimum distance to scene edge points in a joint direction and location space. The DCD tensor Ω is defined as:

$$\Omega(x, \phi(d)) = \min_{x_i, d_i \in \mathcal{S}} (\|x - x_i\| + \lambda \|\phi(d) - \phi(d_i)\|) \tag{13}$$

The DCD tensor can be easily computed by employing a forward and backward recursion to the sequence of distance transform maps [4]. By looking up the DCD

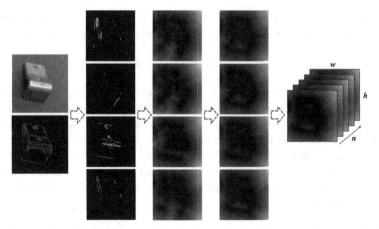

Fig. 5. Computation of the DCD tensor. First column: the input image and extracted edge map. Second column: divided edge subsets according to the quantized edge direction. Third column: distance transform maps of edge subsets. Fourth column: directional distance transform maps. The last: the DCD tensor.

tensor, we can compute the matching cost in linear time. Since the bounding box of the object is detected previously, we can eliminate a large number of possible matching results by verifying the intersection over union (IoU) ratio between the bounding box of the template and the detected bounding box. Only the matching results with IoU ratio bigger than a threshold are selected as pose hypotheses. Finally, we sort these pose hypotheses for increasing matching cost d_{DCM} ($\mathcal{T}$, $\mathcal{S}$), the top-rated 10 poses are selected as candidate poses.

3.4 Pose Optimization Using IRLS

The candidate poses are coarse estimation, we need to recover the accurate pose and remove the outliers. We refine the object pose by solving a non-linear optimization problem that minimizes the DCM matching cost function.

The pose parameters to be optimized are the rotation vector $\mathbf{r} = [r_x, r_y, r_z]^T$ and the translation vector $\mathbf{t} = [t_x, t_y, t_z]^T$. Each candidate template $\mathcal{T}$ contains a set of 3D raster points, we project $\mathbf{x}_i$ onto the 2D image plane and search the edge pixel x_j on the scene image $\mathcal{S}$ that minimizes the directional chamfer distance as the corresponding point. After establishing correspondence, an iterative optimization process is performed to find the optimal pose so that the projection error is minimized. For a 3D raster points $(\mathbf{x}_i, \mathbf{d}_i)$, the projection error e_i is denoted as follows, which can be computed by looking up the DCD tensor.

$$e_i = \Omega[\pi(\mathbf{x}_i, \mathbf{r}, \mathbf{t}),\ \phi(\sigma(\mathbf{d}_i, \mathbf{r}, \mathbf{t}))] \tag{14}$$

However, the foreground occlusion and background clutter will cause mismatches. A simple quadratic error is sensitive to these outliers. Motivated by [22], a re-weighted least-squares (IRLS) method is utilized to suppress the effects of these mismatches.

Each projection error term e_i is multiplied by a weight function w_i. The edge direction consistency score s_i is taken as the weight function and is defined as follows:

$$s_i(\mathbf{x}_i, \mathbf{d}_i) = |\cos(\mathbf{I}_\theta(\pi(\mathbf{x}_i)) - n(\sigma(\mathbf{d}_i)))| \tag{15}$$

The score compares the projected point orientations with the current local image gradient directions. Given a 3D raster point $(\mathbf{x}_i, \mathbf{d}_i)$, the projected image pixel location is $\pi(\mathbf{x}_i)$ and the image normal direction in radians is $n(\sigma(\mathbf{d}_i))$. $\mathbf{I}_\theta(\pi(\mathbf{x}_i))$ is the local image gradient direction at the same pixel $\pi(\mathbf{x}_i)$. In the ideal case of a correct match, these two directions should be consistent and s_i would be close to 1. The edge direction consistency score is robust to outlier matches caused by occlusion, edge missing and disturbing edges, which cannot keep high consistency in edge directions. The IRLS loss function is denoted as follows:

$$L(\mathbf{r}, \mathbf{t}) = \frac{1}{2} \sum_{i=1}^{m} w_i \cdot e_i^2 \tag{16}$$

Starting from the initial value $\mathbf{p}^0 = [\mathbf{r}^0, \mathbf{t}^0]$ given by the candidate pose, the optimal 6D pose is calculated by iteratively minimizing the loss function using the Levenberg-Marquardt (L-M) algorithm. The derivatives of loss function required by the Jacobin matrix is given by:

$$\nabla L = \sum_{i=1}^{m} w_i \cdot \Omega \cdot \nabla\Omega \cdot \nabla[\pi(\mathbf{x}_i), \phi(\sigma(\mathbf{d}_i))] \tag{17}$$

Since the DCD tensor Ω is only defined at discrete pixel frame and direction channels, a numerical and approximate method is used to calculate its derivatives $\nabla\Omega$. The x and y derivatives are computed as the image derivatives of the currently selected distance maps at pixel location $\pi(\mathbf{x}_i)$, the derivative along the orientation direction is given by:

$$\frac{\partial\Omega}{\partial\sigma(\mathbf{d}_i)} = \frac{1}{2}[\Omega(x, \ \phi(\sigma(\mathbf{d}_i)) + 1) - \Omega(x, \ \phi(\sigma(\mathbf{d}_i)) - 1))] \tag{18}$$

3.5 Pose Validation

After the optimization process, we obtain the refined poses for these pose candidates. Due to the existence of local minima after optimization, we need to eliminate these outliers and select the final 6D poses. The direction consistency score S is used to validate these optimization results. The value of S ranges from 0 to 1. We take those refined poses whose score exceeds a threshold as the final results. S is defined as:

$$S(\mathcal{T}, \mathbf{r}, \mathbf{t}) = \frac{1}{m} \sum_{\mathbf{x}_i, \mathbf{d}_i \in \mathcal{T}} s_i(\mathbf{x}_i, \mathbf{d}_i) \tag{19}$$

4 Experimental Results

In this section, experiments are performed to validate the proposed method. We use Intel RealSense D435 camera to collect a dataset of texture-less industrial parts in different light condition and foreground occlusion. The dataset is composed of 640×480 RGB-D images of scenes that contain more than five object instances in arbitrary 6D pose.

In the object detection stage, we fine-tune the YOLOv3 on our synthetic dataset using a Darknet-53 backend pretrained on ImageNet. The synthetic dataset contains 20,000 labelled images. In pose hypotheses generation stage, edge templates are generated at three different scales by setting the virtual camera at depth 0.3 m, 0.36 m and 0.5 m. At each depth, we generate 642 camera viewpoints that evenly distributed on the surface of sphere. The in-plane rotation step is set to $20°$. These 3D template edges are discretized into edge points by rasterization operation at a step of 0.005 m. The final template library contains 13482 templates. As for DCD tensor, we discretize the edges into 60 directions, thus the DCD tensor size is $640 \times 480 \times 60$. The weighting factor λ is set to 6.0. The IoU filter threshold is set to 0.7 and we select the top 10 rated matching results according to their matching cost as the final candidate poses. In pose optimization stage, the edge direction consistency score is set to 0.8 to help select out the final accurate poses. Some experiment results on our dataset are shown in Fig. 6.

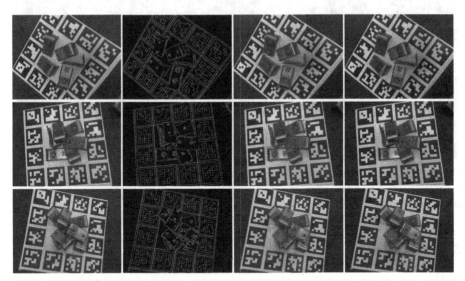

Fig. 6. Experiment results of the proposed method on our dataset. First column: the input image. Second column: edge maps with highlight removal. Third column: initial pose hypothesis. Last column: pose optimization result.

Our pipeline takes an average time of 165 ms for object detection in the input 640×480 RGB image and 75 ms for the DCD tensor calculation. The FDCM takes about 250 ms to generate 10 candidate poses and the final pose optimization takes about 375 ms. All the algorithms run on a computer with Intel i7-9700K CPU, NVidia GTX

Titan 2080Ti GPU, and 64 GB memory. As for object detection, the YOLOv3 network can obtain the bounding boxes of complete or partially occluded parts accurately. As for pose estimation, we compare the estimated pose with the ground truth, only if the total translation distance error is less than 5 mm and rotation angular error is less than 0.2 radians, this pose is considered correct. The proposed method achieves correct 6D pose estimation rate of 85% on our dataset.

To evaluate the effectiveness of the IRLS optimization, we perform the pose refinement using IRLS optimization and direct optimization respectively. For direct optimization, we set the weight w_i of each error term to 1.0 while keeping other parameters the same as IRLS optimization. The results are shown in Fig. 7. The direct optimization gets miss aligned results caused by partial occlusion (Fig. 7(e) and (h)) or disturbing edges (Fig. 7(f) and (g)). However, the IRLS optimization is more robust to these disturbances.

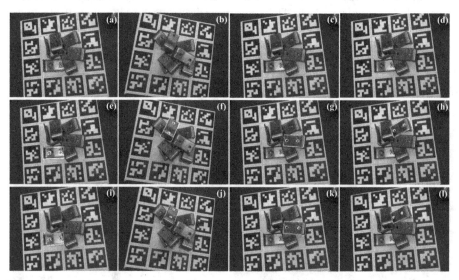

Fig. 7. Comparison between IRLS optimization and direct optimization. First row: the initial candidate poses. Second row: final results using direct optimization. Last row: final results using IRLS optimization.

5 Conclusion

In this paper, we have proposed a 6D pose estimation approach for texture-less industrial part in the crowd. A YOLOv3 network is used to detect the bounding boxes of object instances in the image. We design an efficient approach to automatically synthesize training data. An edge-based template matching method is used to generate coarse pose hypotheses. Refined 6D poses are obtained by employing a non-linear optimization procedure that minimizes an IRLS loss function. Several experimental results demonstrate the capabilities of our proposed method in handling texture-less and partially occluded objects. However, our method still needs improvement to deal with severe occlusion scenarios.

Acknowledgement. This work is jointly supported by National Key R&D program of China (2018YFB1306500) and National Natural Science Foundation of China (51805325, 51675325).

References

1. Lowe, D.G.: Distinctive image features from scale-invariant key-points. Int. J. Comput. Vis. **60**(2), 91–110 (2004)
2. Bay, H., Tuytelaars, T., Van Gool, L.: SURF: speeded up robust features. In: Leonardis, A., Bischof, H., Pinz, A. (eds.) ECCV 2006. LNCS, vol. 3951, pp. 404–417. Springer, Heidelberg (2006). https://doi.org/10.1007/11744023_32
3. Rublee, E., Rabaud, V., Konolige, K., Bradski, G.: ORB: an efficient alternative to SIFT or SURF. In: Proceedings of the 2011 International Conference on Computer Vision, pp. 2564–2571 (2011)
4. Liu, M.-Y., Tuzel, O., Veeraraghavan, A., Taguchi, Y., Chellappa, R.: Fast object localization and pose estimation in heavy clutter for robotic bin picking. Int. J. Robot. Res. **31**(8), 951–973 (2012)
5. Imperoli, M., Pretto, A.: D^2CO: fast and robust registration of 3D textureless objects using the Directional Chamfer Distance. In: Nalpantidis, L., Krüger, V., Eklundh, J.-O., Gasteratos, A. (eds.) ICVS 2015. LNCS, vol. 9163, pp. 316–328. Springer, Cham (2015). https://doi.org/10.1007/978-3-319-20904-3_29
6. Redmon, J., Farhadi, A.: YOLOv3: an incremental improvement. arXiv preprint arXiv:1804.02767 (2018)
7. Hinterstoisser, S., et al.: Gradient response maps for real-time detection of texture-less objects. IEEE Trans. Pattern Anal. Mach. Intell. **34**(5), 876–888 (2012)
8. Rios-Cabrera, R., Tuytelaars, T.: Discriminatively trained templates for 3D object detection: a real time scalable approach. In: ICCV, pp. 2048–2055 (2013)
9. Kehl, W., Tombari, F., Navab, N., Ilic, S., Lepetit, V.: Hashmod: a hashing method for scalable 3D object detection. In: BMVC, pp. 1–12 (2015)
10. Hodan, T., Zabulis, X., Lourakis, M., Obdrzalek, S., Matas, J.: Detection and fine 3D pose estimation of texture-less objects in RGB-D images. In: IROSs, pp. 4421–4428 (2015)
11. Rusu, R.B., Marton, Z.C., Blodow, N.: Persistent point feature histograms for 3D point clouds. In: Proceedings of the 10th International Conference on Intelligent Autonomous Systems (IAS-10), Baden-Baden, Germany (2008)
12. Rusu, R.B., Blodow, N., Beetz, M.: Fast point feature histograms (FPFH) for 3D registration. In: IEEE International Conference on Robotics and Automation, pp. 3212–3217 (2009)
13. Rusu, R.B., Bradski, G., Thibaux, R.: Fast 3D recognition and pose using the viewpoint feature histogram. In: IEEE/RSJ International Conference on Intelligent Robots and Systems, pp. 2155–2162 (2010)
14. Drost, B., Ulrich, M., Navab, N., Ilic, S.: Model globally, match locally: efficient and robust 3D object recognition. In: 2010 IEEE Computer Society Conference on Computer Vision and Pattern Recognition, pp. 998–1005 (2010)
15. Choi, C., Christensen, H.I.: 3D pose estimation of daily objects using an RGB-D camera. In: IEEE/RSJ International Conference on Intelligent Robots and Systems, pp. 3342–3349 (2012)
16. Brachmann, E., Krull, A., Michel, F., Gumhold, S., Shotton, J., Rother, C.: Learning 6D object pose estimation using 3D object coordinates. In: Fleet, D., Pajdla, T., Schiele, B., Tuytelaars, T. (eds.) ECCV 2014. LNCS, vol. 8690, pp. 536–551. Springer, Cham (2014). https://doi.org/10.1007/978-3-319-10605-2_35

17. Doumanoglou, A., Kouskouridas, R., Malassiotis, S., Kim, T.K.: Recovering 6D object pose and predicting next-best-view in the crowd. In: Proceedings of the IEEE Conference on Computer Vision and Pattern Recognition, pp. 3583–3592 (2016)

18. Kehl, W., Milletari, F., Tombari, F., Ilic, S., Navab, N.: Deep learning of local RGB-D patches for 3D object detection and 6D pose estimation. In: Leibe, B., Matas, J., Sebe, N., Welling, M. (eds.) ECCV 2016. LNCS, vol. 9907, pp. 205–220. Springer, Cham (2016). https://doi.org/10.1007/978-3-319-46487-9_13

19. Kehl, W., Manhardt, F., Tombari, F., Ilic, S., Navab, N.: SSD-6D: making RGB-based 3D detection and 6D pose estimation great again. In: Proceedings of the IEEE International Conference on Computer Vision, pp. 1521–1529 (2017)

20. Xiang, Y., Schmidt, T., Narayanan, V.: Pose-CNN: a convolutional neural network for 6D object pose estimation in cluttered scenes. In: Robotics: Science and Systems (RSS) (2018)

21. dos Santos Souza, A.C., et al.: Real-time high-quality specular highlight removal using efficient pixel clustering. In: 2018 31st SIBGRAPI Conference on Graphics, Patterns and Images (SIBGRAPI), Parana, pp. 56–63 (2018)

22. Wang, B., Zhong, F., Qin, X.: Robust edge-based 3D object tracking with direction-based pose validation. Multimed. Tools Appl. **78**, 12307–12331 (2019)

Multi-scale Crack Detection Based on Keypoint Detection and Minimal Path Technique

Nianfeng Wang[(✉)], Hao Zhu, and Xianmin Zhang

Guangdong Province Key Laboratory of Precision Equipment and Manufacturing Technology, South China University of Technology, Guangzhou, Guangdong 510640, People's Republic of China
menfwang@scut.edu.cn

Abstract. In this paper, a method for the detection of multi-scale cracks based on computer vision is introduced. This crack detection method is divided into two parts to extract crack features from the images. In the first part, the original image is mapped to different scale Spaces and the pixels with strong ridge characteristic are detected with a Hessian-matrix in these scales Spaces. Then the detection results in different scales are superimposed. Finally, an evaluation index is designed to select the Keypoints detected in the previous step. In the second part, the cracks are detected based on a modified Fast Marching Mothed which is improved into an iterative algorithm with self-terminating capability. The Keypoints detected and selected in the first part are used as endpoints for the crack detection. Then the burrs are removed from the detection results. The experimental results show that under different lighting and road conditions, the crack feature can be extracted stably by this method.

Keywords: Multi-scale crack detection · Keypoint detection · Minimal path technique · Fast marching mothed

1 Introduction

Surface cracks on objects are often difficult to inspect automatically due to its complex background, varies topology and intensity inhomogeneity. In many cases such as pavement or magnetic components, cracks are even in poor continuity and low contrast. Volumes of papers have contributed to address this challenging problem. Most of these methods are based on three shared assumptions of crack.

(1) **Intensity:** relative low intensity compared to image background.
(2) **Continuity:** pixels belonging to the cracks form continuous paths of arbitrary shape.
(3) **Edge feature:** cracks can be represented by thin edge/ridge/line features.

© Springer Nature Switzerland AG 2020
C. S. Chan et al. (Eds.): ICIRA 2020, LNAI 12595, pp. 429–441, 2020.
https://doi.org/10.1007/978-3-030-66645-3_36

Combinations of the premises listed above generate different approaches that can be categorized as follow.

Intensity Based Only: this photometric property of cracks is the sole basis of most early studies. Therefore threshold-based methods are popular in early researches. But with intensity information only is not sufficient to distinguish crack pixel from background due to the illumination inhomogeneity. Moreover, simply taking brightness into consideration makes these methods be sensitive to noise and thus be far from satisfactory.

Continuity and Intensity Based: for more robust and precise crack detection, continuity and intensity are combined and highly performance algorithms are proposed. This is because continuity raises the possibility of taking geometric information into account that allow for usage of higher-level features. Based on the additional assumption that cracks are locally organized along 4 possible global orientations, Nguyen [1] proposes the "free-form anisotropy" (FFA), by measuring the correlation of 4 minimal paths with same distance generated from one source point for each pixel in an image, FFA is able to extract arbitrary cracks with good continuity and high contrast. In [2], crack extraction problem is formulated as a seed-growing problem and Ford algorithm is used to collect crack strings within a square sub-image centered at seed point. In [3], Image is first partitioned to blocks where endpoints are selected by intensity. Then Dijkstra algorithm is executed to generate minimal paths which are filtered in post-processing to obtain final detection. Zou [4] identify crack seeds through tensor voting and crack probability-map generation, with these crack seeds a graph is constructed then minimum spanning trees are found to obtain the coarse crack segmentation. In [5], Candidate pixels are identified within each sub-window by applying intensity threshold. Dijkstra is adopted to generate candidate crack paths and a statistical hypothesis test is developed to remove false-positive cracks.

Edge Feature Based: Since cracks appear as thin edge/ridge intuitively, edge-detection based methods are proposed. Traditional edge detector like the Laplacian of Gaussian (LoG) algorithm [6], Canny operator [7], Sobel [8] are employed to deal with different kinds of crack detection task including bridge deck maintenance, magnetic tile defect inspection and pavement crack detection. Due to the poor performance in heavy noise of these methods based on traditional edge detector, approaches developed from frequency domain analysis techniques such as Discrete Fourier Transform [9] is proposed. But time information is lost and it is hard to identify the position of the edge. To counter this problem, methods of multi-scale analysis including Gabor filter [10] and wavelet transform [11] are proposed. Curvelet transform [7], shearlet transform [12,13], contourlet transform [14] and ridgelet transform[15] are adopted to address problems like road crack detection, magnetic tile crack inspection and surface detection of continuous casting slabs. The main drawback of these edge-detection base methods is the limitation in distinguishing cracks tangled by complex texture or with low continuity.

It can be observed that minimal path methods show the best performance in the accuracy of crack images segmentation. To exploit the advantage of minimal path technique and to break the limitation of its prior knowledge requiring, the idea is to adopt the minimal path approach with precise initial point detection.

In Sect. 2, the Keypoint detection in scale space is detailed. In Sect. 3, the crack segmentation approach based on Keypoint and minimal path technique is described. In Sect. 4, the experimental results are proposed to demonstrate the performance of introduced method. In Sect. 5, conclusions and future works are discussed.

2   Multi-scale Keypoint Detection and Selection

As for the minimal path method in crack detection issue, an ideal situation is that a least and sufficient pixel set is provided automatically as the initial points of minimal path algorithm, instead of the manual choose or redundant candidates that coarsely selected by intensity threshold. Note that the main motivation of Keypoint detection is to efficiently and accurately localize a proper number of pixels that are right on the crack region as the input to minimal path algorithm which would be described in Sect. 3.

2.1   Multi-scale Keypoint Detection

The first step of Multi-scale Keypoint detection is to map the original image to different scale Spaces and the pixels with strong ridge characteristic are detected with a Hessian-matrix in these scale Spaces. It has been shown by Linderberg [16] that Gaussian function is the only kernel to generate scale space. That is, scale spaces are usually implemented as an image pyramid. The images are iteratively convoluted with a Gaussian kernel and then down-sampled for a higher level of the pyramid. Lowe [17] subtracts these blurred images within each octave to get the Difference of Gaussian (DoG) images and applied non-maximum suppression in a 3×3 neighborhood to localize edges and blobs.

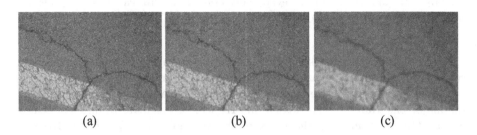

Fig. 1. The original image in different scale Spaces. (a) The 1st scale ($\sigma = 0$) of original image. (b) The 2nd scale ($\sigma = 2$) of original image. (c) The 3rd scale ($\sigma = 4$) of original image.

As shown in Fig. 1, the original image is mapped to different scale Spaces. The construction of the scale space starts with $\sigma = 0$, then the next scale is $\sigma = 2$. From second scale the σ is multiplied with a factor of 2, i.e., the third scale is $\sigma = 4$ and the fourth scale is $\sigma = 8$ and so on. After each convolution with Gaussian kernel, the blurred image is down-sampled to half the size, as shown in Fig. 1(b), (c).

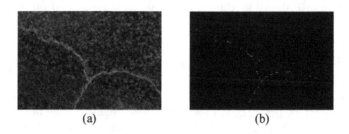

(a)   (b)

Fig. 2. Keypoint detection and selection. The size of the pictures shown here is 480×320. (a) Ridge characteristic extraction. Pixels are displayed in grayscale corresponding to the major eigenvalue of Hessian Matrix. (b) By applying a threshold on ridge.

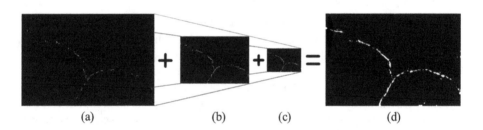

(a)   (b)   (c)   (d)

Fig. 3. The Keypoint detection in different scale Spaces. (a) is the result of Keypoint detection in the 1st scale ($\sigma = 0$). (b) is the result of Keypoint detection in the 2nd scale ($\sigma = 2$). (c) is the result of Keypoint detection in the 3rd scale ($\sigma = 4$). (d) is the superposition result of (a), (b), (c).

According to the prior knowledge, there is a strong correlation between the ridge feature of an image and the major eigenvalue of the Hessian matrix at a certain point. Ridge feature may occur at different scales for different images so the major eigenvalue of the Hessian matrix will be calculate in different scales Spaces. Given an image I, the major eigenvalue, noted as e_{max}, of the Hessian matrix can be computed by:

$$e_{max} = \frac{1}{2}(I_{xx} + I_{yy} + \sqrt{I_{xx}^2 + 4I_{xy}^2 - 2I_{xx}I_{yy} + I_{yy}^2}) \tag{1}$$

where I_{xx}, I_{yy}, I_{xy} are the second derivatives. As shown in Fig. 2(a), taking the original image for an example, the e_{max} values of every points in I are calculated and normalized to the range $[0, 255]$. To remove those pixels with weak ridge feature, all points with a e_{max} value less than 150 were eliminated. Figure 2(b) shows the result of removing those pixels with weak ridge features in the original image. The same operation is carried out for other scale Spaces. In Fig. 3, (a), (b), (c) are the results of Keypoint detection of (a), (b), (c) in Fig. 1 respectively. And Fig. 3 (d) is the superposition result of (a), (b), (c). The points on the ridge feature in the original image are detected.

2.2 Keypoint Selection

The previous step has detected the Keypoints with ridge feature in different scales and superposed the results of detection. This section introduces how to select the Keypoints that have been detected.

As shown in Fig. 4, the gradient vector field of local crack area is established. Figure 4(c) shows that the gradient vectors are distributed in the opposite direction along the crack. And the gradient magnitude of the points at the crack center line are lower, while the gradient magnitude of the points on both sides are higher. According to this property, an evaluation index is designed to measure whether a point is in the crack areas. The gradient vectors in a certain point and its eight neighborhoods are transformed into a discrete point in the 2-D coordinate system, as shown in Fig. 5(a).

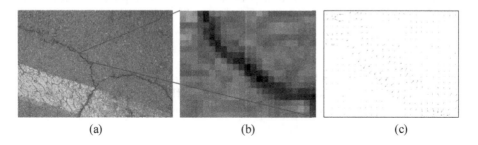

      (a)   (b)   (c)

Fig. 4. (a) Source image. (b) Local crack image by of (a). (c) The gradient vector field of (b).

Algorithm 1. K-meansClusteringAlgorithm

Require: Point set x_i, the number of points n and initial conditions $N = 6, c = 2$.
Ensure: Six points divided into two groups.
 1: Set $U(s) = 0$; $L(s) = 0$ and $Status(s) = Trail$;
 2: **while** $(n > L)$ **do**
 3:     K-means clustering; Getting two class centers μ_1, μ_2;
 4:     Calculating the Euclidean distance d_{ij} between x_{ij} and μ_i in each class;
 5:     Eliminating d_{ijmax}; $n = n - 1$;
 6: **end while**

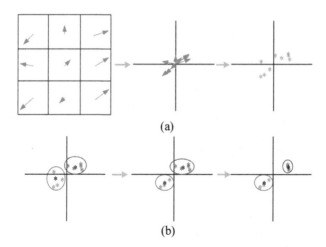

Fig. 5. (a) The transformation of gradient vectors into discrete points. (b) The process of clustering.

As shown in Algorithm 1, K-means algorithm is used to cluster the discrete points. Finally, 6 points are retained and divided into two groups. The clustering process is shown in Fig. 5(b). The evaluation index should express the characteristics of the clustering results as much as possible. Two class centers μ_1, μ_2 and the Euclidean distance d_{ij} between x_{ij} and μ_i in each class have been acquired from Algorithm 1. Therefore, the dispersion of the discrete points in each class can be expressed as follows:

$$S_i = \sum_{j=1}^{m} d_{ij}^2 \tag{2}$$

where m is the number of points in each class, d_{ij} is the Euclidean distance between the jth point and the class center μ_i in the ith class. The Euclidean distance between two classes can be expressed as follows:

$$d = (\mu_1 - \mu_2)^2 \tag{3}$$

Then the evaluation index can be expressed by a ratio of d and weighted sum of S_i, which is expressed as follows:

$$Index = \frac{d}{\sum \omega_i S_i} \tag{4}$$

where ω_i is the weighted of S_i. The Keypoints detected in previous step can be selected by the evaluation index $Index$. The points at the crack areas always have a larger $Index$ value. Based on the $Index$ value, as shown in Fig. 6(a) a proper amount of Keypoints with the largest $Index$ value are chosen as the final Keypoints. The result of Keypoint selection is shown in Fig. 6(b).

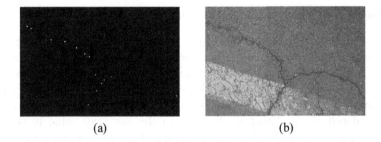

(a)   (b)

Fig. 6. The result of multi-scale Keypoint detection and selection. (a) The white points are the Keypoints that are detected and selected. (b) The Keypoints marked in the original image.

3 Crack Detection Based on Self-terminating Fast Marching Method

By the ordinary Fast Marching Method [18,19], the cracks in the image have been found. However, the detection results depend on the two points (s and p). As shown in Fig. 7(b), the detected crack curve is confined between s and p. In other words, the quality of the detection depends on the selection of two Keypoints and only one crack curve can be detected. Therefore, an optimized method is needed to improve the adaptability of detection algorithm. The self-terminating Fast Marching Methods requires three elements:

(a) one or several source points s on the cracks;
(b) an automatic iteration rule based on FMM;
(c) a suitable termination condition.

In Sect. 2, the Keypoints on the cracks had been detected. A source point set S can be created by the Keypoints. In this section, an automatic iteration rule based on FMM and a suitable termination condition on crack detection will be introduced.

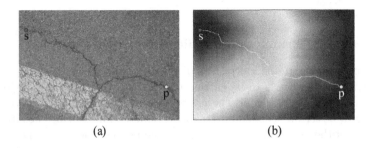

(a)   (b)

Fig. 7. The process of FMM with source point s and terminal point p.

3.1   An Automatic Iteration Rule Based on FMM

As shown in Fig. 8(a), a point s on the crack which is one of the source point set S was selected. Then the geodesic distance map U with s as the initial point was generated in Fig. 8(b). In the geodesic distance map U, the Euclidean distance L values of each point was calculated. When the point whose L reached a preset value λ firstly was obtained, the first step of the iteration terminated and this point was noted as s_1. Theoretically, due to the low gray value in the crack region, the wave propagates faster in the crack. Therefore, at the same time of arrival, the points on the crack region always have larger L values. This is why s_1 is considered a node for the iteration. As shown in Fig. 8(c), the geodesic distance map U with s and s_1 as the initial points was generated. Same as in the previous step, the iteration terminated at a point whose L firstly reached λ. And this point was noted as s_2. In this way, the iterative process is repeated as shown in Fig. 8(d) and Fig. 8(e). As shown in Fig. 8(f), the iterative process terminates when it goes to the 49th.

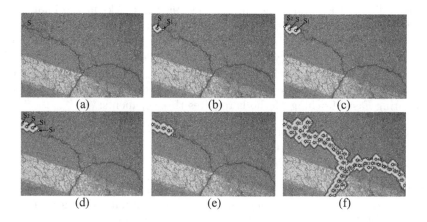

Fig. 8. The process of Self-terminating Fast Marching Method.

3.2   A Suitable Termination Condition on Crack Detection

In order to make the detection results of crack curve as accurately as possible, the iterative process needs to be carried out normally on the crack region, and terminates immediately beyond the crack region. Therefore, it is important to design a suitable termination condition on crack detection. During each iteration, the U value of the detected crack region and the points around it are constantly updated.

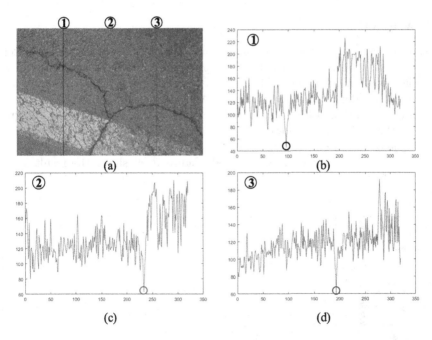

Fig. 9. Single column gray value of the crack image. (a) The original image marked by three columns. (b) The gray value information on columns① in (a). (c) The gray value information on columns② in (a). (d) The gray value information on columns③ in (a).

Figure 9 shows one dimensional gray value information on three different columns of the image. It is easy to find that the gray value of the pixels in the crack region is lower than that of the surrounding pixels. Meanwhile, the U value is the result of summing up all the gray value of points in one iteration. In theory, if the iterative process always goes in the crack region, the U value will be smaller than the product of the step length and the mean gray value of the image. Then the termination condition can be shown as follow:

$$ifU_1(x) > avg * \lambda, x \in \Omega \tag{5}$$

where avg is the average of the gray value of the original image, λ is the step length of every iterative process. However, considering the diversity of image information, variance is introduced. Variance represents the degree of dispersion of gray values in an image.

$$ifU_1(x) > avg * (k_1\lambda - k_2\sigma), x \in \Omega \tag{6}$$

where σ is the standard deviation of the gray value of the original image, k_1, k_2 are the coefficients of λ, σ, respectively. λ and σ specific gravity can be adjusted by adjusting the values of k_1 and k_2. As shown in Fig. 8(f), according to the termination condition, the crack search terminates automatically.

3.3 Burr-Free

As shown in Fig. 10(b), the crack skeleton has been obtained in the previous step. However, there are several burrs on the extraction results. This section describes a method for removing burrs. The idea is to divide the extracted crack into several independent connected domains. And then, each connected domain is mapped to the original image as shown in Fig. 10(a). It is known that the gray value of the crack area is lower than the background. Therefore, the average gray value of each isolated connected domain is calculated, and the connected domain with large gray value is removed. Figure 10(c) shows the result of (b) after burr-free process.

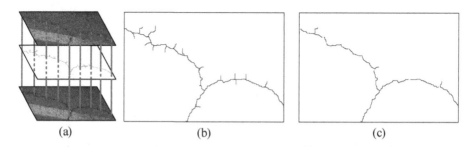

| (a) | (b) | (c) |

Fig. 10. The process of burr-free. (a) The detected crack is mapped to the original image. (b) is the crack detected in the previous step. (c) is the result of (b) after burr-free process.

4   Experimental Results

In order to analyze the self-terminating minimal path algorithm explained in Sect. 3, the algorithm was tested on the CFD dataset [20]. This public dataset is composed of 118 images with a resolution of 480×320 pixels and each image has hand labeled ground truth. As shown in Fig. 11, ten representative images were selected from the data set. Column (a) and column (d) are the original images. Column (b) and column (e) are the results of self-terminating minimal path algorithm. Column (c) and column (f) are the ground truth. By comparing the experimental results and the ground truth, it's obvious that the algorithm works well on multi-scale crack detection. The method can be used to completely extract the crack in a single image without branching. Complete extraction can also be realized in the image with branching but uniform crack gray value. Although incomplete extraction only occurs in some complex images such as the network cracks. The general shape of the crack can often be detected. The crack detection algorithm can basically meet the requirements.

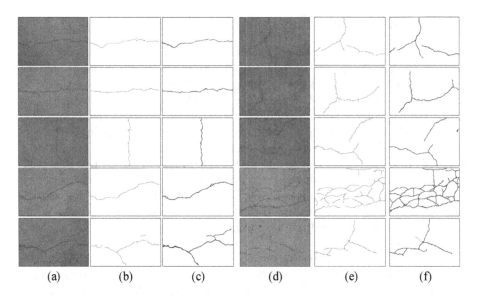

(a)   (b)   (c)   (d)   (e)   (f)

Fig. 11. Some experimental results of CFD dataset. Column (a) and column (d) are the original images. Column (b) and column (e) are the results of self-terminating minimal path algorithm. Column (c) and column (f) are the ground truth.

To eveluate the performance of the crack detection algorithms quantitatively, Precision, Recall and F1-measure are employed. They are computed based on true positive (TP), false negative (FN), and false positive (FP) as follows:

$$\text{Precision} = \frac{\text{TP}}{\text{TP} + \text{FP}} \tag{7}$$

$$\text{Recall} = \frac{\text{TP}}{\text{TP} + \text{FN}} \tag{8}$$

$$\text{F1-measure} = 2 \times \frac{\text{Precision} \times \text{Recall}}{\text{Precision} + \text{Recall}} \tag{9}$$

The crack detection result on the CFD dataset is shown in Table 1. Comparing with the MFCD (multi-scale fusion based crack detection) algorithm proposed in [5], the method in this paper has a better Precision and F1-measure.

Table 1. Crack detection results evaluation on CFD dataset

Method	Precision	Recall	F1-measure
MFCD [5]	0.8990	0.8947	0.8804
Method in this paper	**0.9314**	0.8636	**0.8962**

5   Conclusions and Future Works

In this paper, a multi-scale crack detection based on Keypoint detection and minimal path technique is introduced. The method can be divided into two steps. The first step is multi-scale Keypoint detection. The multi-scale Keypoint detection is to identify pixels with strong ridge characteristic in different scales with a Hessian-matrix and superimpose the detection results of them. And then compute a indicator for the local image region to select Keypoints. The second step is to detect the cracks based on Fast Marching Mothed. The FMM is improved into an iterative algorithm with self-terminating capability. The Keypoints detected and selected in the previous step are used as endpoints for crack detection. Then the burr was removed from the detection results. The experimental results show that the algorithm is effective in pavement crack detection. Under different lighting and road conditions, the crack feature can be extracted stably by this method.

However, incomplete detection of complex cracks sometimes occurs. Although the results of individual images can be improved by adjusting the parameters in the algorithm. The results of other images will be affected. The future works is to improve the algorithm for the complete extraction of complex crack features.

Acknowledgements. This work is supported by National Natural Science Foundation of China (Grant Nos.U1713207 and 52075180), Science and Technology Program of Guangzhou (Grant Nos.201904020020), and the Fundamental Research Funds for the Central Universities.

References

1. Nguyen, T.S., Bégot, S., Duculty, F., Avila, M.: Free-form anisotropy: a new method for crack detection on pavement surface images. In: IEEE International Conference on Image Processing (2011)
2. Li, Q., Zou, Q., Zhang, D., Mao, Q.: Fosa: F* seed-growing approach for crack-line detection from pavement images. Image & Vision Computing (2011)
3. Amhaz, R., Chambon, S., Idier, J., Baltazart, V.: Automatic crack detection on two-dimensional pavement images: an algorithm based on minimal path selection. IEEE Trans. Intell. Transp. Syst. **17**(10), 2718–2729 (2016)
4. Qin, Z., Yu, C., Li, Q., Mao, Q., Wang, S.: Cracktree: automatic crack detection from pavement images. Pattern Recogn. Lett. **33**(3), 227–238 (2012)
5. Li, H., Dezhen, S., Yu, L., Li, B.: Automatic pavement crack detection by multi-scale image fusion. IEEE Trans. Intell. Transp. Syst. **99**, 1–12 (2018)
6. Lim, R.S., La, H.M., Sheng, W.: A robotic crack inspection and mapping system for bridge deck maintenance. IEEE Trans. Autom. Sci. Eng. **11**(2), 367–378 (2014)
7. Li, X., Jiang, H., Yin, G.: Detection of surface crack defects on ferrite magnetic tile. Ndt E Int. **62**, 6–13 (2014)
8. Ayenu-Prah, A., Attoh-Okine, N.: Evaluating pavement cracks with bidimensional empirical mode decomposition. EURASIP J. Adv. Signal Process. **1–7**, 2008 (2008)
9. Paulraj, M.P., Shukry, A.M.M., Yaacob, S., Adom, A.H., Krishnan, R.P.: Structural steel plate damage detection using DFT spectral energy and artificial neural network. In: 2010 6th International Colloquium on Signal Processing & its Applications, pp. 1–6. IEEE (2010)

10. Salman, M., Mathavan, S., Kamal, K., Rahman, M.: Pavement crack detection using the gabor filter. In: 16th International IEEE Conference on Intelligent Transportation Systems (ITSC 2013), pp. 2039–2044. IEEE (2013)
11. Yang, C., Liu, P., Yin, G., Jiang, H., Li, X.: Defect detection in magnetic tile images based on stationary wavelet transform. Ndt E Int. **83**, 78–87 (2016)
12. Yang, C., Liu, P., Yin, G., Wang, L.: Crack detection in magnetic tile images using nonsubsampled shearlet transform and envelope gray level gradient. Optics Laser Technol. **90**, 7–17 (2017)
13. Xie, L., Lin, L., Yin, M., Meng, L., Yin, G.: A novel surface defect inspection algorithm for magnetic tile. Appl. Surf. Sci. **375**, 118–126 (2016)
14. Fang, L.: An image segmentation technique using nonsubsampled contourlet transform and active contours. Soft. Comput. **23**(6), 1823–1832 (2018). https://doi.org/10.1007/s00500-018-3564-4
15. Zhang, D., Shiru, Q., He, L., Shi, S.: Automatic ridgelet image enhancement algorithm for road crack image based on fuzzy entropy and fuzzy divergence. Opt. Lasers Eng. **47**(11), 1216–1225 (2009)
16. Lindeberg, T.: Scale-space theory: a basic tool for analyzing structures at different scales. J. Appl. Statist. **21**(1–2), 225–270 (1994)
17. Lowe, D.G.: Distinctive image features from scale-invariant keypoints. Int. J. Comput. Vis. **60**(2), 91–110 (2004)
18. Osher, S., Sethian, J.A.: Fronts propagating with curvature-dependent speed: algorithms based on hamilton-jacobi formulations. J. Comput. Phys. **79**(1), 12–49 (1988)
19. Tsitsiklis, J.N.: Efficient algorithms for globally optimal trajectories. IEEE Trans. Autom. Control **40**(9), 1528–1538 (1995)
20. Shi, Y., Cui, L., Qi, Z., Meng, F., Chen, Z.: Automatic road crack detection using random structured forests. IEEE Trans. Intell. Transp. Syst. **17**(12), 3434–3445 (2016)

A Self-correction Based Algorithm for Single-Shot Camera Calibration

Shuangfei Yu[1], Jie Hong[2], Tao Zhang[1], Zhi Yang[1], and Yisheng Guan[1(✉)]

[1] Guangdong University of Technology, Guangzhou, Guangdong Province, China
ysguan@gdut.edu.cn
[2] School of Engineering and Computer Science, Australian National University,
Canberra, ACT 2601, Australia

Abstract. Camera calibration is a fundamental task in photogrammetry and computer vision. In view of the requirements in live camera characteristics, we present a novel calibration approach to obtain all the camera optimal parameters and the distortion rectification by using only a single image. The existing automatic calibration approaches inspired by the vanishing point theory or the homography matrix are not capable of entirely deal with the internal, external and the len distortion parameters simultaneously. Compared to the previous works, our approach solves the problem by applying the linear features extracted from the image and thus improves the accuracy and efficiency. Results of experiments in different scenes demonstrate the comparable performance to the traditional methods.

Keywords: Camera calibration · Single image · Distortion · Self-correction

1 Introduction

The traditional methods in camera calibration, such as the classical Zhang's method [1], typically require multiple images taken of the target at different distances and orientations [2], which are cumbersome and time-consuming. They are not applicable to the scenes where the re-calibration needs to be done frequently.

In order to tackle the drawbacks of the traditional methods, a new calibration method which is able to achieve both high accuracy and simple operation is definitely demanding. Recently, many researchers have attempted calibration methods based on a single image. Lihua et al. [3] substituted the image center for the principle point and obtained the homography matrix using the points close to the image center. Richard Strand et al. [4] proved that real straight lines become circular arcs in images under lens distortion. Based on this theory, a lot of similar works [5–8] have been developed. Such methods work well for the

This work is partially supported by the Special Funds for the Frontier and Key Technology Innovation of Guangdong Province (Grant No. 2017B050506008), the Key R&D Program of Guangdong Province (Grant No. 2019B090915001), and the Program of Foshan Innovation Team of Science and Technology (Grant No. 2015IT100072).

C. S. Chan et al. (Eds.): ICIRA 2020, LNAI 12595, pp. 442–455, 2020.
https://doi.org/10.1007/978-3-030-66645-3_37

highly distorted lenses, like fisheye cameras, but they are less stable for industrial cameras. To enhance the robustness, most methods need to filter inappropriate arc elements in the obtained images. In [9], authors estimated image distortion center and corrected radial distortion from the vanishing points via a single image, but it is only based on the first-order radial distortion model which has certain limitation in accuracy. In addition, it is unable to solve the rest of the camera parameters. Alvarez et al. [10] applied the homography matrix from a single image to estimate the pose of a robot's end-effector. Researchers [11–15] have been doing similar work by using one single image to realize calibration of a camera, whereas the lens distortion is neglected.

Although lots of works have been done on camera calibration, how to calibrate efficiently, accurately and robustly is still a challenging problem. The current approaches based on a single image can only obtain part of camera parameters. In other words, the linear and nonlinear model parameters can not be directly solved simultaneously. Hence, in this work, we present an analytical framework for camera full calibration from a single image using the self-correction of lens distortion and the solution of imaging model parameters by the vanishing theory. In addition to the commonly used calibration patterns, for some natural scenes, high-precision full camera calibration can also be achieved.

2   Calibration Parameters

In this section, we describe the task of our work, to solve the linear and nolinear model parameters. These two components constitute the basic process of camera imaging.

2.1   Linear Imaging Mode

The linear model is considered as the pinhole camera model, and it projects a point $P_w(x_w, y_w, z_w)$ in the world coordinate system to a point $P = (u, v)$ in the image coordinate system, which can be expressed by:

$$z_c \begin{bmatrix} u \\ v \\ 1 \end{bmatrix} = \begin{bmatrix} f_x & 0 & u_0 & 0 \\ 0 & f_y & v_0 & 0 \\ 0 & 0 & 1 & 0 \end{bmatrix} \begin{bmatrix} \boldsymbol{R} & \boldsymbol{T} \\ \boldsymbol{0} & 1 \end{bmatrix} \begin{bmatrix} x_w \\ y_w \\ z_w \\ 1 \end{bmatrix} \tag{1}$$

where $\boldsymbol{R}$ is the attitude matrix composed of three unit direction vectors of the world coordinate frame corresponding to the camera coordinate system, and the translation matrix $\boldsymbol{T}$ is the offset vector, which denotes the location and direction of the camera in the world coordinate system. The goal of the linear camera model calibration is to solve the intrinsic parameters (f_x, f_y, u_0, v_0) and the extrinsic parameters $(\boldsymbol{R}, \boldsymbol{T})$.

2.2 Lens Distortion Models

Failing to compensate for lens distortion in cameras can lead to severe errors. The most commonly used distortion model is the radial distortion model, which is standardized as the polynomial model (PM):

$$r_u = T(r_d) = r_d + r_d \sum_{i=0}^{\infty} \kappa_i r_d^{2i} \tag{2}$$

where $r_d = ||P_d O||_2$ and $r_u = ||P_u O||_2$, and $o(u_0, v_0)$ is the distortion center, also the principle point. Then we can obtain the following forms

$$\begin{cases} u_u - u_0 = (u_d - u_0)(1 + \kappa_1 r_d^2 + \cdots + \kappa_n r_d^{2n4}) \\ v_u - v_0 = (v_d - v_0)(1 + \kappa_1 r_d^2 + \cdots + \kappa_n r_d^{2n4}) \end{cases} \tag{3}$$

where (u_u, v_u) is the undistorted point which follows the pinhole camera model. (u_d, v_d) is the original point with distortion in actual image, and (u_0, v_0) presents the image distortion center. Empirical experiences have shown that it is sufficient to take only the two parameters κ_1 and κ_2 into account. Using more parameters brings no major improvement to the approximation of $T(r_d)$ for distortion parameters resolution [16]. In addition, an calibration of more parameters is less robust. Thus, (3) can be simplified as:

$$\begin{cases} u_u - u_0 = (u_d - u_0)(1 + \kappa_1 r_d^2 + \kappa_2 r_d^4) \\ v_u - v_0 = (v_d - v_0)(1 + \kappa_1 r_d^2 + \kappa_2 r_d^4) \end{cases} \tag{4}$$

Compared with PM, the division model is preferred because it takes a more accurate approximation to the true undistortion function of a typical camera [5]:

$$\boldsymbol{P}_u = T(\boldsymbol{P}_d) = \frac{\boldsymbol{P}_d - \boldsymbol{e}}{1 + \lambda r_d^2} + \boldsymbol{e} \tag{5}$$

where $\boldsymbol{e} = (u_0, v_0)^{\mathrm{T}}$.

The camera distortion calibration is mainly to solve (u_0, v_0, λ) of the division model (4) or $(u_0, v_0, \kappa_1, \kappa_2)$ of the polynomial model (5), and further use them to rectify image distortion. This article uses division model as an example.

3 Image Distortion Self-correction

3.1 Identifying Good Lines

To evaluate the effect of image distortion correction, it is necessary to establish a quantitative index to measure the degree of overall image distortion. Here, we propose a dimensionless distortion measurement index to measure the degree of curve distortion as follows:

$$F_{dis} = \frac{\Delta d}{A} \sum_{i=1}^{n} \xi_i \tag{6}$$

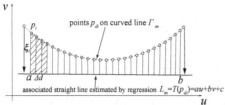

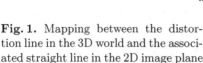

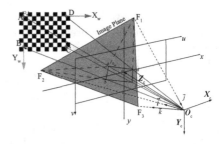

Fig. 1. Mapping between the distortion line in the 3D world and the associated straight line in the 2D image plane

Fig. 2. Vanishing points and their spatial position relation based on projective geometry

where $A \in R^{w \times h}$ ($w \times h$ is the size of the image), and $\Delta d = \frac{L}{N_m}$ (L is the length of a line segment, and N_m is the number of pixels of it, as illustrated in Fig. 1).

Due to distortion, a straight line in the 3D world space projects to a curved line in the 2D image plane. The curved line Γ_m can be detected by edge extraction using Canny operator in sub-pixel range. However, only the good lines can be used to accurately estimate the distortion. To judge whether a line can be used or not, we perform the following steps to select good lines: 1) We discard any contour whose length is less than l_{min} pixels, and link the adjacent edges; 2) We select contours that is very close to straight lines in terms of candidate parameters Root Mean Square Error (RMSE); 3) We compute the distortion index function F_{dis} of every remaining contour and remove contours whose F_{dis} near to 0, then left contours are obtained as the good lines.

3.2 Solution of Distortion Model Parameters

As paper [17] indicates, a straight line in the 3D world must always project to a straight line in the 2D image plane, if the radial lens distortion is compensated. Therefore, the task of distortion correction based on a single image is to find an inverse radial distortion function T that maps all points $p_d(u_d, v_d)$ on the curved lines into points $p_u(u_u, v_u)$ on their corresponding straight lines.

As shown in Fig. 1, if an associated straight line L_m is found, then we can solve the parameters of T by using the correspondence between points on L_m and Γ_m according to (5). In order to obtain the projection straight line L_m, we use total least squares (TLS) method to fit the good lines Γ_m as L_m. For the line L_m, from (5) we can get:

$$\frac{u_u}{v_u} = \frac{u_d + \lambda r_d^2}{v_d + \lambda r_d^2} \tag{7}$$

where (u_d, v_d) is the coordinate of point p_d on Γ_m, and (u_u, v_v) is the correspondingly point on L_m. In this equation, there are 3 unknowns, u_0, v_0, λ, therefore it is able to achieve least squares solution by using points on one line, but more lines might benefit the precision. For m lines selected by the steps above, we can get the initial parameters of T, where cost function of distortion correction for all pixels can be

$$\epsilon_{est} = \sum_{k=1}^{m} \sum_{i=1}^{n} \| p_u - T(p_{di}) \|_2 \tag{8}$$

To minimiz ϵ_{est}, we use an improved particle swarm optimization (PSO) algorithm explained in Sect. 4.3 to obtain the best estimation of distortion parameters.

4   Estimation of Camera Linear Model Parameters

4.1   Internal Parameters Estimation

Most calibration methods [10,18–23] based on single image replace the principle point, also the distortion center, with image center. However, this crude simplification is not safe, especially for cases where cameras are with severe distortion [4]. In contrast to this, we realized distortion correction and estimated principle point to achieve the higher accuracy of camera linear model parameters than the simplified substitution. In order to facilitate the evaluation, we used calibration board to solve the camera linear model parameters in (1) by using the vanishing point theory after implementing distortion self-correction.

In the vanishing point theory, the vanishing point is defined as the intersection point of parallel line group projected on the image plane, like F_1, F_2 and F_3 in Fig. 2.

As illustrated in Fig. 2, two sets of orthogonal parallel lines $AD//BC$ and $AB//CD$, which are on the world coordinate frame, consist of A, B, C and D on the calibration pattern. These four points are projected on the image plane as a, b, c and d. Let the principal point be O, $F_1(u_1, v_1)$ is the vanishing point corresponding to the parallel lines AB and CD, and $F_2(u_2, v_2)$ is the one corresponding to the parallel lines AD and BC. After obtaining the coordinate of the principal point (u_0, v_0), the pixel coordinate of the third vanishing point $F_3(u_3, v_3)$ can be obtained by any two equations:

$$\begin{cases} \overrightarrow{F_1F_2} \cdot \overrightarrow{F_3O} = 0 \\ \overrightarrow{F_1F_3} \cdot \overrightarrow{F_2O} = 0 \\ \overrightarrow{F_2F_3} \cdot \overrightarrow{F_1O} = 0 \end{cases} \tag{9}$$

As illustrated in Fig. 3(a), the three vanishing points and the camera coordinate system origin O_c form a right-angle tetrahedron with three right-angle triangles $\triangle O_c F_1 F_2$, $\triangle O_c F_1 F_3$ and $\triangle O_c F_2 F_3$. Then we can obtain:

$$\begin{cases} \| \overrightarrow{F_1F_2} \|^2 = \| \overrightarrow{O_cF_1} \|^2 + \| \overrightarrow{O_cF_2} \|^2 \\ \| \overrightarrow{F_1F_3} \|^2 = \| \overrightarrow{O_cF_1} \|^2 + \| \overrightarrow{O_cF_3} \|^2 \\ \| \overrightarrow{F_2F_3} \|^2 = \| \overrightarrow{O_cF_2} \|^2 + \| \overrightarrow{O_cF_3} \|^2 \end{cases} \tag{10}$$

For a certain point $p(u, v)$ in the pixel plane, its corresponding camera coordinate can be obtained by:

$$\begin{cases} x_c = (u - u_0)d_x \\ y_c = (v - v_0)d_y \\ z_c = f \end{cases} \tag{11}$$

where f is the focal length of camera, d_x and d_y represent the physical size of each pixel on the CCD of a camera in horizontal and vertical directions, respectively. By substituting (11) into (10), we can get the parameters about focal length in the following equations:

$$\begin{cases} f_x = \sqrt{\dfrac{(u_1-u_0)[(v_2-v_0)(u_3-u_2)-(v_3-v_2)(u_2-u_0)]}{v_3-v_2}} \\ f_y = \sqrt{-\dfrac{(v_1-v_0)[(v_2-v_0)(u_3-u_2)-(v_3-v_2)(u_2-u_0)]}{u_3-u_2}} \end{cases} \tag{12}$$

4.2   External Parameters Estimation

The external parameter matrix in (1) can be expressed as:

$$\boldsymbol{R} = \begin{bmatrix} i_x & j_x & k_x \\ i_y & j_y & k_y \\ i_z & j_z & k_z \end{bmatrix} \qquad \boldsymbol{T} = \begin{bmatrix} T_x & T_y & T_z \end{bmatrix}^{\mathrm{T}} \tag{13}$$

As illustrated in Fig. 2, for fair comparison with Zhang's method [1], point A is taken as the origin O_w of the world coordinate system. The direction vectors of axes X_w and Y_w are the same with the direction vector $\overrightarrow{AD}$ and $\overrightarrow{AB}$, respectively. According to the right-hand rule, the direction vector of axis Z_w is the same with vector $\overrightarrow{AD} \times \overrightarrow{AB}$. In the camera coordinate system, the unit direction vectors $\boldsymbol{i} = \begin{bmatrix} i_x & i_y & i_z \end{bmatrix}^{\mathrm{T}}$, $\boldsymbol{j} = \begin{bmatrix} j_x & j_y & j_z \end{bmatrix}^{\mathrm{T}}$ and $\boldsymbol{k} = \begin{bmatrix} k_x & k_y & k_z \end{bmatrix}^{\mathrm{T}}$ are the same with $\overrightarrow{O_cF_2}$, $\overrightarrow{O_cF_1}$ and $\overrightarrow{O_cF_3}$, respectively. Then the matrix $\boldsymbol{R}$ can be obtained by:

$$\boldsymbol{i} = \begin{bmatrix} i_x \\ i_y \\ i_z \end{bmatrix} = \frac{\overrightarrow{O_cF_2}}{\| \overrightarrow{O_cF_2} \|}, \boldsymbol{j} = \begin{bmatrix} j_x \\ j_y \\ j_z \end{bmatrix} = \frac{\overrightarrow{O_cF_1}}{\| \overrightarrow{O_cF_1} \|}, \boldsymbol{k} = \begin{bmatrix} k_x \\ k_y \\ k_z \end{bmatrix} = \frac{\overrightarrow{O_cF_3}}{\| \overrightarrow{O_cF_3} \|} \tag{14}$$

Figure 3(b) further illustrates the spatial geometric constraints in camera projective space. Let aB' be a space line parallel to AB, as $\triangle O_c aB' \sim \triangle O_c AB$, then

$$\frac{O_c a}{O_c O_w} = \frac{aB'}{AB} \tag{15}$$

and then

$$\boldsymbol{T} = \overrightarrow{O_c O_w} = \frac{AB}{aB'}\overrightarrow{O_c a} \tag{16}$$

Let $AB = m$, which can be obtained by the size of calibration board. aB' can be obtained through geometric relationship analysis. Finally, we can obtain the external parameters $\boldsymbol{T}$ from (15) and (16).

4.3   Global Optimization for Calibration Parameters

In the previous subsections, we have solved all the initial camera parameters. It is effective to conduct nonlinear optimization for the internal parameters to improve the accuracy and robustness. The optimization of internal parameters $(u_0, v_0, \lambda, f_x, f_y)$ is essentially a multi-objective optimization problem. Compared with the traditional Levenberg-Marquardt (LM) optimization strategy, which is easy to fall into the local optimal situation, particle swarm optimization (PSO) has the better ability to solve multi-objective optimization problems [24]. For the classic PSO, on account of the particle number is fixed, when the number is small, it tends to fall into a local optimum solution. However, with the increasing particle number, the computation time becomes longer. In other words, the particle number is inversely proportional to the solution efficiency.

Therefore, we use the improved algorithm which combines the natural selection method with the PSO to improve algorithm efficiency for large-scale particle swarm. The core ideal of this strategy can be referred to [24]. We initialize the position and velocity of each particle in the population with the value obtained by least square method rather than random numbers adopted in conventional method, which can greatly speed up convergence. We design the fitness function, namely, the objective function as follows:

$$Fitness = \sum_{i=1}^{N} \parallel p_i - \hat{p}_i(\lambda, v_0, v_0, f_x, f_y, \boldsymbol{R}, \boldsymbol{T}) \parallel_2 \qquad (17)$$

where p_i is the actual coordinates of a point in the world coordinate system projected on the pixel coordinate system, and $\hat{p}_i$ is the pixel coordinate calculated by the calibration parameters of the point. As it can be seen from Fig. 7(a) in the experimental part, after using the least squares solution as the initial value, the optimization of the camera internal parameters basically reach global convergence after about 200 iterations.

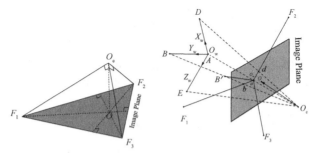

(a) Space relation between vanishing points and principle point

(b) Space geometric constraints of vector $\boldsymbol{T}$

Fig. 3. Signal of projective space geometric constraints

5 Experiments and Evaluation

In this section, we describe a detailed quantitative study of our method on synthetic images and experiments were also carried out on a real scene.

5.1 Experiment on Synthetic Image

We now present experiments on a synthetic image to compare the five algorithms in evaluating the accuracy of distortion model and the ability of distortion rectification. Since there is no standardized dataset for radial distortion correction, for fair comparison, we selected a synthetic image as our source data as shown in Fig. 4(a), which was also adopted by the latest works [5,20,21,25]. Results of these works are presented in Table 1. The main process of our method is shown in Fig. 4. By using particular ground truth values of $(u_0, v_0) = (320, 240)$ and $\lambda = -10^{-6}$ to distort the source image in Fig. 4(a), we get the distorted image, as shown in Fig. 4(b). Extracted edge curves by Canny operator in sub-pixel range are illustrated in Fig. 4(c). As shown in Fig. 4(d), the edge segmentation is adopted in this experiment, then we can identify good lines among the contours to estimate the distortion parameters. Results are shown in Table 1. We can see that the our approach produces convincing distortion parameters which are very close to the true distortion parameters. Compared with other papers, the result of our approach is competitive. In Table 1, Dis is the Euclidean distance between $(u_0, v_0)_{true}$ and $(u_0, v_0)_{estimate}$. Rel is the relative error for λ, i.e., $|(\lambda_{estimate} - \lambda_{true})/\lambda_{true}|$. In order to evaluate the eventual correction effect of each algorithm on distorted images, we employ the corrected source images in their papers and calculate their overall distortion function F_{dis}. By comparing the value of F_{dis}, we can see that our method achieves the best performance in distortion correction. From index Dis, we can conclude that our method has advantages in the estimation of distortion parameters and distortion center over other methods. It should be noted that since the method of Luis Alvarez et al. [20] does not implement the estimation of the distortion center of the image, but simply replaces it with image center, we cannot know the estimation result of the distortion center, which is not placed in the table.

Table 1. Comparison of results on the synthetic image

Parameter	True value	Method				
		Faisal Bukhari [5]	Luis Alvarez [20]	Faisal Bukhari [25]	Santanacedres Daniel [21]	Our method
(u_0, v_0)	(320, 240)	(319.352, 238.009)	–	(319.632, 247.75)	(319.526, 239.6648)	(319.63, 239.864)
λ	-10^{-6}	$1.00419e^{-6}$	$-1.000129e^{-6}$	$-9.80979e^{-7}$	$-1.0416667e^{-6}$	$-1.00328e^{-6}$
F_{dis}	0.0151563	0.128697	0.138528	0.0135797	0.346973	0.0104668
Dis	–	2.0934	–	7.7589	0.7946	0.3942
Rel	–	$4.19e^{-3}$	$1.29e^{-4}$	$1.9021e^{-2}$	$4.16667e^{-2}$	$3.28e^{-3}$

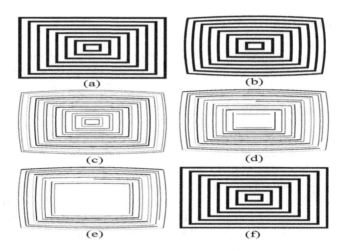

Fig. 4. Example experiment with synthetic image size of 640×480. (a) Original image. (b) Distorted image with $\lambda = -10^{-6}$ and $(u_0, v_0) = (320, 240)$ (the distortion center). (c) Extracted edges. (d) Contours smoothing. (e) Identifying Estimated lines

5.2 Experiment on a Real Scene

For the experiments on real scenes, the classical Zhang's method [1] is used for comparison. The implementation process of Zhang's method is as follows: (1) obtaining 20 pictures by the IDS UI-1220LE-C-HQ industry camera with less lens distortion; (2) implementing the calibration with MATLAB toolbox for camera calibration, Single Camera Calibrator App, which is based on Zhang's method whose results can be used as criteria for evaluation. In addition, we use the results achieved by encapsulated functions in OpenCV as a reference, wich is also based on Zhang's method.

Our calibration method only needs to take a single image. We took one of images obtained in process above to conduct our method, and the result is

Table 2. Calibration results by the two methods

Calibration method	Internal parameters			Distortion parameters	External parameters		
	f_x/mm	f_y/mm	(u_0, v_0)/pixel		R $\quad \bullet$		T/mm
Zhang's Method (Matlab)	843.6230	843.8606	(323.8685, 221.4643)	$\kappa_1\,naturalscene$ $imag = -0.1317$ $\kappa_2 = 0.1897$	$\begin{bmatrix} 0.7167 & -0.07538 & -0.6933 \\ 0.2749 & 0.9442 & 0.1815 \\ 0.6409 & -0.3207 & 0.6974 \end{bmatrix}$		$\begin{bmatrix} -65.44 \\ -55.49 \\ 275.26 \end{bmatrix}$
Zhang's Method (OpenCV)	864.06199	865.1194	(336.4875, 216.5105)	$\kappa_1 = 0.02317$ $\kappa_2 = -0.02183$	$\begin{bmatrix} 0.7113 & -0.07690 & -0.6987 \\ 0.2772 & 0.9441 & 0.1783 \\ 0.6460 & -0.3205 & 0.6928 \end{bmatrix}$		$\begin{bmatrix} -70.07 \\ -54.22 \\ 286.18 \end{bmatrix}$
Our Method	843.4618	843.4618	(348.6460, 233.4913)	$\lambda = -6.1266^{-11}$	$\begin{bmatrix} 0.7146 & -0.07316 & -0.6957 \\ 0.2712 & 0.9457 & 0.17906 \\ 0.6449 & -0.3166 & 0.6956 \end{bmatrix}$		$\begin{bmatrix} -70.21 \\ -59.97 \\ 273.15 \end{bmatrix}$

shown in Table 2. In this experiment, we used corners on the calibration pattern to generate lines for estimation, instead of extracting the checkerboard edges directly, because it can greatly enhance the robustness with the help of automatic corner detection method [26]. The corners' position in image coordinate which map those on the planar checkerboard in the real world coordinate have been distorted because of lens distortion, as shown in Fig. 5(d). In the distortion correction process, we used the corners' position to fit lines, and thought that the corrected positions of these points should be located on this straight line see Fig. 5(a–c). Then we can solve the distortion parameters in (5) by the corners' position before and after correction.

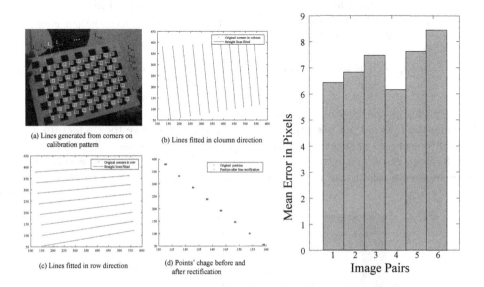

(a) Lines generated from corners on calibration pattern

(b) Lines fitted in cloumn direction

(c) Lines fitted in row direction

(d) Points' chage before and after rectification

Fig. 5. Calibration of our method

Fig. 6. Calibration results of the household camera

As it can be seen from the Table 2, the results of our method are very close to the evaluation criteria conducted by MATLAB. We projected the world coordinates of the corner points in space onto the image plane according to the parameters calibrated by our method in Table 2, and we can see that they can also be accurately restored for all the points, which is shown in Fig. 7(b). Parameters optimization is important for distortion model parameters estimation. We use the least squares solution as the initial value and different particle numbers to achieve the convergence of the objective function value. Although increasing the number of particles can accelerate the convergence rate, it can reach convergence within only 200 times for all tests. This proves that our optimization strategy is successful.

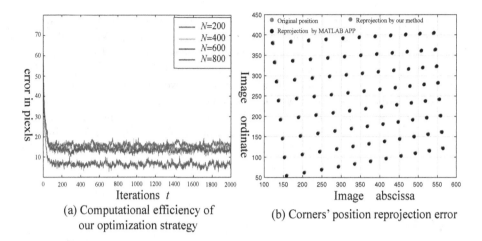

(a) Computational efficiency of our optimization strategy

(b) Corners' position reprojection error

Fig. 7. The total error of our method for an image calibration

In addition, we conducted experiments without manual calibration plate. Keeping the internal parameters of the camera unchanged, we took a picture in a natural scene to calibrate the camera, and the main process is illustrated in Fig. 8. From the edges extracted from the original image, we reserve good lines to estimate the distortion model parameters and implemented rectification as

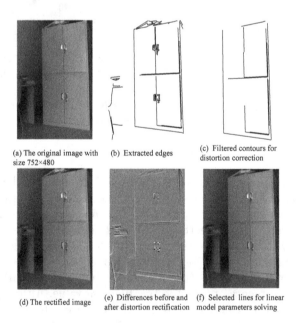

(a) The original image with size 752×480

(b) Extracted edges

(c) Filtered contours for distortion correction

(d) The rectified image

(e) Differences before and after distortion rectification

(f) Selected lines for linear model parameters solving

Fig. 8. Calibration process in natural scene

explained in Sect. 3. We subtracted images before and after rectification, and highlighted the difference, as shown in Fig. 8(e). The internal parameters result is shown in Table 3, and the rotation matrix of the external parameter is represented by a vector of RPY. We can summarize that, the distortion rectification is important for camera calibration. For the case without distortion correction, the principle point of the image is replaced by the image center. As we can see that, after the simply replacement, the obtained internal parameters (f_x, f_y, u_0, v_0) are almost unreliable compared to the evaluation criterion. Furthermore, we chose a household camera, i.e. Cannon PowerShot G7X which is usually equipped with a wide-angle lens and consequently has large image distortion, for further testing. We take 6 images with large size 2736 × 2736 pixels to calibrate it six times. The reprojection error of every time is shown in Fig. 6. It can be seen that, for the household camera, the average error of our method is within only 9 pixels.

Table 3. Calibration result of the natural scene image

Parameters	Without distortion rectification	With distortion rectification
(f_x, f_y)	(510.712,510.712)	(859.573,859.573)
(u_o, v_0)	(376, 240)	(327.558, 216.980)
Distortion	–	$\lambda = -1.1554 \times 10^{-7}$
R	(0.031,0.236,-1.546)	(0.417,0.052,-1.527)
T	$(-17.782, 39.967, 261.077)^T$	$(-6.194, 52.305, 446.536)^T$

6 Conclusion

This article provides a one-shot camera calibration method which can replace the traditional multi-view method in some application scenarios. The strategy is to realize the self correction of lens distortion through the line estimation in images, and then use the vanishing point to solve the linear model parameters of the camera. Experiments show that this strategy is effective. In order to evaluate the performance, we used manual calibration board to achieve the full calibration of the camera by using the minimum number of straight lines and the most basic two sets of parallel lines. The result is very competitive compared with the traditional methods implemented by MATLAB and OpenCV. In addition, all parameters can be calibrated by our method in natural scenes without manual calibration board.

References

1. Zhang, Z.: A flexible new technique for camera calibration. IEEE Trans. Pattern Anal. Mach. Intell. **22**(11), 1330–1334 (2000)
2. Corke, P.: Robotics, Vision and Control: Fundamental Algorithms in MATLAB® Second, Completely Revised, vol. 118. Springer, Cham (2017). https://doi.org/10.1007/978-3-319-54413-7
3. Hu, L., Zhang, J., Zhang, S., Li, X.: A method of estimating the camera distortion parameters from a single image. J. Comput.-Aided Des. Comput. Graph. (2015)
4. Hartley, R., Kang, S.B.: Parameter-free radial distortion correction with center of distortion estimation. IEEE Trans. Pattern Anal. Mach. Intell. **29**(8), 1309–1321 (2007)
5. Bukhari, F., Dailey, M.N.: Automatic radial distortion estimation from a single image. J. Math. Imaging Vis. **45**(1), 31–45 (2013)
6. Wang, A., Qiu, T., Shao, L.: A simple method of radial distortion correction with centre of distortion estimation. J. Math. Imaging Vis. **35**(3), 165–172 (2009)
7. Wu, F., Wei, H., Wang, X.: Correction of image radial distortion based on division model. Opt. Eng. **56**(1), 013108 (2017)
8. Lucas, L., Loscos, C., Remion, Y.: Camera Calibration: Geometric and Colorimetric Correction. Wiley, Hoboken (2013)
9. Liu, D., Liu, X., Wang, M.: Camera self-calibration with lens distortion from a single image. Photogram. Eng. Remote Sens. **82**(5), 325–334 (2016)
10. Santana-Cedrés, D., Gomez, L., Alemán-Flores, M., Salgado, A., Esclarín, J., Mazorra, L., Alvarez, L.: An iterative optimization algorithm for lens distortion correction using two-parameter models. Image Process. On Line **6**, 326–364 (2016)
11. Boby, R.A., Saha, S.K.: Single image based camera calibration and pose estimation of the end-effector of a robot. In: 2016 IEEE International Conference on Robotics and Automation (ICRA), pp. 2435–2440 (2016)
12. Espuny, F.: A new linear method for camera self-calibration with planar motion. J. Math. Imaging Vis. **27**(1), 81–88 (2007)
13. Lee, J.: Camera calibration from a single image based on coupled line cameras and rectangle constraint. In: Proceedings of the 21st International Conference on Pattern Recognition (ICPR 2012), pp. 758–762 (2012)
14. Grammatikopoulos, L., Karras, G., Petsa, E.: An automatic approach for camera calibration from vanishing points. ISPRS J. Photogram. Remote Sens. **62**(1), 64–76 (2007)
15. Li, B., Peng, K., Ying, X., Zha, H.: Simultaneous vanishing point detection and camera calibration from single images. In: Bebis, G., et al. (eds.) ISVC 2010. LNCS, vol. 6454, pp. 151–160. Springer, Heidelberg (2010). https://doi.org/10.1007/978-3-642-17274-8_15
16. Broszio, H.: Automatic Line-Based Estimation of Radial Lens Distortion. IOS Press, Amsterdam (2005)
17. Devernay, F., Faugeras, O.D.: Automatic calibration and removal of distortion from scenes of structured environments. Invest. Trial Image Process. **2567**, 62–72 (1995)
18. Xianyu, X.J.S.: Camera calibration with single image based on two orthogonalonedimensional objects. Acta Optica Sinica **32**(1), 145–151 (2012)
19. Thormählen, T., Broszio, H.: Automatic line-based estimation of radial lens distortion. Integr. Comput.-Aided Eng. **12**(2), 177–190 (2005)

20. Alemán-Flores, M., Alvarez, L., Gomez, L., Santana-Cedrés, D.: Automatic lens distortion correction using one-parameter division models. Image Process. On Line **4**, 327–343 (2014)
21. Santana-Cedrés, D., et al.: Invertibility and estimation of two-parameter polynomial and division lens distortion models. SIAM J. Imaging Sci. **8**(3), 1574–1606 (2015)
22. Alvarez, L., Gomez, L., Sendra, J.R.: Algebraic lens distortion model estimation. Image Process. On Line **1**, 1–10 (2010)
23. Alvarez, L., Gómez, L., Sendra, J.R.: An algebraic approach to lens distortion by line rectification. J. Math. Imaging Vision **35**(1), 36–50 (2009)
24. Engelbrecht, A.P.: Computational Intelligence: An Introduction. Wiley, Hoboken (2007)
25. Bukhari, F., Dailey, M.N.: Robust radial distortion from a single image. In: International Symposium on Visual Computing (2010)
26. Geiger, A., Moosmann, F., Car, O., Schuster, B.: Automatic camera and range sensor calibration using a single shot. In: Proceedings IEEE International Conference on Robotics and Automation, pp. 3936–3943 (2012)

An Automated View Planning Method of Robot 3D Measurement

Yilin Yang[1,2], Lin Zhang[2], Zeping Wu[1,2], and Xu Zhang[1,2(✉)]

[1] School of Mechatronic Engineering and Automation, Shanghai University,
No.99 Shangda Road, Baoshan District, Shanghai 200444, China
`xuzhang@shu.edu.cn`
[2] HUST-Wuxi Research Institute, No 329 Yanxin Road, Huishan District,
Wuxi 214100, China

Abstract. Precision measurement of large-scale workpiece is one of the most difficult problems in industrial measurement. In order to select suitable viewpoints to obtain precise measurement data, a lot of research has been done in the past decades. This paper proposes an automated view planning method of robot 3D measurement for large-scale workpiece. A viewpoint constraint model is used for landmarks inspection. The viewpoints are generated based on the visual cone theory and checked for validity. A cost function for robot motion is designed based on path length minimization, after what the optimal viewpoint can be determined. The simulated annealing algorithm is used to calculate the optimal sequence for robot to pass through all optimal viewpoints with the shortest total distance. Experimental result shows that the view planning method proposed in this paper increases the automation and efficiency of the whole system.

Keywords: View planning · 3D measurement · Robot · Measuring system

1 Introduction

In aerospace, military and automotive industry, the machining and measurement of large surface has always been a difficult problem. For large-scale workpieces, for example propeller or body-in-white, precision measuring instruments such as laser tracker, 3D scanner and vision system are widely used to get the shape and dimension information of high accuracy.

Due to the limited range of single measurement, the integrated result is composed of data collected from multiple positions, which can also be called as viewpoints. During the measurement, the viewpoint directly determines the pose of the sensors and influences the final result. It's a hard work to determine appropriate viewpoints to meet the needs of different measurement tasks: on the one hand, when faced with curved surfaces with sparse features, only by finding reliable viewpoints can such precision measuring system bring its function into full play; objects with complex structures, on the other hand, bring about

© Springer Nature Switzerland AG 2020
C. S. Chan et al. (Eds.): ICIRA 2020, LNAI 12595, pp. 456–468, 2020.
https://doi.org/10.1007/978-3-030-66645-3_38

superfluous viewpoints that increase the measuring time and complexity.

Traditional viewpoint planning relies on moving the sensors manually by experience to reach each viewpoint. This method is not only very time-consuming, but also difficult to obtain the optimal measuring viewpoints and path. In addition, the measurement task will become absolutely redundant especially when repeated measurement is required. Novel viewpoint planning draws a significant research interest.

According to the type of the target application, the research can be divided into two direction: multi-point detection and active inspection. In the former, the problem usually involves multiple sensors, while in the latter the problem often involves single range sensor.

View planning systems that import spatial geometric constraints were developed by Cowan et al. [1–3]. Factors such as resolution, field-of-view and visibility are considered to restrict sensor placement.

Tarbox [4] proposed a measurability matrix that measures visual coverage. The solution is found by a greedy search that operates over the entries of the matrix, where the rows represent points in the scene and the columns represent viewpoints. The solution is then to cover as many rows with as few columns as possible. Scott [5] and Guo [6] carried out further researches about global optimization on the basis.

Zhang [7] deployed a multi-camera network with the goal of maximizing the coverage area under a limited number of viewpoints. The model consists of a triangle mesh and visual frustums, where Frobenius distance and Euclidean distance between a triangle unit and a visual frustum are introduced to reflect the camera-to-triangle coverage strength. On this basis, the optimal viewpoints are determined by minimizing the sum of squares of these two distances.

However, multiple sensors contribute to increasing cost and difficulty in location, and the system is lack of flexibility on different targets as well. Robot measuring system, as an automatic measuring equipment, makes data acquisition from different positions more maneuverable and economical. Compared with system consisting of handheld sensors [8], automatic measuring system based on robot provides not only higher stability but also satisfying accuracy and efficiency [9].

In some researches, the view planning is simplified to the robot motion planning of multi-target tracking based on multiple constraints [10–12]. Shi proposed a robot assisted 3D scanning system, which can automatically plan the viewpoint of the probe by adding more viewpoints in iterations and finally generate the minimum number of viewpoints and the shortest scanning path [13].

Methods mentioned above partially solve the planning problem but leave some shortcomings such as too long optimization time and low efficiency, which lead to the failure to reach an acceptable solution in application. The originality of this paper is to solve the view planning problem from the perspective of robot motion based on the constraint model. Firstly, some important concepts and the robot measuring system are introduced in Sect. 2. In Sect. 3 we illustrate the method of generating credible viewpoints and finding an optimal measuring path based on the robot motion distance theory. The experimental results are presented in Sect. 4. Finally, there are some concluding remarks in Sect. 5.

2   Overview of Robot Measuring System

In this paper we make use of a constraint model to describe the viewpoints. This model considers four factors: visual distance, field of view, visibility and accessibility. The first two among them are determined by the type and parameters of sensors in the 3D measuring system.

Visibility distinguishes eligible viewpoints from those that existing occlusion during inspection. Viewpoints that do not satisfy the expectation of data fusion will also be abandoned. The requirement of accessibility relies on the motion of robot in space. The first thing to pay attention to is the interference between the robot and the environment (including the measuring system, the measured object and external environment). Secondly, in the working range of the robot, the viewpoints should be reachable for the robot reasonably without singularity problem. The optimal viewpoint model is determined by the constraints above.

Active multi-view inspection focuses on single sensor and multiple targets. The measuring system controlled by the robot moves to these viewpoints in turn and measures specific areas. A large-scale measurement task is carried out in this way: The multi-DOF robot moves to first viewpoint and locates the measuring system, then the 3D data is obtained by the sensors. After finishing the measurement of current position, the robot controls the measuring system to move to the next viewpoint and continues the process above. The measuring process can be summarized into four main steps:

Step 1: Displaying feature points and building the landmark map. The feature points are pasted reasonably on the appropriate positions of the object to be tested. Adjust the sensor position to record feature data, then establish a full map of the landmarks.

Step 2: View planning. Plan viewpoints and measuring path.

Step 3: 3D measurement. By measuring the object from all viewpoints, the data can be matched and spliced according to the landmark map.

Step 4: Quality assessment. Analyze the deviation and geometric tolerance between the measurements and the CAD model, then generate a report as final result.

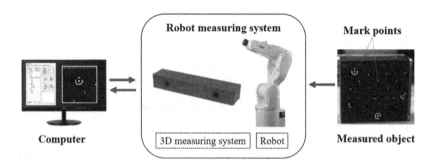

Fig. 1. Measuring system and hardware equipment

As shown in Fig. 1, the robot measuring system consists of a 3D measuring system and a robot. The robot can be placed on an AGV further. Other auxiliary measuring instruments include a host computer, coded and non-coded mark points. Next section will focus on how to carry out view planning.

3   View Planning

The process of view planning (Fig. 2) is as follows: firstly, for each coded mark pasted, a set of viewpoints is generated by visual cone theory. The visual cone is determined by the measuring system's visual distance and field of view. Then the validity of all viewpoints is checked to ensure that each viewpoint meets the requirement of visibility, which means that these viewpoints are qualified for data acquisition and 3D reconstruction. According to the principle of robot accessibility, the optimal viewpoint with shortest robot moving distance is determined. Finally, starting from the current position of the end-effector of the robot, the shortest path that linked all optimal viewpoints is found.

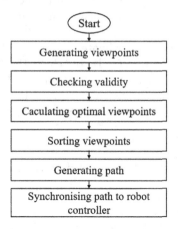

Fig. 2. View planning procedure

3.1   Generating Viewpoints

As shown in Fig. 3(a), for each coded mark pasted on the object, we use a virtual cone based on Spyridi's visual cone theory [14] to represent the space where all viewpoints can observe the coded mark. 4 to 8 points around the bottom of the cone are evenly selected as a set of possible viewpoints.

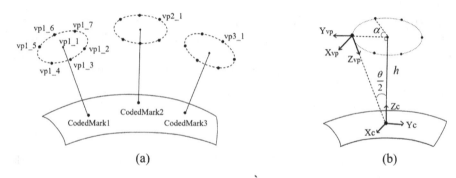

Fig. 3. Generation of viewpoints (a) Viewpoints generated by visual cone theory (b) Generation of viewpoint model

As shown in Fig. 3(b), we record the coded mark pasted on the surface as CodedMark and construct coded mark coordinate system $\{C\}$ here with the normal vector of the coded mark as the axis Z_C. The height of the visual cone is the offset value h between the viewpoint plane and the coded mark, and it can be related to visual distance. In order to ensure the effectiveness of scanning, the cone angle of the visual cone θ is usually within 90. The radius of the visual cone bottom is $r = h \tan \frac{\theta}{2}$.

We construct the viewpoint coordinate system $\{VP\}$ at the generated viewpoint ViewPoint[i] position respectively, of which the Z-axis direction always points to CodedMark. Taking the first viewpoint vp-1 as the benchmark, the angle between every two adjacent viewpoints in the bottom plane is α. The pose transformation of the viewpoint in the two coordinate systems can be described as follows:

$$^{c}P = {}^{c}_{vp}T \; {}^{vp}P \tag{1}$$

The mapping from $\{VP\}$ to $\{C\}$ is obtained by the following transformation:

$$^{c}_{vp}T = \begin{bmatrix} 1 & 0 & 0 & 0 \\ 0 & -1 & 0 & 0 \\ 0 & 0 & 1 & 0 \\ 0 & 0 & 0 & 1 \end{bmatrix} \begin{bmatrix} \cos\alpha & -\sin\alpha & 0 & r\sin\alpha \\ \sin\alpha & \cos\alpha & 0 & r\cos\alpha \\ 0 & 0 & 1 & 0 \\ 0 & 0 & 0 & 1 \end{bmatrix} \begin{bmatrix} 1 & 0 & 0 & 0 \\ 0 & \cos\frac{\theta}{2} & -\sin\frac{\theta}{2} & -r \\ 0 & \sin\frac{\theta}{2} & \cos\frac{\theta}{2} & h \\ 0 & 0 & 0 & 1 \end{bmatrix} \tag{2}$$

After the possible viewpoints are generated, the camera moves to all the viewpoints to check whether occlusion exists and whether the camera's field of view contains enough landmarks one by one, so as to screen out all eligible viewpoints for subsequent work.

3.2   Calculating Optimal Viewpoints

After initial screening, for the qualified viewpoints, the viewpoint with the shortest moving distance is selected as the optimal viewpoint to the coded mark by calculating the motion distance of the robot moving from the initial position to here.

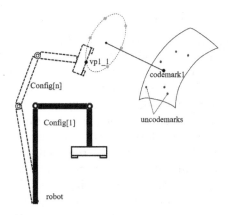

Fig. 4. Moving distance from initial pose to the viewpoint

An obstacle avoidance path can be generated between two points by means of a bidirectional RRT-connect method [15]. The robot will generate a continuous set of poses on the path to approximate the optimal solution. As shown in Fig. 4, the poses of the robot can be determined by the status Config[1], Config[2]..., Config[n], and each pose consists of m joint angles joint[1], joint[2],...joint[m]. In this paper, an algorithm of robot motion distance is defined to calculate the distance between the starting and ending joint configurations of the robot. The specific content of the algorithm is shown in the figure.

Algorithm 1 Algorithm of robot motion distance

Input: Robot joint configurations at n points on the path Config[1],...,Config[n]
     , Length←0
Output: Length;
 1: **def** PathLength(Config[n], Config[n-1])
 2:     d=0;
 3: **for** i = 1; i <= m; i++ **do**
 4:        dx = Config[n](joint[i]) - Config[n-1](joint[i]);
 5:        d = d + dx * dx;
 6: **return** math.sqrt(d)
 7:
 8: **for** j = n; j > 0; j−− **do**
 9:        Length ← Length + PathLength(Config[j], Config[j-1]);
10: **return** Length

If all viewpoints of a coded mark are unqualified, or the position is unreachable to the robot, the viewpoint can be added manually by teaching robot.

3.3 Finding Measuring Path

After the optimal viewpoint model extracted, starting from the robot initial pose, we solve this problem optimally by reducing it to generalized traveling salesman problem. By means of simulated annealing (SA) algorithm [16], the optimal sequence that passes through all viewpoints without repeat is calculated, and the path with the shortest total distance is finally generated.

$F(y)$ is used to represent the value of the evaluation function in state S, which is also known as the path length. $S(i)$ represents the current sequence, $S(i + 1)$ represents the new sequence, and r is the coefficient used to control the speed of cooling. T0 and Tmin respectively represent initial temperature and minimum acceptable temperature. After searching for the local optimal solution, SA algorithm will continue to update the solution with a certain probability. If $F(S(i + 1)) >= F(S(i))$, that is, a better solution is obtained after updating, it will always accept the update. If $F(S(i + 1)) < F(S(i))$, that is, the updated solution is worse than the current solution, the updated solution is accepted a certain probability, and this probability decreases gradually over time until the global optimal solution is obtained.

Algorithm 2 Simulated Annealing Algorithm

Input: T_0, Tmin, S(0)
Output: S(i);
1: **while**(T > Tmin)
2:      dE = F(S(i+1)) - F(S(i)) ;
3:      **if** (dE >=0)
4:          S(i+1) = S(i);
5:      **else**
6:          **if** (exp(dE/T) > random(0, 1))
7:              S(i+1) = S(i) ;
8:      T = r * T ;
9:      i++;
10: **return** S(i)

The result of this algorithm is the optimal measuring path that passes through all viewpoints without repeat.

4   Experiment

Industrial photogrammetry, as one of the common measuring methods, is widely used in product quality inspection and process control. The principle is to paste the mark points on the object and take multiple photos to form a landmark map of the mark points. Then point cloud registration is carried out on the basis of the landmark map to reconstruct an accurate point cloud model. This paper will take the photogrammetric system along with a 6-DOF robot as an example to introduce the automated view planning method.

The photogrammetric and reconstruction system adopted in the experiment is LaserVision700 and the robot is ABB IRB1200. An Aluminum blade with the size of 1200 * 400 * 300 is taken as the object to be tested, of which the surface is evenly arranged with coded and non-coded circular marks.

The simulation software VREP based on digital twin technology helps digitize the environment, measured object and measurement process so as to verify the generated viewpoints.

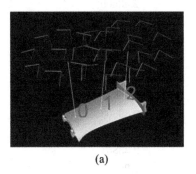   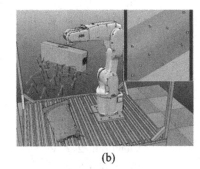

(a)   (b)

Fig. 5. The viewpoint constraint model (a) Viewpoints generated by the visual cone theory (b) Checking validity of all generated viewpoints in VREP

Figure 5(a) shows the viewpoints generated according to the method described in this paper. The Aluminum blade is pasted with three coded marks and several non-coded marks. The coded marks are represented by points in green and the generated viewpoints are represented by frames above. Each coded mark possesses 9 viewpoints as a set with 8 around the bottom of the visual cone as well as 1 at the center of the circle.

As shown in Fig. 5(b), the virtual vision sensor simulates taking photos at each viewpoint to identify that there are adequate non-coded marks(at least three non-coded marks) in the camera's field of view.

Eligible viewpoints are marked in the column of Validity in Table 1. Those viewpoints that the robot has appropriate joint configurations to get to are marked in the column of Robot Config, and the path lengths of the robot are shown in the last column.

Table 1. Viewpoints inspection

ViewPoint	Validity	Robot config	Path length
ViewPoint 1-1	✓	✗	/
ViewPoint 1–2	✓	✗	/
ViewPoint 1–3	✓	✗	/
ViewPoint 1–4	✓	✗	/
ViewPoint 1–4	✓	✗	/
ViewPoint 1–5	✓	✓	2.35294
ViewPoint 1–6	✓	✓	1.8535
ViewPoint 1–7	✓	✓	2.00873
ViewPoint 1–8	✓	✓	2.43609
ViewPoint 1–9	✓	✗	/
ViewPoint 2-1	✓	✓	1.81131
ViewPoint 2-2	✓	✓	1.84061
ViewPoint 2–3	✓	✓	1.87477
ViewPoint 2–4	✓	✓	2.08033
ViewPoint 2–5	✓	✓	2.39856
ViewPoint 2–6	✓	✓	2.4834
ViewPoint 2–7	✓	✓	2.33851
ViewPoint 2–8	✓	✓	2.29635
ViewPoint 2–9	✓	✓	2.35314
ViewPoint 3-1	✓	✓	1.73256
ViewPoint 3-2	✓	✓	1.76363
ViewPoint 3-3	✓	✓	1.80438
ViewPoint 3–4	✓	✓	2.07109
ViewPoint 3–5	✓	✓	2.3933
ViewPoint 3–6	✓	✓	2.41715
ViewPoint 3–7	✓	✓	2.28236
ViewPoint 3–8	✓	✓	2.28799
ViewPoint 3–9	✓	✓	2.30963

The optimal viewpoints and the sequence that passes through all of them are calculated, and the path with the shortest total distance is indicated in Table 2.

Table 2. The optimal viewpoints and the shortest distance

Order	Position	Total distance
0	Starting Point	0
1	ViewPoint 1–6	396
2	ViewPoint 2-1	580
3	ViewPoint 3-1	660.3

As shown in Table 3, the overall measurement planning time for the blade in the experiment is 35.8 s while traditional manual planning takes more than 4 min. Although it takes more time to calculate the optimal path than to plan the path directly according to the viewpoints, due to the time-consuming of the robot space motion, it takes shorter time in overall measurement.

Table 3. Comparison of experimental results

Duration	Traditional	Normal path	Optimal path
Planning time(s)	287	24.1	35.8
Total time(s)	308	154.4	117.3

As the dimensions of measured object increase, taking large wind turbine blade as an example, traditional manual planning takes more than 15 min or even longer [17]. This method can greatly improve the time efficiency, and there is no random problems such as redundant view or insufficient view coverage. As Fig. 6 shown, take a 5-meter-long blade as the measured object, the overall time of planning and measurement can be controlled within 8 min. In addition, the method based on robot can give full play to its advantages and stability, especially in the case of repeated measurement.

Fig. 6. Experiment of large-scale blade

The measurement process is shown in Fig. 7. The planning results are synchronized to ABB controller to drive LaserVision700 to scan and reconstruct. The process can be fully automated without human intervention.

Fig. 7. Robot 3D measurement in practical

Figure 8 is the point cloud scanned from three viewpoints and the final result of data fusion. The accuracy of the measurement system is 0.02 mm. According to our previous matching algorithm, we set the threshold of distance error at 0.20 mm, and the success rate of point cloud registration under the error is about 87.8%.

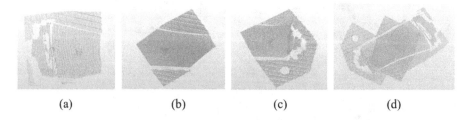

(a)   (b)   (c)   (d)

Fig. 8. Experimental result (a) Point cloud scanned from first viewpoint (b) Point cloud scanned from second viewpoint (c) Point cloud scanned from third viewpoint (d) The final result of data fusion

5   Conclusion

In this paper, an automated view planning method based on robot is presented to improve the flexibility and efficiency of measuring system. The method uses a viewpoint constraint model to represent the sensor position, which consists of generating viewpoints, checking visuality and calculating the optimal viewpoint by the robot motion distance algorithm. The optimal path that passes through all viewpoints is calculated as well. This method provides accurate and efficient viewpoints and optimal measuring path for robot 3D measurement. In addition, the optimal viewpoint reachable for robot can be determined in an acceptable time to improve the time efficiency with high reliability. Experimental result shows that the view planning method based on robot expands the measurement range of 3D measurement flexibly and increases the automation and efficiency of the whole system, providing favorable conditions to fully automatic measurement.

Acknowledgement. This research was partially supported by the key research project of the Ministry of Science and Technology (Grant No. 2018YFB1306802) and the National Natural Science Foundation of China (Grant No. 51975344).

References

1. Cowan, C.K.: Model-based synthesis of sensor location. In: Proceedings 1988 IEEE International Conference on Robotics and Automation, pp. 900–905 (1998)
2. Tarabanis, K., Tsai, R.Y., Abrams, S.: Planning viewpoints that simultaneously satisfy several feature detectability constraints for robotic vision. In: Fifth International Conference on Advanced Robotics' Robots in Unstructured Environments, pp. 1410–1415. IEEE (1991)
3. Yang, C.C., Ciarallo, F.W.: Optimized sensor placement for active visual inspection. J. Field Robot. **18**(1), 1–15 (2015)
4. Tarbox, G.H., Gottschlich, S.N.: Planning for complete sensor coverage in inspection. Comput. Vis. Image Underst. **61**(1), 84–111 (1995)
5. Scott, W.R.: Model-based view planning. Mach. Vis. Appl. **20**(1), 47–69 (2009). https://doi.org/10.1007/s00138-007-0110-2
6. Guotian, Y., Ruifang, D., Hua, W.U., et al.: Viewpoint optimization using genetic algorithm for flying robot inspection of electricity transmission tower equipment. Chin. J. Electron. **02**, 213–218 (2014)
7. Zhang, X., Alarcon-Herrera, J.L., Chen, X.: Coverage enhancement for deployment of multi-camera networks, pp. 909–914 (2015)
8. Zhe, C., Fumin, Z., Xinghua, Q., et al.: Fast measurement and reconstruction of large workpieces with freeform surfaces by combining local scanning and global position data. Sensors **15**(6), 14328–14344 (2015)
9. Wang, J., Tao, B., Gong, Z., et al.: A mobile robotic measurement system for large-scale complex components based on optical scanning and visual tracking. Robot. Comput. Integr. Manuf. **67**, 102010 (2020)
10. Budhiraja, A.K.: View point planning for inspecting static and dynamic scenes with multi-robot teams. Virginia Polytechnic Institute and State University (2017)

11. Huang, Y., Gupta, K.: An adaptive configuration-space and work-space based criterion for view planning. In: Intelligent Robots and Systems. IEEE (2005)
12. Li, L., Xiao, N.: Volumetric view planning for 3D reconstruction with multiple manipulators. Ind. Robot **42**(6), 533–543 (2015)
13. Shi, Q., Xi, N., Spagnuluo, C.: A feedback design to a CAD-guided area sensor planning system for automated 3D shape inspection. Comput. Aided Des. Appl. **4**(1–4), 209–218 (2017)
14. Spyridi, A.J., Requicha, A.A.G.: Accessibility analysis for the automatic inspection of mechanical parts by coordinate measuring machines. Robot. Autom. **15**(4), 714–727 (1990)
15. Kuffner, J.J., Lavalle, S.M.: RRT-connect: an efficient approach to single-query path planning. In: Robotics and Automation. IEEE (2002)
16. Kirkpatrick, S., Gelatt, C.D., Vecchi, M.P.: Optimization by simulated annealing. Science **220**(4598), 671–680 (1983)
17. Han, P.: Research on key technique of viewpoint generation and path planning for automated surface structured-light 3D measurement. Huazhong University of Science & Technology (2018)

Research on Point Cloud Processing Algorithm Applied to Robot Safety Detection

Nianfeng Wang[(✉)], Jingxin Lin, Kaifan Zhong, and Xianmin Zhang

Guangdong Province Key Laboratory of Precision Equipment and Manufacturing Technology, South China University of Technology, Guangzhou 510640, Guangdong, People's Republic of China
menfwang@scut.edu.cn

Abstract. In this paper, a method of point cloud recognition and segmentation based on neural network is introduced. This method will be applied to the specific industrial scene to detect whether there are sudden obstacles around the robot during the working process. This method is mainly divided into two parts. The first part is to design an efficient neural network structure, which achieves modification from state of art methods. The second part is to generate the corresponding neural network point cloud training data set for the specific scene. A simulation model is used to generate scene point cloud, and a large number of data are generated randomly. Simulation results verify the effectiveness and practicability of this method.

Keywords: Point cloud processing · Neural network · Obstacle recognition

1 Introduction

In recent years, with the development of computer vision and artificial intelligence, more and more attention has been paid to 3D discrete point cloud, which can provide more spatial information. Point cloud can give the detailed information of objects and environment, including their 3D position, even color, surface normal, etc. Now it is widely used in reverse engineering, measurement, robot and other applications. In the field of robot, point cloud can be used to identify the pose and position of the target object or obstacle, and assist the robot to make movement and operation judgment.

The order of points in point cloud data does not affect the overall shape of the whole point set in space. The same point cloud can be represented by different matrices, which is called the disorder of point cloud. Therefore, the neural network used in point cloud processing can not be as simple as that of 2D image. The disorder is a difficult point in the research of point cloud processing. In the early deep learning of point cloud processing, inspired by the convolution neural network [1,2] of 2D image, many networks make the point

© Springer Nature Switzerland AG 2020
C. S. Chan et al. (Eds.): ICIRA 2020, LNAI 12595, pp. 469–479, 2020.
https://doi.org/10.1007/978-3-030-66645-3_39

cloud object structured or voxelized by operating rasterization preprocessing, such as VoxNet [3] and 3D ShapeNets [4]. However, the amount of data required by these methods is very large, and there are a large number of blank voxels in the sparse point cloud, which leads to low efficiency. In addition, there are methods [5,6] to directly transform the point cloud into multiple images under multiple perspectives, and then process the multiple images through neural network. Such methods reduce most of the data demand, but still large.

Then came the point cloud processing algorithm of fully end-to-end architecture designs. Pointnet [7], a pioneering work, applies symmetric function to solve the disorder problem of point cloud. The point cloud data matrix can be directly used as input, which greatly improves the efficiency of neural network processing. After that, many scholars develop a variety of neural network structures based on inputting the original points of point cloud. For example, Pointnet++ [8], which considers the local features of point cloud, is the improved version of Pointnet, but its structure is very complicated. RSNets [9], which is recurrent neural networks (RNN)-based, slices the point cloud in space and input them into a bidirectional recurrent neural network. PointCNN [10] proposes an x-transform to solve the disorder problem, which improves the performance, but the effect is not as good as expected. DGCNN [11] uses edge convolution to extract local features, but the network structure is also complicated. On the basis of DGCNN, LDGCNN [12] connects the hierarchical features extracted from different dynamic graphs, and reduces the network size by using MLP [1] instead of the conversion network. The above three are deep convolutional neural networks (ConvNets)-based. And other deep learning neural networks based on tree structure and automatic encoder are proposed.

This paper studies the neural network point cloud processing algorithm applied to robot safety detection. Traditional point cloud processing algorithm is applied to robot security detection [13], but the algorithm based on neural network is rarely proposed. In this paper, the neural network processes the scene point cloud information while a robot or manipulator is working. It is used to automatically extract the features of the point cloud and recognize the points of robot and the points of obstacle parts (if there are obstacles) in the point cloud, and then the point cloud is segmented. Various cases should be regarded as obstacles intrusion: other workpieces suddenly flying in, personnel mistakenly entering the working range of the robot or the freight car accidentally entering, etc. On this premise, the point cloud of the extracted obstacle part is processed to calculate the information of the obstacle, such as the shape, position and posture, etc. So as to judge whether it has influence on the robot movement, and then let the robot decide whether to stop or dodge. The research can also be applied to the safety detection in human-robot interaction collaborative work.

In Sect. 2, the structure of neural network is introduced. In Sect. 3, point cloud data generation method for neural network training are described. In Sect. 4, the experimental results are proposed to demonstrate the performance of introduced method. In Sect. 5, conclusions and future works are discussed.

2 Point Cloud Segmentation Based on Neural Network

According to the results of the classical point cloud recognition and classification training set of various structures in [14,15], linked dynamic graph CNN (LDGCNN) [12] has good segmentation performance for different objects in the point cloud. In this study, LDGCNN is used as the prototype, and then simplified and modified. This section describes the important module and the overall structure of the neural network used in this study.

2.1 Edge Convolution

Edge convolution (EdgeConv) is an important feature recognition module used in the neural network, which is an operator proposed by DGCNN [11]. It can capture the local geometric information while keeping the permutation invariance.

The operation of EdgeConv is as follows. A F-dimension point cloud with n points, denoted by $\mathbf{X} = \{x_1, \ldots, x_n\} \subseteq R^F$. F is the number of features of each point. A k-nearest neighbor (k-NN) graph $\mathcal{G} = (\mathcal{V}, \mathcal{E})$ represents the local structure of the point cloud, where $\mathcal{V} = \{1, \ldots, n\}$ and $\mathcal{E} \subseteq \mathcal{V} \times \mathcal{V}$ are the vertices and edges, respectively. The graph $\mathcal{G}$ contains directed edges of the form $(i, j_{i1}), \ldots, (i, j_{ik})$ such that points $x_{j_{i1}}, \ldots, x_{j_{ik}}$ are the closest to x_i. Edge features are defined as $e_{ij} = h_\Theta(x_i, x_j)$, where $h_\Theta : \mathbb{R}^F \times \mathbb{R}^F \to \mathbb{R}^{F'}$ is some parametric non-linear function. Finally, the EdgeConv operation is defined by applying a channel-wise symmetric aggregation operation $\square$ (max, sum or $average$) on the edge features associated with all the edges emanating from each vertex [11].

In this paper, $h_\Theta(x_i, x_j) = h_\Theta(x_i, x_j - x_i)$ is chosen as the edge function and $\square = max$ as the aggregation operation, and h_Θ is instantiated with some convolution layers, which are the same as in [11]. The whole edge convolution is shown in Fig. 1.

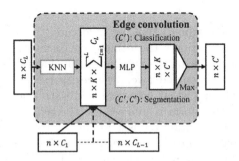

Fig. 1. Edge convolution operation

2.2   Neural Network Structure Design

The structure diagram of neural network used in this paper is shown in Fig. 2. It uses the original 3D point cloud information as input. After several EdgeConv operations, the extracted point cloud features are inputted into one or several convolution layers, and the features obtained after convolution are saved. These features include the characteristics of each point itself and the local features around it. They are pooled to get a global feature. These are shown in the blue box in Fig. 2. The global feature is reproduced into N copies, and then spliced with the local features of each point saved before, so that the features of each point include its own features, local features and global features. Finally, all the features of each point are inputted into MLP [1], and are integrated and compressed. A corresponding result is obtained for each point to determine whether the point is an obstacle point. These are shown in the green box in Fig. 2.

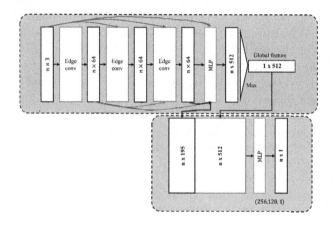

Fig. 2. Structure diagram of neural network (Color figure online)

Compared with the original LDGCNN structure, the neural network structure used in this paper reduces the complexity, including reducing one edge convolution operation, reducing the number of global features extracted by half, and reducing the fully-connected layer by one layer. The simplified structure proposed in this paper has great advantages in data calculation and variable storage. When the output form is unified, the two networks are tested on the classic point cloud segmentation dataset ModelNet40. The number of weights and threshold variables required by all neurons of LDGCNN is 886974, while the number of variables required by the simplified neural network structure is 359358, which brings great convenience to the training process of neural network. Finally, after 100 epochs of training, their accuracy is 85.11% and 85.06% respectively, which shows little disadvantage of the accuracy of the proposed structure.

In this paper, the collected scenes point clouds is inputted into the neural network after down sampling, and the output result of the neural network is limited between 0 and 1. The closer the output value to 1, the greater the probability that the point corresponding to the value is the obstacle point. Finally, the obstacle point recognized is segmented by thresholding.

3   Data Set Generation

The training of neural network needs a lot of data. In this section, the generation method of simulation scenarios in the neural network training data set is described.

At present, there are few neural network training sets for point cloud processing. Some universities or other point cloud research institutions in the world have training data sets for point cloud processing, including point cloud of single object, point cloud of indoor scene, etc., but there is a lack of specific working scene for industrial robots. The point clouds studied in this paper is obtained by a structured light system [16] in a industrial scene, so the point cloud data that simulate this case is generated.

Figure 3 shows the process of generating data set and the corresponding example diagram. The main steps are modeling, transforming the model into point cloud of object surface, synthesizing point cloud of each part of scene, and post processing.

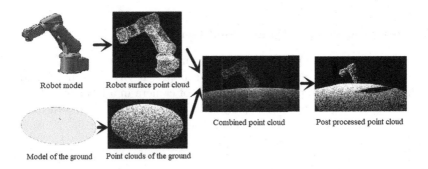

Robot model   Robot surface point cloud

Combined point cloud   Post processed point cloud

Model of the ground   Point clouds of the ground

Fig. 3. Point cloud generation process and example

3.1   Preliminary Generation and Combination of Point Clouds

In this study, the models of robot working scene are mainly divided into three categories: robot or manipulator, background and possible obstacles. In the simulation scenarios of the training set, these three categories is produced separately. Other some scenarios used to test the accuracy of neural network are to directly integrate the three parts in the model. The robot model uses a certain type of manipulator and it generates several different poses. In this study, the ground where the robot is located is the only background element. The ground in the simulated view is quite large for collecting the point cloud. So only part of the ground can be displayed in the simulation scene, and the shape of this part of ground will include regular shape and irregular shape. The obstacle models include some regular simple geometries, some models of complex parts, AGV models and pedestrian models.

After obtaining the model, according to the vertex and patch information, the surface point cloud of the model is generated. For a patch on the model, three vertices V_1, V_2, V_3 are assumed on the this patch but not in the same line. Their 3D coordinates are $V_i(x_i, y_i, z_i), i = 1, 2, 3$. Let O be the origin of the coordinate system, and the coordinates of point V_i can also be expressed by vector $\overrightarrow{OV_i}$. A new point V generated by these three points can be given by:

$$\overrightarrow{OV} = a \cdot \overrightarrow{OV_1} + b \cdot \overrightarrow{OV_2} + c \cdot \overrightarrow{OV_3} \tag{1}$$

where a, b, c are three random values and they satisfy $a, b, c \in [0, 1]$ and $a+b+c = 1$. Select different patches and vertices and repeat the above steps until enough points are generated. Finally, the surfaces in the model are covered with random points and the surface point cloud of the model is obtained.

The parts of the scene are combined in random position relationship within the allowed position range, and they need to be labeled appropriately. In this paper, the points of the robot and the background are labeled as 0, and the points of the obstacles are labeled as 1.

3.2   Post-processing of Point Cloud Generation

In order to simulate the point cloud as a result of a depth vision system, such as the structured light system used in this paper, a virtual camera model is established by using the projection principle of camera in a random view point. The schematic diagram of camera projection transformation is shown in Fig. 4.

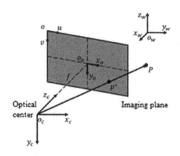

Fig. 4. Camera projection transformation

o_w-$x_w y_w z_w$ is the world coordinate system. At this time, the origin is set at the center of the bottom surface of the robot base. The ground is taken as the o_w-$x_w y_w$ plane, and the z_w axis is upward. o_c-$x_c y_c z_c$ is the camera coordinate system. The origin of the camera coordinate system is at the optical center of the camera, and the z_c axis direction is the optical axis direction of the camera, pointing vertically to the imaging plane. o-uv is the image coordinate system, the upper left corner of the image is the origin, and the unit of image coordinate system is pixel. Taking the intersection point of z_c axis of camera coordinate system and imaging plane as the origin, the imaging plane coordinate system o-$x_o y_o$ is established.

The coordinate value of the point cloud in the world coordinate system is transformed into the image coordinate system under the view point. The z_c axis coordinate value in the camera coordinate system is taken as the image pixel value. Each point is projected into the image coordinate system, and the hidden back points are removed by comparing the z_c axis coordinate value of each point. In this process, occasionally some points on the back are not covered due to the gap between the points in front. So some special filtering methods are needed to remove these points as much as possible. For example, setting a threshold, the distance value of a point is not only less than the original value in the pixel, but also less than the value obtained by subtracting a threshold value from the surrounding value, then the point is accepted. After removing the back points, the original point cloud is restored from the original depth map. At this time, the point cloud is represented in the camera coordinate system. The scene point cloud is transformed back to the world coordinate system by coordinate transformation. The simulation results can be seen in Fig. 3.

4   Experiment

In this paper, simulation experiments are designed to verify the effectiveness of the above algorithm, mainly including the recognition and segmentation of obstacle points in continuous scenes based on neural network.

Using the training set generated by the method described in Sect. 3, the neural network is trained. The experimental environment of neural network training is as follows: python 3.6 and tensorflow 2.0 are used in windows 10. The detailed experimental parameters are as follows: the number of point cloud points in each scene is sampled to 2048 points. The learning method of neural network is stochastic gradient descent (SGD) optimization algorithm. The learning parameters are set as the default value of the program, that is, the learning rate is 0.001, the momentum factor is 0.9. Then the training batch number is 4, and the training epochs is 100. In the training process, the data enhancement technology is used. Before each epoch of training, the point cloud data in the training set will rotate around the z-axis randomly and add random noise to enhance the robustness of the neural network.

The simplified neural network structure used in this paper is compared with the original LDGCNN in [12]. They use the same parameters. The comparison results are shown in Table 1.

Table 1. Comparison of experimental results of neural network

Structure	Amount of data	Training time for each batch	Accuracy in training set	Accuracy in test set
LDGCNN [12]	880653	1.042 s	99.36%	99.51%
Structure in this paper	353037	0.564 s	99.18%	99.42%

The accuracy of the two different network structures have little difference, but there is a huge difference in the amount of data, resulting in a huge gap in training time. It can be seen that the neural network with simplified structure has great practicability in this study.

The trained neural network is used to recognize obstacles in continuous scenes. One of the consecutive scenarios is shown in Fig. 5. These scenes simulate an object flying into the robot. Assuming that the time interval $\Delta t = 0.5$ s.

Fig. 5. Point clouds of continuous scenes simulating obstacles flying in

Each point cloud is down sampled to 2048 points, inputted into the neural network, and the output results are thresholded. The recognition results are shown in Fig. 6. The obstacles recognized in the figure are marked in black, and the robot and background are marked in red.

The recognition accuracy of obstacles in these scenes is 100%. In addition, there is also the recognition and segmentation of the scenes where pedestrian approach the robot. The segmentation results are shown in Fig. 7, and the segmentation accuracy is shown in Table 2.

From the above figures and table, it can be seen that the proposed method performs a stable segmentation in most cases. In the case of occlusion, especially when the obstacle is close to the robot, the neural network is likely to misjudge the robot point as the obstacle point. However, this kind of misjudgment is not fatal to the overall project task, because the obstacles can be correctly recognized and segmented when they are still far away from the robot. Therefore, the motion track of the obstacle will be captured at an earlier time when the obstacles approach the robot. Based on this, the action of the robot can be predicted and adjusted.

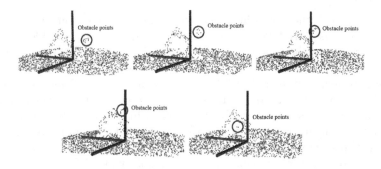

Fig. 6. Recognition results of point clouds of continuous scenes (Color figure online)

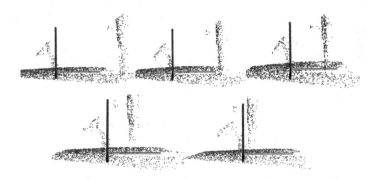

Fig. 7. Segmentation results of continuous scene point clouds with human (half body) approaching

Table 2. Segmentation accuracy results of human approaching scenes

Scene	Number of obstacle points recognized	Number of false positive points	Number of false negative points	Overall error rate
1st moment	209	1	0	0.48%
2nd moment	253	0	7	2.69%
3rd moment	238	0	0	0%
4th moment	244	26	0	11.93%
5th moment	230	42	0	22.34%

5   Conclusions and Future Works

In this paper, a method of point cloud recognition and segmentation based on neural network is introduced, which is used to detect the possible obstacles during operation of the robot. The method can be divided into two parts. The first part is to design an efficient neural network structure. The main part of the structure comes from LDGCNN, which is simplified and modified in this paper. The second part is to generate the corresponding neural network point cloud training data set for the specific scene, which makes the segmentation performance of the neural network greatly improved. The main process is object modeling, transforming into point clouds and post-processing. The simulated experiment proves the effectiveness and practicability of this method.

However, in some special cases, the recognition accuracy of neural network is poor. The future work is to further modify the structure of neural network to improve its recognition accuracy in these special cases.

Acknowledgements. The authors would like to gratefully acknowledge the reviewers comments. This work is supported by National Key R&D Program of China (Grant Nos. 2019YFB1310200), National Natural Science Foundation of China (Grant Nos. U1713207 and 52075180), Science and Technology Program of Guangzhou (Grant Nos. 201904020020), and the Fundamental Research Funds for the Central Universities.

References

1. Fei-Yan, Z., Lin-Peng, J., Jun, D.: Review of convolutional neural network. Chin. J. Comput. **40**(6), 1229–1251 (2017). (in Chinese)
2. Goodfellow, I., Bengio, Y., Courville, A.: Deep Learning. MIT press, Cambridge (2016)
3. Maturana, D., Scherer, S.: VoxNet: a 3D convolutional neural network for real-time object recognition. In: 2015 IEEE/RSJ International Conference on Intelligent Robots and Systems (IROS), pp. 922–928. IEEE (2015)
4. Wu, Z., et al.: 3D ShapeNets: a deep representation for volumetric shapes. In: Proceedings of the IEEE Conference on Computer Vision and Pattern Recognition, pp. 1912–1920 (2015)
5. Su, H., Maji, S., Kalogerakis, E., Learned-Miller, E.: Multi-view convolutional neural networks for 3D shape recognition. In: Proceedings of the IEEE International Conference on Computer Vision, pp. 945–953 (2015)
6. Kalogerakis, E., Averkiou, M., Maji, S., Chaudhuri, S.: 3D shape segmentation with projective convolutional networks. In: Proceedings of the IEEE Conference on Computer Vision and Pattern Recognition, pp. 3779–3788 (2017)
7. Qi, C.R., Su, H., Mo, K., Guibas, L.J.: PointNet: deep learning on point sets for 3D classification and segmentation. In: Proceedings of the IEEE Conference on Computer Vision and Pattern Recognition, pp. 652–660 (2017)
8. Qi, C.R., Yi, L., Su, H., Guibas, L.J.: PointNet++: deep hierarchical feature learning on point sets in a metric space. In: Advances in Neural Information Processing Systems, pp. 5099–5108 (2017)
9. Huang, Q., Wang, W., Neumann, U.: Recurrent slice networks for 3D segmentation of point clouds. In: Proceedings of the IEEE Conference on Computer Vision and Pattern Recognition, pp. 2626–2635 (2018)

10. Li, Y., Bu, R., Sun, M., Wu, W., Di, X., Chen, B.: PointCNN: convolution on X-transformed points. In: Advances in Neural Information Processing Systems, pp. 820–830 (2018)
11. Wang, Y., Sun, Y., Liu, Z., Sarma, S.E., Bronstein, M.M., Solomon, J.M.: Dynamic graph CNN for learning on point clouds. ACM Trans. Graph. (ToG) **38**(5), 1–12 (2019)
12. Zhang, K., Hao, M., Wang, J., de Silva, C.W., Fu, C.: Linked dynamic graph CNN: learning on point cloud via linking hierarchical features. arXiv preprint arXiv:1904.10014 (2019)
13. Yao, X., Xu, P., Wang, X.: Design of robot collision avoidance security scheme based on depth image detection. Control Eng. China **24**(7), 1514–1518 (2017). (in Chinese)
14. Liu, W., Sun, J., Li, W., Ting, H., Wang, P.: Deep learning on point clouds and its application: a survey. Sensors **19**(19), 4188 (2019)
15. Zhang, J., Zhao, X., Chen, Z., Zhejun, L.: A review of deep learning-based semantic segmentation for point cloud. IEEE Access **7**, 179118–179133 (2019)
16. Chen, Y.J., Zuo, W.M., Wang, K.Q., Wu, Q.: Survey on structured light pattern codification methods. J. Chin. Comput. Syst. **9**, 1856–1863 (2010)

A Brief Simulation Method for Coded Structured Light Based 3D Reconstruction

Nianfeng Wang$^{(\boxtimes)}$, Weiyong Xie, Kaifan Zhong, and Xianmin Zhang

Guangdong Province Key Laboratory of Precision Equipment and Manufacturing Technology, South China University of Technology, Guangzhou, Guangdong 510640, People's Republic of China
menfwang@scut.edu.cn

Abstract. Coded structured light based 3D reconstruction is one of the most reliable methods to recover a scene. The errors in the reconstruction are mainly caused by indirect illumination. So it is necessary to suppress the effect of indirect illumination to improve the reconstruction accuracy. In this paper, a simulation method of coded structured light based 3D reconstruction is presented. A simulated point cloud with high accuracy is generated for the noise reduction for the actual point cloud. A down-sampling method based on edge detection is also proposed to improve the efficiency of loading and transporting a point cloud. The experiment results verify the performance of simulation and down-sampling.

Keywords: 3D reconstruction · Coded structured light · Simulation · Down-sampling

1 Introduction

Coded structured light based 3D reconstruction is based on the assumption that the object only receives the illumination directly from the light source. However, various indirect illuminations effects the reconstruction result in reality, such as inter-reflections, subsurface scattering, ambient illumination, etc. There has been a considerable amount of research on this problem. Seitz [1] illuminates each point of the scene at a time and then computes its contribution to the other points according to the captured images. But this method becomes expensive when the scene is large or complicated. In [2], the indirect illumination received by each point in the scene is assumed as a smooth function with respect to the frequency of the lighting. Then high frequency illumination patterns are projected to the scene and the images are captured to separating the direct and indirect components of the illumination. This approach needs only two images in theory. However, 25 images are used in actual because of the resolution limits. Gupta [3] shows that different kinds of indirect illumination lead to decoding errors for different patterns and construct codes with only high-frequency binary patterns using conventional gray code patterns and logical XOR operation. They also propose a depth recovery algorithm to improve the decoding accuracy by

© Springer Nature Switzerland AG 2020
C. S. Chan et al. (Eds.): ICIRA 2020, LNAI 12595, pp. 480–488, 2020.
https://doi.org/10.1007/978-3-030-66645-3_40

projecting four different kinds of patterns onto the scene. Xu [4] needs an iteration to progressively reduce the indirect component of the illumination. They reduce the illumination area progressively using a mask which is created based on the pixels with correct depth values. The correct pixels are not illuminated in next iteration. These methods attempt to detect the indirect illumination and then reduce it. But the indirect part is usually not able to removed completely and the direct part remained may be too low in actual environment. On the contrary, illumination is completely controllable in virtual environment which results in accurate reconstruction.

After reconstruction, the obtained point cloud is usually down-sampled to accelerate the loading and transporting operation. However, some feature information is also lost after down-sampling. So it is necessary to save feature information while down-sampling.

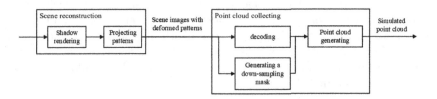

Fig. 1. Simulation of coded structured light vision system.

This paper focuses on achieving a reliable simulation of coded structured light vision system, and it can be divided into two parts: scene reconstruction and point cloud collecting, as shown in Fig. 1. In scene reconstruction, the shadows are rendered correctly while the indirect illumination is removed completely. In point cloud collecting, the down-sampling method is applied to simplify the point cloud.

2   Scene Reconstruction

The virtual reconstruction system is implemented in a robot software platform. The system consists of a projector and a camera. The projector throws coded patterns onto the scene and encode each scene point. Then the images with deformed patterns are captured by the camera. The camera and the projector are placed vertically so that the stripes in the captured images are vertical or horizontal. This kind of stripes is required in the down-sampling method which are explained in next chapter. The simulation of the scene can be divided to two parts: shadow rendering and simulation of projecting.

2.1   Shadow Rendering

In real condition, the depth information of scene points in shadows cannot be acquired because they are not coded. As a result, this part of point data is lost

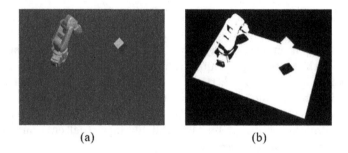

(a)   (b)

Fig. 2. The shadow generated in simulation environment. (a) The scene to be rendered. (b) The shadow generated.

in the actual point cloud. Therefore, rendering shadow correctly is necessary to ensure that the simulated point cloud does not contain points which are not existed in the actual one. Shadow mapping [5] is an effective way of shadow rendering and is applied in this paper. The depth map which stores the depth values of scene points in the image is rendered at first. Then each scene point is determined to be in shadow or not through a depth test. The shadows generated in the scene are shown in Fig. 2.

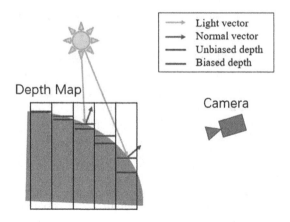

Fig. 3. The needed depth bias depends on the angle between the light vector and normal vector.

A depth bias is also used to cope with incorrect self-shadowing cause by undersampling or errors when storing depth information in the depth map. As shown in Fig. 3, the greater the angle between the normal vector of the point and the light vector, the greater the depth value required. However, an excessively value leads to artifact called peter panning. So it is important to set a suitable bias. The depth bias value in this paper can be computed by:

$$bias = c + f_s \cdot \cos < \overrightarrow{l}, \overrightarrow{n} > \tag{1}$$

where c is a constant based on the parameters of the projector, f is a coefficient determined by experiment, $\vec{l}$ is the light vector and $\vec{n}$ is the normal vector of the surface.

2.2 Simulation of Projecting

The coordinates of each point in the projector coordinate system are calculated according to the parameters of the projector. Each patterns to be projected are stored as textures. Then each scene points are rendered using the color at the corresponding pixel of the patterns. The patterns have a great influence on the reconstruction result. For example, when the minimum stripe width of the patterns is too small to distinguish in the captured images, the decoding error occurs. In this paper, min-SW Gray codes [6] patterns which maximize the minimum stripe width is used to avoid this problem. One of the projected patterns and the scene with pattern are shown in Fig. 4. In real condition, indirect illumination occurs when projecting structured light and leads to decoding errors. However, in simulation, the lights received by each point are all calculated. That is, indirect illumination can be completely eliminated in this step and leads to no decoding errors.

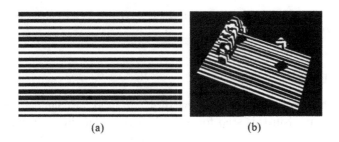

(a)   (b)

Fig. 4. The projected pattern and the scene with pattern. (a) One of the min-SW Gray codes patterns. (b) The scene with pattern.

3 Point Cloud Collecting

The virtual camera is simplified as a pinhole camera. The captured images are then decoded to determine the position of scene points in the projector coordinate system. Then the position of the point in world coordinate system can be calculated by triangulation. In order to improve the accuracy of triangulation, an efficient algorithm based on minimizing the L_2 reprojection error proposed by Lindstorm [7] is applied. To simplify the obtained point cloud while retaining feature information of the object, a down-sampling method based on edge detection is presented. This method down-samples the point cloud by generating a mask which is used in point cloud generating. The edge detection method proposed in [8] is applied. In this method, stripe patterns, which depart at edges, are projected onto the scene and a Gabor filter is used to detect the discontinuity.

3.1   Shadow Regions Filling

Shadow regions where light can not reach lead to double edges because the junctions of coded stripes and shadow regions lead to large deviations of stripes. So shadow regions should be removed to maintain only the correct edge. To locate the shadow regions, a white light is projected onto the scene and then the shadow region images are captured. According to the relative position between camera and projector, stripes on certain side are extended to fill the shadow regions. The images before and after filling are shown in Fig. 5.

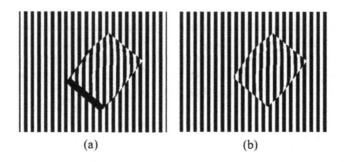

(a)   (b)

Fig. 5. The filling of shadow regions. (a) Shadow regions in an image. (b) the image after shadow regions filling

3.2   Edge Detection

The captured images with deformed patterns are used for edge detecting. A 2D Gabor filter is used in this step for edge detection. The imaginary part of the Gabor filter is:

$$g(x, y; \lambda, \theta, \psi, \sigma, \gamma) = exp(-\frac{x^{'2} + \gamma^2 y^{'2}}{2\sigma^2}) \sin(2\pi\frac{x^{'}}{\lambda} + \psi)$$
$$x^{'} = x\cos\theta + y\sin\theta \qquad\qquad (2)$$
$$y^{'} = -x\sin\theta + y\cos\theta$$

where λ is the wavelength of the sinusoidal factor, θ is the orientation of the normal to the stripes, ψ is the phase offset of the sinusoidal factor, σ is the standard deviation of the Gaussian factor and γ determines the ellipticity of Gabor function. According to the function, a Gabor kernel whose parameters are determined by the stripes is created. The amplitudes after filtering are different according to the deviation of stripes. Therefore, a threshold operation is applied to extract the contours of the objects.

3.3   Down-Sampling

As shown in Fig. 6, two images with coded stripes in different directions are detected to extract edges in different directions. The results are composed to one edge image as shown in Fig. 6(b). Then white noise is added into the edge image to generate a down-sampling mask as shown in Fig. 6(d) which includes two components, the white noise part and the contour part. The mask is used in the process of generating the point cloud. The points whose corresponding pixel is white will be generated while the points whose corresponding pixel is black will not. In this way, the white noise part applies a random down-sampling to the point cloud to reduce the size of it while the contour part can maintain the points at the contours of the objects where contains most of the feature information.

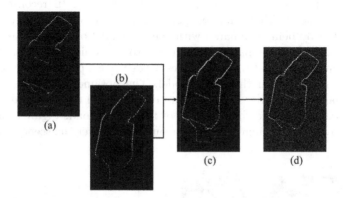

Fig. 6. The process of generating a down-sampling mask. (a) Edges detected using image with horizontal stripes. (b) Edges detected using image with Vertical stripes. (c)Edge image composed using (a) and (b). (d) Mask with edges and white noise.

4   Experimental Results

The simulation of 3D reconstruction is tested based on our robot software platform. The test scene consists of an IRB120 robot and a cubic object. A virtual reconstruction system including one camera and one projector is built based on a real system. The poses of camera and projector in world coordinate are shown in Table 1.

Table 1. The poses of camera and projector in world coordinate

Device	Pose
camera	$\begin{bmatrix} -0.995 & -0.010 & -0.099 & 300 \\ 0 & 0.995 & -0.100 & 300 \\ 0.100 & -0.099 & -0.990 & 2500 \\ 0 & 0 & 0 & 1 \end{bmatrix}$
projector	$\begin{bmatrix} 1 & 0 & 0 & 0 \\ 0 & -1 & 0 & 0 \\ 0 & 0 & -1 & 3000 \\ 0 & 0 & 0 & 1 \end{bmatrix}$

The whole process of the coded structured light based 3D reconstruction is simulated to generate the simulated point cloud. The obtained point cloud is then evaluated by being compared with the 3D model of the scene using open source software CloudCompare [9]. The cloud-to-mesh distance function offered by CloudCompare calculates the distances between each point of the point cloud to the nearest surface in the mesh model. The obtained distances are fitted with Gaussian distribution, and the mean distance and the standard deviation are computed to evaluate the accuracy of the point cloud. Figure 7 shows the comparison result. The simulated point cloud can recover the scene efficiently.

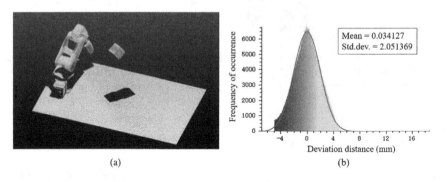

Fig. 7. The evaluation result of the simulated point cloud. (a) The cloud-to-mesh distance of the point cloud. (b) The Gaussian fitting result of the distribution of the distances.

The point cloud after downsampling is shown as Fig. 8. The down-sampling method based on edge detection does have a certain effect on maintaining the contour of the objects. However, some regions on the ground are still misidentified as contours.

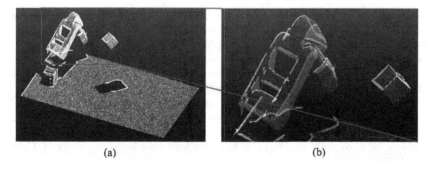

(a)   (b)

Fig. 8. The down-sampling result acquired by our method. (a) The point cloud after down-sampling. (b) Local image of (a).

5   Conclusions and Future Works

In this paper, a simulation of coded structured light based 3D reconstruction is proposed. The proposed method achieves a virtual 3D reconstruction system in simulation environment. In the environment, shadows are rendered correctly using shadow mapping method to ensure the authenticity of the simulated point cloud and the projecting of patterns is also realized. A down-sampling method based on edge detection is also be proposed. Shadow regions in the images are fulfilled to cope with double edges. A Gabor filter is used to detect the contours of the objects in the scene. The generated simulated point cloud is evaluated using the 3D model of the scene. The experiment result verifies the accuracy of the generated point cloud and the performance of the proposed down-sampling method in data simplification. However, the simulated point cloud has not been used to denoise the actual point cloud and misidentifications still occurs in edge detection. The future works are to improve the edge detection effect, finish comparing experiment between the simulated and the actual point cloud and realize the application of using simulated point cloud to denoise the actual one.

Acknowledgements. The authors would like to gratefully acknowledge the reviewers comments. This work is supported by National Key R&D Program of China (Grant Nos. 2019YFB1310200), National Natural Science Foundation of China (Grant Nos. U1713207 and 52075180), Science and Technology Program of Guangzhou (Grant Nos. 201904020020), and the Fundamental Research Funds for the Central Universities.

References

1. Seitz, S.M., Matsushita, Y., Kutulakos, K.N.: A theory of inverse light transport. In: Tenth IEEE International Conference on Computer Vision (ICCV 2005), vol. 1, vol. 2, pp. 1440–1447. IEEE (2005)
2. Nayar, S.K., Krishnan, G., Grossberg, M.D., Raskar, R.: Fast separation of direct and global components of a scene using high frequency illumination. In: ACM SIG-GRAPH 2006 Papers, pp. 935–944 (2006)

3. Gupta, M., Agrawal, A., Veeraraghavan, A., Narasimhan, S.G.: A practical approach to 3D scanning in the presence of interreflections, subsurface scattering and defocus. Int. J. Comput. Vis. **102**(1–3), 33–55 (2013)
4. Yi, X., Aliaga, D.G.: An adaptive correspondence algorithm for modeling scenes with strong interreflections. IEEE Trans. Vis. Comput. Graph. **15**(3), 465–480 (2009)
5. Williams, L.: Casting curved shadows on curved surfaces. In: Proceedings of the 5th Annual Conference on Computer Graphics and Interactive Techniques, pp. 270–274 (1978)
6. Goddyn, L., Gvozdjak, P.: Binary gray codes with long bit runs. Electr. J. Comb., R27–R27 (2003)
7. Lindstrom, P.: Triangulation made easy. In: 2010 IEEE Computer Society Conference on Computer Vision and Pattern Recognition, pp. 1554–1561. IEEE (2010)
8. Park, J., Kim, C., Yi, J., Turk, M.: Efficient depth edge detection using structured light. In: Bebis, G., Boyle, R., Koracin, D., Parvin, B. (eds.) ISVC 2005. LNCS, vol. 3804, pp. 737–744. Springer, Heidelberg (2005). https://doi.org/10.1007/11595755_94
9. CloudCompare. http://www.danielgm.net/cc

Human Gait Analysis Method Based on Kinect Sensor

Nianfeng Wang[✉], Guifeng Lin, and Xianmin Zhang

Guangdong Province Key Laboratory of Precision Equipment and Manufacturing Technology, South China University of Technology, Guangzhou 510640, Guangdong, People's Republic of China
menfwang@scut.edu.cn

Abstract. In this paper, a new method for human gait analysis based on the Kinect Sensor is introduced. Such method based on Kinect sensor can be divided into three steps: data acquisition, pre-processing and gait parameter calculation. First, a GUI (Graphical User Interface) was designed to control the Kinect sensor and get the required raw gait data. In the pre-processing, abnormal frames are removed first. Afterwards,the influence of Kinect's installation error is eliminated by coordinate system transformation. What's more,the noise is eliminated by using moving average filtering and median filtering. Finally, gait parameters are obtained by the designed algorithm which composed of gait cycle detection, gait parameter calculation, and gait phase extraction. The validity of the gait analysis method based on Kinect v2 was verified by experiments.

Keywords: Gait analysis · Kinect · Motion capture

1 Introduction

Human gait is the locomotion achieved through the movements of human limbs, which needs the harmony of bones, muscles and nervous systems. Gait analysis is the systematic study of human gait involving measurement of kinematics and kinetics, and assessment of the subject in health, age, state, speed and so on. Gait analysis is an effective tool used for a wide range of applications including diagnosis neurological diseases [1], rehabilitation evaluation [2], studies of exoskeleton [3] and so on. Clinical gait assessment was carried out by medical staff through various measures, scales and movement patterns in the past.

Although there are many kinds of gait analysis systems on the market, most of them are high priced and requiring technical expertise. What's more, these technologies have certain limitations. For example, wearable dynamic capture equipment consisting of pressure sensors and inertial sensors [4] is troublesome to wear. Noise may be introduced easily and signal drift[5] may appear. Additionally, the data is counterintuitive which requires data analysis by professionals.

Camera sensor is also a kind of device for gait analysis. Now commonly used cameras are RGB camera, infrared camera, and Kinect [5]. RGB camera

© Springer Nature Switzerland AG 2020
C. S. Chan et al. (Eds.): ICIRA 2020, LNAI 12595, pp. 489–502, 2020.
https://doi.org/10.1007/978-3-030-66645-3_41

is the most common camera and most of the gait datasets are built based on RGB camera such as Georgia [6], CASIA(A) [7], and USDC [8]. But the data acquisition of RGB camera is easily affected when illumination changing. Infrared camera can avoid the influence of light conditions but requires multiple cameras and pre-placing markers on the subjects' body.

Alternatively, Kinect v2 is the second generation of motion sensing input devices produced by Microsoft [9]. Kinect v2 incorporates an RGB camera, infrared projectors and detector. It can extract 3D-coordinates of skeleton joints from the images of human directly without markers on the human body. The release of the Kinect software development kit (Kinect SDK) , makes it simple for developers to call Kinect for research and development [10]. Kinect is expected to be used to develop a cheap and convenient gait analysis method based on its advantages of high-cost performance, simple equipment, non-contact data acquisition, and intuitive data.

In this paper, a gait analysis method based on Kinect is proposed. The proposed method contains the following step: (1) a GUI (Graphical User Interface) was designed to control the Kinect sensor and get its raw data we need. (2) the pre-processing of Kinect's raw data contains three steps: removal of abnormal frames, coordinate system transformation and data filtering. (3) gait parameters are obtained by three steps: gait cycle detection, gait parameter calculation, and gait phase extraction. Finally, the validity of the gait analysis method based on Kinect v2 was verified experiments.

2   Data Acquisition

The human gait raw data are acquired from Microsoft Kinect at this stage.

2.1   Required Data

Kinect provides a variety of data including color images, depth images, skeleton joints, etc. In this paper,three kinds of Kinect's raw data are needed.They are skeleton joints, timestamps, and the floor clip plane (FCP).

Kinect V2 can provide the 3D coordinates of 25 skeleton joints of up to 6 human bodies at 30 frames per second (fps) [10]. The name, number and position which is based on the Kinect coordinate of each skeleton joint are shown in Fig. 1.

The origin of Kinect coordinate system is the Kinect sensor. As shown in Fig. 2, the coordinate system is set as following: z-axis is vertical to the front face of the Kinect, the y-axis is always vertical to the bottom of Kinect, and the x-axis is parallel with the front surface of Kinect.

Timestamp sequence represents the time of each captured frame.Since most of the gait analysis is carried out in time domain, it is necessary to record the timestamps data.

Floor clip plane is a set of parameters that used to determine the floor plane recognized by Kinect sensor. The floor plane is given by:

$$Ax + By + Cz + D = 0 \tag{1}$$

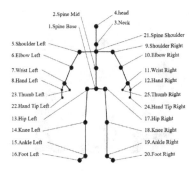

Fig. 1. Name, number, and position of 25 skeleton joints tracked by Kinect v2.

Fig. 2. Kinect coordinate system.

where A, B, C, and D are the elements of Floor clip plane, and D also equals to the distance between two coordinate in the y-axis direction.

2.2 Data Acquisition Procedure

Human gait data is designed to be acquired in the lateral view in this paper. The distance range that Kinect can track human body is from 0.8 m to 4 m. The vertical view angle of Kinect is 60° and the horizontal perspective is 70°. So the sensor should be placed 2.5m from the footpath, at a height of 0.5 m.

The connection between Kinect v2 and the computer is built via Kinect SDK, Kinect toolbox for Matlab [11] and Matlab. With Matlab GUI, a user interface for the data acquisition module is built which can realize two functions: displaying real-time image captured by Kinect and recording lateral view human gait data sequences.

As shown in Fig. 3, the user can preview the depth image captured by Kinect by press the start button. When the subject is successfully tracked by Kinect, the skeleton image of the subject will be drawn on the depth image. The function enables the tracking state of subjects to be checked directly. Then the user can press the record button or stop button to start or stop recording gait data. Skeleton joints, timestamps and floor clip plane will be acquired and saved as TXT files finally.

Fig. 3. The user interface of the data acquisition procedure.

3   Pre-processing

Because of the installation error of Kinect, data loss, and noisy data, the pre-processing of Kinect's raw data consists of three steps: removal of abnormal frames, coordinate system transformation, and data filtering.

3.1   Removal of Abnormal Frames

In course of the data acquisition, if any part of the subject is obscured or out of Kinect field of view, Kinect would fail to capture the subject and return data. In this case, the skeleton joint coordinate value of this frame is "NaN". Thus, the abnormal frames are detected by traversing data frames and replace their value with linear interpolation.

3.2   Coordinate System Transformation

Because of the Kinect installation error, the y-axis of Kinect coordinate system is not parallel but at a certain angle to the y-axis of the floor coordinate system [12] (see Fig. 4). In this case, the skeleton joint coordinate values in such a coordinate system that cannot accurately reflect the human gait. So it is necessary to transform the Kinect coordinate system to a standard one which is the floor coordinates.

It is assumed that there is movement in the y-axis direction and rotation around the x-axis between two coordinate systems. The equation of floor based on Kinect coordinate system can be obtained from the floor clip plane. Consider the equation of floor plane (see Eq.(1)), the vector of the y-axis of the floor coordinate system can be expressed as :

$$F = (A, B, C) \tag{2}$$

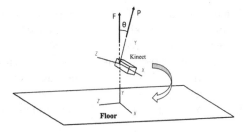

Fig. 4. Comparison of Kinect coordinate system and the floor coordinate system.

And the vector of the y-axis of the Kinect coordinate system is:

$$P = (0, 1, 0) \tag{3}$$

The rotation angle between two vectors can be computed as:

$$\theta = \arccos\left(\frac{F \cdot P}{|F| \cdot |P|}\right) \tag{4}$$

The vector of the rotation axis of two vectors can be calculated as:

$$C = \begin{pmatrix} c1 \\ c2 \\ c3 \end{pmatrix} = F \times P = \begin{pmatrix} -C \\ 0 \\ A \end{pmatrix} \tag{5}$$

According to the Rodrigues' rotation formula:

$$R_{\hat{\omega}}(\theta) = I + \hat{\omega}\sin\theta + \hat{\omega}^2(1 - \cos\theta)$$

$$\begin{bmatrix} \cos\theta + \omega_x^2(1 - \cos\theta) & \omega_x\omega_y(1 - \cos\theta) - \omega_z\sin\theta & \omega_y\sin\theta + \omega_x\omega_z(1 - \cos\theta) \\ \omega_z\sin\theta + \omega_x\omega_y(1 - \cos\theta) & \cos\theta + \omega_y^2(1 - \cos\theta) & -\omega_x\sin\theta + \omega_y\omega_z(1 - \cos\theta) \\ -\omega_y\sin\theta + \omega_x\omega_z(1 - \cos\theta) & \omega_x\sin\theta + \omega_y\omega_z(1 - \cos\theta) & \cos\theta + \omega_z^2(1 - \cos\theta) \end{bmatrix} \tag{6}$$

where $\hat{\omega} = (\omega_x, \omega_y, \omega_z) \in \mathbb{R}^3$ is the normalized vector of the rotation axis, ω is the rotation angle and I is a unit matrix.

The rotation matrix R between two y-axis' vectors can be calculated by substituding the rotation angle θ and the rotation axis vector C into the Rodrigues' Rotation formula, which is also the rotation matrix between two coordinates system. And the transformation of the coordinate system can be conducted as follow:

$$X_{floor} = RX + T \tag{7}$$

where X is a $3 \times N$ matrix consist of N coordinates in Kinect coordinate system, X_{floor} is based on the floor coordinate, and $T = (0, D, 0)^T$ defines the translation from Kinect coordinate to the floor coordinate.

3.3 Data Filtering

Due to the influence of many factors like light distribution, distance change between the human body and Kinect, wearing of the subjects and blocking of skeleton joints, and the quantization noise from the process of data quantization of Kinect SDK, there are mainly two kinds of noise in Kinect's raw data [13].

One is the jitter noise with high frequency and lower excitation. As shown in Fig. 5(a), jitter noise is manifested in the data curve with continuous slight jitter. This noise can be eliminated by using a moving average filter (MA). In this paper, a moving average filter with the window size of 5 is used to smooth data and eliminate jitter noise. The principle of the filter is shown as follow:

$$yy(n) = \frac{y(n-2) + y(n-1) + y(n) + y(n+1) + y(n+2)}{5} \tag{8}$$

for all $n=[3,N]$, where N equal total number of frames for a gait data sequence, $yy(n)$ is frame n of the filtered data and $y(n)$ is frame n of the raw data.

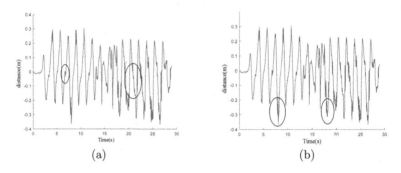

(a)   (b)

Fig. 5. Two types of noise in the gait data collected by Kinect.(a)the jitter noise in the gait data curve.(b)the impulse noise in the gait data curve.

The other one is impulse noise with short duration and large amplitude. As shown in Fig. 5(b), the impluse noise appears as a pulse mutation at the peak of the data. To eliminate thec impulse noise, a median filter (MF) with the window size of 5. A median filter (MF) with the window size of 5 is used for eliminating the impulse noise.

Combining the moving average filter and the median filter, the noise can be removed effectively. Comparison of the data curve before and after filtering is shown in Fig. 6.

4   Gait Parameters Calculation

4.1   Gait Cycle Detection

Most of the gait parameters are carried out in time domain [14], and it is important to complete the detection of gait cycle. An automatic algorithm for the detection of half gait cycle in lateral view was proposed in [15]. The half gait cycle is considered as the frames between two consecutive local minima of distance vector. In this paper, a new method by computing two adjacent local

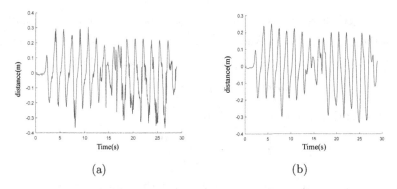

Fig. 6. Comparison of the data curve before and after filtering.(a)the data curve before filtering.(b)the data curve after filtering.

maximum of the distance vector is used to detect the gait cycle. The distance vector is calculated as follow:

$$distance = x_{ankleLeft} - x_{ankleRight} \tag{9}$$

where $x_{ankleLeft}$ and $x_{ankleRight}$ is the x-coordinate value of the left and right ankle joints. All the transition from positive slope to negative slope is detected to identify the local maximums. The frames between two adjacent local maximums belong to a single gait cycle.

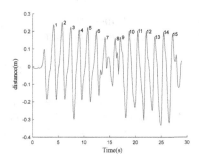

Fig. 7. The curve of distance vector and the marked peak frames.

Figure 7 shows a section of the distance vector curve and the local maximums are marked with numbers.

4.2 Gait Parameters Calculation

Gait parameters are first calculated in the divided gait cycle, and then the calculated values are averaged to get the final results.

(1) **The length of gait cycle**: $time_{peak}^{[n]}$ is the timestamp of the maxima frame n of the distance vector. And $\Delta T_{gaitCycle}^{[n]}$ is the length of the n gait cycle of the sequence of gait data. $\Delta T_{gaitCycle}^{[n]}$ can be computed as:

$$\Delta T_{gaitCycle}^{[n]} = time_{peak}^{[n+1]} - time_{peak}^{[n]} \tag{10}$$

(2) **Cadence**: Cadence can be computed as:

$$Cadence^{[n]} = \frac{60}{\Delta T_{gaitCycle}^{[n]}} \times 2 \tag{11}$$

(3) **Step length and stride length**: Step length contains left step length and right step length [16]. Left step length is given by:

$$d_{stepLeft}^{[n]} = distance_{peak}^{[n]} \tag{12}$$

Right step length is obtained by capturing the absolute value of the trough of the distance vector:

$$d_{stepRight}^{[n]} = \left| distance_{trough}^{[n]} \right| \tag{13}$$

Stride length is the distance human moves in a gait cycle. It is the sum of left and right step length.

(4) **Pace**: Pace is obtained by dividing the step length by the gait period:

$$pace^{[n]} = \frac{d_{stride}^{[n]}}{\Delta T_{gaitclycle}^{[n]}} \tag{14}$$

(5) **Step width**: Step width is obtained by calculating the average distance in the z-axis direction between two ankle joints:

$$d_{width} = |z_{ankleLeft} - z_{ankleRight}| \tag{15}$$

(6) **Joint angle**: Joint angle is given by calculating the angle of joint vector. Take the right knee joint angle as an example, as shown in Fig. 8, P is the vector from the right knee joint to the right hip joint, and Q is the vector from the right knee joint to the right ankle joint. P and Q can be given by:

$$P = \begin{pmatrix} X_{hipRight} - X_{kneeRight} \\ Y_{hipRight} - Y_{kneeRight} \\ Z_{hipRight} - Z_{kneeRight} \end{pmatrix} \tag{16}$$

$$Q = \begin{pmatrix} X_{hipRight} - X_{kneeRight} \\ Y_{hipRight} - Y_{kneeRight} \\ Z_{hipRight} - Z_{kneeRight} \end{pmatrix} \tag{17}$$

And the right knee joint angle $\theta_{kneeRight}$ can be computed as:

$$\theta_{kneeRight} = \arccos\left(\frac{P \cdot Q}{|P| \cdot |Q|}\right) \tag{18}$$

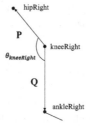

Fig. 8. Constructing vectors with adjacent skeleton joints for joint angle calculating.

4.3   Gait Phase Extraction

There are two main phases in the gait cycle:during stance phase,the foot is on the ground,whereas in the swing phase that same foot is no longer in contact with the floor and the leg is swinging through in preparation for the next foot strike [17]. Based on the principle above, a gait phase division method based on the change of coordinate gradient of ankle joint in x-axis direction was adopted.

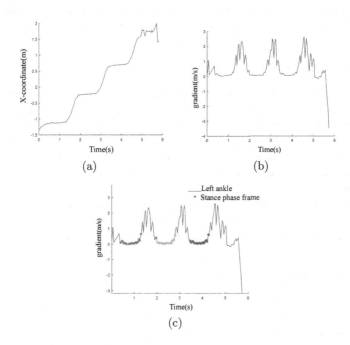

Fig. 9. (a)X-coordinate curve of left ankle joint.(b)the gradient of x-coordinate of left ankle joint.(c)the stance phase frames are marked with dots.

The x coordinate curve of the left ankle joint is shown in Fig. 9(a). The coordinate gradient sequence (see Fig. 9(b))is calculated by the formula as follow:

$$gradient^{[n]}_{ankleLeft} = \frac{x^{[n]}_{ankleLeft} - x^{[n-1]}_{ankleLeft}}{time^{[n]} - time^{[n-1]}} \tag{19}$$

During the stance phase, the foot is on the ground and the gradients of x-coordinate of this ankle joint are close to zero. Therefore, the stance and swing phases can be extracted from the gradient value sequence by using a threshold. As shown in Fig. 9(C), frames with gradient below the threshold are marked with dots, which are the frames in the stance phase of the gait cycle. The ratio of swing phase and stance phase can be calculated from the proportion of frames below or above the threshold in gait cycle.

5   Experiments

In this section, the validity of the gait analysis method based on Kinect v2 is verified with experiments. First, gait parameters were calculated in two ways. One was obtained by the method described above. The other set of gait parameters was calculated manually through ground markers and video recordings. The verification experiment was carried out by contrasting two sets of parameters.

5.1   Experimental Setup

Lateral view gait data were collected concurrently using a single Kinect v2 sensor (Microsoft Corp. Redmond, WA) and a cellphone camera (Redmi Note8 pro, Xiaomi Corp. Beijing, China). As shown in Fig. 8, two sensors were both positioned on the side of the data acquisition area, 2.5 m from the area and 0.5 m high. Kinect was connected with a computer to collect data using the compiled program. During the experiment, the cellphone recorded a video of the entire course of walking.

A footpath (see Fig. 10) of 2.52 m long by 0.36 m wide was marked with step length of 0.42 m and step width of 0.18 m to limit the subject. A healthy subject (man, age: 21 years, height: 160 cm, mass: 56 kg) took part in this experiment.

5.2   Experimental Procedures

Subject began by standing on the origin of the area until experimenter confirmed Kinect was capturing the subject through the interactive interface of the data acquisition program. To ensure that subjects walk in accordance with the specified gait parameters (step width and step length), the subject walked following ground markers. To ensure the data integrity, each trial lasted 5s after the subject walking out of the footpath. Ten sets of valid data were collected finally.

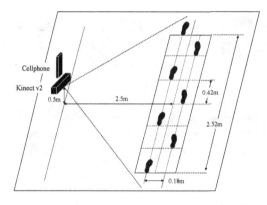

Fig. 10. Schematic diagram of experimental site layout.

5.3   Data Processing

Gait data from Kinect was processed and analyzed by using preceding method to calculate subject's gait parameters.The recorded video was played frame by frame with Adobe Premiere Pro (Adobe Inc. San Jose, California) to find and record the timing of gait events. The gait events within a single gait cycle that need to be recorded are shown in Table 1. The video snapshots of gait events are shown in Fig. 11.

Table 1. Gait events in a single gait cycle.

Number	Gait events	time
1	Left Heel Contact with ground	t_{HCL}^1
2	Right Heel Contact with ground	t_{HCR}^1
3	Left Toe Off the ground	t_{TOL}^1
4	Right Toe Off the ground	t_{TOR}^1
5	Left Heel Contact with ground again	t_{HCL}^2

The length of gait cycle and support phase can be calculated as follow:

$$\Delta T_{gaitCycle} = t_{HCL}^2 - t_{HCL}^1 \tag{20}$$

$$\Delta T_{leftStancePhase} = t_{TOL}^1 - t_{HCL}^1 \tag{21}$$

$$\Delta T_{rightStancePhase} = t_{TOR}^1 - t_{HCR}^1 \tag{22}$$

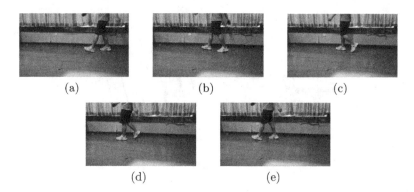

Fig. 11. The video snapshots of gait events.(a)Left heel contact with ground.(b)Right heel contact with ground.(c)Left toe off the ground.(d)Right toe off the ground(e)Left heel contact with ground again.

5.4 Statistical Analysis

To quantify the magnitude of the difference between two sets of data, absolute error and relative error were calculated as follow:

$$AbsoluteError = |KinectData - MeasurementData| \qquad (23)$$

$$RelativeError = \frac{KinectData - MeasurementData}{MeasurementData} \times 100\% \qquad (24)$$

Relativity between two sets of data was assessed by using correlation coefficient given as follow:

$$r = \frac{\sum(x - \bar{x})(y - \bar{y})}{\sqrt{\sum(x - \bar{x})^2 \sum(y - \bar{y})}} \qquad (25)$$

And the degree of dispersion of various data was evaluated by the standard deviation coefficient:

$$V_\sigma = \frac{\sigma}{\bar{x}} \times 100\% \qquad (26)$$

where σ is the standard deviation and x is the average value of data.

5.5 Experimental Results

The analyses of experiment are reported in Table 2.

According to the result of comparison, the relative error of all gait spatial parameters is below 10%. The standard deviation coefficient indicates that the dispersion degree of step width data is larger than other spatial parameters.

The error of gait spatiotemporal parameters from Kinect is slight as compared with the measurement one, while the error of left support phase ratio is relatively large. Additionally, correlation coefficients of gait cycle, pace and phase ratio are all above 0.85 which means a high consistency between the gait parameters from Kinect and measurement.

Table 2. Comparison of gait parameters from Kinect and measurement.

	Measurement Mean ± SD	Kinect Mean ± SD	Standard deviation coefficient	Absolute Error	Relative Error	Correlation coefficient
Left step length(m)	0.42	0.385 ± 0.014	3.64%	0.035±0.009	8.33%	—
Right step length(m)	0.42	0.393 ± 0.012	3.05%	0.027 ± 0.014	6.43%	—
Stride length (m)	0.84	0.778 ± 0.020	2.57%	0.062 ± 0.010	7.38%	—
Step width (m)	0.18	0.178 ± 0.043	24.16%	0.013 ± 0.015	7.22%	—
Gait cycle (s)	1.338 ± 0.060	1.321 ± 0.050	3.79%	0.027 ± 0.022	2.02%	0.85
Pace (m/s)	0.629 ± 0.027	0.590 ± 0.021	3.56%	0.039 ± 0.012	6.20%	0.91
Left Support Phase Ratio (%)	0.684 ± 0.021	0.621 ± 0.037	5.96%	0.064 ± 0.050	9.35%	0.96
Right Support Phase Ratio (%)	0.638 ± 0.028	0.624 ± 0.038	6.09%	0.034 ± 0.026	5.33%	0.84

Considering the error of the ground marked itself, the analysis result above indicates that Kinect is acceptable for measuring gait spatial parameters. At the same time, the accuracy of Kinect measurement in the depth direction is weaker than in the other two directions.related to the different principles of measurement.

Additionally, Kinect performs well in measuring spatiotemporal parameters, but poorly in gait phase division due to the imprecise threshold, filters and the precision of Kinect.

6   Conclusion

In conclusion, the gait analysis method based on Kinect v2 in this paper is efficient for gait parameters and analysis. In view of the advantages of simple external connection, low price and convenient operation, it can be a substitute for expensive motion-capture equipment in some cases with low precision requirement. Additional investigations will focus on comparison with other gait analysis methods and improvement of the algorithm for more accurate results.

Acknowledgements. The authors would like to gratefully acknowledge the reviewers comments. This work is supported by National Natural Science Foundation of China (Grant Nos. 52075180 and U1713207), Science and Technology Program of Guangzhou (Grant Nos. 201904020020), and the Fundamental Research Funds for the Central Universities.

References

1. Schlachetzki, J.C.M., et al.: Wearable sensors objectively measure gait parameters in parkinson's disease. PloS One **12**(10), e0183989 (2017)
2. LeMoyne, R., Mastroianni, T.: Wearable and wireless gait analysis platforms: smartphones and portable media devices. In: Wireless MEMS Networks and Applications, pp. 129–152. Elsevier, 2017
3. Long, Y., Zhijiang, D., Cong, L., Wang, W., Zhang, Z., Dong, W.: Active disturbance rejection control based human gait tracking for lower extremity rehabilitation exoskeleton. ISA Trans. **67**, 389–397 (2017)
4. Bouten, C.V.C., Koekkoek, K.T.M., Verduin, M., Kodde, R., Janssen, J.D.: A triaxial accelerometer and portable data processing unit for the assessment of daily physical activity. IEEE Trans. Biomed. Eng. **44**(3), 136–147 (1997)
5. Li, Y., Guo, J., Zhang, Q.: Methods and technologies of human gait recognition. J. Jilin Univ.(Eng. Technol. Ed.) **50**(1), 1–18 (2020)
6. Tanawongsuwan, R., Bobick, A.: Gait recognition from time-normalized joint-angle trajectories in the walking plane. In: Proceedings of the 2001 IEEE Computer Society Conference on Computer Vision and Pattern Recognition, CVPR 2001, volume 2, pages II-II. IEEE, 2001
7. Wang, L., Tan, T., Ning, H., Weiming, H.: Silhouette analysis-based gait recognition for human identification. IEEE Trans. Pattern Anal. Mach. Intell. **25**(12), 1505–1518 (2003)
8. Little, J., Boyd, J.: Recognizing people by their gait: the shape of motion. Videre: J. Comput. Vis. Res. **1**(2), 1–32 (1998)
9. Kinect for xbox one, (2020). https://en.wikipedia.org/wiki/Kinect
10. Rahman, M.: Beginning Microsoft Kinect for Windows SDK 2.0: Motion and Depth Sensing for Natural User Interfaces. Apress, Montreal (2017)
11. Terven, J.R., Córdova-Esparza, D.M.: Kin2. a kinect 2 toolbox for matlab. Sci. Comput. Program. **130**, 97–106 (2016)
12. Huang Q.: Design and implementation of gait analysis system based on rgb-d information. Master's thesis, Huazhong University of Science and Technology (2019)
13. Leijie, L.: Real-time optimization based on kinect v2 skeleton data. Electron. World **6**, 145–146 (2018)
14. Sahak, R., Zakaria, N.K., Tahir, N.M., Yassin, A.I.M. and Jailani, R.: Review on current methods of gait analysis and recognition using kinect. In: 2019 IEEE 15th International Colloquium on Signal Processing & Its Applications (CSPA), pp. 229–234. IEEE, 2019
15. Sinha, A., Chakravarty, K., Bhowmick, B., et al.: Person identification using skeleton information from kinect. In: Proceedings of International Conference on Advances in Computer-Human Interactions, pp. 101–108 (2013)
16. Qing, J.G., Song, Y.W., Ye, Q., Li, Y.Q., Tang, X.: The biomechanics principle of walking and analysis on gaits. J. Nanjing Inst. Phys. Educ. (Nat. Sci.). **04**, 1–7+39 (2006)
17. Vaughan C.L., Brian, L.D., Jeremy, C.O.: Dynamics of Human Gait, Human Kinetics Publishers (1992)

A Method for Welding Track Correction Based on Emulational Laser and Trajectory

Nianfeng Wang[✉], Jialin Yang, Kaifan Zhong, and Xianmin Zhang

Guangdong Province Key Laboratory of Precision Equipment and Manufacturing Technology, South China University of Technology, Guangzhou, Guangdong 510640, People's Republic of China
{menfwang,zhangxm}@scut.edu.cn
814846784@qq.com

Abstract. In this paper, a method for welding track correction based on emulational laser and trajectory are introduced. The proposed method is divided into two parts: seam tracking and trajectory correction. In the seam tracking method, by using the prior information of emulational laser stripes which are generated by the simulation software and affine transformation of emulational laser stripes, the real seam point can be detected. And then the trajectory correction method mainly consists of three steps: pre-processing, coarse matching and presice matching by using Iterative Closest Point Matching (ICP). For various conditions and workpieces, the corresponding experiments are conducted in this paper. Experimental results demonstrate that the method can meet the requirement of the internal seams tracking of workpiece and accuracy of welding track correction.

Keywords: Structured light · Track correction · Affine transformation · Emulational laser

1 Introduction

Nowadays, intelligent robotic welding is widely used in lots of industries such as automobile, ship, aerospace, petrochemical and electronics industry. And welding track correction based on vision sensor is one of the most well-explored areas. In order to overcome the shortcomings of traditional teach-and-playback method such as being susceptible to harsh environmental conditions and various disturbance, lots of vision sensors with different modules and structures and seam tracking methods have been proposed previously. Generally, optical sensors can be categorised as passive and active vision. And the main distinction between these two sensors is whether if using the auxiliary light.

Passive Vision Sensor: Vision sensors are used to extract the seam point with natural light. Also the passive vision sensor can be further categorised

© Springer Nature Switzerland AG 2020
C. S. Chan et al. (Eds.): ICIRA 2020, LNAI 12595, pp. 503–511, 2020.
https://doi.org/10.1007/978-3-030-66645-3_42

into monocular vision sensor or binocular vision sensor by using one camera or two camera. Xu [1] proposed a real-time seam tracking method based on passive monocular vision sensor which consists one CCD camera and corresponding optical fitters for GTAW. Ma [2] proposed a passive vision sensor for seam tracking of thin plate, which consists of a CCD camera, a wideband filter, a neural density filter and reflective mirrors.

Active Vision Sensor: Vision sensors are used to extract the seam point with auxiliary light. In [3], an active vision system was proposed to acquire the images of molten pool for evaluating the welding quality. And a laser vision sensor was proposed for an adaptive feature extraction algorithm which consists of neural network and edges extraction method in [4].

Comparing these two vision sensors above-mentioned, passive vision sensor is mostly used for detecting of workpiece and seam tracking based on molten pool detection, which is susceptible to disturbance such as spatter, arc light and the changes of natural light. However, active vision sensor has the advantages of high stability with the use of auxiliary light sources. And structured light sensor is widely studied mostly regarded as a typical representative of active light vision because of the features of relatively high precision and anti-interference of the laser light.

Structured light sensor can be catagorised into surface structured light sensor and line structured light sensor. 3D surface imaging system which consists of a structured light projector and a camera was proposed in [5] for acquiring 3D surface profiles. However, for welding seam tracking, line structured light sensor is mostly used. In [6], a precise seam tracking method for narrow butt seams based on a structured light sensor is proposed and in [7], a correction method of weld pass position is proposed based on line structured light sensor for the large and thick workpieces, which must be welded by MLMPW technology.

Furthermore, lots of image process method have been proposed previously by many researchers to extract the weld positions based on line structured light sensors. And the image process method usually comprises of three steps: pre-processing, laser stripe extraction and welding joint feature extraction. For image pre-processing method, in [8], average filtering process was proposed to smoothen high intensity saturated pixels as a software-based colour filtration and Gu [9] used the median filter to eliminate the noises such as salt and pepper noise. And for laser stripe extraction method, some conventional edge detection methods such as Canny edge detection [10], Roberts, Sobel, Prewitt, Robinson, and Kirsch, are employed to determine the edge of the laser stripes. Besides, Huang [11] proposed a method extracting laser by searching for all the pixels in each column with the maximum light intensity and Yang [12] proposed the adaptive Hough transformation which has made up for the time-consuming disadvantages of the Hough transformation as the weld line detection method.

After image process, next step is the visual control methods. Visual control methods consists of image-based visual serving, position-based visual serving and the combination method of two previous method regard as 2.5D visual control methods. But it's difficult to achieve the accuracy control due to the nonlinear and time-varying characteristics of welding process. Gu [9] proposed a control method which contains fuzzy controller and proportion controller and in [6] Fuzzy-PID controller is proposed which adopted to achieve seam tracking control, Also in [12] the neural work is utilized to predict the proper weld variables in nonlinear system.

However, the algorithms mentioned above mainly are susceptible to the disturbance during welding and the structured light sensors mainly focused on the external of workpiece, so a more compact sensor and robust method for meeting the need of weld track correction in the internal of workpieces need to be proposed.

In Sect. 2, the seam tracking method based on affine transformation of emulational laser stripe is introduced. In Sect. 3, the trajectory correction method based on ICP of emulational trajectory in simulation software is proposed. In Sect. 4, corresponding experiments are validated under various welding conditions. In Sect. 5, conclusions and future works are discussed.

2   Welding Track Correction Based on Structured Light Sensor

Various algorithms proposed for welding track correction have been proposed by many researchers previously [13–15]. Whereas, the image pre-processing methods for denoising such as using average gray value of all pixedls and taking the least gray value between sequent images in [13] or using median filter, adaptive threshold segmentation and selecting the maximum area as laser stripe in [14] cannot remove the disturbance such as splash and arc light during welding completely. Meanwhile, laser stripe extraction method such as searching the maximum grayscale position depending on the pre-processing methods. Therefore, a robust method for meeting the need of weld track correction are proposed as follows. The proposed method is divided into two part: seam tracking and trajectory correction.

2.1   Emulational Laser of Different Postures of Workpiece

As shown in section Fig. 1, image of emulational laser can be generated in emulation software by using boundary of workpiece and the laser plane geometrical intersection method when the welding robot is set to a certain position. So when the model of workpiece, sensor and robot are the same as the real condition, it's convenient to regard the seam point of emulational laser image as the real welding track point directly. However, the position and shape of the real workpieces are different from the emulational model slightly.

As shown in Fig. 1, the emulational laser changes as the posture of the same workpiece changes, therefore, a transformation method should be used to correction the emulational laser in order to find the accurate seam point of real workpiece.

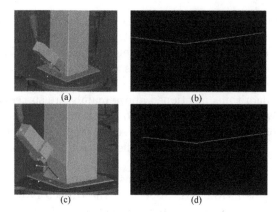

Fig. 1. The emulational laser images generated when the posture of workpiece are different.

2.2 Affine Transformation for the Correction of Emulational Laser

Affine transformation is one kind of non-rigid body transformation, which consists of rotation transformation, translation transformation, shear transformation and scaling transformation.

In order to make the emulational laser as accurate as possible, the partial data of real laser stripe should be extracted by image process, which can utilize the prior information from emulational laser stripe such as grayscale and slope, before the correction of emulation laser stripe. After getting the data from real laser stripe, the next step is pre-process the data of real laser stripe and emulational laser stripe as shown in Fig. 2.

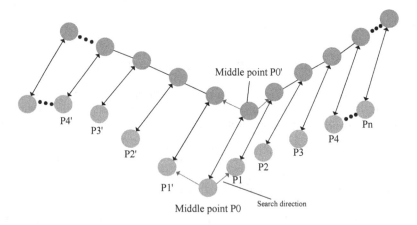

Fig. 2. The process of searching for corresponding points among emulational laser stripe and actual laser stripe.

As shown in Fig. 2, the pre-processing of data consists of two step: In the first step, the middle point $P0$ of emulational laser stripe, which is generally referred to the emulational seam point, should be matched with its corresponding point called the middle point $P0'$. And $P0'$ can be the approximate point such as the intersection of the fitting line of real laser stripe. In the second step, respectively search correspondent points with a distance k on the left and right of the point $P0$ and the point $P0'$ which can be computed by:

$$Pi = P0 + ki, Pi' = P0 - ki, i = 1, 2, ...n \qquad (1)$$

After the preprocessing of data, the next step is using the affine transformation on the points of emulational laser with corresponding the points of real laser as follows:

$$\begin{bmatrix} u' \\ v' \\ 1 \end{bmatrix} = \mathbf{M} \cdot \begin{bmatrix} u \\ v \\ 1 \end{bmatrix} = \begin{bmatrix} a_1 & a_2 & t_x \\ a_2 & a_2 & t_y \\ 0 & 0 & 1 \end{bmatrix} \cdot \begin{bmatrix} u \\ v \\ 1 \end{bmatrix} \qquad (2)$$

where $(u, v), (u', v')$ are respectively the coordinates of the partial points of emulational laser and real laser, and $\mathbf{M}$ is the matrix of affine transformation. In order to calculate the matrix of affine transformation by using least square method, it's convenient to form the equation as follows:

$$\mathbf{E} \cdot \mathbf{H} = \mathbf{R}, \mathbf{E} = \begin{bmatrix} u_1 & v_2 & 1 \\ u_2 & v_2 & 1 \\ ... & ... & ... \\ u_n & v_n & 1 \end{bmatrix}, \mathbf{H} = \begin{bmatrix} a_{11} & a_{12} \\ a_{21} & a_{22} \\ t_x & t_y \end{bmatrix}, \mathbf{R} = \begin{bmatrix} u_1' & v_2' \\ u_2' & v_2' \\ ... & ... \\ u_n' & v_n' \end{bmatrix} \qquad (3)$$

And then according to the principle of least square method, the matrix $\mathbf{H}$ can be computed by:

$$\mathbf{H} = (\mathbf{E}^T \mathbf{E})^{-1} \mathbf{E}^T \mathbf{R} \qquad (4)$$

After affine transformation of emulational laser, the middle point $P0$ can be regard as the seam point of workpiece.

3   Robot Trajectory Correction Based on Emulational Trajectory for Different Conditions

After features of weld seam extracted by the affine transformation of emulational laser stripe, the data are used by the controller module of seam tracking system. First of all, the emulational trajectory is needed to be generated in simulated program as the yellow line shown in Fig. 3. And then the data of actual seam points are needed to be processed as shown in Fig. 4: The pre-processing consists of denoising method by using DBSCAN [16] to remove outliers and using the average of k points in the region to smooth the original trajectory.

$$\begin{bmatrix} x_i \\ y_i \\ z_i \end{bmatrix} = \frac{1}{k} \begin{bmatrix} x_{i-\frac{k}{2}} ... + x_{i-1} + x_i + x_{i+1} ... + x_{i+\frac{k}{2}} \\ y_{i-\frac{k}{2}} ... + y_{i-1} + y_i + y_{i+1} ... + y_{i+\frac{k}{2}} \\ z_{i-\frac{k}{2}} ... + z_{i-1} + z_i + z_{i+1} ... + z_{i+\frac{k}{2}} \end{bmatrix}, i = 1 + \frac{k}{2}, ...n - \frac{k}{2} \qquad (5)$$

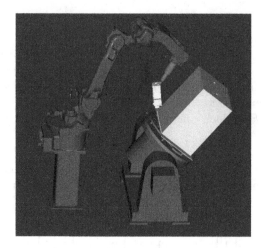

Fig. 3. The emulational trajectory of workpiece generated in simulation software

After pre-processing, the next step is using Iterative Closest Point Matching(ICP) [17] for the set of emulational points which computed as follows:

$$\mathbf{M} = \mathbf{R} \cdot \mathbf{D} + \mathbf{T}, \mathbf{D} = \mathbf{R}^{-1} \cdot (\mathbf{M} - \mathbf{T}) \tag{6}$$

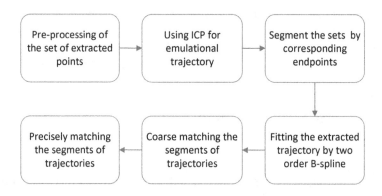

Fig. 4. The flow chart of trajectory correction method in this paper.

Where $\mathbf{M}$ is the set of emualtional points, $\mathbf{D}$ is the set of detected points, $\mathbf{R}$ and $\mathbf{T}$ are respectively rotation and translation transformation matrixs. And after using ICP, the sets of emulational and detected points are segmented respectively by corresponding endpoints of trajectory and fitting the points of detected trajectory by two order B-spline. In the end, the rotation transformation matrix $\mathbf{R}$ and translation transformation matrix $\mathbf{T}$ of coarse matching can be computed according to the geometry of the two segment of trajectories and the ICP is used as precisely matching method for each segment of corresponding trajectory.

4 Experimental Results

As shown in Fig. 5(a), the two workpieces of different shapes are fixed on the welding robotic positioner, so the emulational laser stripe generated of these workpiece are different from each other. And as shown in Fig. 5(b), the green laser stripe is the original laser stripe when the workpiece is fixed with the first position and the blue laser stripe is the target laser stripe when the workpiece is fixed with the second position in order to simulate the situation that the pose of the emulational model is different from the real pose of workpiece.

And the result demonstrate that for different types of laser stripes, the stripes after affine transformation are both accurately matched the target laser stripes.

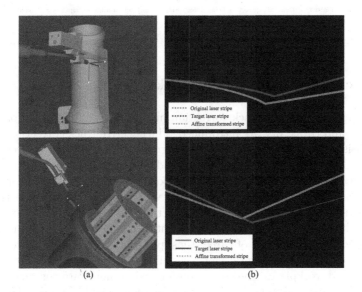

(a)   (b)

Fig. 5. The experiments based on affine transformation for different types of seam. (a) The different types of seam of workpieces. (b) The results of affine transformation of emulational laser for different workpieces.

For various different conditions of welding, the corresponding experiments need to be conducted to validate the trajectory correction method in Fig. 6. As shown in Fig.6, (a), (d), (g) respectively are corresponding to three conditions. For example, (a) the workpiece on the left is 0.9 times the size of the workpiece on the right, (d) the workpiece on the left has partial shape different from the right one and (g) these two workpieces with perpendicular insertion weld are both different in size and fixed pose. (b), (e), (h) are the result of welding track correction method and (c), (f), (i) are the partially enlarged views.

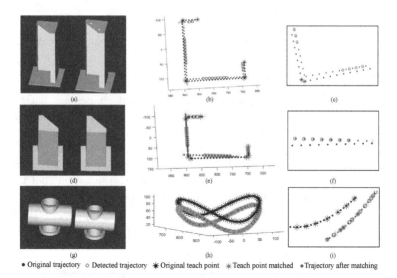

• Original trajectory ○ Detected trajectory ✳ Original teach point ✳ Teach point matched • Trajectory after matching

Fig. 6. The experiments of the robot trajectory correction method.

5   Conclusions and Future Works

In this paper, an effective track correction method for welding based on emulational laser and trajectory are introduced. The effective track correction method on the basis of the sensor consists of seam tracking method and trajectory correction method. In the seam tracking method, the seam point of the emulational laser stripes generated by the simulation software have deviation from actual laser stripe because of shape and pose of workpiece, so after affine transformation which based on the data of the actual laser stripe, emulational seam point can be regard as the real seam point. And then the trajectory correction method mainly consists of three steps: pre-processing, coarse matching and presice matching by using Iterative Closest Point Matching (ICP). For different conditions and workpieces, the corresponding experiments are conducted in this paper. Experimental results demonstrate that the method can meet the requirement of the internal seams tracking of workpiece and accuracy of welding track correction.

However, wrong detection of complex laser images sometimes happens and the length of trajectories after transformation sometimes will be different from the detected trajectories. The future works is to improve the algorithm for the complete extraction of complex laser images.

Acknowledgements. The authors would like to gratefully acknowledge the reviewers comments. This work is supported by National Natural Science Foundation of China (Grant Nos. U1713207 and 52075180), Science and Technology Program of Guangzhou (Grant Nos. 201904020020), and the Fundamental Research Funds for the Central Universities.

References

1. Yanling, X., Huanwei, Yu., Zhong, J., Lin, T., Chen, S.: Real-time seam tracking control technology during welding robot GTAW process based on passive vision sensor. J. Mater. Process. Technol. **212**(8), 1654–1662 (2012)
2. Ma, H., Wei, S., Sheng, Z., Lin, T., Chen, S.: Robot welding seam tracking method based on passive vision for thin plate closed-gap butt welding. Int. J. Adv. Manuf. Technol. **48**(9), 945–953 (2010). https://doi.org/10.1007/s00170-009-2349-4
3. Zhang, Y., Gao, X.: Analysis of characteristics of molten pool using cast shadow during high-power disk laser welding. Int. J. Adv. Manuf. Technol. **70**(9), 1979–1988 (2013). https://doi.org/10.1007/s00170-013-5442-7
4. Xiao, R., Yanling, X., Hou, Z., Chen, C., Chen, S.: An adaptive feature extraction algorithm for multiple typical seam tracking based on vision sensor in robotic arc welding. Sensors Actuat. A Phys. **297**, 111533 (2019)
5. Geng, J.: Structured-light 3D surface imaging: a tutorial. Adv. Optics Photon. **3**(2), 128–160 (2011)
6. Fan, J., Jing, F., Yang, L., Long, T., Tan, M.: A precise seam tracking method for narrow butt seams based on structured light vision sensor. Opt. Laser Technol. **109**, 616–626 (2019)
7. Zeng, J., et al.: A weld position recognition method based on directional and structured light information fusion in multi-layer/multi-pass welding. Sensors **18**(1), 129 (2018)
8. Muhammad, J., Altun, H., Abo-Serie, E.: A robust butt welding seam finding technique for intelligent robotic welding system using active laser vision. Int. J. Adv. Manuf. Technol. **94**(1), 13–29 (2016). https://doi.org/10.1007/s00170-016-9481-8
9. Gu, W.P., Xiong, Z.Y., Wan, W.: Autonomous seam acquisition and tracking system for multi-pass welding based on vision sensor. Int. J. Adv. Manuf. Technol. **69**(1), 451–460 (2013). https://doi.org/10.1007/s00170-013-5034-6
10. Chen, X.Z., Chen, S.B.: The autonomous detection and guiding of start welding position for arc welding robot. Ind. Robot Int. J. **37**(1), 70–78 (2010)
11. Huang, W., Kovacevic, R.: Development of a real-time laser-based machine vision system to monitor and control welding processes. Int. J. Adv. Manuf. Technol. **63**(1), 235–248 (2012). https://doi.org/10.1007/s00170-012-3902-0
12. Yang, S.-M., Cho, M.-H., Lee, H.-Y., Cho, T.-D.: Weld line detection and process control for welding automation. Meas. Sci. Technol. **18**(3), 819–826 (2007)
13. Xu, D., Jiang, Z., Wang, L., Tan, M.: Features extraction for structured light image of welding seam with arc and splash disturbance. In: ICARCV 2004 8th Control, Automation, Robotics and Vision Conference, 2004, vol. 3, pp. 1559–1563 (2004)
14. Chu, H.-H., Wang, Z.-Y.: A vision-based system for post-welding quality measurement and defect detection. Int. J. Adv. Manuf. Technol. **86**(9), 3007–3014 (2016). https://doi.org/10.1007/s00170-015-8334-1
15. Kiddee, P., Fang, Z., Tan, M.: An automated weld seam tracking system for thick plate using cross mark structured light. Int. J. Adv. Manuf. Technol. **87**, 3589–3603 (2016). https://doi.org/10.1007/s00170-016-8729-7
16. Ester, M., Kriegel, H.P., Sander, J., Xu, X.: A density-based algorithm for discovering clusters in large spatial databases with noise. In: Proceedings 1996 International Conference Knowledge Discovery and Data Mining (KDD 1996), pp. 226–231 (1996)
17. Besl, P.J., McKay, N.D.: Method for registration of 3-D shapes. Sensor Fusion IV Control Paradigms Data Struct. **1611**, 586–606 (1992)

A Novel Edge Detection and Localization Method of Depalletizing Robot

Weihong Liu[1], Yang Gao[2], Yong Wang[2], Zhe Liu[3], and Diansheng Chen[3,4(✉)]

[1] School of Mechanical Engineering and Automation, Beihang University, Beijing, China
[2] Zhejiang Cainiao Supply Chain Management Co., Ltd., Hangzhou, Zhejiang, China
[3] Robotics Institution, Beihang University, Beijing, China
chends@buaa.edu.cn
[4] Beijing Advanced Innovation Center for Biomedical Engineering, Beihang University, Beijing, China

Abstract. The application of intelligent robots to perform the depalletizing task is a common requirement in warehouse automation. To solve the problem of identification and localization caused by the disorderly stacking of boxes in pallet, and to eliminate the interference of the reflective material contained in the stacks, this paper proposes an edge extraction algorithm that combines 3D and 2D data. The algorithm firstly obtains the plane position data through three-dimensional point cloud, secondly uses an edge detection algorithm to extract edges in the two-dimensional image. Finally, an optimal segmentation strategy is performed, which is based on the results of point cloud segmentation, edge extraction, and the size information of boxes. Therefore, we can determine the position of each box in the space accurately. Compared with algorithms that only use 2D and 3D data, our method can effectively filter interference. The accuracy rate is close to 100%, which meets the requirements of industrial applications.

Keywords: Mixed-load palletizing · Edge detection · Image processing · Point cloud segmentation

1 Introduction

The function of the depalletizing robot is to disassemble boxes from pallet. The working process is as follows: Firstly the vision system use a camera to collect pictures of boxes, and obtain the sequence of the detachable target through the detection and analysis of the data, then plan the path and use the end-effector to complete the depalletizing task. Collecting the position information of the target box through machine vision is the core part of the system. It is necessary to develop an effective algorithm to detect the edge of the target box.

As early as the 20th century, scientists began to study edge detection algorithms for 2D images. Canny [1] proposed a multi-stage algorithm to detect edges in an image, Martin [2] defined a boundary detection algorithm based on image features such as brightness, color, and texture. However, they are unable to distinguish useless texture or

© Springer Nature Switzerland AG 2020
C. S. Chan et al. (Eds.): ICIRA 2020, LNAI 12595, pp. 512–521, 2020.
https://doi.org/10.1007/978-3-030-66645-3_43

identify the contours required for specific objects in the scene. With the application of deep learning technology in contour detection, Bertasius proposed DeepEdge [3] that uses object-related features to assist edge detection. Nonetheless, the above algorithm is only based on a partial area, and its computational cost and test cost are high.

Scholars have developed a detection algorithm based on global feature learning by constructing an end-to-end deep convolutional neural network (CNN). For example, Holistically-Nested Edge Detection (HED) [4] proposed by Xie adopts the pre-trained trimmed Visual Geometry Group (VGG) Net architecture and uses the final convolutional layers before the pooling layers of neural networks, which greatly improves the edge detection result compared with traditional methods. Richer Convolutional Features for Edge Detection (RCF) [5] is also based on the VGG network, but can fully exploit the CNN features from all the convolutional layers to perform prediction for edge detection. In terms of the test results of the BSD 500 data set [6], RCF achieves better results in edge extraction accuracy than HED.

However, we live in a three-dimensional world. During the process of mapping a 3D object to a plane to obtain a 2D image, it is susceptible to lose data due to factors such as illumination and distance, which makes it difficult for 2D images to accurately and effectively express 3D information. In recent years, the 3D vision has been applied to the edge detection of an object in depalletizing [8]. Chen [7] set the search conditions according to the overall size of the material bag to conduct a neighborhood search after obtaining the 3D information, and determine the position and posture of the material bag respectively; Zuo [9] proposed a point cloud segmentation method based on statistical distance and spatial clustering of target parts and extracted the plane structure from the whole cloud to achieve the goal segmentation. Nonetheless, The accuracy of the target point cloud obtained only by referring to the 3D data is related to the material and shape of the target object, sundries, reflection, transparency, and hollow on the object will also lead to the problem of missing measurement data.

Therefore, this paper proposes a new segmentation strategy that can improve the edge detection performance. Considering both the 3D point cloud data and the edge detection obtained from the 2D image as well as the size information of boxes, the optimal segmentation strategy can locate each grasping target effectively. It can be applied to the depalletizing robot.

2 Edge Detection Algorithm Based on 2D and 3D Data

2.1 Overall Framework

The algorithm proposed in this paper is based on 3D data point cloud segmentation and 2D data edge detection principle to identify the edge of target boxes, which are of the same size. The steps are as follows.

- Step 1: Pallet in place, the vision system obtains the size information of boxes (including length, width, and height). Then the camera fixed above the depalletizing port is triggered to collects both the 3D point cloud and 2D image data of the boxes;
- Step 2: Analysis 3D point cloud data and use normal vector clustering to obtain point cloud segmentation plane (Output as Depth_Output in Fig. 4);

- Step 3: Based on the segmentation plane gained in Step 2, extract an edge map in the 2D image by edge detection algorithm (Output as Edge_Output in Fig. 4);
- Step 4: Based on Depth_output and Edge_output, combining with the size information in Step 1, segment the image by the optimal segmentation strategy;
- Step 5: Compared with the segmentation result with the edge map, output the final result (Output as Seg_Output in Fig. 4).

The overall algorithm framework is shown in Fig. 1 and the detailed steps of each part will be explained in turn.

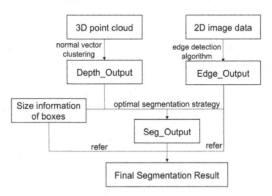

Fig. 1. Flow chart of our algorithm.

2.2   Acquisition and Preliminary Segmentation of 3D Point Cloud

3D point cloud data can be obtained by the depth camera. We use the segmentation algorithm based on point cloud normal vector clustering. Firstly, the least square plane is established by searching K-Neighborhood points. Then calculate the normal vector of the plane as the normal vector estimation of the data point. Secondly, the points with the same normal vector are extracted and the plane is generated by clustering the similar points according to the Euclidean distance, and the normal vector of the plane is calculated. Finally, the corresponding point cloud segmentation plane is determined by refitting the point cloud data. An example of a point cloud segmentation plane is shown in Fig. 2, which is referred to as Depth_Output in Fig. 1:

2.3   Edge Detection Algorithm

This part refers to two edge detection algorithms, HED [4] and RCF [5]. During the training process, we manually annotated 800 images, of which 500 were simulated data in the laboratory and 300 were actual production data shot at the production site). Then performed a certain amount of training data augmentation during the preprocessing of the original annotated images (rotation and cropping). Besides, due to the instability of the ambient light in the warehouse, we convert the image from RGB space to HSV

(a)   (b)   (c)

Fig. 2. Illustration of the preliminary segmentation process of 3D point cloud: (a) the original test image; (b) 3D point cloud image analyzed by normal vector clustering; (c) point cloud segmentation plane (Depth_Output).

space and change its V-channel (brightness value), thereby increasing the breadth of the sample.

During the training and testing process, we used 600 of the 800 manually labeled samples for training, the remaining 200 for model evaluation. HED and RCF networks are used for edge detection respectively. The result of automatic detection is considered to be correct if it is within 2 pixels of the manual calibration position. The Precision/Recall curve of the 200 test set images is shown in Fig. 3, showing that when the recall rate is the same, RCF can extract edges more accurately than HED.

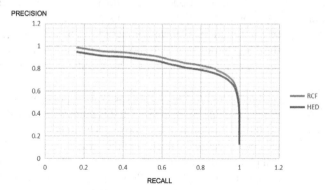

Fig. 3. Comparison of Precision/Recall curve between HED and RCF edge detection algorithm. It can be seen that when the recall rate is the same, RCF is better than HED in terms of the precision of the extracted edge map.

Whereby the edge detection based on 2D data and 3D point cloud is obtained respectively. It can be seen in Fig. 4 that the Depth_Output attained by the 3D point cloud is not suitable for segmentation when boxes are closely spaced, and the Edge_Output detected by the RCF is also easily affected by the reflections of the tape. Therefore, we designed an optimal segmentation strategy that comprehensively considers these two sets of data and the size information of boxes.

(a)Origin Img   (b)Depth_Output   (c)Edge_Output   (d)Seg_Output

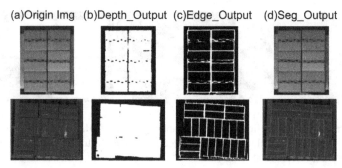

Fig. 4. (a) are the original test image; (b) are the Depth_Output obtained by 3D point cloud data, which is unsuitable for segmentation due to the close distance between boxes of the same height; (c) are the Edge_Output obtained by RCF edge detection algorithms, which contains interference information of tape reflection. Those interfering lines can easy be mistaken for the edge lines of the boxes; (d) are the final segmentation results obtained by the optimal segmentation strategy proposed in our paper, which is referred to as Seg_Output in Fig. 1, it can avoid the influence of interference information such as tape, and achieve accurate identification of the edge of the box.

3   Optimal Segmentation Strategy

As Fig. 4 shows, Depth_Output and Edge_Output do not meet the requirements of direct segmentation. Therefore, we designed an optimal segmentation strategy, which can divide a pallet plane composed of a known number (N) of boxes of the same size. It uses Depth_Output and Edge_Output as well as the size information (including Length, Width, and Height) of boxes as known conditions. The specific steps are as follows:

- Step 1: According to Depth_Output (Fig. 4), we use convex polygon fitting to obtain the top-left vertex as the seed point of the plain. On this point we establish a rectangle template according to the size information of boxes. Theoretically, there should be three templates for each box because the posture of the three-dimensional box mapped to the two-dimensional plane is unknown, so there are three different combinations of side lengths for the template (that is (L, W), (W, H), (L, H)). Besides, each template needs to be discussed in two cases according to the different selection of vertex as the rotation center. Figure 5 shows the situation of a template with Length and Width as the side length. The two cases have selected different vertex.
  The template rotates around the seed point. We need to find the optimal rotation angle for each template based on the overlapping area of the rectangle template and the segmentation plane of Depth_Output, record the angle that maximizes the overlapping area.
- Step 2: After determining the rotation angles of different template in Step 1, we establish the Range of Interest (ROI) of each boundary based on the side length of the template (as shown in Fig. 6). The ROI is used to calculate the 'edge pixels density' in the subsequent steps, the width of ROI is a set value, and the length is the side length of the rectangular template;

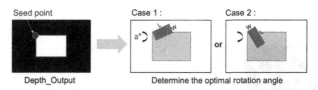

Depth_Output   Determine the optimal rotation angle

Fig. 5. The convex polygon vertex in the upper left corner is the seed point (blue dot). The red box represents one of the templates with Length (L) and Width (W) as the side length. Case 1 and 2 are used to discuss the selection of different vertices as the center of rotation. The process in Step 1 is: We establish a template and take the seed point as the rotation center, calculate the optimal rotation angle to maximize the overlapping area between the template and point cloud segmentation plane. The above pictures are all schematic diagrams drawn by the computer as the simulation of the Edge_Output. (Color figure online)

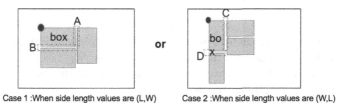

Case 1 :When side length values are (L,W)   Case 2 :When side length values are (W,L)

Fig. 6. According to different cases, the ROI regions of the two cases can be calculated (as shown in the red dotted box in the figure). We can name the above ROI regions as A, B, C, D respectively. (Color figure online)

- Step 3: Based on the Edge_Output (Fig. 4) and ROI, we calculate the joint edge density of the template at the different case sequentially, and then take the case with the highest density of edges as a node to save the value of its side length.
  For instance, in Fig. 7, P(Edge|A) means the edge pixel density of ROI A. It represents the ratio of pixels occupied by the edge to the total pixels of A. Similarly, we can calculate the P(Edge|B), P(Edge|C), P(Edge|D) as the edge pixel density of ROI B, C and D. Then we multiply P(Edge|A) and P(Edge|B) to get P(Edge|A, B) as the joint edge density of Case 1 and calculate P(Edge|C, D) similarly. Lastly, we can compare the joint edge density between the two cases, and record the higher one as a node.
- Step 4: This step is used to judge the posture of the box on the pallet since the side length of the box mapped to the two-dimensional plane is unknown. Based on the above steps, the optimal case is determined for each template, then three possible posture of each box could be saved (that is (L, W), (W, H), (L, H)). We selected the more likely postures by their joint edge density (P(Edge|ROI 1, ROI 2)). If it is less than a set threshold, the node stops searching; otherwise, the node is saved and return to Step 1 to determine the posture of the next box as the sub-nodes until the location of N boxes is determined. Thus the point cloud surface segmentation ends.

Multiple results may be obtained in the last iteration, it is necessary to compare the results of each sub-node concerning the actual size of the box. We set a loss function as:

$$loss = \sum_{i=1}^{N} |L\_cal_i - L\_gt_i| + |W\_cal_i - W\_gt_i| \tag{1}$$

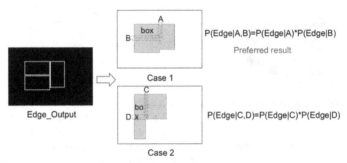

Fig. 7. The joint percentage of edge pixels in the two ROI of the template in a different case is calculated since the box posture of Case 1 is more similar to the edge detected in Edge_Output, the joint edge density of Case 1 is higher than Case 2. Therefore Case 1 is selected as a preferred result and saved as a node.

Where N represents the number of boxes, L_cal_i, W_cal_i represents the length and width of the i-th box output in the final plane segmentation (Seg_Output), L_gt_i, W_gt_i represents the side length determined in Step 3. Finally, we save the segmentation result with the smallest value of loss function.

The schematic diagram of the overall optimal segmentation search algorithm is shown in the following figure (Fig. 8):

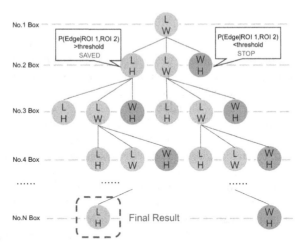

Fig. 8. The schematic diagram of the optimal segmentation strategy is based on joint edge density. Nodes in different colors represent the different postures of the box. When the edge density is less than the threshold, the node stops searching, otherwise, it starts looping to search the position of the next box. (Color figure online)

4   Experiment and Results

In this section, we discuss the implementation and the performance of our proposed algorithm.

4.1   Implementation

We set up a depth camera (HV-1000, SmartEye) at a height of 4 m from the bottom of the pallet to acquire 2D images and 3D depth image of boxes (keeping the accuracy of the Z-direction within 5 mm), and collect a total of 879 sets of actual production data shot at the production site as the Test Set. Four algorithms are used to test the recognition of the edge of the box respectively: (i) edge segmentation algorithm based on HED, (ii) edge segmentation algorithm based on RCF, (iii) HED + optimal segmentation strategy, (iv) RCF + optimal segmentation strategy.

4.2   Results

Figure 9 shows a comparison of edge detection result calculated using RCF and RCF combined with the optimal segmentation strategy. It can be seen that the latter method can better avoid the interference of sundries (such as paper, reflective tape) on the boxes, then accurately segment the stack. Figure 10 represents more results of edge detection using RCF combined with optimal segmentation strategy.

| Origin Img | Depth_Output | Edge_Output | Seg_Output |

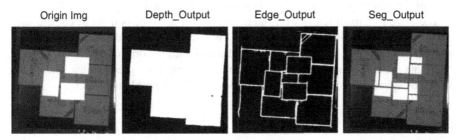

Fig. 9. From left to right are the original image, the point cloud segmentation plane, the edge detection result of RCF, and the final segmentation result based on the optimal segmentation search strategy. It can be seen that the point cloud segmentation plane is affected by the close distance between the boxes, and the box boundaries cannot be displayed. The edge detection is affected by the covering objects on the box surface, and the reflective block on the box surface is incorrectly identified as the boundary of the boxes. The algorithm that using RCF combined with the optimal segmentation strategy accurately perform segmentation.

Besides, not only the precision of the edge detection has been improved, but the overall recognition accuracy has also been effectively enhanced. The accuracy of the four algorithms is shown in the table, it proves that combining with the optimal segmentation algorithm can better the edge detection result obtained by HED and RCF by helping filter the interference information (Table 1).

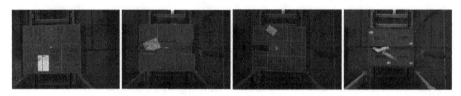

Fig. 10. Results of edge detection using RCF combined with the optimal segmentation strategy.

Table 1. Accuracy of edge recognition with different algorithms

Algorithm	Correct detection	Incorrect detection	Correct rate
HED	814	65	93%
RCF	823	56	94%
HED + optimal segmentation search	879	0	100%
RCF + optimal segmentation search	879	0	100%

Compared with the visual recognition algorithm currently used by manufacturers in the depalletizing system, our algorithm can use a fixed camera position without an external light source to achieve segmentation recognition for arbitrary stacking and complex surface texture with accuracy close to 100%. The processing time is about 3–3.5 s (the time to generate the point cloud is about 1.5 s.), which is twice the processing speed of the other visual algorithm only based on 3D point data.

There are two main reasons for the increase in speed: one is that our camera uses GPU to accelerate point cloud generation. The other is that in the actual production situation, we find that the identification and positioning algorithm based on 3D point cloud data generally uses multi-threshold segmentation. The speed is slow when there is obvious interference, our algorithm can overcome such problems.

5   Conclusion

This paper presents an edge extraction algorithm that can be applied to the process of depalletizing. Combining the plane position data obtained through the three-dimensional point cloud and the edge detection algorithm, an optimal segmentation strategy is proposed, which can accurately segment the same type of boxes when its size is known. This algorithm can reduce the influence of the close stacking of boxes and the interference of surface reflections. Therefore it can improve the accuracy of edge detection. In general, this paper provides a novel edge detection and localization method for palletizing robots.

Acknowledgement. This research was supported by the National Key R&D Program of China 2018YFB1309300. The authors would like to personally thank all the team members.

References

1. Canny, J.: A computational approach to edge detection. IEEE Trans. Pattern Anal. Mach. Intell. PAMI **8**, 679–698 (1986)
2. Martin, D.R., Fowlkes, C.C., Malik, J.: Learning to detect natural image boundaries using local brightness, color, and texture cues. IEEE Trans. Pattern Anal. Mach. Intell. **26**, 530–549 (2004)
3. Bertasius, G., Shi, J., Torresani, L.: DeepEdge: a multi-scale bifurcated deep network for top-down contour detection. In: Proceedings of the IEEE Computer Society Conference on Computer Vision and Pattern Recognition, 07–12 June, pp. 4380–4389 (2015)
4. Xie, S., Tu, Z.: Holistically-nested edge detection. Int. J. Comput. Vis. **125**, 3–18 (2017)
5. Liu, Y., et al.: Richer convolutional features for edge detection. IEEE Trans. Pattern Anal. Mach. Intell. **41**, 1939–1946 (2019)
6. Arbeláez, P., Maire, M., Fowlkes, C., Malik, J.: Contour detection and hierarchical image segmentation. IEEE Trans. Pattern Anal. Mach. Intell. **33**, 898–916 (2011)
7. Chen, X., Shi, J., Jia, R.: Application of 3D machine vision in intelligent robot destacking. Electrotech. Appl. **S1**, 31–35 (2019)
8. Xu, K., Liang, L., Wang, F., et al.: Research on depalletizing analysis and localization method of mixed-loaded pallet based on three-dimensional vision technology. Harbin Institute of Technology (2019)
9. Zuo, L.: Research on automatic identification and location of scattered parts for robot picking. Harbin Institute of Technology (2015)

Soft Actuators

Control of a Series Elastic Actuator Based on Sigmoid-Proportional-Retarded (SPR) with Online Gravity Compensation

Feng Jiang, Jiexin Zhang, and Bo Zhang[✉]

State Key Laboratory of Mechanical System and Vibration, School of Mechanical Engineering,
Shanghai Jiao Tong University, Shanghai 200240, China
{f.jiang,zhangjiexin,b_zhang}@sjtu.edu.cn

Abstract. In this paper, a rotary series elastic actuator (SEA) based on torsion spring is designed. A novel position control law (Sigmoid-proportional-retarded) with online gravity compensation (OGC) is presented in order to reduce the residual vibration of the link and shorten the response time. Moreover, the stability of SPR control law is proved by the Lyapunov method. Some comparative experiments were implemented. It is concluded that SPR control based on OGC can reach the target position accurately and quickly. Meanwhile, the results show that the method is also effective in eliminating residual vibration.

Keywords: Series elastic actuator · Sigmoid-Proportional-Retarded · Online gravity compensation · Residual vibration

1 Introduction

In present times, modeling and control of flexible joint robot have been considered increasingly [1]. Flexible transmissions such as harmonic reducers, belts, and cables are widely used in robots [2]. Since Pratt proposed a series elastic drive (SEA) [3], more scholars actively add flexibility to the joints to improve certain performance [4], such as increasing the torque control accuracy [5], storing energy [6], improving safety performance [7]. However, modeling and control of the flexible joint compared to the rigid joint is more complicated because of the following reasons:

1. The order of dynamics model and the degrees of freedom for the SEA are significantly increased [8]. The rigid joint is described via second-order nonlinear equations, but the SEA system is described via fourth-order nonlinear equations.
2. The number of inputs is less than the degrees of system freedom, so the system is underactuated [9].
3. The ectopic control problem inspires the residual vibration of the link [10].

The SEA model mostly used is the simplified model proposed by SPONG [11]. Furthermore, many control methods are proposed, such as singular perturbation [12],

C. S. Chan et al. (Eds.): ICIRA 2020, LNAI 12595, pp. 525–537, 2020.
https://doi.org/10.1007/978-3-030-66645-3_44

feedback linearization [13], adaptive control [14], etc. And currently, the potential difference (PD) control method has been widely used. Tomei proposed a PD controller with static gravity compensation to solve the SEA adjustment problem [15]. On the basis of static compensation, DeLuca et al. proposed an online gravity compensation control strategy and proved it is a better control effect, but there is still residual vibration [16]. Mo et al. of Shanghai Jiao Tong University proposed using Proportional-Integral-Retarded (PIR) controller to apply to joint position loop control [17], this method can effectively reduce residual vibration compared to conventional PD control. However, the residual vibration problem and improvement of control accuracy are still the focus of research at present.

In this paper, an independent torsion spring based on the SEA system is designed, and a Sigmoid-Proportional-Retarded (SPR) controller is proposed. At the same time, the SPR control method with online gravity of compensation is proposed. The contributions made in this paper are as follows:

1. A SEA system based on torsion spring is designed.
2. The Sigmoid-Proportional-Retarded (SPR) controller is proposed to effectively solve the problem of residual vibration in the SEA system. The Lyapunov method is used to verify the stability of the system.
3. The OCG-based SPR control is proposed to solve the SEA joint control problem. And the integral time squared error criterion (ITSE) is used as a parameter adjustment criterion for the control method.

The remaining part of this paper is organized as follows. Joint design and modeling are introduced in section 2. In Sect. 3, a joint control algorithm is proposed. And stability analysis is detailed in Sect. 4. In Sect. 5, Some experiments are carried out on the platform to verify the correctness of the algorithm. Section 6 concludes paper findings.

2 Design and Modeling

In this paper, the authors have designed the SEA system based on torsion spring. Figure 1 shows the structure of the SEA. Figure 2 shows the SEA system platform. The structure includes motor, harmonic reducer, control board, elastic link, and absolute encoder of motor and link.

The servo motor input is connected in series to the harmonic reducer, transmitted to the torsion spring, and then connected to the output. An absolute encoder is connected to the output of the servo motor side, and inner axles transfer the rotation motion from the link to an encoder, so the other absolute encoder can be connected to the link side. Thereby real-time position detection of the motor side and the link side is completed.

The torsion spring is made of maraging steel (18Ni type, alloy 350) with a yield stress of up to 2400 MPa and good elasticity and ductility. The design principle of the Archimedes spiral double spring is proposed. The double helix design can offset the radial force generated by the spring winding in the center of the spring when the deformation occurs. Figure 3 shows the results of the torsion spring under finite element analysis (ANSYS). It can be found that the maximum stress of the torsion spring is

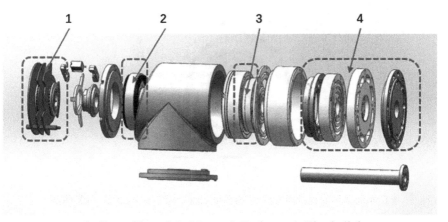

1. Control board; 2. Motor; 3. Reducer; 4. Elastic link.

Fig. 1. Overall view of SEA system.

1.Control board; 2. Emulator; 3.CAN receiver; 4. SEA system; 5. Link and load.

Fig. 2. SEA system platform.

958 MPa, the total strain is 1.15 cm, and the spring stiffness is 239 Nm/rad. Under load conditions, the real spring stiffness is 220 Nm/rad.

The dynamics model of a single flexible joint can be divided into a gravity dynamics model and a gravity-less dynamic model depending on whether there is load.

In the case of no-load gravity, the difference between the motor and link angle can be ignored, and the driving torque of the motor can be accurately controlled for the SEA system [18]. In this paper, the torsion spring with stiffness K is used instead of the joint flexible part to connect the link end and the motor end. The figure of the schematic image of the simplified physical model are shown in Fig. 4.

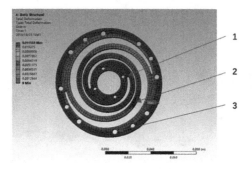

1.Connect to motor side; 2. Torsion spring; 3. Connect to link side.

Fig. 3. Static simulation for stress distribution.

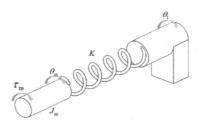

Fig. 4. SEA model.

The effects of friction are ignored so that joint dynamics models can be simplified. the ideal dynamic model of the flexible joint is as follows without considering the gravity term [11],

$$J_l \ddot{\theta}_{l0} + K(\theta_{l0} - \theta_{m0}) = 0 \tag{1}$$

$$J_m \ddot{\theta}_{m0} + K(\theta_{m0} - \theta_{l0}) = \tau_0 \tag{2}$$

Where θ_{l0} is the angle of the link without gravity, θ_{m0} is the angle of the motor without gravity (after passing through the reducer), τ_{m0} is the output torque of the motor without gravity, J_l, J_m is the inertia of the joint and the motor, K is the joint stiffness.

However, the effect of load gravity is not negligible in the presence of load. The flexible link will be affected, and the motor output angle will deviate from the output angle of the link. Therefore, it is necessary to establish a model with gravity compensation. The influence of gravity is added in (1)(2). In summary, the dynamic equation of the flexible joint with load gravity added is shown in (3)(4) [19].

$$J_l \ddot{\theta}_l + g(\theta_l) + K(\theta_l - \theta_m) = 0 \tag{3}$$

$$J_m \ddot{\theta}_m + K(\theta_m - \theta_l) = \tau_m = Pi \tag{4}$$

Where $g(\theta_l)$ is the gravity compensation term in the system, θ_l is the rotation angle of the link, θ_m is the angle of the motor (passing through the reducer), τ_m is the torque applied by the motor, P is the proportional coefficient between torque and current, i is the current of the motor.

3 Control Method

In order to eliminate residual vibration, we propose a control method Proportional-Retarded (PR), which adds a time-delay link to the p-control. For an n-order linear time-invariant system, it can be expressed as (5)(6):

$$x^{(n)}(t) + \sum_{i=0}^{n-1} a_i x^{(i)}(t) = \sum_{j=0}^{n-1} a_j u^{(j)}(t) \tag{5}$$

$$u(t) = k_P e(t) - k_R e(t - t_r), e(t < 0) = 0 \tag{6}$$

Where $u(t)$ is the input of the system, $x(t)$ is the output of the system, a_i, b_j are constant, k_R the time lag coefficient, t_r is the error of the time lag, $e(t)$ is the error, and k_P, k_R are the proportional term and the hysteresis term coefficient respectively. It has been proven that PR control can control flexible joint better, but there are still some shortcomings: First, faster convergence speeds can be obtained with large gains, but it is easy to overshoot and enhance vibration. Second, small gains result in slow convergence and insensitivity to small errors. For the parameters of P control, the sigmoid function is applied to realize the nonlinear control of k_P. This paper proposes an improved sigmoid-proportional-retarded (SPR) controller.

$$u(t) = k_P * c_1 \left(\frac{2}{1 + |e^{-c_2 x}|} - 1 \right) - k_R e(t - t_r), e(t < 0) = 0 \tag{7}$$

Where c_1, c_2 are the adjustable normal number. Compared with the traditional PD control, SPR controller do nonlinear changes to the error input of proportional control: when the input error is small, the system can still obtain a large gain, which can improve sensitivity to small errors, when the errors are large, it can effectively avoid excessive control input of the system, thereby suppressing overshoot and reducing output oscillation caused by overshoot. Increasing the hysteresis section can effectively eliminate the residual vibration of the link caused by the torsion spring when the motor is stable. Therefore, overshoot and residual vibration on the link side can also be effectively suppressed while ensuring rapid convergence.

However, it can be found that the joint angle will not reach the predetermined position due to the influence of the load when the flexible joint model with gravity is established. Moreover, the torque of the load will affect the control accuracy of the motor torque during the control process. Hence, an on-line gravity compensation method is proposed. The torque applied by the motor is expressed as follows in consideration of gravity:

$$i = i_0 + i_g \tag{8}$$

Where i_0 is the current value corresponding to the torque applied by the motor in the control request without considering the gravity of the load. i_g is the current value corresponding to the additional compensation torque of the motor caused by the gravity.

At present, the most common of i_g computing applications are fixed-value operations. For constant position control, there is

$$\tau_g = Pi_g = g(\theta_d) \tag{9}$$

Where θ_d is the link position setting. The above control rate can be applied to obtain a good compensation effect in the final stable state, but the method contributes less to the improvement of the real-time control accuracy. In summary, an online gravity compensation method is proposed as follows.

$$\tau_g = Pi_g = g(\theta_l) + J_m K^{-1} \ddot{g}(\theta_l) \tag{10}$$

The configuration of the joint and load is considered in this paper, and the following relationship is obtained,

$$g(\theta_l) = mg\sin(\theta_l) \times r \tag{11}$$

$$\ddot{g}(\theta_l) = mg\cos(\theta_l)\ddot{\theta}_l \times r - mg\sin(\theta_l)\dot{\theta}_l^2 \times r \tag{12}$$

Where r is the distance from the axis of the shaft to the center of mass of the link.

The angle of the joint has a slight difference from the angle of the motor due to the influence of the flexible link. In order to ensure the correct position of the load, i_0 control law can be considered as

$$i_0 = K_p e(t) + K_r(e(t - t_r)) \tag{13}$$

$$e(t) = \theta_d - \theta_m + \theta_g \tag{14}$$

$$\theta_g = K^{-1} g(\theta_l) \tag{15}$$

Where θ_g is the value of the joint angle that is deviated from the load.

Finally, the S-PRG control law is obtained by synthesizing the SPR control law described above and the gravity compensation control described in this subsection. The motor angle is used as a feedback signal for PR control. After the gravity term compensation term is added, the control block diagram is shown in Fig. 5.

4   Stability Analysis

In this section, the Lyapunov method is used to judge the stability of the SPR control system. In the previous section, the gravity compensation algorithm was implemented, so the gravity term can be ignored in the stability analysis.

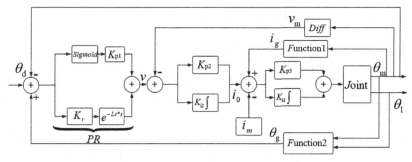

Fig. 5. The method of joint control is three-loop control, which is the position loop, speed loop, and current loop. The motor angle is used as a feedback signal to control the position loop. The position loop uses the SPR control law and adds a gravity compensation term. The speed loop adopts PI control. After passing through the speed loop, the gravity compensation current is added to the current to obtain an accurate real-time current value. Finally, the PI control law is used to control motor rotation.

Equation (3)(4) is combined with the SPR control law to obtain the following expression.

$$\tau_m = k_p * c_1\left(\frac{2}{1+\left|e^{-c_2 e(t)}\right|} - 1\right) - k_R e(t - t_r) \tag{16}$$

$$J_l \ddot{\theta}_l + K(\theta_l - \theta_m) = 0,\, J_l \ddot{\theta}_l + K(\theta_l - \theta_m) = 0 \tag{17}$$

$$J_m \ddot{\theta}_m + K(\theta_m - \theta_l) = \tau_m,\, J_m \ddot{\theta}_m + K(\theta_m - \theta_l) = \tau_m \tag{18}$$

To facilitate the calculation, set:

$$e(t) = \theta_d - \theta_m(t) \tag{19}$$

Where θ_d is the expected angle.

Next, the above controller proved to be able to stabilize the closed loop.

It is known that the energy of the system consists of three parts, namely the kinetic energy of the motor, the kinetic energy of the joint link and the elastic potential energy of the flexible link.

$$V = \frac{1}{2}J_m \dot{\theta}_m^2 + \frac{1}{2}J_l \dot{\theta}_l^2 + \frac{1}{2}K(\theta_m - \theta_l)^2 \tag{20}$$

Derivation, we get

$$\dot{V} = J_m \dot{\theta}_m \ddot{\theta}_m + J_l \dot{\theta}_l \ddot{\theta}_l + K(\theta_m - \theta_l)(\dot{\theta}_m - \dot{\theta}_l) \tag{21}$$

Therefore, the dynamic model brought into the joint and the motor. Where,

$$\dot{V} = k_p * c_1\left(\frac{2}{1 + e^{-c_2 e(t)}} - 1\right) - k_R e(t - t_r)\dot{\theta}_m \tag{22}$$

In order to make $\dot{V}$ negative, add V_m, V_d to V.
Where,

$$V_m = k_p c_1 \left(-e(t) - \frac{2}{c_2} \ln(e^{-c_2 e(t)} + 1) \right) + \frac{2}{c_2} \ln 2), V_m 0 = 0 \tag{23}$$

$$V_d = -\frac{1}{2} k_R (e(t))^2 + \frac{1}{2} k_R (\sup|e(t)|)^2 \tag{24}$$

Where,

$$\frac{\partial V_m}{\partial e(t)} = k_p * c_1 \left(\frac{2}{1 + e^{-c_2 e(t)}} - 1 \right) \tag{25}$$

$$\dot{V}_d = k_R e(t) \dot{\theta}_m(t) \tag{26}$$

Obviously,

$$\frac{\partial V_m}{\partial \Delta \theta} \{ \begin{array}{l} > 0 \text{if } e(t) > 0 \\ = 0 \text{if } e(t) = 0 \\ < 0 \text{if } e(t) < 0 \end{array} \tag{27}$$

Therefore, it can be judged that V_m, V_d are positive. We get the positive-definite function V.

$$V = \frac{1}{2} J_m \dot{\theta}_m^2 + \frac{1}{2} J_l \dot{\theta}_l^2 + \frac{1}{2} K (\theta_m - \theta_l)^2 + V_m + V_d \tag{28}$$

V is the Lyapunov function. Derivation, we get,

$$\dot{V} = -k_R e(t - t_r) \dot{\theta}_m(t) + k_R e(t) \dot{\theta}_m(t) \tag{29}$$

Assuming,

$$f(x) = e(x) \tag{30}$$

The Taylor expansion is shown in (31):

$$f(x) = \frac{f(x_0)}{0!} + \frac{\dot{f}(x_0)}{1!}(x - x_0) + R_n(x) \tag{31}$$

This function can perform Taylor expansion at $x = t$. (32) can be obtained with ignoring high-order terms.

$$e(t - t_r) = e(t) - \dot{e}(t) t_r \tag{32}$$

Put (32) in (29) and you can get:

$$\dot{V} = -k_R t_r \dot{\theta}_m(t)^2 \tag{33}$$

Therefore, it can be concluded that $\dot{V}$ is less than zero, and further, the control system is stable.

Combined with closed-loop dynamics, it can be seen that on $\{(\theta_m, \theta_l, \dot{\theta}_m, \dot{\theta}_l) | \dot{V} = 0\}$, the only equilibrium point of the system is

$$\theta_m = \theta_l = \theta_d, \dot{\theta}_m = \dot{\theta}_l = 0 \tag{34}$$

Therefore, it can be obtained from the Russell invariant theorem that the system is asymptotically stable at the above equilibrium point.

5 Hardware Experiments

The design control method is verified on a self-built SEA platform shown in Fig. 2, the control experiment is performed in the following four cases:

1. Comparing PD, PR, SPR control method without considering gravity.
2. Comparing PD, PD with OGC, SPR with OGC with considering gravity.
3. Comparing Irregular trapezoid wave tracking with considering gravity.
4. Comparing sine wave tracking with considering gravity.

The control gains of each method are repeatedly adjusted to make integral time squared error (ITSE) as small as possible, the nominal parameters of SEA are K = 180 Nm/rad, $B_m = 0.0525$ kgm^2, $J_m = 0.1254$ kgm^2.

Case1: Position Control Without Gravity
The platform is placed horizontally, and the Experiment results are shown in Fig. 6. The subfigure on the right side is a partially enlarged view of the left side. It is obvious that under the PD control method, a small gain (purple line) allows the link to reach the desired position with a slow response. Although a large gain (blue line) can respond quickly, there is a large residual vibration. And we can find that PR control (brown line) can effectively reduce residual vibration, but still cannot be eliminated entirely. The proposed method (red line) is used to drive the link to the desired position quickly and to suppress residual vibration well.

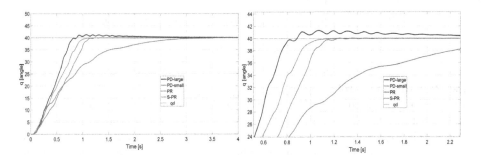

Fig. 6. Position control without gravity. (Color figure online)

- Control parameters are chosen as:
- **Pd control(large):** p = 0.25 d = 0.01.
- **Pd control(small):** p = 0.015 d = 0.0015.
- **PR control:** P = 0.25 kr = 0.01 t = 440.
- **Sigmoid-PR control**: P = 0.3 kr = 0.01 t = 440 c1 = 2 c2 = 0.005.

In this system, errors in the later stages of the response process should be considered. Therefore, it is very suitable to use the integral time squared error criterion (ITSE) as a parameter adjustment criterion for the system. The ITSE error criteria can be expressed as:

$$ITSE = \int_0^T t|e(t)|^2 dt \tag{35}$$

Where t is the response time, T is the time when the response is stable, and e(t) is the systematic error at time t.

The exhaustive method is used to search for the optimal solution near the empirical value, and the ITSE criterion is used as the objective function. The maximum overshoot of the control methods under the optimal parameters, the time from the start of the motion to steady-state, the stable error, The ITSE values from the initial arrival to the specified position to the steady-state are shown in Table 1, it shows that the SPR controller has a better control effect than other control methods.

Table 1. Experiment results of case 1.

Position Loop Controller	Maximum Overshoot(^{o})	Response time(s)	Stable error	ITSE
PD	1.28	4.200	0.15	10030
PD(small)	0.01	3.6	0.10	
PR	0.07	3.004	0.01	8.8943
SPR	0.01	1.197	0.01	1.9975

Case2: Position Control with Gravity

The platform is placed vertically, and three sets of compare experiment results were carried out in Fig. 7, PD control, PD control with OGC, and SPR control with OGC, respectively. It is obvious that online gravity compensation (OGC) can compensate well for gravity errors by comparing the blue line and the yellow line. The maximum overshoot, the response time, and stable error are shown in Table 2. It can be seen that SPR control with online gravity can perform gravity compensation well, and the link can reach the desired position without the residual vibration.

Case3: Comparing Irregular Trapezoid Wave Tracking With Gravity

The platform is placed vertically and the link tracking the desired position, experiment result under three method are detailed in Fig. 8. under PD control, there are obvious steady-state error and the residual vibration, and under PD control with OGC, link side can track the desired position but experience violent oscillation. The online gravity link is driven to track the desired position without residual vibration under SPR control, but we can also find that the control method has a very small tracking delay.

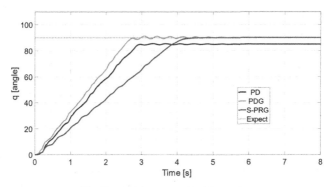

Fig. 7. Position control with gravity.

Table 2. Experiment results of case 2

Position Loop Controller	Maximum Overshoot(^{o})	Response time(s)	Stable error
PD	1.16	6.185	0.16
PDG	0.53	6.225	−4.92
S-PRG	0.01	4.648	0.04

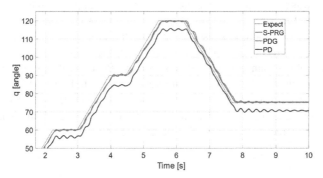

Fig. 8. Irregular trapezoid wave tracking with gravity.

Case4: Comparing Sine Wave Tracking With Gravity

The platform is placed vertically, and the link tracking sine wave with different ampli-tudes and frequency. The result under S-PRG and PDG control method are detailed in Fig. 9, and the tracking error under different experiments is also shown in the figure. In the case of tracking different sinusoids, it can be seen that although the S-PRG control method can reduce the fluctuation, there is a phenomenon of tracking lag, resulting in no significant improvement in tracking error compared to PDG control.

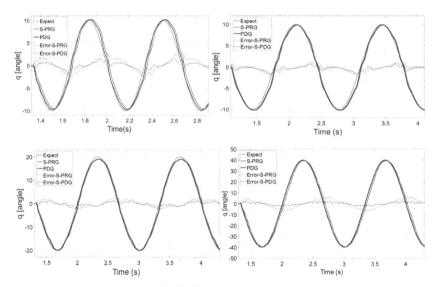

Fig. 9. Sine wave tracking.

6 Conclusion

In this paper, a SEA system based on torsion springs was designed. SPR control method based on OGC was proposed to solve the SEA joint control problem effectively. Moreover, ITSE is used as a parameter adjustment criterion for the control method. Subsequently, The Lyapunov method is used to analyze the stability of the control algorithm. At last, four cases of the experiment were carried out, and it proves that SPR control has a better control effect than classical PD control and PR control in terms of overshoot response time and the residual vibration suppression of the link. In addition, this control method can perform simple trajectory tracking. However, when tracking complex tracks, there will be a tracking delay, and it is no better effect than the general control method. This will be considered as the focus of research in the future.

References

1. Tadele, T.S., de Vries, T., Stramigioli, S.: The safety of domestic robotics: a survey of various safety-related publications. IEEE Robot. Autom. Mag. **21**(3), 134–142 (2014)
2. Good, M., Sweet, L., Strobel, K.: Dynamic models for control system design of integrated robot and drive systems. J. Dynam. Syst. Measure. Control **107**(1), 53–59 (1985)
3. Pratt, G.A., Williamson, M.M.: Series elastic actuators. In: Proceedings of International Conference on Intelligent Robots and Systems. Human Robot Interaction and Cooperative Robots. IEEE/RSJ. pp. 399–406 (1995)
4. Roy, N., Newman, P., Srinivasa, S.: CompActTM Arm: A compliant manipulator with intrinsic variable physical damping, p. 504. MIT Press, Massachusetts (2013)
5. Kashiri, N., Laffranchi, M., Tsagarakis, N.G., et al.: Physical interaction detection and control of compliant manipulators equipped with friction clutches. In: Proceedings of International Conference on Robotics and Automation. IEEE, pp. 1066–1071 (2014)

6. Garofalo, G., Englsberger, J., Ott, C.: On the regulation of the energy of elastic joint robots: Excitation and damping of oscillations. In: Proceedings of American Control Conference. IEEE, pp. 4825–4831 (2015)

7. Sariyildiz, E., Chen, G., Yu, H.: A unified robust motion controller design for series elastic actuators. IEEE/ASME Trans. Mechatronics 22(5), 2229–2240 (2017)

8. Kostarigka, A.K., Doulgeri, Z., Rovithakis, G.A.: Prescribed performance tracking for flexible joint robots with unknown dynamics and variable elasticity. *Automatica*, *49*(5),.1137–1147 (2013)

9. Fantoni, I., Lozano, R.: Non-linear control for underactuated mechanical system. Springer Science & Business Media, Dordrecht (2002)

10. Petit, F., Lakatos, D., Friedl, W., et al.: Dynamic trajectory generation for serial elastic actuated robots. Int. Federation Autom. Control 45(22), 636–643 (2012)

11. Spong, M.W.: Modeling and control of elastic joint robots. J. Dynam. Syst. Measure. Control 109(4), 310–319 (1987)

12. Spong, M.W., Khorasani, K., Kokotovic, P.: An integral manifold approach to the feedback control of flexible joint robots. IEEE J. Robot. Autom. 3(4), 291–300 (1987)

13. Spong, M.W., Hung, J.Y., Bortoff, S.A., et al.: A comparison of feedback linearization and singular perturbation techniques for the control of flexible joint robots. In: Proceedings of American Control Conference. IEEE, pp. 25–30 (1989)

14. Ghorbel F, Spong, M.W.: Stability analysis of adaptively controlled flexible joint manipulators. In: Proceedings of Conference on Decision and Control. IEEE, pp. 2538–2544 (1990)

15. Tomei, P.: A simple PD controller for robots with elastic joints. IEEE Trans. Autom. Control 36(10), 1208–1213 (1991)

16. De Luca, A., Siciliano, B., Zollo, L.: PD control with on-line gravity compensation for robots with elastic joints: Theory and experiments. Automatica 41(10), 1809–1819 (2005)

17. Xixian, Mo., Feng, J.: Control of a mechanically compliant joint with proportional-integral-retarded (PIR) Controller. In: ICIRA, pp. 379–390 (2018)

18. Readman, M.C., Belanger, P.R.: Analysis and control of a flexible joint robot. In: IEEE Conference on Decision & Control. IEEE (1990)

19. Luca, A.D., Flacco, F.: A PD-type regulator with exact gravity cancellation for robots with flexible joints. In: IEEE International Conference on Robotics & Automation. IEEE (2011)

A Flexible Mechanical Arm Based on Miura-Ori

Xiu Zhang, Meng Yu, Weimin Yang, Yuan Yu, Yumei Ding, and Zhiwei Jiao[✉]

Beijing University of Chemical Technology, Beijing 100029, China
jiaozw@buct.edu.cn

Abstract. In this paper, a flexible mechanical arm (FMA) based on Miura ori is designed and its driving performance is studied. Through inflating or pumping air passage of the FMA, the relationship between the air pressure and the bending angle of the FMA is obtained; Through the inflation or extraction of the three air passages of the FMA, the relationship between the air pressure and the elongation or compression of the robotic arm is obtained. The results show that the FMA is flexible and can realize the bending, elongation and compression of the FMA in the two situations of inflation and extraction.

Keywords: Miura-ori · FMA · Bending · Elongation

1 Introduction

Traditional rigid industrial mechanical arms perform well in regular environment, but in some chaotic and crowded environments, such as mining accident site and scientific detection, due to the limitation of the working space of rigid mechanical arms, good control effect cannot be achieved [1]. Compared with rigid mechanical arm, FMA is lighter in structure and can be adapted to a variety of operating spaces, and have important research value in the fields of aerospace, medical equipment and robotics. The material of the FMA is soft, and the stiffness needs to be changed to meet the requirements of grasping a certain weight target. The FMA makes use of the elastic deformation of the body to make the arm continuously bend along the length direction, so as to realize the movement of animals like elephant trunk and octopus tentacles [2]. At present, the driving methods of FMA can be divided into pneumatic control, hydraulic control, rope control, shape memory, etc. [3], among which the FMA designed pneumatically is the lightest and simplest [4]. In 2019, scholars such as Guan Qinghua of Harbin Institute of Technology [5], inspired by the elephant trunk, proposed and analyzed curved and spiral pneumatic artificial muscles on the basis of pneumatic artificial muscles (PAM). The team connected the bent and helical PAM in series to form a FMA, which proved its adaptability, functionality and flexibility through testing. In recent years, origami structure has been increasingly applied in aerospace, flexible electronics, medicine, robotics and other fields. Origami has important research value and scientific significance [6]. With the development of science and technology, the combination of origami structure

Z. Jiao — Supported by Beijing Nova program(Z201100006820146). Zhuhai Industrial Core and Key Technology Research Project (ZH01084702180085HJL).

C. S. Chan et al. (Eds.): ICIRA 2020, LNAI 12595, pp. 538–544, 2020.
https://doi.org/10.1007/978-3-030-66645-3_45

and modern science and technology makes the practical value of origami far beyond its aesthetic value [6–8]. The introduction of the origami concept establishes the connection between the rigid and FMA, which can better design the flexible manipulator with certain stiffness. In 2017, Donghwa Jeong of Case Western Reserve University and other scholars [9] designed a new three-finger FMA based on Twisted Tower origami structure. The arm is made of a ten-story twisted tower, driven by four cables, and each finger is made of a smaller eleven-story tower. The team conducted kinematic modeling and stiffness and durability tests on the arm, Experiments show that the origami structure of the arm absorbs the excessive force exerted on the object through force distribution and mechanism deformation, and can be used to grasp the fragile object. In this paper, a FMA is designed based on the Miura-ori, and the performance of the robotic arm is evaluated by studying its movement characteristics.

2   Miura-Ori Mechanical Arm Structure

2.1   The Miura-Ori

In the science of origami, there are many typical folding mechanisms, among which the most significant is the "Miura folding mechanism" invented by Japanese scholar Miura. "The Miura-ori" is shown in Fig.1 below [4]. The Miura-ori can change the mechanical properties and material properties of paper, and can make humble paper have a certain degree of stiffness, compressibility, and shrinkability after folding, and the Miura-ori has certain memory characteristics; After being folded, the paper can be restored from the unfolded state to the folded state under a small force.

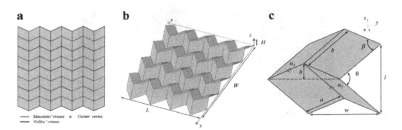

Fig. 1. Configuration of Miura-ori.

2.2   Structural Design of Miura-Ori FMA

Miura-ori mechanical arm is composed of multiple identical modules, and the number of modules can be adjusted according to requirements. Each module consists of two upper and lower connecting plates and three pneumatic foldable actuators located between the connecting plates. The office paper is inserted into the folding portion of the pneumatic foldable actuators to improve deformation stability. The pneumatic foldable actuators with paper skeleton is shown in Fig. 2. The 3D model of Miura-ori FMA unit is shown in Fig. 3. The actual Miura-ori FMA is shown in Fig. 4, and (a), (b), (c) are the initial state, the inflated drive bending state and the folded and compressed state, respectively.

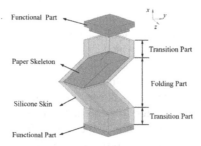

Fig. 2. Pneumatic foldable actuators with paper skeleton.

Fig. 3. 3D model of Miura-ori FMA unit.

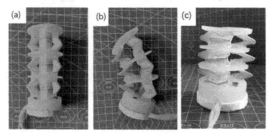

Fig. 4. Physical Picture of Miura-ori FMA (a) initial state (b) inflatable drive state (c) folded compression state.

3   Research on the Characteristics of Miura-Ori FMA

3.1   Selection of Materials

The pneumatic foldable actuator is made of two kinds of silicone, Ecoflex-0030 and Dragon Skin 30 produced by Smooth-On. Both of the two rubber materials have non-linear behavior, which can produce large deformation under small stress, and can recover their original shape after unloading. They are super-elastic materials.

3.2 Inflatable Bending Test

The motion of the Miura FMA depends on the folding, contraction and expansion of the arm, which is mainly realized by gas charging and discharging. In order to explore the influence of air pressure and gas volume on the pneumatic foldable actuator, and accurately test the relationship between air pressure and displacement as well as the relationship between air pressure and air volume, a set of displacement air pressure-air volume measuring device were built. The device is mainly composed of stepping motor, syringe, laser displacement sensor (DP101A, Panasonic), air pressure sensor (HG-C1050, Panasonic), Raspberry Pi control board (Raspberry 3B +) and AD converter (ADS1115) (Fig. 5) [10].

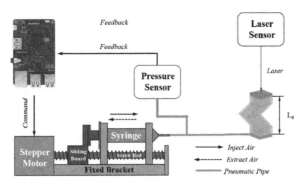

Fig. 5. The experimental setup: (a) schematic diagram of the experimental setup for measuring the pneumatic foldable actuator; (b) the experimental setup for measuring the pneumatic foldable actuator.

In order to understand the performance of the pneumatic foldable actuator in detail, so as to facilitate the control of the Miura-ori mechanical arm, the relationship curve between displacement, air pressure and gas volume of the pneumatic foldable actuator within −15 kPa ~ 15 kPa is measured, specifically speaking, to test the relationship curve between the bending angle, air pressure and gas volume of the foldable gas driver during the change process of the pneumatic foldable actuator from the normal pressure state, the folded state, and the unfolded state. The measurement process includes three steps (Inflate and extract the gas from any airway of the Miura-ori FMA, and record the bending angles at different pressure and gas volumes):

(1) Air is extracted from the Miura-ori FMA through a syringe until $P_A = -15$ kPa. The folded part is folded along the crease, and the pneumatic foldable actuator shrinks from the initial state to the folded state.
(2) Inject the gas through a syringe into a Miura-ori FMA. The folding part of the Miura-ori FMA is unfolded along the crease from the folded state until $P_A = 15$ kPa;
(3) The air is extracted from the Miura-ori FMA through a syringe until $P_A = 0$ kPa, and the foldable gas driver is restored to the initial state.

The measurement results are shown in Fig. 6. The air pressure can be conveniently controlled according to requirements and the deformation of the Miura-ori FMA can be controlled. Considering the low precision of the arm, the three airway volumes of the Miura-ori FMA have errors, and the bending angle is different to some extent when the gas volume is the same. Similarly, under the same pressure, there is also bending angle error.

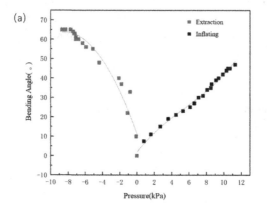

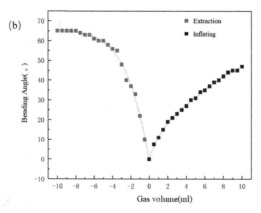

Fig. 6. The testing of Miura-ori FMA (a) gas pressure-bending angle curve; (b) gas volume-bending angle curve.

3.3 Inflatable Elongation Test

The three air passages of Miura-ori FMA are simultaneously inflated, and arm is axially extended under the action of air pressure. When the free end displacement is 15 mm, the mechanical arm is exhausted and the original length is restored to 90 mm. The pneumatic drive process is repeated for three times, and the pressure and the length of the Miura-ori FMA are monitored in real time to obtain the relation curve between the pressure and the length, as shown in Fig. 7. When the Miura-ori FMA is shrunk by extracting gas, the length of the arm will not change when it reaches 75 mm. It can be seen from the test results that during the process of increasing inflation pressure, the Miura-ori FMA stretches in the axial direction. When the length of the manipulator arm reaches 105 mm, the air pressure is about 15 kPa. During the exhaust process, the air pressure gradually decreases, and the Miura-ori FMA gradually returns to its original length; When extracting gas from the robotic arm, the length of the arm will be shortened to a fixed value, and the air pressure in the arm will no longer change.

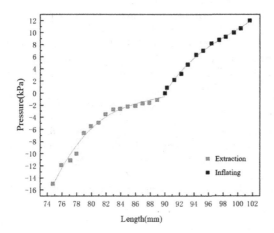

Fig. 7. The relationship between the length of Miura-ori FMA and air pressure.

4   Conclusions

This paper designs a Miura-ori mechanical arm based on the research of the pneumatic foldable actuator. A Miura-ori mechanical arm with a simple structure is innovatively realized, which has broad application prospects in flexible control tasks. The research in this article is a part of the research on foldable actuator. The future research needs are:

Due to the complex preparation process, the three airways of Miura-ori FMA are not exactly the same, which has a certain influence on the experimental results. Then material and structural design will need to be optimized to improve the unit consistency of the Miura-ori FMA. In the process of bending and elongation of inflation and bending and compression of extracted gas, it can be found that the air pressure-elongation curves of the inflatable section and the pumped section show nonlinear changes. As the pressure increases, the elongation of the arm increases. The research shows that the Miura-ori FMA can achieve inflatable bending and elongation, extraction gas bending and compression. The current work is limited to the driving test of the solid model, and subsequent simulations and experiments can be carried out by selecting different materials, and materials and models with more extensive adaptability can be selected for exploration; Optimizing the motion performance of the arm, using sensor technology to improve the degree of automation of the arm and laying a good technical foundation for further research on more intelligent flexible Miura-ori FMA.

References

1. Wang, C.: Method Of 3D shape detection of soft manipulator based on fiber grating sensor. Chem. Indus. Autom. Instrum. **42**(10), 1130–1133 (2015)
2. Tian, J., Wang, T., Shi, Z., Luo, R.: Kinematics analysis and experiment of imitating elephant trunk manipulator. Robot **39**(05), 585–594 (2017)
3. Yao, L., Li, J., Dong, H.: Analysis of variable stiffness performance of pneumatic soft manipulator module. Chin. J. Mech. Eng. **56**(09), 36–44 (2020)
4. Guan, Q., Sun, J., et al.: Novel bending and helical extensile/contractile pneumatic artificial muscles inspired by elephant trunk. Soft Robotics **00**(00), 1–18 (2020)
5. Li, X., Li, M.: A review of research on origami and its crease design. Chin. J. Theoret. Appl. Mech. **50**(11672056), 467–476 (2018)
6. Feng, H., Yang, M., Yao, G., Yan, C., Dai, J.: Origami robot. Sci. China: Techn. Sci. **48**(12), 1259–1274 (2018)
7. Ming, X.: The neglected ancient art—origami art. Art Technol. **6**, 377–378 (2014)
8. Jeong, D., Lee, K.. Design and analysis of an origami-based three-finger manipulator. Robotica, 1–14 (2017)
9. Meng, Y., Yang, W., Yuan, Y., Cheng, X., Jiao, Z.: A crawling soft robot driven by pneumatic foldable actuators based on Miura-Ori. Actuators **9**(2), 26 (2020)

Author Index

Printed in the United States
By Bookmasters